T0234628

Lecture Notes in Computer Science　　10396

Commenced Publication in 1973
Founding and Former Series Editors:
Gerhard Goos, Juris Hartmanis, and Jan van Leeuwen

More information about this series at http://www.springer.com/series/7407

Émilie Charlier · Julien Leroy
Michel Rigo (Eds.)

Developments in Language Theory

21st International Conference, DLT 2017
Liège, Belgium, August 7–11, 2017
Proceedings

 Springer

Editors
Émilie Charlier
Université de Liège
Liège
Belgium

Julien Leroy
Université de Liège
Liège
Belgium

Michel Rigo
Université de Liège
Liège
Belgium

ISSN 0302-9743 ISSN 1611-3349 (electronic)
Lecture Notes in Computer Science
ISBN 978-3-319-62808-0 ISBN 978-3-319-62809-7 (eBook)
DOI 10.1007/978-3-319-62809-7

Library of Congress Control Number: 2017946697

LNCS Sublibrary: SL1 – Theoretical Computer Science and General Issues

Printed on acid-free paper

This Springer imprint is published by Springer Nature
The registered company is Springer International Publishing AG
The registered company address is: Gewerbestrasse 11, 6330 Cham, Switzerland

Preface

The 21st International Conference on Developments in Language Theory (DLT 2017) was organized by the Department of Mathematics of the University of Liège, Belgium, during August 7–11, 2017.

The DLT conference series is one of the major international conference series in language theory and related areas. The DLT conference was established by G. Rozenberg and A. Salomaa in 1993. Since then, the DLT conferences have been held on every odd year: Magdeburg, Germany (1995), Thessaloniki, Greece (1997), Aachen, Germany (1999), and Vienna, Austria (2001). Since 2001, a DLT conference has been taking place in Europe on every odd year and outside Europe on every even year. The locations of DLT conferences since 2002 were: Kyoto, Japan (2002), Szeged, Hungary (2003), Auckland, New Zealand (2004), Palermo, Italy (2005), Santa Barbara, California, USA (2006), Turku, Finland (2007), Kyoto, Japan (2008), Stuttgart, Germany (2009), London, Ontario, Canada (2010), Milan, Italy (2011), Taipei, Taiwan (2012), Marne-la-Vallée, France (2013), Ekaterinburg, Russia (2014), Liverpool, UK (2015), and Montréal, Canada (2016). This 21st edition was thus the first time that the conference was organized in Belgium.

The series of International Conferences on Developments in Language Theory provides a forum for presenting current developments in formal languages and automata. Its scope is very general and includes, among others, the following topics and areas: combinatorial and algebraic properties of words and languages; grammars, acceptors and transducers for strings, trees, graphs, arrays; algebraic theories for automata and languages; codes; efficient text algorithms; symbolic dynamics; decision problems; relationships to complexity theory and logic; picture description and analysis; polyominoes and bidimensional patterns; cryptography; concurrency; cellular automata; bio-inspired computing; quantum computing.

The papers submitted to DLT 2017 were from 19 countries including Belgium, Canada, Czech Republic, France, Germany, Hungary, India, Italy, Japan, The Netherlands, Poland, Portugal, Republic of Korea, Russia, Slovakia, South Africa, Thailand, and USA.

This volume of *Lecture Notes in Computer Science* contains the papers that were presented at DLT 2017. There were 47 qualified submissions. Each submission was handled by three Program Committee members and received at least three reviews. The committee decided to accept 24 papers. The volume also includes the abstracts or full papers of the invited speakers:

- Véronique Bruyère (University of Mons): "Computer-Aided Synthesis: A Game-Theoretic Approach"
- Sergei Kitaev (University of Strathclyde): "A Comprehensive Introduction to the Theory of Word-Representable Graphs"
- Robert Mercaş (Loughborough University): "On the Number of Factors with Maximal-Exponent in Words"

- Balasubramanian Ravikumar (Sonoma State University): "Language Approximation: Asymptotic and Non-asymptotic Results"
- Eric Rowland (Hofstra University): "Binomial Coefficients, Valuations, and Words"
- Michał, Skrzypczak (University of Warsaw): "Connecting Decidability and Complexity for MSO Logic"

We warmly thank all the invited speakers and all the authors of the submitted papers. We also would like to thank all the members of the Program Committee and all the external reviewers (listed in the proceedings) for their excellent work in evaluating the papers. We finally thank all the members of the Organizing Committee at the University of Liège.

The organization of the conference benefited from the support of the F.R.S.-FNRS, the Faculty of Sciences of the University of Liège and the Research Unit in Mathematics of the University of Liège. The reviewing process was organized using the EasyChair conference system created by Andrei Voronkov. We would like to acknowledge that this system greatly helped to improve the efficiency of the committee work. Finally, we wish to thank the editors of the *Lecture Notes in Computer Science series* and Springer.

May 2017

Émilie Charlier
Julien Leroy
Michel Rigo

Organization

Steering Committee

Marie-Pierre Béal	University Paris-Est Marne-la-Vallée, France
Cristian S. Calude	University of Auckland, New Zealand
Volker Diekert	University of Stuttgart, Germany
Juraj Hromkovic	ETH Zürich, Switzerland
Oscar H. Ibarra	UCSB, Santa Barbara, USA
Masami Ito	Kyoto Sangyo University, Japan
Natasha Jonoska	University of South Florida, USA
Juhani Karhumaki (Chair)	University of Turku, Finland
Martin Kutrib	Universität Giessen, Germany
Michel Rigo	University of Liège, Belgium
Antonio Restivo	University of Palermo, Italy
Grzegorz Rozenberg	Leiden University, The Netherlands
Arto Salomaa	University of Turku, Finland
Kai Salomaa	Queen's University, Canada
Wojciech Rytter	Warsaw University, Poland
Mikhail Volkov	Ural Federal University, Russia
Takashi Yokomori	Waseda University, Japan

Program Committee

Srecko Brlek	Université du Québec à Montréal, Canada
Emilie Charlier	University of Liège, Belgium
Henning Fernau	Universität Trier, Germany
Emmanuel Filiot	ULB Brussels, Belgium
Dora Giammarresi	Università degli Studi di Roma Tor Vergata, Italy
Yo-Sub Han	Yonsei University, Republic of Korea
Markus Holzer	Universität Giessen, Germany
Oscar H. Ibarra	UCSB, Santa Barbara, USA
Artur Jeż	University of Wroclaw, Poland
Lila Kari	University of Western Ontario, Canada
Julien Leroy	University of Liège, Belgium
K. Narayan Kumar	Chennai Mathematical Institute, India
Giovanni Pighizzini	Università degli Studi di Milano, Italy
Jean-Eric Pin	CNRS and University Paris-Diderot, France
Michel Rigo	University of Liège, Belgium
Aleksi Saarela	University of Turku, Finland
Shinnosuke Seki	The University of Electro-Communications, Japan
Mikhail Volkov	Ural Federal University, Russia

Organizing Committee

Danielle Bartholomeus	University of Liège, Belgium
Valérie Berthé	CNRS and University Paris-Diderot, France
Emilie Charlier	University of Liège, Belgium
Vinciane Godfrind	University of Liège, Belgium
Raphaël Jungers	UCLouvain, Belgium
Julien Leroy	University of Liège, Belgium
Victor Marsault	University of Liège, Belgium
Adeline Massuir	University of Liège, Belgium
Michel Rigo	University of Liège, Belgium
Manon Stipulanti	University of Liège, Belgium

Additional Reviewers

Ananichev, Dmitry	Haase, Christoph	Masopust, Tomas
Andrew Ryzhikov	Hashimoto, Kenji	McQuillan, Ian
Badkobeh, Golnaz	Hoffmann, Stefan	Monmege, Benjamin
Beier, Simon	Huynh, Dung	Monteil, Thierry
Berglund, Martin	Jamet, Damien	Moutot, Etienne
Birget, Jean-Camille	Jecker, Ismaël	Okhotin, Alexander
Boigelot, Bernard	Jolivet, Timo	Okubo, Fumiya
Cadilhac, Michaël	Kim, Hwee	Otop, Jan
Carayol, Arnaud	Ko, Sang-Ki	Paperman, Charles
Caron, Pascal	Kufleitner, Manfred	Parreau, Aline
Carton, Olivier	Kuske, Dietrich	Pelantova, Edita
Cho, Da-Jung	Kutrib, Martin	Prigioniero, Luca
Choffrut, Christian	Labbé, Sébastien	Rampersad, Narad
Currie, James	Lavado, Giovanna	Ravikumar, Bala
Czarnetzki, Silke	Lecomte, Pierre	Rogers, Trent
Dartois, Luc	Lhote, Nathan	Saivasan, Prakash
Delacourt, Martin	Löding, Christof	Salo, Ville
Diekert, Volker	Lombardy, Sylvain	Salomaa, Kai
Dolce, Francesco	Madonia, Maria	Sebej, Juraj
Don, Henk	Mahajan, Meena	Shur, Arseny
Endrullis, Joerg	Mainz, Isabelle	Starosta, Štěpán
Fazekas, Szilard Zsolt	Malcher, Andreas	Stipulanti, Manon
Gańczorz, Michał	Maletti, Andreas	Szykuła, Marek
Ganty, Pierre	Manea, Florin	Törmä, Ilkka
Golovach, Petr	Maneth, Sebastian	Vandomme, Elise
Guillon, Pierre	Mantaci, Sabrina	Wendlandt, Matthias
Gusev, Vladimir	Marsault, Victor	Winslow, Andrew

Abstract of Invited Talks

Computer Aided Synthesis: A Game-Theoretic Approach

Véronique Bruyère

Computer Science Department, University of Mons, 20 Place du Parc,
7000 Mons, Belgium
Veronique.Bruyere@umons.ac.be

Abstract. In this invited contribution, we propose a comprehensive introduction to game theory applied in computer aided synthesis. In this context, we give some classical results on two-player zero-sum games and then on multi-player non zero-sum games. The simple case of one-player games is strongly related to automata theory on infinite words. All along the article, we focus on general approaches to solve the studied problems, and we provide several illustrative examples as well as intuitions on the proofs.

A Comprehensive Introduction to the Theory of Word-Representable Graphs

Sergey Kitaev

Department of Computer and Information Sciences, University of Strathclyde,
26 Richmond Street, Glasgow G1 1XH, UK
sergey.kitaev@cis.strath.ac.uk
https://personal.cis.strath.ac.uk/sergey.kitaev/

Abstract. Letters x and y alternate in a word w if after deleting in w all letters but the copies of x and y we either obtain a word $xyxy\cdots$ (of even or odd length) or a word $yxyx\cdots$ (of even or odd length). A graph $G = (V, E)$ is word-representable if and only if there exists a word w over the alphabet V such that letters x and y alternate in w if and only if $xy \in E$.

Word-representable graphs generalize several important classes of graphs such as circle graphs, 3-colorable graphs and comparability graphs. This paper offers a comprehensive introduction to the theory of word-represent-able graphs including the most recent developments in the area.

On the Number of Factors
with Maximal-Exponent in Words

Robert Mercaş

Department of Computer Science, Loughborough University, LE11 3TU,
Loughborough, UK
R.G.Mercas@lboro.ac.uk

A *repetition* (unary pattern) is represented as concatenations of several instances of the same factor. A word contains a repetition if it has one as a factor, and it is said to be repetition-free, otherwise. The word shshsh is an example of a *cube*, that is three consecutive repetitions of the factor sh. If, for a given alphabet, every infinite word contains an instance of the repetition, then such a repetition is called unavoidable for an alphabet of that size. Otherwise, it can be avoided by an infinity of words constructed over such an alphabet [5].

The investigation of repetitions has been around from the beginnings of the Combinatorics on Words research area. People were interested in how it is possible to avoid them [30, 36, 37], what bounds exist on the unavoidable ones [11, 12, 31, 34], and extended the notion of avoidability to other settings, such as the abelian one [14–16, 25, 26, 32], the case of words with "don't cares" [7, 22, 28], as well as other similar extensions [13, 18, 23, 33, 35]. Counting different types of repetitions has also been a highly investigated topic [6, 18, 19, 21, 24, 29]. Moreover, most of these topics have been accompanied by algorithmical results related to the identification and counting of repetitions [8, 10, 20, 27].

Similar to the way that repetitions are defined, one can extend the notion by allowing a fractional exponent. A *period* of a word is an interval at which each of it's previous symbols repeat. In the case of repetitions, given that these are concatenations of the same factor, we immediately conclude that the length of the factor represents a period of the word. Whence, we can say that every word has an *exponent* equal to its length divided by a period. The word alfalfa has a period of 3 and its exponent is 7 / 3. As a direct consequence, considering the prefix of the word that determines its minimal period, will give us the highest exponent for that word. Maximal-exponent factors are those factors of a word that have the highest exponent among all factors of the word. For the word abaca, its maximal-exponent factors are aba and aca, respectively, each of a 3 / 2 exponent. Bounds on the maximal-exponent that a factor of a word can have been thoroughly investigated [11, 12, 31, 34]. Even more, recently it has been proven that every word has at most as many runs as its length [4, 9, 17]. A run represents a factor of a word that is a repetition of exponent at least two, such that extending it to either left or right (considering the letter before, or the letter following the factor) breaks its periodicity (renders a smaller exponent for the newly obtained factor). The factor anana represents a run in the word bananas since its exponent is 2.5 and both banana and ananas have exponent 1, therefore smaller.

The runs conjecture, which stood open for over 15 years, contributed to the investigation of the number of maximal-exponent factors in a word. In particular, after being proven that the number of such factors is no more than 3.11 times the size of the word [3] when looking at square-free words (the upper bound for words that are not square-free is the length of the word and is a direct consequence of the runs result) this bound was already improved in the extended version of the former paper to 2.25 times the length of the word [1]. Furthermore, in the latter, the authors also provide a lower bound on the number of such factors, showing that there exist words that have at least 2/3 times the length of the word maximal-exponent factors. An important observation is that in this particular case, the exponents that are investigated are in fact obtained with the help of the longest border of the respective factor. A border of a word is any prefix that also occurs as a suffix of the word.

This talk will focus on work carried out together with Golnaz Badkobeh and Maxime Crochemore [2] and will have as its main focus both the lower and the upper bounds on the number of factors with maximal-exponent.

References

1. Badkobeh, G., Crochemore, M.: Computing maximal-exponent factors in an overlap-free word. J. Comput. Syst. Sci. **82**(3), 477–487 (2016)
2. Badkobeh, G., Crochemore, M., Mercaş, R.: Counting maximal-exponent factors in words. Theor. Comput. Sci. **658**, 27–35 (2017)
3. Badkobeh, G., Crochemore, M., Toopsuwan, C.: Computing the maximal-exponent repeats of an overlap-free string in linear time. In: Calderón-Benavides, L., González-Caro, C., Chávez, E., Ziviani, N. (eds.) SPIRE 2012. LNCS, vol. 7608, pp. 61–72. Springer, Heidelberg (2012). doi:10.1007/978-3-642-34109-0_8
4. Bannai, H., I, T., Inenaga, S., Nakashima, Y., Takeda, M., Tsuruta, K.: The "runs" theorem. CoRR abs/1406.0263v7 (2014)
5. Bean, D., Ehrenfeucht, A., McNulty, G.: Avoidable patterns in strings of symbols. Pac. J. Math. **85**, 261–294 (1979)
6. Blanchet-Sadri, F., Mercaş, R., Scott, G.: Counting distinct squares in partial words. In: International Conference on Automata and Formal Languages. AFL, pp. 122–133 (2008)
7. Blanchet-Sadri, F., Mercaş, R., Scott, G.: A generalization of Thue freeness for partial words. Theor. Comput. Sci. **410**(8–10), 793–800 (2009)
8. Crochemore, M.: An optimal algorithm for computing the repetitions in a word. Inf. Process. Lett. **12**(5), 244–250 (1981)
9. Crochemore, M., Mercas, R.: On the density of lyndon roots in factors. Theor. Comput. Sci. **656**, 234–240 (2016)
10. Crochemore, M., Rytter, W.: Squares, cubes, and time-space efficient string searching. Algorithmica **13**(5), 405–425 (1995)
11. Currie, J.D., Rampersad, N.: A proof of Dejean's conjecture. Math. Comput. **80**(274), 1063–1070 (2011)
12. Dejean, F.: Sur un théorème de Thue. J. Comb. Theor. **13**(1), 90–99 (1972)
13. Dekking, F.: On repetitions of blocks in binary sequences. J. Comb. Theor. **20**(3), 262–299 (1976)

14. Erdös, P.: Some unsolved problems. Magyar Tudományos Akadémia Matematikai Kutató Intézete **6**, 221–254 (1961)
15. Evdokimov, A.: Strongly asymmetric sequences generated by a finite number of symbols. Doklady Akademii Nauk SSSR **179**, 1268–1271 (1968). Russian. English translation in Soviet Mathematics Doklady **9**(1968), 536–539 (1968)
16. Evdokimov, A.: The existence of a basis that generates 7-valued iteration-free sequences. Diskretnyĭ Analiz **18**, 25–30 (1971)
17. Fischer, J., Holub, Š., I, T., Lewenstein, M.: Beyond the runs theorem. In: Iliopoulos, C., Puglisi, S., Yilmaz, E. (eds.) SPIRE 2015. LNCS, vol. 9309, pp. 277–286. Springer, Cham (2015). doi:10.1007/978-3-319-23826-5_27
18. Fraenkel, A.S., Simpson, R.J.: How many squares must a binary sequence contain? Electron. J. Comb. **2**, R2 (1995)
19. Fraenkel, A.S., Simpson, R.J.: How many squares can a string contain? J. Comb. Theor. **82** (1), 112–120 (1998)
20. Gusfield, D.: Algorithms on Strings, Trees, and Sequences: Computer Science and Computational Biology. Cambridge University Press, New York (1997)
21. Halava, V., Harju, T., Kärki, T.: On the number of squares in partial words. RAIRO - Theor. Inf. Appl. **44**(1), 125–138 (2010)
22. Halava, V., Harju, T., Kärki, T., Séébold, P.: Overlap-freeness in infinite partial words. Theor. Comput. Sci. **410**(8–10), 943–948 (2009)
23. Huova, M., Karhumäki, J., Saarela, A.: Problems in between words and abelian words: k-abelian avoidability. Theor. Comput. Sci. **454**, 172–177 (2012)
24. Ilie, L.: A note on the number of squares in a word. Theor. Comput. Sci. **380**(3), 373–376 (2007)
25. Justin, J.: Characterization of the repetitive commutative semigroups. J. Algebra **21**, 87–90 (1972)
26. Keränen, V.: Abelian squares are avoidable on 4 letters. In: Kuich, W. (ed.) ICALP 1992. LNCS, vol. 623, pp. 41–52. Springer, Heidelberg (1992). doi:10.1007/3-540-55719-9_62
27. Kolpakov, R., Kucherov, G.: Finding maximal repetitions in a word in linear time. In: Annual Symposium on Foundations of Computer Science. FOCS, pp. 596–604 (1999)
28. Manea, F., Mercaş, R.: Freeness of partial words. Theor. Comput. Sci. **389**(1–2), 265–277 (2007)
29. Manea, F., Seki, S.: Square-density increasing mappings. In: Manea, F., Nowotka, D. (eds.) WORDS 2015. LNCS, vol. 9304, pp. 160–169. Springer, Cham (2015). doi:10.1007/978-3-319-23660-5_14
30. Morse, M., Hedlund, G.: Unending chess, symbolic dynamics and a problem in semigroups. Duke Math. J. **11**(1), 1–7 (1944)
31. Pansiot, J.J.: A propos d'une conjecture de F. Dejean sur les répétitions dans les mots. In: Diaz, J. (ed.) ICALP 1983. LNCS, vol. 154, pp. 585–596. Springer, Heidelberg (1983). doi:10.1007/BFb0036939
32. Pleasants, P.: Non repetitive sequences. Proc. Camb. Philos. Soc. **68**, 267–274 (1970)
33. Rampersad, N., Shallit, J., Wang, M.W.: Avoiding large squares in infinite binary words. Theor. Comput. Sci. **339**(1), 19–34 (2005)
34. Rao, M.: Last cases of Dejean's conjecture. Theor. Comput. Sci. **412**(27), 3010–3018 (2011)
35. Rao, M.: On some generalizations of abelian power avoidability. Theor. Comput. Sci. **601**, 39–46 (2015)

36. Thue, A.: Über unendliche Zeichenreihen. Norske Vid. Selsk. Skr. I, Mat. Nat. Kl. Christiana **7**, 1–22 (1906). (Reprinted in Selected Mathematical Papers of Axel Thue, T. Nagell, editor, Universitetsforlaget, Oslo, Norway (1977), pp. 139–158)

37. Thue, A.: Über die gegenseitige Lage gleicher Teile gewisser Zeichenreihen. Norske Videnskabers Selskabs Skrifter, I Mathematisch-Naturwissenschaftliche Klasse Christiana **1**, 1–67 , (Reprinted in Selected Mathematical Papers of Axel Thue, T. Nagell, editor, Universitetsforlaget, Oslo, Norway (1977), pp. 413–478) (1912)

Language Approximation: Asymptotic and Non-asymptotic Results

Bala Ravikumar

Department of Computer and Engineering Science, Sonoma State University,
Rohnert Park, CA 94928, USA
ravikuma@sonoma.edu

Abstract. Approximation is a central concept in computational problem solving and an effective way to deal with intractable problems as well as in contexts with limited resources. Our specific focus will be on using finite automaton as a computational model, and study how well various languages can be approximately recognized by finite automata (FA). In [1] and [10], a notion of approximation was introduced that measures the proportion of inputs of a problem (or language L) that are correctly processed by a machine M. In [2–7] and other works, several regular and non-regular languages were considered and the question how well they can be approximately recognized by a finite automaton was addressed. Specifically, in [5], it was shown that Majority language over $\{0, 1\}$ (the set of strings with more 1's than 0's) can't be approximated by a FA of any size much better than a 1-state FA that accepts (or rejects) all strings. In this presentation, in addition to theoretical results, we will also explore approximation by finite automaton as a tool for algorithm design - for recognition, optimization as well as counting problems. We also consider some decision questions related to approximations - such as computing the success ratio (defined as the fraction of the inputs correctly processed in the limit) of a given FA, relative to specific, nonregular languages.

Most of the prior work listed above deals with approximation in an asymptotic sense - how well is a language approximated by a machine when the length of the string is arbitrarily large. For the notion of approximation to be useful in practice, we should consider the case of fixed (or bounded) length inputs. Specifically, we consider for various languages L (including Majority language and Center = the set of strings w of length n over $\{0, 1\}$ such that $[n/2]$-th bit is 1 where $|w| = n$), and for a given integer n, which FA best approximates the language L over strings of length at most n. We provide a general algorithm to find efficient, and in some cases, an optimal approximating finite automaton with a specified number of states. Finally, we will attempt to present some case studies in which approximate computation enables the solution of hard counting problems.

References

1. Cai, J.-Y.: With probability one, a random oracle separates PSPACE from the polynomial-time hierarchy. In: Proceedings of 18th ACM Symposium on Theory of Computing, pp. 21–29 (1986)

2. Combs, J., Ravikumar, B.: DFA for approximate integer multiplication. In: Presentation at SIAM Conference on Discrete Mathematics, Atlanta, Georgia, June 2016
3. Cordy, B., Salomaa, K.: On the existence of regular approximations. Theoret. Comput. Sci. **387**, 125–135 (2007)
4. Eisman, G., Ravikumar, B.: Approximate recognition of non-regular languages by finite automata. In: Australasian Symposium on Computer Science, pp. 219–228 (2005)
5. Eisman, G., Ravikumar, B.: On approximating non-regular languages by regular languages. Fundamenta Informaticae. **110**(1–4), pp. 125–142 (2011)
6. Kappes, M., Kintala, C.M.R.: Tradeoffs between reliability and conciseness of deterministic finite automata. J. Automata Lang. Comb. **9**, 281–292 (2004)
7. Kappes, M., Niener, F.: Succinct representations of languages by DFA with different levels of reliability. Theoret. Comput. Sci. **330**, 299–310 (2005)
8. Nrederhof, M.-J.: Practical experiments with regular approximations of context-free languages. In: Proceedings of Information Processing and Management of Uncertainty on Knowledge Based Systems, pp. 891–895 (1996)
9. Ponitz, A., Tittmann, P.: Improved upper bounds for self-avoiding walks in Z^d. Electron. J. Comb. **7**(Research Paper R-21) (2000)
10. Wilber, R.E.: Randomness and the density of hard problems. In: Proceedings of 24th IEEE Foundations of Computer Science, pp. 335–342 (1983)

Binomial Coefficients, Valuations, and Words

Eric Rowland

Department of Mathematics, Hofstra University, Hempstead, NY 11549, USA
eric.rowland@hofstra.edu

Abstract. The study of arithmetic properties of binomial coefficients has a rich history. A recurring theme is that p-adic statistics reflect the base-p representations of integers. We discuss many results expressing the number of binomial coefficients $\binom{n}{m}$ with a given p-adic valuation in terms of the number of occurrences of a given word in the base-p representation of n, beginning with a result of Glaisher from 1899, up through recent results by Spiegelhofer–Wallner and Rowland.

Connecting Decidability and Complexity for MSO Logic (Extended Abstract)

Michał Skrzypczak

University of Warsaw, Banacha 2 Warsaw, Poland

mskrzypczak@mimuw.edu.pl

Abstract. This work is about studying reasons for (un)decidability of variants of Monadic Second-order (MSO) logic over infinite structures. Thus, it focuses on connecting the fact that a given theory is (un)decidable with certain measures of complexity of that theory.

The first of the measures is the topological complexity. In that case, it turns out that there are strong connections between high topological complexity of languages available in a given logic, and its undecidability. One of the milestone results in this context is the Shelah's proof of undecidability of MSO over reals.

The second complexity measure focuses on the axiomatic strength needed to actually prove decidability of the given theory. The idea is to apply techniques of reverse mathematics to the classical decidability results from automata theory. Recently, both crucial theorems of the area (the results of Büchi and Rabin) have been characterised in these terms. In both cases the proof gives strong relations between decidability of the mso theory with concepts of classical mathematics: determinacy, Ramsey theorems, weak Konig's lemma, etc...

The author was supported by the Polish National Science Centre grant no. UMO-2016/21/D/ST6/00491.

Contents

Invited Talks

Computer Aided Synthesis:
A Game-Theoretic Approach

Véronique Bruyère[✉]

Computer Science Department, University of Mons,
20 Place du Parc, 7000 Mons, Belgium
Veronique.Bruyere@umons.ac.be

Abstract. In this invited contribution, we propose a comprehensive introduction to game theory applied in computer aided synthesis. In this context, we give some classical results on two-player zero-sum games and then on multi-player non zero-sum games. The simple case of one-player games is strongly related to automata theory on infinite words. All along the article, we focus on general approaches to solve the studied problems, and we provide several illustrative examples as well as intuitions on the proofs.

Keywords: Games played on graphs · Boolean objective · Quantitative objective · Winning strategy · Nash Equilibrium · Synthesis

1 Introduction

Game theory is a well-developed branch of mathematics that is applied to various domains like economics, biology, computer science, etc. It is the study of mathematical models of interaction and conflict between individuals and the understanding of their decisions assuming that they are rational [54,70].

The last decades have seen a lot of research on algorithmic questions in game theory motivated by problems from *computer aided synthesis*. One important line of research is concerned with *reactive systems* that must continuously react to the uncontrollable events produced by the environment in which they evolve. A *controller* of a reactive system indicates which actions it has to perform to satisfy a certain objective against any behavior of the environment. An example in air traffic management is the autopilot that controls the speed of the plane, but have no control on the weather conditions. Such a situation can be modeled by a *two-player game played on a graph*: the system and the environment are the two players, the vertices of the graph model the possible configurations, the infinite paths in the graph model all the continuous interactions between the system and the environment. In this game, the system wants to achieve a certain objective while the environment tries to prevent it to do so. The objectives of the two players are thus *antagonistic* and we speak of *zero-sum* games. In this framework, checking whether the system is able to achieve its objective reduces to the existence of a *winning strategy* in the corresponding game, and

© Springer International Publishing AG 2017
E. Charlier et al. (Eds.): DLT 2017, LNCS 10396, pp. 3–35, 2017.
DOI: 10.1007/978-3-319-62809-7_1

building a controller reduces to computing such a strategy [38]. Whether such a controller can be automatically designed from the objective is known as the *synthesis problem.*

Another, more recent, line of research is concerned with the modelization and the study of *complex systems.* Instead of the simple situation of a system embedded in a hostile environment, we are faced with systems/environments formed of several components each of them with their own objectives that are not necessarily conflicting. Imagine the situation of several users behind their computers on a shared network. In this case, we use the model of *multi-player non zero-sum games played on graphs*: the components are the different players, each of them aiming at satisfying his objective. In this context, the synthesis problem is a little different: winning strategies are no longer appropriate and are replaced by the concept of *equilibrium*, that is, a strategy profile where no player has an incentive to deviate [39]. Different kinds of equilibria have been investigated among which the famous notion of *Nash equilibrium* [53].

A lot of study has been done about *Boolean* objectives, in particular about the class of ω-*regular* objectives, like avoiding a deadlock, always granting a request, etc [38]. An infinite path in the game graph is either winning or losing depending on whether the objective is satisfied or not. To allow richer objectives, such as minimizing the energy consumption or guaranteeing a limited response time to a request, existing models have been enriched with quantitative aspects in a way to associate a payoff (or a cost) to all paths in the game graph [19]. In this setting, we speak of *quantitative* objectives, and a classical *decision problem* in two-player zero-sum games is whether there exists a winning strategy for the system that ensures a payoff satisfying some given *constraints* no matter how the environment behaves. For instance we would like an energy consumption lying within a certain given interval. The same kind of question is also considered for multi-player non zero-sum games, that is, whether there exists an equilibrium such that the payoff of each player satisfies the constraints.

Decidability of those problems is not enough. Indeed in case of positive answer, it is important to know the exact *complexity class* of the problem and how complex are the strategies used to solve it. Given past interactions between the players, a strategy for a player indicates the next action he has to perform. The *amount of memory* on those past interactions is one of the ways to express the complexity of the strategy. The simplest strategies are those that require no memory at all. When all these characteristics are known and indicate practical applicability of the models, the final step is the *implementation* of the solving strategies into a program (like for instance a controller for a reactive system) by using adequate data structures and possibly heuristics.

In this article, we propose a *comprehensive introduction* to classical algorithmic solutions to the synthesis problem for two-player zero-sum games and for multi-player non zero-sum games. A complementary survey can be found in [9], and detailed expositions in the case of Boolean objectives are provided in [38,39]. We study the existence of winning strategies (in two-player zero-sum games) and equilibria (in multi-player non zero-sum games) satisfying some given constraints, in particular the complexity class of the decision problem and the memory required for the related strategies. We provide several illustrative examples as well as intuitions on some proofs. We do not intend to present an exhaustive survey, but rather focus on some lines of research, with an emphasis on *general approaches*. In particular, we only consider *(i) turned-based* (and not concurrent) games such that the players choose their actions in a turned-based way (and not concurrently), *(ii) deterministic* (and not stochastic) games such that their edges are deterministic and not labeled by probabilities *(iii) pure* (and not randomized) strategies such that the next action is chosen in a deterministic way (and not according to a probability distribution).

Our approach is as follows. We begin with a general definition of game that includes the class of games with Boolean objectives and the class of games with quantitative objectives. For two-player zero-sum games, we present a criterium [36] that implies, for several large families of games, the existence of memoryless winning strategies ensuring a payoff satisfying some given constraints. For non zero-sum multi-player games, we present a characterization of plays (used for instance in [14,65]) that are the outcome of a Nash equilibrium. The existence of Nash equilibrium in many different families of games is derived from this characterization, as well as results on the existence of a Nash equilibrium satisfying some constraints. We also present two other well-studied equilibria: the secure equilibria [23] and the subgame perfect equilibria [61]. For the studied decision problems, in addition to the results derived from our general approaches, we provide in this survey an overview of known results for games with Boolean and quantitative objectives.

The article is organized in the following way. In Sect. 2, we introduce the concepts of game and strategy, we then present the studied decision problems, and we finally recall the Boolean and quantitative objectives that are classically studied. In Sect. 3 devoted to two-player zero-sum games, we begin with the simple case of one-player games, and show how the decision problems are connected to problems in automata theory and numeration systems. We then present the general criterium mentioned before, and then the solutions to the decision problems for the classes of games with Boolean and quantitative objectives. Finally, we present several recent extensions of those classes of games, where for instance the single objective is replaced by a Boolean intersection of several objectives. The case of multi-player non zero-sum games is investigated in Sect. 4 by starting with the characterization of outcomes of Nash equilibrium. Derived results on the existence of Nash equilibrium (under some given constraints) are then detailed, followed by a study of other kinds of equilibria like secure and subgame perfect equilibria. We provide a short conclusion in Sect. 5.

2   Terminology and Studied Problems

We consider multi-player turn-based games played on finite directed graphs. The set of vertices are partitioned among the different players. A play is an infinite sequence of vertices obtained by moving an imaginary pebble from vertex to vertex according to existing edges. The owner of the current vertex decides what is the next move of the pebble according to some strategy. Each player follows a strategy in a way to achieve a certain objective. This objective depends on a preference relation that the player has on the payoffs assigned to plays. In this section, we introduce all these notions and state the problems studied in this article.

2.1 Preliminaries

Games. We begin with the notions of arena and game.

Definition 1. *An* arena *is a tuple* $A = (\Pi, V, (V_i)_{i \in \Pi}, E)$ *where:*

- *Π is a finite set of* players,
- *V is a finite set of* vertices *and $E \subseteq V \times V$ is a set of* edges, *such that each vertex has at least one outgoing edge[1],*
- *$(V_i)_{i \in \Pi}$ is a partition of V, where V_i is the set of vertices owned[2] by player $i \in \Pi$.*

A *play* is an infinite sequence $\rho = \rho_0\rho_1 \ldots \in V^\omega$ of vertices such that $(\rho_k, \rho_{k+1}) \in E$ for all $k \in \mathbb{N}$. *Histories* are finite sequences $h = h_0 \ldots h_n \in V^*$ defined in the same way. We often use notation hv to mention the last vertex $v \in V$ of the history. The set of plays is denoted by *Plays* and the set of non empty histories (resp. ending with a vertex in V_i) by *Hist* (resp. by *Hist$_i$*). A *prefix* (resp. *suffix*) of a play $\rho = \rho_0\rho_1 \ldots$ is a finite sequence $\rho_{\leq n} = \rho_0 \ldots \rho_n$ (resp. infinite sequence $\rho_{\geq n} = \rho_n\rho_{n+1} \ldots$). We often use notation $h\rho$ for a play of which history h is prefix. Given a play ρ, we denote by $inf(\rho)$ the set of vertices visited infinitely often by ρ. We say that ρ is a *lasso* if it is equal to hg^ω with h, g being two histories. This lasso is called *simple* if hg has no repeated vertices.

Definition 2. *A* game *G is an arena $A = (\Pi, V, (V_i)_{i \in \Pi}, E)$ such that each player i has:*

- *a payoff function $f_i : Plays \to P_i$ where P_i is a set of* payoffs,
- *a preference relation $\prec_i \subseteq P_i \times P_i$ on his set of payoffs.*

A *preference* relation $\prec_i$ is a strict total order[3]. It allows player i to compare two plays $\rho, \rho' \in Plays$ with respect to their payoffs: $f_i(\rho) \prec_i f_i(\rho')$ means that player i prefers ρ' to ρ. Given $p, p' \in P_i$, we write $p \preceq_i p'$ when $p \prec_i p'$ or $p = p'$; notice that $p \not\prec_i p'$ iff $p' \preceq_i p$ since $\prec_i$ is total.

A payoff function f_i is *prefix-independent* if $f_i(h\rho) = f_i(\rho)$ for all $h\rho \in Plays$. It is *prefix-linear* if for all $h\rho, h\rho' \in Plays$,

$$f_i(\rho) \preceq_i f_i(\rho') \Rightarrow f_i(h\rho) \preceq_i f_i(h\rho'), and \tag{1}$$

$$f_i(\rho) \prec_i f_i(\rho') \Rightarrow f_i(h\rho) \prec_i f_i(h\rho'). \tag{2}$$

---

[1] This condition guarantees that there is no deadlock. It can be assumed w.l.o.g. for all the problems considered in this article.
[2] We also say that player i *controls* the vertices of V_i.
[3] that is, an irreflexive, transitive and total binary relation.

Any prefix-independent function f_i is prefix-linear.

When an initial vertex $v_0 \in V$ is fixed, we call (G, v_0) an *initialized* game. In this case, plays and histories are supposed to start in v_0, and we then use notations $Plays(v_0)$, $Hist(v_0)$, and $Hist_i(v_0)$ (instead of $Plays$, $Hist$, and $Hist_i$).

Example 3. Consider the initialized two-player game (G, v_0) in Fig. 1 such that player 1 (resp. player 2) controls vertices v_0, v_2, v_3 (resp. vertex v_1).[4] Both players use the same set P of payoffs equal to $\{p_1, p_2, p_3\}$, and the same payoff function f that is prefix-independent: $f((v_0 v_1)^\omega) = p_1$, $f(v_2^\omega) = p_2$, and $f(v_3^\omega) = p_3$. The preference relation for player 1 (resp. player 2) is $p_1 \prec_1 p_2 \prec_1 p_3$ (resp. $p_2 \prec_2 p_3 \prec_2 p_1$).

Fig. 1. A two-player game with payoff functions $f = f_1 = f_2$, preference relations $p_1 \prec_1 p_2 \prec_1 p_3$ and $p_2 \prec_2 p_3 \prec_2 p_1$, such that $f((v_0 v_1)^\omega) = p_1$, $f(v_2^\omega) = p_2$, and $f(v_3^\omega) = p_3$

Strategies. Let (G, v_0) be an initialized game. A *strategy* σ_i for player i in (G, v_0) is a function $\sigma_i : Hist_i(v_0) \to V$ assigning to each history $hv \in Hist_i(v_0)$ a vertex $v' = \sigma_i(hv)$ such that $(v, v') \in E$. Thus $\sigma_i(hv)$ is the next vertex chosen by player i (that controls vertex v) after history hv has been played. A play $\rho \in Plays(v_0)$ is *consistent* with σ_i if $\rho_{n+1} = \sigma_i(\rho_{\leq n})$ for all n such that $\rho_n \in V_i$.

A strategy σ_i for player i is *positional* if it only depends on the last vertex of the history, i.e., $\sigma_i(hv) = \sigma_i(v)$ for all $hv \in Hist_i(v_0)$. More generally, it is *finite-memory* if $\sigma_i(hv)$ needs only a finite information out of the history hv. This is possible with a finite-state machine that keeps track of histories of plays. The strategy chooses the next vertex depending on the current state of the machine and the current vertex in the game.[5] The previous definition of positional strategy σ_i for player i is given for an initialized game (G, v_0). We call it *uniform* if it is defined for all $hv \in Hist_i$ (instead of $Hist_i(v_0)$), that is, when σ_i is a positional strategy in all initialized games (G, v), $v \in V$.

A *strategy profile* is a tuple $(\sigma_i)_{i \in \Pi}$ of strategies, where each σ_i is a strategy of player i. It is called *positional* (resp. *uniform, finite-memory*) if all σ_i, $i \in \Pi$, are positional (resp. uniform, finite-memory). Given an initial vertex v_0, such a strategy profile determines a unique play of (G, v_0) that is consistent with all strategies σ_i. This play is called the *outcome* of $(\sigma_i)_{i \in \Pi}$ in (G, v_0) and is denoted by $\langle (\sigma_i)_{i \in \Pi} \rangle_{v_0}$.

---

[4] In all examples of this article, circle (resp. square) vertices are controlled by player 1 (resp. player 2).

[5] This informal definition is enough for this survey. See for instance [38] for a definition.

Example 3 (continued). An example of strategy profile (σ_1, σ_2) in (G, v_0) is the following one:

- the positional strategy σ_2 for player 2 is defined such that $\sigma_2(hv_1) = v_3$ for all $hv_1 \in Hist(v_0)$,
- the finite-memory strategy σ_1 for player 1 is defined such that $\sigma_1(v_0) = v_1$ and $\sigma_1(hv_0) = v_2$ for all $hv_0 \in Hist(v_0) \setminus \{v_0\}$.[6] Hence player 1 chooses to move to v_1 (resp. to v_2) at the first visit (resp. next visits) to v_0. The needed memory is whether the current history has visited v_0 once or more time.

The outcome $\langle (\sigma_1, \sigma_2) \rangle_{v_0}$ is equal to $v_0 v_1 v_3^\omega$ with payoff p_3.

2.2 Studied Problems

In this paper, we want to study *two problems*. In the first problem, one designated player, say player 1, wants to apply a strategy that *guarantees* certain constraints on the payoffs of the plays (with respect to his preference relation) *against* any strategy of the other players. The other players can thus be considered as one player, say player 2, being the *opponent* of player 1. This is the class of so-called *two-player zero-sum games*.

Problem 4. Let (G, v_0) be an initialized two-player zero-sum game and $\mu, \nu \in P_1$ be two bounds. Decide whether player 1 has a strategy σ_1 such that $\mu \preceq_1 f_1(\rho)$ (resp. $\mu \preceq_1 f_1(\rho) \preceq_1 \nu$) for all plays $\rho \in Plays(v_0)$ consistent with σ_1.[7]

Case $\mu \preceq_1 f_1(\rho)$ is called the *threshold problem* whereas case $\mu \preceq_1 f_1(\rho) \preceq_1 \nu$ is called the *constraint problem*. When a strategy σ_1 as required in Problem 4 exists, it is called *winning* and a play ρ consistent with σ_1 is also called *winning*; we also say that player 1 can *ensure* a payoff $f_1(\rho)$ such that $\mu \preceq_1 f_1(\rho)$ (resp. $\mu \preceq_1 f_1(\rho) \preceq_1 \nu$). When this problem is decidable, we are interested in finding its complexity class and the simplest winning strategies σ_1, like positional or finite-memory ones when they exist.

In a two-player zero-sum game G, the opposition between player 1 and player 2 is most often described in terms of objectives. An *objective* Ω for player 1 is a subset of *Plays*, here the set of plays ρ such that $\mu \preceq_1 f_1(\rho)$ (resp. $\mu \preceq_1 f_1(\rho) \preceq_1 \nu$). Player 1 wants to ensure a play in Ω against any strategy of player 2. As an opponent, player 2 wants wants to avoid plays in Ω, that is, to ensure the opposite objective $Plays \setminus \Omega$. We say that the game G with objective Ω is *determined* if for each initial vertex v_0, either player 1 has a winning strategy to ensure Ω in (G, v_0) or player 2 has a winning strategy to ensure $Plays \setminus \Omega$. *Martin's theorem* [51] states that every two-player zero-sum game with *Borel objectives* is determined. Nevertheless, it gives no information on which player has a winning strategy and on the shape of such a winning strategy. This motivates studying Problem 4.

---

[6] As player 1 can only loop on vertices v_2 and v_3, we do not formally define σ_1 on histories ending with v_2 or v_3.

[7] This problem is focused on Player 1, the payoff function f_2 and preference relation $\prec_2$ of Player 2 do not matter.

Example 5. Let us come back to the game of Fig. 1 seen as a two-player zero-sum game (we thus focus on player 1). In (G, v_0), player 1 has a winning strategy σ_1 for the threshold problem with $\mu = p_2$, that is, for the objective $\Omega = \{\rho \mid f(\rho) \in \{p_2, p_3\}\}$: take the positional strategy σ_1 such that $\sigma_1(v_0) = v_2$. However he has no winning strategy for the threshold problem with $\mu = p_3$. Indeed with the positional strategy σ_2 such that $\sigma_2(v_1) = v_0$, player 2 has a winning strategy for the opposite objective since he can ensure a payoff equal to p_1 or p_2.

In the second problem studied in this article, we come back to multi-player games where each player has his own payoff function and preference relation. Here, the players are not necessarily antagonistic: this is the class of so-called *multi-player non zero-sum games*. Instead of looking for a strategy ensuring a certain objective for one designated player, we are now interested in strategy profiles, called *solution profiles*, that provide payoffs satisfactory to all players with respect to their own objectives. A classical example of solution profile is the notion of *Nash equilibrium* (NE) [53]. Informally, a strategy profile is an NE if no player has an incentive to deviate (with respect to his preference relation) when the other players stick to their own strategies. In other words, an NE can be seen as a *contract* that makes every player satisfied in the sense that nobody wants to break the contract if the others follow it.

Definition 6. *Given an initialized game* (G, v_0), *a strategy profile* $(\sigma_i)_{i \in \Pi}$ *is a Nash equilibrium if* $f_i(\langle (\sigma_i)_{i \in \Pi} \rangle v_0) \nprec_i f_i(\langle \sigma_i', \sigma_{-i} \rangle v_0)$ *for all players* $i \in \Pi$ *and all strategies* σ_i' *of player* i.

In this definition, notation (σ_i', σ_{-i}) means the strategy profile such that all players stick to their own strategy except player i who shifts from strategy σ_i to strategy σ_i'. We say that σ_i' is a *deviating* strategy from σ_i. When $f_i(\langle (\sigma_i)_{i \in \Pi} \rangle v_0) \prec_i f_i(\langle \sigma_i', \sigma_{-i} \rangle v_0)$, σ_i' is called a *profitable deviation* for player i with respect to $(\sigma_i)_{i \in \Pi}$.

Example 3 (continued). Let us reconsider the non zero-sum game G of Fig. 1 and the strategy profile (σ_1, σ_2) given previously in (G, v_0) ($\sigma_1(v_0) = v_1$, $\sigma_1(hv_0) = v_2$ for all $hv_0 \in Hist(v_0) \setminus \{v_0\}$, and $\sigma_2(hv_1) = v_3$ for all $hv_1 \in Hist(v_0)$). This strategy profile is an NE with outcome $\langle (\sigma_1, \sigma_2) \rangle_{v_0} = v_0 v_1 v_3^\omega$. Indeed, player 1 has no incentive to deviate since the payoff p_3 of $\langle \sigma_1, \sigma_2 \rangle_{v_0}$ is the best possible with respect to $\prec_1$. If player 2 uses the deviating strategy σ_2' from σ_2 such that $\sigma_2'(v_0 v_1) = v_0$, then the resulting outcome $\langle \sigma_1, \sigma_2' \rangle_{v_0} = v_0 v_1 v_0 v_2^\omega$ has a less preferable payoff for him since $p_2 \prec_2 p_3$. So player 2 has no profitable deviation.

Other kinds of solution profiles will be studied in Sect. 4.

Problem 7. Let (G, v_0) be an initialized multi-player non zero-sum game and $(\mu_i)_{i \in \Pi}, (\nu_i)_{i \in \Pi} \in (P_i)_{i \in \Pi}$ be two tuples of bounds. Decide whether there exists a solution profile $(\sigma_i)_{i \in \Pi}$ such that $\mu_i \preceq_i f_i(\langle (\sigma_i)_{i \in \Pi} \rangle v_0)$ (resp. $\mu_i \preceq_i f_i(\langle (\sigma_i)_{i \in \Pi} \rangle v_0) \preceq_i \nu_i$) for all players $i \in \Pi$.

Similarly to Problem 4, the two cases are respectively called *threshold problem* and *constraint problem*, and we want to compute the complexity class and the simplest solution profiles in case of decidability.

In Sects. 3 and 4, we present some known results about solutions to Problems 4 and 7 respectively with an emphasis on *general approaches*. Before, we end Sect. 2 with a list of payoff functions that are classically studied.

2.3 Classical Payoff Functions

In the classes of games that are *classically* studied, each player $i \in \Pi$ uses a real-valued payoff function $f_i : Plays \to \mathbb{R}$ and a preference relation $\prec_i$ equal to the usual ordering $<$ on $P_i = \mathbb{R}$. Hence, player i prefers to maximize the payoff $f_i(\rho)$ of a play ρ.[8] In this classical setting, we focus on two particular subclasses: the *Boolean* payoff functions and the *quantitative* payoff functions.

Boolean Payoff Functions. A particular subclass of games G are those equipped with *Boolean* functions $f_i : Plays \to \{0, 1\}$, for all $i \in \Pi$, where payoff 1 (resp. payoff 0) means that the play is the most (resp. the less) preferred by player i. Particularly interesting related objectives are $\Omega_i = \{\rho \in Plays \mid f_i(\rho) = 1\}$, $i \in \Pi$. Classical such objectives Ω_i are ω-*regular objectives* like the following ones [38,39,55].

Definition 8. – *Let $U \subseteq V$,*
  - Reachability : $\Omega_i = \{\rho \in Plays \mid \rho$ *visits a vertex of U at least once*$\}$,
  - Safety: $\Omega_i = \{\rho \in Plays \mid \rho$ *visits no vertex of U*$\}$,
  - Büchi: $\Omega_i = \{\rho \in Plays \mid inf(\rho) \cap U \neq \emptyset\}$,
  - Co-Büchi: $\Omega_i = \{\rho \in Plays \mid inf(\rho) \cap U = \emptyset\}$.
– *Let $c : V \to \mathbb{N}$ be a coloring of the vertices by integers,*
  - Parity: $\Omega_i = \{\rho \in Plays \mid$ *the maximum color seen infinitely often along* $c(\rho_0)c(\rho_1) \ldots$ *is even*$\}$.
– *Let $(F_k, G_k)_{1 \le k \le l}$ be a family of pairs of sets $F_k, G_k \subseteq V$,*
  - Rabin: $\Omega_i = \{\rho \in Plays \mid \exists k, 1 \le k \le l,$ *such that* $inf(\rho) \cap F_k = \emptyset$ *and* $inf(\rho) \cap G_k \neq \emptyset\}$,
  - Streett: $\Omega_i = \{\rho \in Plays \mid \forall k, 1 \le k \le l, inf(\rho) \cap F_k \neq \emptyset$ *or* $inf(\rho) \cap G_k = \emptyset\}$.
– *Let $\mathcal{F} \subseteq 2^V$ be a family of subsets of vertices,*
  - Muller[9] : $\Omega_i = \{\rho \in Plays \mid inf(\rho) \in \mathcal{F}\}$.

Notice that reachability and safety (resp. Büchi and co-Büchi, Rabin and Streett) are dual objectives. The complement of a parity (resp. Muller) objective is again a parity (resp. Muller) objective: from the coloring function $c : V \to \mathbb{N}$, define the new function c' such that $c'(v) = c(v) + 1$ for all $v \in V$ (resp. from the family $\mathcal{F} \subseteq 2^V$, define the new family $\mathcal{F}' = 2^V \setminus \mathcal{F}$). A Büchi (resp. co-Büchi) objective is a particular case of a parity objective: assign color 2 to vertices of

---

[8] Alternatively, $\prec_i$ can be the ordering $>$ meaning that player i prefers to minimize the payoff of a play.

[9] A *colored variant* of Muller objective is defined from a coloring $c : V \to \mathbb{N}$ of the vertices: the family $\mathcal{F}$ is composed of subsets of $c(V)$ (instead of V) and $\Omega_i = \{\rho \in Plays \mid inf(c(\rho_0)c(\rho_1) \ldots) \in \mathcal{F}\}$ [39]. See [42] for several variants of Muller games.

U and 1 to vertices of $V \setminus U$ (resp. color 1 to U and 0 to $V \setminus U$). Similarly, one can easily prove that a parity objective is both a Rabin and a Streett objective which are themselves a Muller objective [38].

In the previous definition, the payoff function f_i is prefix-independent in each case except for reachability and safety where only condition (1) of prefix-linearity is satisfied.

Example 9. Suppose that in the game of Fig. 1, player 1 wants to achieve the Büchi objective with $U = \{v_2, v_3\}$ whereas player 2 wants to achieve the Muller objective with $\mathcal{F} = \{\{v_0, v_1\}, \{v_3\}\}$. Then the play $\rho = (v_0 v_1)^\omega$ has payoff $(0, 1)$, that is a payoff 0 for player 1 and a payoff 1 for player 2.

Quantitative Payoff Functions. Classical *quantitative* payoff functions f_i : *Plays* $\rightarrow \mathbb{R}$ are defined from a *weight function* $w_i : E \rightarrow \mathbb{Q}$ as follows [19] (each edge of the game G is thus labeled by a $|\Pi|$-tuple of weights).

Definition 10. *Let $w_i : E \rightarrow \mathbb{Q}$ be a weight function and $\lambda \in {]0, 1[}$ be a rational discount factor. Then f_i : Plays $\rightarrow \mathbb{R}$ is defined as one among the following payoff functions: let $\rho = \rho_0 \rho_1 \ldots \in$ Plays,*

- Supremum: $\mathsf{Sup}_i(\rho) = \sup_{n \in \mathbb{N}} w_i(\rho_n, \rho_{n+1})$,
- Infimum: $\mathsf{Inf}_i(\rho) = \inf_{n \in \mathbb{N}} w_i(\rho_n, \rho_{n+1})$,
- Limsup: $\mathsf{LimSup}_i(\rho) = \limsup_{n \to \infty} w_i(\rho_n, \rho_{n+1})$,
- Liminf: $\mathsf{LimInf}_i(\rho) = \liminf_{n \to \infty} w_i(\rho_n, \rho_{n+1})$,
- Mean-payoff $\overline{\mathsf{MP}}_i$: $\overline{\mathsf{MP}}_i(\rho) = \limsup_{n \to \infty} \frac{1}{n} \sum_{k=0}^{n-1} w_i(\rho_k, \rho_{k+1})$,
- Mean-payoff $\underline{\mathsf{MP}}_i$: $\underline{\mathsf{MP}}_i(\rho) = \liminf_{n \to \infty} \frac{1}{n} \sum_{k=0}^{n-1} w_i(\rho_k, \rho_{k+1})$,
- Discounted sum: $\mathsf{Disc}_i^\lambda(\rho) = \sum_{n=0}^{\infty} w_i(\rho_n, \rho_{n+1}) \lambda^n$.

Some of these payoff functions provide natural generalizations of the previous ω-regular objectives. Indeed the supremum (resp. infimum, limsup, liminf) function is a quantitative generalization of the reachability (resp. safety, Büchi, co-Büchi) objective. The mean-payoff and discounted sum functions are much studied in classical game theory [33].

There are two variants of mean-payoff functions because the limit may not exist. Nevertheless in case of a lasso $\rho = hg^\omega$, both payoffs $\overline{\mathsf{MP}}_i(\rho)$ and $\underline{\mathsf{MP}}_i(\rho)$ coincide and are equal to the average weight of the cycle g (with respect to the weight function w_i).

In Definition 10, the payoff function f_i is prefix-independent in limsup, liminf and mean-payoff cases, prefix-linear in discounted sum case, and satisfies condition (1) of prefix-linearity in supremum and infimum cases.

Example 11. We equip the game of Fig. 1 with two weight functions w_1, w_2, leading to the game of Fig. 2. Suppose that $f_1 = \mathsf{LimSup}_1$ and $f_2 = \overline{\mathsf{MP}}_2$. The preferences of the players with respect to plays $(v_0 v_1)^\omega$ and $v_0 v_2^\omega$ are opposed since

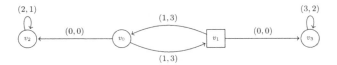

Fig. 2. A quantitative two-player game

$f_1((v_0 v_1)^\omega) = 1 < f_1(v_0 v_2^\omega) = 2$ for player 1, and $f_2(v_0 v_2^\omega) = 1 < f_2((v_0 v_1)^\omega) = 3$ for player 2.

In the sequel, games with the Boolean payoff functions of Definition 8 are called *Boolean games*. Similarly games with the quantitative payoff functions of Definition 10 are called *quantitative games*. We also speak about *reachability game*, *supremum game*, etc., when we want to refer to a game where *all* the players use the *same* type of payoff function. The complexity results mentioned later depend on the number of vertices, edges and players, as well as on the number of colors (resp. pairs, elements of $\mathcal{F}$) for parity (resp. Rabin/Streett, Muller) games, and on numerical rational values (of weights, discount factor, and bounds) given in binary for quantitative games.

3   Two-Player Zero-Sum Games

In two-player zero-sum games, players 1 and 2 have opposite objectives. This class of games has been much studied. In particular solutions to Problem 4 are well established for Boolean games and quantitative games as introduced in Sect. 2.3. Before presenting them, we begin with the simplest situation of games played by a *unique player* and we show that the problems studied in this article are connected to problems in automata theory and numeration systems.

3.1   One-Player Games

In *one-player games*, player 1 has no opponent, he is the only player to choose the next vertex at any moment of a play. In other words, a strategy σ_1 for player 1 is nothing else than a play ρ in the game. The statement of Problem 4 thus simplifies as follows:[10]

Problem 12. Let (G, v_0) be an initialized one-player game. Let $\mu, \nu \in P$ be two bounds. Decide whether there exists a play $\rho \in Plays(v_0)$ such that $\mu \preceq f(\rho)$ (resp. $\mu \preceq f(\rho) \preceq \nu$)?

---

[10] In Sect. 3.1, we omit index 1 everywhere since player 1 is the unique player of the game.

Boolean Games. For Boolean games, this problem is interesting only with bounds $\mu = \nu = 1$. Indeed recall that the payoff function f is Boolean and that player 1 prefers plays ρ such that $f(\rho) = 1$. This is the classical well-known *non emptiness problem for automata* [55]. For instance, Problem 12 for one-player reachability (resp. Büchi) games with $\mu = \nu = 1$ is the non emptiness problem for automata accepting finite words (resp. Büchi automata accepting infinite words).

Theorem 13. *Let (G, v_0) be an initialized one-player Boolean game. Then Problem 12 (with $\mu = \nu = 1$) is decidable in polynomial time with positional winning strategies, except for Streett and Muller games where finite-memory strategies are necessary and sufficient.*

Let us comment this theorem. Notice that a winning strategy for player 1 that is finite-memory (resp. positional) means that the corresponding winning play ρ, or in terms of automata the accepted word, is a (resp. simple) lasso. It is well-known that positional strategies are sufficient for Büchi objectives. This also happens for the other objectives except for Streett and Muller objectives (we will discuss this point in more details in Sect. 3.2, see Theorem 21). Example 14 illustrates that finite-memory strategies are necessary for Streett and Muller games. In cases where positional strategies are sufficient, an algorithm for Problem 12 has thus to concentrate on the existence of winning simple lassos, which can be easily done in polynomial time. The case of Streett and Muller games can also be solved in polynomial time [31,41]. Problem 12 is NL-complete for reachability and Büchi games [45,67] as well as for safety, co-Büchi, Rabin, parity, and Muller games, and it is P-complete for Streett games [31,60].

Example 14. Consider the initialized one-player game (G, v_0) of Fig. 3 with $V = \{v_0, v_1, v_2\}$. For the Muller objective with $\mathcal{F} = \{V\}$ (or the Streett objective with the two pairs $(F_1, G_1), (F_2, G_2)$ such that $F_1 = \{v_1\}$, $G_1 = V$ and $F_2 = \{v_2\}$, $G_2 = V$), a winning play $\rho \in Plays(v_0)$ cannot be a simple lasso as it has to alternate between v_1 and v_2.

Fig. 3. A one-player game

Quantitative Games. Let us turn to quantitative games. The existence of plays ρ with $\mu \leq f(\rho)$ in one-player quantitative games (threshold problem) have been studied in [19].

Theorem 15 *[19]. Let (G, v_0) be an initialized one-player quantitative game, and $\mu \in \mathbb{Q}$ be a rational threshold. Then deciding whether there exists a play $\rho \in Plays(v_0)$ such that $\mu \leq f(\rho)$ is solvable in polynomial time with positional strategies.*

Let us comment this theorem. Dealing with functions Sup, Inf, LimSup, and LimInf is equivalent to respectively consider reachability, safety, Büchi, and co-Büchi objectives (studied in Theorem 13). For instance, satisfying $\mu \leq \mathsf{Sup}(\rho)$ is equivalent to visiting an edge with a weight $\geq \mu$ along ρ. For functions $\overline{\mathsf{MP}}$, $\underline{\mathsf{MP}}$, and Disc^λ, once one knows that positional strategies are sufficient (we will discuss this point in more details in Sect. 3.2), the problem again reduces to the existence of a simple lasso $\rho = hg^\omega$ with maximum payoff $f(\rho)$. In case of mean-payoff function, recall that both payoffs $\overline{\mathsf{MP}}(\rho)$, $\underline{\mathsf{MP}}(\rho)$ coincide and are equal to the average weight of the cycle g. A polynomial algorithm is proposed in [47] to compute a cycle in a weighted graph with maximum average weight. The case of function Disc^λ is polynomially solved by a linear programming approach in [2].

We now discuss the existence of a play ρ such that $\mu \leq f(\rho) \leq \nu$, given two rational bounds $\mu, \nu \in \mathbb{Q}$ (constraint problem). The problem is more involved, in particular it is currently unsolved for function Disc^λ.

Theorem 16 *[43,66]. Let (G, v_0) be an initialized one-player quantitative (except discounted sum) game, and $\mu, \nu \in \mathbb{Q}$ be two rational bounds. Then deciding whether there exists a play $\rho \in Plays(v_0)$ such that $\mu \leq f(\rho) \leq \nu$ is solvable in polynomial time. Positional strategies are sufficient for supremum, infimum, limsup, and liminf games, whereas finite-memory is necessary and sufficient for mean-payoff $\overline{\mathsf{MP}}$ and $\underline{\mathsf{MP}}$ games.*

Let us comment this theorem. If we focus on function LimSup, looking for a play ρ such that $\mu \leq \mathsf{LimSup}(\rho) \leq \nu$ reduces to the non emptiness problem for Rabin automata (studied in Theorem 13). Indeed the required play ρ is such that at least one weight seen infinitely often along ρ is $\geq \mu$ and none of them is $> \nu$. A similar approach exists for functions Sup, Inf and LimInf. Whereas positional winning strategies are sufficient in all these cases, finite-memory is needed for mean-payoff functions as indicated in Example 17. Since finite-memory strategies are sufficient [43], the problem in both cases $\overline{\mathsf{MP}}$, $\underline{\mathsf{MP}}$ reduces to the existence of a lasso ρ satisfying the constraints. This can be checked in polynomial time by solving a linear program [66].

Example 17. Consider the game of Fig. 3 equipped with the weight function w that labels the two left edges by 0 and the two right edges by 2. A winning play ρ for $\mu = \nu = 1$ cannot be a simple lasso (with payoff either 0 or 2). However the non simple lasso $\rho = (v_0 v_1 v_0 v_2)^\omega$ is winning.

Concerning function Disc^λ, Problem 12 is open. It is closely related to the following open problem, called *target discounted-sum problem* in [6].

Problem 18. Given three rational numbers a, b and t, and a rational discount factor $\lambda \in]0, 1[$, does there exist an infinite sequence $u = u_0 u_1 \ldots \in \{a, b\}^\omega$ such that $\sum_{n=0}^\infty u_n \lambda^n$ is equal to t?

The authors of [6] show that Problem 18 is related to several open questions in mathematics and computer science. In particular it is related to *numeration systems* and more precisely to β-representations of real numbers [4,50]. Given

$\beta > 1$ a real number (the base) and $A \subseteq \mathbb{N}$ a finite alphabet (the set of digits), a β-*representation* of a real number $x \geq 0$ is an infinite sequence $(x_n)_{n \leq k} \in A^\omega$, also written $x_k \ldots x_0.x_{-1}x_{-2} \ldots$, such that $x = \sum_{n \leq k} x_n \beta^n$. A well-known result [58] is that every $x \geq 0$ has a β-representation using $A = \{0, 1, \ldots, \lceil \beta - 1 \rceil\}$. It follows that Problem 18 asks whether t has a β-representation $x_0.x_{-1}x_{-2} \ldots$ (with $k = 0$) using $\beta = \frac{1}{\lambda}$ and $A = \{a, b\}$. This problem is therefore decidable when $a = 0, b = 1$ and $\lambda \geq \frac{1}{2}$. Indeed using the result of [58], either $t > \frac{1}{\beta - 1}$ and it has no β-representation $x_0.x_{-1}x_{-2} \ldots \in \{0, 1\}^\omega$, or $t \leq \frac{1}{\beta - 1}$ and it has such a β-representation. Other partial results to Problem 18 can be found in [6].

3.2   Two-Player Games

We now turn to two-player zero-sum games. In Problem 4, the objective of player 1 is the set Ω of plays ρ such that $\mu \preceq_1 f_1(\rho)$ (resp. $\mu \prec_1 f_1(\rho) \preceq_1 \nu$), whereas player 2 has the opposite objective $Plays \setminus \Omega$. Examples of the threshold problem are the following ones: in a reachability game, player 1 aims at reaching some target set of vertices whereas player 2 tries to prevent him from reaching it; in a limsup game, player 1 aims at maximize the payoff $\mathsf{LimSup}(\rho)$ of the play ρ (in a way to be $\geq \mu$) whereas player 2 tries to minimize it. Recall that by Martin's theorem, every two-player zero-sum games with Borel objectives is determined. This large class of games includes the objectives Ω of player 1 in Problem 4 for the Boolean and quantitative games introduced in Sect. 2.3. A lot of research has been developed to solve Problem 4 that we present in this section. In Sects. 3.2 and 3.3, as the objectives Ω and $Plays \setminus \Omega$ of players 1 and 2 only depend on f_1, $\prec_1$, and P_1, we simplify the used notation by omitting index 1.

Criterium for Uniform Optimal Strategies We begin by studying the winning strategies that player 1 can use for the threshold problem in Problem 4. This is related to the notion of *value* and *optimal* strategy.

Definition 19. *Let (G, v_0) be an initialized two-player zero-sum game. If there exists $val(v_0) \in P$ such that*

- *player 1 has a strategy σ_1 such that $val(v_0) \preceq f(\rho)$ for all plays ρ in $Plays(v_0)$ consistent with σ_1, and*
- *player 2 has a strategy σ_2 such that $f(\rho) \preceq val(v_0)$ for all plays ρ in $Plays(v_0)$ consistent with σ_2,*

then $val(v_0) = f(\langle \sigma_1, \sigma_2 \rangle_{v_0})$ is the value *of v_0 and σ_1 (resp. σ_2) is an optimal strategy for player 1 (resp. player 2).*

Intuitively, $val(v_0)$ is the highest threshold μ for which player 1 can ensure (with an optimal strategy) a payoff $f(\rho)$ such that $\mu \preceq f(\rho)$. In this definition, the antagonistic player 2 behaves in the opposite way. When the value $val(v_0)$ exists and is computable, the threshold problem is easily solved: we just check whether the given threshold μ satisfies $\mu \preceq val(v_0)$. Moreover both players can limit themselves to use optimal strategies, that is, if player 1 has a winning strategy (resp. no winning strategy) for the threshold problem, then player 1 (resp. player 2) can use an optimal strategy as winning strategy (resp. for the opposite objective).

Example 5 (continued). Let us come back to the two-player zero-sum game of Example 5. Recall that in (G, v_0), player 1 has a winning strategy for the threshold problem with $\mu = p_2$ but not with $\mu = p_3$, meaning that $val(v_0) = p_2$. Indeed, one can check that $val(v_0) = val(v_1) = val(v_2) = p_2$ and $val(v_3) = p_3$, and that both players have optimal strategies that are positional, and even more uniform. The values are indicated under the vertices in Fig. 4, and the two uniform optimal strategies are given as thick edges.

Fig. 4. Values and uniform optimal strategies for a two-player zero-sum game with $f((v_0v_1)^\omega) = p_1 \prec f(v_2^\omega) = p_2 \prec f(v_3^\omega) = p_3$

We will see later in this section that Boolean and quantitative games often have uniform optimal strategies (see Theorems 21 and 22). In [36], the authors propose a *unified approach* to all these results: they give a general criterium on the payoff function that guarantees uniform optimal strategies for both players.

Theorem 20 *[36].*[11] *Let G be a two-player zero-sum game with a preference relation $\prec$ on P such that each subset of P has an infimum and a supremum. If the payoff function f is* fairly mixing, *that is,*

1. *$\forall h\rho, h\rho' \in Plays$, if $f(\rho) \preceq f(\rho')$ then $f(h\rho) \preceq f(h\rho')$,*
2. *$\forall h\rho, h\rho' \in Plays$, $\min\{f(\rho), f(h^\omega)\} \preceq f(h\rho) \preceq \max\{f(\rho), f(h^\omega)\}$,*
3. *$\forall h_k \in Hist, k \in \mathbb{N}$,*
   $\min\{f(h_0h_2h_4\ldots), f(h_1h_3h_5\ldots), \inf_k f(h_k^\omega)\}$
   $\preceq f(h_0h_1h_2h_3\ldots)$
   $\preceq \max\{f(h_0h_2h_4\ldots), f(h_1h_3h_5\ldots), \sup_k f(h_k^\omega)\},$

then both players have uniform optimal strategies.

Let us comment this theorem. The first condition is condition (1) of prefix-linearity. If f is prefix-independent, then the first and the second conditions are trivially satisfied. The third condition is concerned with shuffles of histories. Let us apply this theorem to quantitative games, for instance to function LimSup (see Definition 10). This function is prefix-independent and satisfies the third

---

[11] The hypotheses of this theorem are those given in the full version of [36] available at http://www.labri.fr/perso/gimbert/.

condition since $\inf_k \mathsf{LimSup}(h_k^\omega) \leq \mathsf{LimSup}(h_0 h_1 h_2 h_3 \ldots) \leq \sup_k \mathsf{LimSup}(h_k^\omega)$. One can check that the payoff functions of all quantitative games are fairly mixing, as well as the payoff functions of the Boolean games with reachability, safety, Büchi, co-Büchi, and parity objectives [36] (but not with Streett and Muller objectives).

The proof[12] of Theorem 20 is simple and elegant; it is by induction on $|E| - |V|$. If $|E| = |V|$ then there is exactly one outgoing edge for each vertex and thus both players have a unique possible strategy that is therefore uniform and optimal. Suppose that $|E| > |V|$ and let us focus on player 1 (a symmetric argument is used for player 2). If all vertices $v \in V_1$ have only one outgoing edge, then player 1 has a unique strategy, and it is uniform and optimal. Suppose that some $v \in V_1$ has at least two outgoing edges. We partition this set of edges into two non empty subsets E'_v and E''_v. From G we define two smaller games G' and G'' with the same vertices and edges except that the set of outgoing edges from v is restricted to E'_v in G' and to E''_v in G''. By induction hypothesis, v has a value $val'(v)$ in G' and $val''(v)$ in G'', and both players have uniform optimal strategies, respectively σ'_1, σ'_2 in G' and σ''_1, σ''_2 in G''. W.l.o.g. $val''(v) \preceq val'(v)$, we then choose σ'_1 as optimal strategy for player 1 in G and for all $u \in V$, we take their value $val'(u)$ in G' as their value in G. Clearly σ'_1 is optimal and uniform in G. The rest of the proof consists in defining a strategy for player 2 (from σ'_2 and σ''_2) that is optimal in G. This is possible thanks to the three conditions of Theorem 20 applied on plays decomposed according to occurrences of v.

Further results can be found in [37]: a characterization of payoff functions is given guaranteeing the existence of uniform optimal strategies for both players. From this characterization, it follows that if both players have uniform optimal strategies when playing solitary in one-player games, then they also have uniform optimal strategies in zero-sum two-player games.

Boolean Games. Let us now focus on Boolean games. As for one-player games, we limit the study of Problem 4 (threshold and constraint problems) to the only interesting case $\mu = \nu = 1$. The following theorem for two-player games is the counterpart of Theorem 13 for one-player games.

Theorem 21. *Let (G, v_0) be an initialized two-player zero-sum Boolean game. Then Problem 4 (with $\mu = \nu = 1$) is*

- P-*complete with uniform winning strategies for reachability, safety, Büchi, and co-Büchi objectives [3, 30, 38, 44],*
- P-*complete with finite-memory winning strategies for Muller[13] objective [41],*
- NP-*complete with uniform winning strategies for Rabin objective [29, 30],*
- co-NP-*complete with finite-memory winning strategies for Streett objective [16, 30],*
- *in* NP $\cap$ co-NP *with uniform winning strategies for parity objectives [30].*

---

[12] Theorem 20 is given in [36] for real-valued payoff functions $f : Plays \to \mathbb{R}$ and the usual ordering $<$, but its proof is easily generalized to the statement given here.

[13] It is PSPACE-complete for the colored variant of Muller objective [42, 52].

Let us comment this theorem. The existence of uniform winning strategies (for all objectives except Rabin and Muller objectives) was previously mentioned as a consequence of Theorem 20 [36]. Notice that here a value $val(v_0) = 1$ is equivalent to say that player 1 has a winning strategy for Problem 4. In case player 1 has no winning strategy ($val(v_0) = 0$), it follows that player 2 has a winning strategy for the opposite objective by Martin's theorem. Hence Theorem 21 also gives information for player 2 by considering the opposite objective. In [48], the author gives general conditions on Boolean objectives that guarantee the existence of a uniform winning strategy for one of the players (and not necessarily for both players). This includes the case of Rabin games where player 1 has a uniform winning strategy (whereas player 2 needs to use a finite-memory strategy to win the opposite Streett objective).

Problem 4 is decidable in $O(|V| + |E|)$ time for reachability and safety games [38], and the current best algorithm for Büchi and co-Büchi games is in $O(|V|^2)$ time [22]. For Muller games with $\mathcal{F} \subseteq 2^V$, the complexity is in $O(|\mathcal{F}| \cdot (|\mathcal{F}| + |V| \cdot |E|)^2)$ time [41], whereas for Rabin and Streett games with l pairs (F_k, G_k), it is in $O(|V|^{l+1}l!)$ time [56]. Concerning parity games, the complexity class of Problem 4 is refined to UP ∩ co-UP in [46] and a major open problem is whether it can be solved in polynomial time. Very recently, a breakthrough quasi-polynomial time algorithm has been proposed in [17] for parity games.

Quantitative Games Let us turn to quantitative games for which we first give results for the threshold problem, and then for the constraint problem. The following theorem provides the known results to the threshold problem. It describes the simplest form of winning strategies for player 1 (resp. player 2) when he has a winning strategy for this problem (resp. for ensuring the opposite objective when player 1 has no winning strategy).

Theorem 22. *Let (G, v_0) be an initialized two-player zero-sum quantitative game, and $\mu \in \mathbb{Q}$ be a rational bound. Then the threshold problem (in Problem 4) is*

- *P-complete for supremum, infimum, limsup, and liminf games with uniform winning strategies for both players,*
- *in NP ∩ co-NP for mean-payoff and discounted sum games with uniform winning strategies for both players [71].*

Let us comment this theorem. We already know the existence of uniform winning strategies from Theorem 20 [36]. The P-completeness for supremum, infimum, limsup, and liminf games follows from the P-completeness for reachability, safety, Büchi, and co-Büchi games in Theorem 21. Parity games are polynomially reducible to mean-payoff games [46] which are themselves polynomially reducible to discounted sum games [71]. For these three classes of games, from the existence of uniform winning strategies, we get a threshold problem in NP as follows: guess a uniform strategy σ_1 for player 1 (by choosing one outgoing edge (v, v') for all $v \in V_1$), fix this strategy σ_1 in the game G to get a one-player game G_{σ_1}, apply the related polynomial time algorithm of Theorems 13 or 15 (from the point of view of player 2 who controls G_{σ_1}). The co-NP membership is symmetrically obtained with player 2.

Concerning the constraint problem, recall that it is more complex already for one-player games (see Sect. 3.1) with no known solution for discounted sum games (see Problem 18).

Theorem 23. *Let (G, v_0) be an initialized two-player zero-sum quantitative (except discounted sum) game, and $\mu, \nu \in \mathbb{Q}$ be rational bounds. Then the constraint problem (in Problem 4) is*

- *P-complete for supremum, infimum, limsup, and liminf games with uniform winning strategies for both players [13, 43],*
- *in NP ∩ co-NP for mean-payoff games with finite-memory (resp. uniform) winning strategies for player 1 (resp. player 2) [43].*

Discounted sum games are studied with bounds μ, ν such that $\mu < \nu$ (to avoid the case $\mu = \nu$ of Problem 18) in [43] where it is proved that the constraint problem is PSPACE-complete with finite-memory winning strategies for both players.

3.3   Variants of Preferences

Several extensions[14] of two-player zero-sum Boolean and quantitative games have been studied in the literature, by using preferences that are irreflexive and transitive but *not necessarily total*, or more generally by using *preorders* $\preceq$ that are reflexive and transitive binary relations (hence, $\preceq$ is not supposed to be total and one can have $p \preceq p'$ and $p' \preceq p$ such that $p \neq p'$).

Such variants naturally appear when we study *intersection of objectives* instead of a single objective as in Sect. 2.3:

- *Intersection of homogeneous objectives.* For instance player 1 has l reachability objectives $U_1, \ldots, U_l$ (instead of just one), and he wants to visit all the sets $U_1, \ldots, U_l$.
- *Intersection of heterogeneous objectives.* In this more general case, player 1 has several objectives not necessarily of the same type. Let us imagine a situation where he has two quantitative objectives depending on two weight functions on the graph, like ensuring a threshold for the liminf of weights with respect to the first weight function and another threshold for the mean-payoff with respect to the second weight function.

In this context, for player 1, we consider a tuple $\bar{f}$ of payoff functions and a tuple $\bar{w}$ of weight functions (instead of a single payoff function f defined from a single weight function w) such that each function $f_k : Plays \rightarrow \mathbb{R}$ is defined from

---

[14] The reader who prefers to know classical solutions to Problem 7 for multi-player non zero-sum games can skip this section and go directly to Sect. 4.

$w_k : E \to \mathbb{Q}$.[15] Tuples of payoffs $\bar{p} = \bar{f}(\rho)$ and $\bar{p}' = \bar{f}(\rho')$ are then compared using the usual ordering on tuples of reals: $\bar{p} \prec_{\mathsf{ord}} \bar{p}'$ iff $p_k \leq p'_k$ for all components k and there exists k such that $p_k < p'_k$ (the preference relation $\prec_{\mathsf{ord}}$ is not total). Let us mention some results first for quantitative objectives and then for Boolean objectives.

Combination of Quantitative Objectives. The threshold problem takes the following form: given a tuple $\bar{\mu}$ of rational thresholds, decide whether player 1 has a strategy σ_1 that ensures a payoff $\bar{f}(\rho)$ such that $\bar{\mu} \prec_{\mathsf{ord}} \bar{f}(\rho)$ for all plays ρ consistent with σ_1.

Theorem 24 *[69]. Let (G, v_0) be an initialized two-player zero-sum game with homogeneous intersections of mean-payoff objectives. Then the threshold problem (in Problem 4) is*

- *in* NP $\cap$ co-NP *for functions* $\overline{\mathsf{MP}}$,
- *is* co-NP-*complete for functions* $\underline{\mathsf{MP}}$.

In both cases, infinite memory is required for winning strategies of player 1 whereas uniform winning strategies are sufficient for player 2.

This theorem indicates different behaviors for the functions $\overline{\mathsf{MP}}$ and $\underline{\mathsf{MP}}$. This is illustrated with the example of the initialized one-player game (G, v_0) depicted in Fig. 5, where player 1 wants to ensure the intersection of two homogeneous objectives. It is shown in [69] that for a pair of functions $\underline{\mathsf{MP}}$, player 1 can ensure a threshold $(1, 1)$, and that for a pair of functions $\overline{\mathsf{MP}}$, he can ensure a threshold $(2, 2)$ (which is impossible with $\underline{\mathsf{MP}}$). In both cases infinite memory is necessary. Indeed recall that with a finite-memory strategy the produced play is a lasso $\rho = hg^\omega$ such that $\overline{\mathsf{MP}}(\rho) = \underline{\mathsf{MP}}(\rho)$ is the average weight of the cycle g. Here this average weight has the form $a \cdot (2, 0) + b \cdot (0, 0) + c \cdot (0, 2) = (2a, 2c)$, with $a + b + c = 1$ and $b > 0$. Clearly $(1, 1) \not\prec_{\mathsf{ord}} (2a, 2c)$ showing that player 1 is losing for threshold $(1, 1)$ with finite-memory strategies.

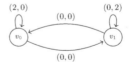

Fig. 5. A one-player game with a pair of weight functions $\bar{w}$

In [68], the author studies objectives equal to *Boolean combinations* of inequalities $f_k(\rho) \sim \mu_k$, with $\sim \in \{\leq, \geq\}$ and $f_k \in \{\overline{\mathsf{MP}}, \underline{\mathsf{MP}}\}$: deciding whether player 1 has a winning strategy in (G, v_0) becomes undecidable. However, this

---

[15] This tuple of payoff functions is used by player 1 contrarily to Definition 2 where function f_i is used by player i for all $i \in \Pi$.

problem remains decidable and is EXPTIME-complete for CNF/DNF Boolean combinations of functions taken among $\{\mathsf{Sup}, \mathsf{Inf}, \mathsf{LimSup}, \mathsf{LimInf}, \mathsf{WMP}\}$ [13], where WMP is an interesting window variant of mean-payoff introduced in [20]. The threshold problem is P-complete (resp. EXPTIME-complete) for a single WMP objective (resp. an intersection of WMP objectives) [20]. Recall that it is in NP ∩ co-NP for a single function $\overline{\mathsf{MP}}$ or $\underline{\mathsf{MP}}$ (see Theorem 22).

Combination of Boolean Objectives. Concerning one-player games, Boolean combinations of Büchi and co-Büchi objectives are introduced in [31] as a generalization of Rabin and Streett objectives. It is proved that the non emptiness problem for this class of automata is NP-complete (for a comparison see Theorem 13). Concerning two-player games, the intersection of homogeneous objectives is simple for safety, co-Büchi, Streett, and Muller cases. Indeed the intersection of safety (resp. co-Büchi, Streett, Muller) objectives is again a safety (resp. co-Büchi, Streett, Muller) objective. In the other cases, we have the following results to be compared with those of Theorem 21.

Theorem 25. *Let (G, v_0) be an initialized two-player zero-sum game with an intersection of homogeneous objectives. Then Problem 4 is*

- PSPACE-*complete for reachability objectives with finite-memory winning strategies for both players [32],*
- P-*complete for Büchi objectives with finite-memory (resp. uniform) winning strategies for player 1 (resp. player 2) [21],*
- co-NP-*complete for parity objectives with finite-memory (resp. uniform) winning strategies for player 1 (resp. player 2) [24],*
- PSPACE-*complete for Rabin objectives with finite-memory winning strategies for both players.*[16]

Problem 4 is PSPACE-complete for heterogeneous intersections of reachability and Büchi objectives [13] as well as for Boolean combinations of Büchi objectives [1,42].

Lexicographic and Secure Preferences. For a tuple $\bar{f}$ of payoff functions defined from a tuple $\bar{w}$ of weight functions for player 1, let us mention two other natural preference relations $\prec$.

Definition 26. *Let $\bar{p}, \bar{p}'$ be two tuples of real payoffs.*

- *lexicographic preference: $\bar{p} \prec_{\mathsf{lex}} \bar{p}'$ iff there exists k such that $p_k < p'_k$ and $p_j = p'_j$ for all $j \leq k$. That is, player 1 prefers to first maximize the first component, then the second, then the third, etc. (see for instance [5]).*

---

[16] We found no reference for this result. The PSPACE membership (resp. the finite memory of the strategies) follows from [1] (resp. [13]). In [1], games with a union of a Streett objective and a Rabin objective are shown to be PSPACE-hard. It is thus also the case for games with a union of Streett objectives. By Martin's theorem, it follows that games with an intersection of Rabin objectives are PSPACE-hard.

- secure preference: $\bar{p} \prec_{\mathsf{sec}} \bar{p}'$ iff either $p_1 < p_1'$ or $\{p_1 = p_1', p_k \geq p_k'$ for all components $k > 1$, and there exists $k > 1$ such that $p_k > p_k'\}$. That is, player 1 prefers to first maximize the first component, and then to minimize all the other components (see for instance [28]).

The lexicographic preference is total whereas the secure preference is total only for *pairs* (instead of tuples) of payoffs. In the latter case, we get a preference which is close to the lexicographic ordering: player 1 prefers to maximize the first component, and then to minimize the second one. The secure preference is used in the notion of secure equilibrium discussed later in Sect. 4.3.

Theorem 27. *Let (G, v_0) be an initialized two-player zero-sum game.*

- *Suppose that $\prec$ is the lexicographic preference $\prec_{\mathsf{lex}}$. Then the threshold problem (in Problem 4) for function $\underline{\mathsf{MP}}$ is in $\mathsf{NP} \cap \mathsf{co\text{-}NP}$ with uniform winning strategies for both players [5].*
- *Suppose that $\prec$ is the secure preference $\prec_{\mathsf{sec}}$ on pairs of payoffs. Then the threshold problem (in Problem 4) is in $\mathsf{NP} \cap \mathsf{co\text{-}NP}$ (resp. P-complete) for functions $\overline{\mathsf{MP}}$, $\underline{\mathsf{MP}}$, and Disc^λ (resp. for functions Sup, Inf, LimSup, and LimInf). Moreover both players have uniform (resp. positional) winning strategies for functions LimSup, LimInf, $\overline{\mathsf{MP}}$, $\underline{\mathsf{MP}}$, and Disc^λ (resp. for functions Sup and Inf) [14].*

In this theorem, it is supposed that the components f_k of $\bar{f}$ are all of the same type (for instance they are all limsup functions); and some results stating the existence of uniform winning strategies can be established thanks to Theorem 20. Notice that the secure preference is limited to pairs of payoffs in a way to be total, which is a necessary condition when dealing with values. Notice also that the authors in [5] consider liminf average of the weight *vector* under lexicographic ordering whereas the authors in [14] consider the secure ordering of components where each *component* is the liminf average value.

The threshold problem is studied in [7] in a general context: the players can use various preorders (like the lexicographic preference, a preorder given by a Boolean circuit, etc.), the players play concurrently and not in a turned-based way, and the objectives are Boolean as in Definition 8.

4   Multi-player Non Zero-Sum Games

In multi-player non zero-sum games, the different players $i \in \Pi$ are not necessarily antagonistic, they have their own payoff functions f_i and preference relations $\prec_i$. Each of them follows a strategy σ_i, the resulting strategy profile $(\sigma_i)_{i \in \Pi}$ induces a play that should be satisfactory to all players. As explained in Sect. 2.2 (see Definition 6), a classical solution profile is the notion of NE, where no player has an incentive to deviate when the other players stick to their own strategies. It is proved in [25,39] that there exists an NE in every initialized multi-player non zero-sum game with Borel Boolean objectives. We go further by presenting in this section additional existence results for quantitative games and some known results for NEs as a solution to Problem 7 (threshold problem and constraint problem). As in Sect. 3, we focus on general approaches.

4.1   Characterization of Outcomes of NE

Given a multi-player non zero-sum game G and an initial vertex v_0, we begin by a *characterization* of plays $\rho \in Plays(v_0)$ that are the outcome of an NE $(\sigma_i)_{i \in \Pi}$ in (G, v_0). It will imply the existence of NE in large classes of games (see Corollaries 29 and 30), and will be useful for the study of Problem 7 (see Theorems 31 and 32). This characterization is related to a *family of two-player zero-sum games* G_i, one for each $i \in \Pi$, associated with G and defined as follows. *(i)* The game G_i has the same arena as G, *(ii)* the two players are player i (player 1) and player $-i$ (player 2) formed by the *coalition* of the other players $j \in \Pi \setminus \{i\}$, *(iii)* the payoff function of player i is equal to f_i and his preference relation is equal to $\prec_i$[17]. For all $v \in V$, when it exists, we denote by $val_i(v)$ the value of vertex v in game G_i, and by τ_i^v, τ_{-i}^v the related optimal strategies for players $i, -i$ respectively in (G_i, v) (see Definition 19).

Proposition 28. *Let G be a multi-player non zero-sum game such that for all $i \in \Pi$,*

- *the payoff function f_i is prefix-linear, and*
- *in the game G_i, all vertices has a value.*

Then $\rho = \rho_0 \rho_1 \ldots \in Plays(v_0)$ is the outcome of an NE in (G, v_0) iff $val_i(\rho_k) \preceq_i f_i(\rho_{\geq k})$ for all $i \in \Pi$ and all $k \in \mathbb{N}$ such that $\rho_k \in V_i$.

The condition of this proposition asks that for all k, if vertex ρ_k is controlled by player i, then in the two-player zero-sum game G_i, its value is less preferred or equal to the payoff of the suffix $\rho_{\geq k}$. The proposed characterization appears under various particular forms, for instance in [14,39,65]. It is here given under two general conditions already studied in Sect. 3.2. Recall that almost all the payoff functions considered in Sect. 2.3 are prefix-linear and that for all the related two-player zero-sum games G_i, the vertices have a value. Notice that when f_i is prefix-independent, condition $val_i(\rho_k) \preceq_i f_i(\rho_{\geq k})$ for all $k \in \mathbb{N}$ with $\rho_k \in V_i$ simplifies in $\max\{val_i(\rho_k) \mid k \in \mathbb{N}, \rho_k \in V_i\} \preceq_i f_i(\rho)$ (the maximum exists since V_i is finite).

Example 3 (continued). An example of NE with outcome $\rho = v_0 v_1 v_3^\omega$ was given in Example 3 for the initialized game (G, v_0) of Fig. 1. Let us verify that ρ satisfies the characterization of Proposition 28. Recall that both players use the same payoff function f that is prefix-independent. The values of G_1 were computed in Example 5: $val_1(v_0) = val_1(v_1) = val_1(v_2) = p_2$ and $val_1(v_3) = p_3$. Similarly one can compute the values of G_2: $val_2(v_0) = val_2(v_2) = p_2$ and $val_2(v_1) = val_2(v_3) = p_3$. One checks that $\max\{val_i(\rho_k) \mid k \in \mathbb{N}, \rho_k \in V_i\} \preceq_i f(\rho) = p_3$, for $i = 1, 2$.

The proof of Proposition 28 is easy to establish.

Firstly suppose that ρ is the outcome of an NE $(\sigma_i)_{i \in \Pi}$ and that there exist $i \in \Pi$ and $k \in \mathbb{N}$ with $\rho_k \in V_i$ such that $f_i(\rho_{\geq k}) \prec_i val_i(\rho_k)$. Let us show that

---

[17] Recall that the payoff function and the preference relation of the second player do not matter in two-player zero-sum games.

player i has a profitable deviation σ_i' with respect to $(\sigma_i)_{i \in \Pi}$ in contradiction with $(\sigma_i)_{i \in \Pi}$ being an NE. The strategy σ_i' consists in playing according to σ_i until producing $\rho_{\leq k}$ and from ρ_k in playing according to his optimal strategy $\tau_i^{\rho_k}$ (in (G_i, ρ_k)). The payoff of the resulting play π from ρ_k is such that $val_i(\rho_k) \preceq_i f_i(\pi)$ by optimality of $\tau_i^{\rho_k}$, and thus $f_i(\rho_{\geq k}) \prec_i f_i(\pi)$. From prefix-linearity of f_i it follows that $f_i(\rho) = f_i(\rho_{<k}\rho_{\geq k}) \prec_i f_i(\rho_{<k}\pi)$ as required.

Secondly suppose that $val_i(\rho_k) \preceq_i f_i(\rho_{\geq k})$ for all $i \in \Pi$ and all $k \in \mathbb{N}$ such that $\rho_k \in V_i$. We are going to construct an NE by using a well-known method in classical game theory that is used in the proof of the Folk Theorem in repeated games [54]. We define a strategy profile $(\sigma_i)_{i \in \Pi}$ that produces ρ as outcome, and as soon as some player i deviates from ρ, say at vertex ρ_k, all the other players (as a coalition) punish him by playing from ρ_k the optimal strategy $\tau_{-i}^{\rho_k}$ (in (G_i, ρ_k)). Let us show that $(\sigma_i)_{i \in \Pi}$ is an NE. Let σ_i' be a deviating strategy from σ_i for player i, and let ρ' be the outcome of the strategy profile (σ_i', σ_{-i}). Consider the longest common prefix $\rho_{\leq k}$ of ρ and ρ'. Then $\rho_k \in V_i$ and by optimality of $\tau_{-i}^{\rho_k}$, we get $f_i(\rho'_{\geq k}) \preceq_i val_i(\rho_k)$ and thus $f_i(\rho'_{\geq k}) \preceq_i f_i(\rho_{\geq k})$. From prefix-linearity of f_i it follows that $f_i(\rho') \preceq_i f_i(\rho)$ showing that σ_i' is not a profitable deviation for player i.

Notice that in this proof, the first (resp. second) implication only requires condition (2) (resp. (1)) of prefix-linearity of f_i. The next corollary follows from this observation and Proposition 28.

Corollary 29 *[27]. Let G be a multi-player non zero-sum game such that for all $i \in \Pi$,*

- *the payoff function f_i satisfies $f_i(\rho) \preceq_i f_i(\rho') \Rightarrow f_i(h\rho) \preceq_i f_i(h\rho')$ for all $h\rho, h\rho' \in Plays$, and*
- *each game G_i has uniform optimal strategies for both players.*

Then there exists a finite-memory NE in each initialized game (G, v_0).

This corollary is a generalization of a theorem[18] given in [12,27] for the existence of NEs in games equipped with payoff functions $f_i : Plays \rightarrow \mathbb{R}$, $i \in \Pi$. The proof of Corollary 29 is as follows. Let us consider the play $\rho \in Plays(v_0)$ produced by the players when each player i plays according to his optimal strategy τ_i in (G_i, v_0) ($\tau_i^v = \tau_i$ for all vertices v since it is uniform). By construction, ρ is the outcome of an NE because it satisfies the characterization of Proposition 28. Notice that ρ is a simple lasso since each τ_i, $i \in \Pi$, is uniform. Therefore the strategies of the constructed NE are finite-memory with a small memory size bounded by $|V| + |\Pi|$ to remember this lasso and the first player who deviates from ρ.

The existence of an NE is also guaranteed in the following corollary that does not require optimal strategies that are uniform, but in counterpart requires payoff functions that are prefix-independent.

Corollary 30 *[27]. Let G be a multi-player non zero-sum game such that for all $i \in \Pi$,*

---

[18] In [12], one hypothesis is missing: the required optimal strategies must be uniform.

- *the payoff function f_i is prefix-independent, and*
- *each game G_i has (resp. finite-memory) optimal strategies for both players.*

Then there exists an (resp. finite-memory) NE in each initialized game (G, v_0).

This is a generalization of a result given in [27] for games equipped with payoff functions $f_i : Plays \to \mathbb{R}$, $i \in \Pi$. The proof is as follows: under the hypotheses of Corollary 30, one can show that there exist optimal strategies $\tau_i^{v_0}$ in (G_i, v_0), $i \in \Pi$, such that for all plays $\rho \in Plays(v_0)$ consistent with $\tau_i^{v_0}$, we have $\max\{val_i(\rho_k) \mid k \in \mathbb{N}, \rho_k \in V_i\} \preceq_i f_i(\rho)$. Then as in Corollary 29, we consider the play $\rho \in Plays(v_0)$ obtained when each player i plays according to his optimal strategy $\tau_i^{v_0}$. As each f_i is prefix-independent, ρ satisfies the characterization of Proposition 28.

From Corollaries 29 and 30, it follows that there exists an NE (which can be constructed) in every game of Sect. 2.3; the case of Boolean (resp. quantitative) game is proved in [25,39] (resp. in [12,27]). The existence of an NE in discounted sum games can be obtained in a second way: the function Disc^λ is continuous and all games with real-valued continuous payoff functions always have an NE [35,40]. Notice that the two previous corollaries allow mixing the types of functions f_i, like for instance f_1 associated with a Büchi objective, a limsup function f_2, a mean-payoff function f_3, etc.

Conditions generalizing those of Corollaries 29 and 30 are given in [59] that guarantee the existence of a finite-memory NE. Moreover, for most of the given conditions counterexamples are provided that show that they cannot be dispensed with.

4.2 Solution to Problem 7

In this section we study how to solve Problem 7 for NEs (threshold problem and constraint problem). The characterization given in Proposition 28 provides a general approach to solve this problem. Indeed consider the case of initialized games (G, v_0) with prefix-independent payoff functions f_i and such that the vertices of each game G_i, $i \in \Pi$, has a value. Then given two tuples of bounds $(\mu_i)_{i \in \Pi}, (\nu_i)_{i \in \Pi}$, we simply have to check whether there exists a play $\rho \in Plays(v_0)$ such that for all $i \in \Pi$,

$$\max\{val_i(\rho_k) \mid \rho_k \in V_i\} \preceq_i f_i(\rho) \text{ and } \mu_i \preceq_i f_i(\rho) \quad (\text{resp. } \mu_i \preceq_i f_i(\rho) \preceq_i \nu_i).(3)$$

Thanks to this general approach or variations based on Proposition 28, Problem 7 is solved for Büchi, co-Büchi, Streett, and parity games in [64], and for the other Boolean games in [26].

Theorem 31 *[26, 64]. Let (G, v_0) be an initialized multi-player non zero-sum Boolean game. Then Problem 7 is*

- *is P-complete for Büchi and Muller[19] games,*

---

[19] We found no reference for Muller objectives. A sketch of proof is given in the appendix. Problem 7 is PSPACE-complete for the colored variant of Muller objectives [26].

- NP-*complete for reachability, safety, co-Büchi, parity and Streett games,*
- *in* P^{NP}, *and* NP-*hard,* co-NP-*hard for Rabin games.*

Let us explain the proof of NP membership for parity games and the constraint problem with bounds $(\mu_i)_{i \in \Pi}, (\nu_i)_{i \in \Pi}$. As each G_i is a two-player zero-sum parity game, recall that the constraint problem is in NP $\cap$ co-NP with uniform winning strategies for both players (see Theorem 21). The required algorithm in NP is as follows. *(i)* For all $i \in \Pi$, in the game G_i, guess a subset $U_i \subseteq V$ of vertices and two uniform strategies τ_i, τ_{-i} for players $i, -i$ respectively (intuitively we guess $U_i = \{v \in V \mid val_i(v) = 1\}$ and $V \setminus U_i = \{v \in V \mid val_i(v) = 0\}$). Check in polynomial time[20] that τ_i is a winning strategy for player i for the constraint problem in each (G_i, v) with $v \in U_i$ and that τ_{-i} is a winning strategy for player $-i$ for the opposite objective in each (G_i, v) with $v \in V \setminus U_i$. *(ii)* Then for all $i \in \Pi$, we guess $r_i \in V_i$ (intuitively we guess r_i such that $val_i(r_i) = \max\{val_i(\rho_k) \mid k \in \mathbb{N}, \rho_k \in V_i\}$ for the required play ρ). Construct in polynomial time a one-player game G' from G such that each set V_i of vertices is limited to $\{v \in V_i \mid val_i(v) \le val_i(r_i)\}$ and the unique player is formed by the coalition of all players $i \in \Pi$. *(iii)* By (3) it remains to check whether there exists a play ρ in (G', v_0) such that for all $i \in \Pi$, $val_i(r_i) \le f_i(\rho)$ and $\mu_i \le f_i(\rho) \le \nu_i$. Recall that the existence of plays satisfying certain constraints in one-player games was studied in Sect. 3.1, see Theorem 13. Here we are faced with the existence of a play in game with an intersection of parity objectives which can be checked in polynomial time by [31].

Problem 7 can be similarly solved for quantitative games.

Theorem 32 *[26, 64, 65]. Let (G, v_0) be an initialized multi-player non zero-sum quantitative (except discounted sum) game. Then Problem 7 is*

- P-*complete for limsup games,*
- NP-*complete for supremum, infimum, liminf, mean-payoff* $\underline{\mathsf{MP}}_i$, *and mean-payoff* $\overline{\mathsf{MP}}_i$ *games.*

The case of supremum, infimum, limsup and liminf games is equivalent to the case of reachability, safety, Büchi and co-Büchi games presented in Theorem 31, whereas the case of mean-payoff games is studied in [65]. The proof of NP membership for mean-payoff games is based on the approach (3), and is similar to the one given above for parity games. The case of discounted sum games is open. Indeed it is proved in [14] that Problem 18 reduces to Problem 7 with the discounted sum function.[21]

Problem 7 is studied in [7] in a general context: the players can use various preorders, they play concurrently and not in a turned-based way, and the objectives are Boolean as in Definition 8. The general approach proposed in [7] is different from the one of Proposition 28.

---

[20] Recall our comment after Theorem 22.
[21] The reduction is given for another kind of solution profile but it also works for NEs.

4.3 Other Solution Profiles

In this section, we present some other solution profiles. Indeed the notion of NE has several drawbacks: *(i)* Each player is selfish since he is only concerned with his own payoff, and not with the payoff of the other players. *(ii)* An NE does not take into account the sequential nature of games played on graphs. We illustrate these drawbacks in the following two examples of quantitative game.

Example 33. Consider the two-player quantitative game of Fig. 6 such that $f_i = \mathsf{LimSup}_i$ for $i = 1, 2$. The strategy profile depicted with thick edges is an NE. Notice that player 1 could decide to deviate at v_0 by moving to v_2. Indeed he then keeps the same payoff of 1 but also decreases the payoff of player 2 (from 2 to 1) which is bad for player 2. To avoid such a drawback, we will introduce hereafter the concept of secure equilibrium, where each player take cares of his own payoff as well as the payoff of the other players (but in a negative way).

Consider now the game of Fig. 7 where the weights of the loops have been modified. The depicted strategy profile is again an NE. Player 1 has no incentive to deviate at v_0 due to the threat of player 2: player 1 will receive a payoff of $0 < 1$. Such a threat of player 2 is non credible because in the subgame induced by v_2, v_3, v_4, at vertex v_2, it is more rational for player 2 to move to v_4 to get a payoff of 2 instead of going to v_3 where he only receives a payoff of 1. To avoid such a drawback, we will introduce hereafter the concept of subgame perfect equilibrium that takes into account rational behaviors of the players in all subgames of the initial game.

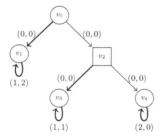

Fig. 6. An NE that is not a secure equilibrium

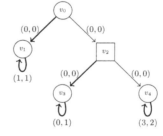

Fig. 7. An NE that is not a subgame perfect equilibrium

Secure Equilibria. The notion of *secure equilibrium* (SE) is introduced in [23] for two-player non zero-sum games. The idea of an SE is that no player has an incentive to deviate in the following sense: he will not be able to increase his payoff, and keeping the same payoff he will not be able to decrease the payoff of the other player. An SE can thus be seen as a contract between the two players which strengthens cooperation: if a player chooses another strategy that is not harmful to himself, then this cannot harm the other player if the latter follows the contract.

The definition of an SE is given in the context of games equipped with payoff functions $f_i : Plays \rightarrow \mathbb{R}$, $i \in \Pi$. It uses the notion of secure preference introduced in Sect. 3.3 (see Definition 26[22]). Let us recall the *secure preference* $\prec_{\mathsf{sec},i}$ for player i: given $\bar{p} = (f_i(\rho))_{i\in\Pi}, \bar{p}' = (f_i(\rho'))_{i\in\Pi}$, we have $\bar{p} \prec_{\mathsf{sec},i} \bar{p}'$ iff either $p_i < p'_i$ or $\{p_i = p'_i, p_k \geq p'_k$ for all components $k \neq i$, and there exists $k \neq i$ such that $p_k > p'_k\}$. Hence player i prefers to increase his own payoff, and in case of equality to decrease the payoffs of all the other players. This preference relation is not total except when there are only two players.

The definition of an SE is very close to the one of NE (see Definition 6). The only difference is that it uses the secure preference:

Definition 34. *Given an initialized game* (G, v_0), *a strategy profile* $(\sigma_i)_{i\in\Pi}$ *is a secure equilibrium if*

$$(f_i(\langle(\sigma_i)_{i\in\Pi}\rangle v_0))_{i\in\Pi} \not\prec_{\mathsf{sec},i} (f_i(\langle\sigma'_i, \sigma_{-i}\rangle v_0))_{i\in\Pi}$$

for all players $i \in \Pi$ *and all strategies* σ'_i *of player* i.

Example 33 (continued). The strategy profile of Fig. 6 is not an SE because player 1 has a profitable deviation if at v_0 he chooses to move to v_2: $(1,2) \prec_{\mathsf{sec},1} (1,1)$.

By definition, every SE is an NE but the converse is false as shown in the previous example. It is proved in [23] that every two-player non zero-sum game with Borel Boolean objectives has an SE; this result is generalized to multi-player games in [28].

Let us turn to quantitative games such that all the players have the same type of payoff function f_i. General hypotheses are provided in [28] that guarantee the existence of an SE in quantitative games, except for functions $\overline{\mathsf{MP}}_i$ and $\underline{\mathsf{MP}}_i$. Thanks to Corollary 29 and Theorem 27, for two-player[23] quantitative games (now including functions $\overline{\mathsf{MP}}_i$ and $\underline{\mathsf{MP}}_i$), there exists such an SE that is finite-memory [14]. Moreover, with the same general approach (3) described previously for NEs, Problem 7 is solved as follows for SEs.

Theorem 35 *[14].* *Let* (G, v_0) *be an initialized two-player non zero-sum quantitative (except discounted sum) game. Then Problem 7 for SEs is*

- *P-complete for supremum, infimum, limsup, and liminf functions,*
- *in* NP $\cap$ co-NP *for functions* $\overline{\mathsf{MP}}_i$ *and* $\underline{\mathsf{MP}}_i$.

The case of discounted sum function is open since it is proved in [14] that Problem 18 reduces to Problem 7 with Disc$^\lambda$. The complexity class of the problem of deciding whether, in an initialized two-player parity game (G, v_0), there exists an SE with payoff respectively equal to $(0,0)$, $(0,1)$, $(1,0)$, and $(1,1)$, is studied in [23, 39].

---

[22] The definition was given for player 1.

[23] A restriction to two-player games is necessary to deal with a secure preference that is total.

Subgame Perfect Equilibria. A solution profile that avoids incredible threats by taking into account the sequential nature of games played on graphs is the notion of *subgame perfect equilibrium* (SPE) [61]. For being an SPE, a strategy profile is not only required to be an NE from the initial vertex but after every possible history of the game.

Before giving the definition of an SPE, we need to introduce the following concepts for an initialized game (G, v_0) with payoff functions f_i and preference relations $\prec_i$, for all $i \in \Pi$. Given a history $hv \in Hist(v_0)$, the *subgame* $(G_{|h}, v)$ of (G, v_0) is the initialized game with payoff functions $f_{i|h}$, $i \in \Pi$, such that $f_{i|h}(\rho) = f_i(h\rho)$ for all plays $\rho \in Plays(v)$ (the preference relation of player i is his preference $\prec_i$ in G). Given a strategy σ_i for player i in (G, v_0), the strategy $\sigma_{i|h}$ in $(G_{|h}, v)$ is defined as $\sigma_{i|h}(h') = \sigma_i(hh')$ for all $h' \in Hist_i(v)$.

Definition 36. *Given an initialized game (G, v_0), a strategy profile $(\sigma_i)_{i \in \Pi}$ is a* subgame perfect equilibrium *if $(\sigma_{i|h})_{i \in \Pi}$ is an NE in each subgame $(G_{|h}, v)$ of (G, v_0) with $hv \in Hist(v_0)$.*

Example 33 (continued). The strategy profile (σ_1, σ_2) of Fig. 7 is not an SPE because in the subgame $(G_{|v_0}, v_2)$, player 2 has a profitable deviation with respect to $(\sigma_{1|v_0}, \sigma_{2|v_0})$ if at v_2 he chooses to move to v_4.

By definition, every SPE is an NE but the converse is false as shown in the previous example. A well-known result is the existence of an SPE in every initialized game (G, v_0) such that its arena is a tree rooted at v_0[24] [49]. The SPE is constructed backwards from the leaves to the initial vertex v_0 in the following way. Suppose that the current vertex v is controlled by player i, and that for each son v' of v one has already constructed an SPE $(\sigma_i^{v'})_{i \in \Pi}$ in the subtree rooted at v'. Then player i chooses the edge (v, v') such that $(\sigma_i^{v'})_{i \in \Pi}$ has the best outcome with respect to his preference relation $\prec_i$. The resulting strategy profile $(\sigma_i^v)_{i \in \Pi}$ is an SPE in the subtree rooted at v.

It is proved in [63] that there exists an SPE in every multi-player non zero-sum game with Borel Boolean objectives, and that in case of ω-regular objectives there exists one that is finite-memory. Existence of an SPE also holds for games with continuous real-valued payoff functions [35, 40] (this is also holds when the functions are upper-semicontinuous (resp. lower-semicontinuous) and with finite range [34] (resp. [57])).

For subgame perfect equilibria, we are not aware of a characterization like the one in Proposition 28. Therefore a solution to Problem 7 for SPEs needs a different approach. Few solutions are known: this problem is in EXPTIME for Rabin games [63] and is NP-hard for co-Büchi games [39].

Whereas NEs exist for large classes of games, see Corollaries 29 and 30, SPEs fail to exist even in simple games like the one of Fig. 1 [62]. Variants of SPE, *weak SPE* and *very weak SPE*, have thus been proposed in [11] as interesting alternatives. In a weak SPE (resp. very weak SPE), a player who deviates from a strategy σ is allowed to use deviating strategies that differ from σ on a *finite number* of histories only (resp. only on the *initial vertex*). Deviating strategies

---

[24] In this particular context, plays are finite paths.

that only differ on the initial vertex are a well-known notion that for instance appears in the proof of Kuhn's theorem [49] with the one-step deviation property. By definition, every SPE is a weak SPE, and every weak SPE is a very weak SPE. Weak SPE and very weak SPE are equivalent notions, but this is not true for SPE and weak SPE [11].

The following theorem gives two general conditions such that each of them separately guarantees the existence of a weak SPE.

Theorem 37 *[15]. Let G be a multi-player non zero-sum game such that*

- *either each payoff function f_i, $i \in \Pi$, is prefix-independent,*
- *or each f_i, $i \in \Pi$, has a finite range.*

Then there exists a weak SPE in each initialized game (G, v_0).

This theorem has to be compared with Corollary 30 that gives general conditions for the existence of an NE, one of them being prefix-independence of f_i, $i \in \Pi$. This latter condition is here enough to guarantee the existence of a weak SPE (the existence of an SPE is not possible as mentioned before with the game of Fig. 1 [62]). It follows from Theorem 37 that there exists a weak SPE in all the Boolean and quantitative games of Sect. 2.3 (except for the case of discounted sum payoff that is neither prefix-independent nor with finite range).

In addition to SEs and (weak) SPEs, other solution profiles have been recently proposed, like Doomsday equilibria in [18], robust equilibria in [8], and equilibria using admissible strategies in [10]. We also refer the reader to the survey [9].

5   Conclusion

In this invited contribution, we gave an overview of classical as well as recent results about the threshold and constraint problems for games played on graphs. Solutions to these problems are winning strategies in case of two-player zero-sum games, and equilibria in case of multi-player non zero-sum games. We tried to present a unified approach through the notion of games equipped with a payoff function and a preference relation for each player, in a way to include classes of Boolean games and quantitative games that are usually studied. We also focussed on general approaches from which one can derived several different results: a criterium that guarantees the existence of uniform optimal strategies in two-player zero-sum games, and a characterization of plays that are the outcome of an Nash equilibrium in multi-player non zero-sum games. Several illustrative examples were provided as well as some intuition on the proofs when they are simple.

Acknowledgments. We would like to thank Patricia Bouyer, Thomas Brihaye, Emmanuel Filiot, Hugo Gimbert, Quentin Hautem, Mickaël Randour, and Jean-François Raskin for their useful discussions and comments that helped us to improve the presentation of this article.

Appendix

In this appendix, we give a sketch of proof for Muller games in Theorem 31. Recall that each player i has the objective $\Omega_i = \{\rho \in Plays \mid inf(\rho) \in \mathcal{F}_i\}$ with $\mathcal{F}_i \subseteq 2^V$, and that the values $val_i(v)$, $v \in V$, in each game G_i can be computed in polynomial time (Theorem 21). To prove P membership for the constraint problem with bounds $(\mu_i)_{i \in \Pi}$, $(\nu_i)_{i \in \Pi}$, we apply the approach (3). Notice that for the required play $\rho \in Plays(v_0)$ in (3), the set $U = inf(\rho)$ must be a strongly connected component that is reachable from the initial vertex v_0. Moreover if for some i, $f_i(\rho) = 0$ then $val_i(\rho_k) = 0$ for all $\rho_k \in V_i$, and if $\nu_i = 0$, then $f_i(\rho) = 0$. We thus proceed as follows. *(i)* For each i such that $\nu_i = 1$, for each $U \in \mathcal{F}_i$ (seen as a potential $U = inf(\rho)$), the following computations are done in polynomial time for all $j \in \Pi$:

- if $\mu_j = 1$ ((3) imposes $f_j(\rho) = 1$), test whether $U \in \mathcal{F}_j$,
- if $\nu_j = 0$ ((3) imposes $f_j(\rho) = 0$), test whether $U \notin \mathcal{F}_j$ and whether each $v \in U \cap V_j$ has value $val_j(v) = 0$,
- if $\mu_j = 0$ and $\nu_j = 1$ ((3) allows either $f_j(\rho) = 0$ or $f_j(\rho) = 1$), then if $U \notin \mathcal{F}_j$, test whether each $v \in U \cap V_j$ has value $val_j(v) = 0$.

Finally, construct in polynomial time the game G' from G such that each V_j is limited to $\{v \in V_j \mid val_j(v) = 0\}$ whenever $U \notin \mathcal{F}_j$, and test whether U is a strongly connected component that is reachable from v_0 in G'. As soon as this sequence of tests is positive, there exists ρ satisfying (3). *(ii)* It may happen that step *(i)* cannot be applied (because there is no j such that $\mu_j = 1$, and for j such that $\mu_j = 0$ and $\nu_j = 1$, there is no potential $U = inf(\rho)$). In this case, we construct in polynomial time a two-player game G' from G such that each V_i is limited to $\{v \in V_i \mid val_i(v) = 0\}$, player 1 controls no vertex and player 2 is formed by the coalition of all $i \in \Pi$, and the objective is a Muller objective with $\mathcal{F} = \cup_{i \in \Pi} \mathcal{F}_i$. We then test in polynomial time whether player 1 has no winning strategy from v_0 in this Muller game.

References

1. Alur, R., La Torre, S., Madhusudan, P.: Playing games with boxes and diamonds. In: Amadio, R., Lugiez, D. (eds.) CONCUR 2003. LNCS, vol. 2761, pp. 128–143. Springer, Heidelberg (2003). doi:10.1007/978-3-540-45187-7_8
2. Andersson, D.: An improved algorithm for discounted payoff games. In: ESSLLI Student Session, pp. 91–98 (2006)
3. Beeri, C.: On the membership problem for functional and multivalued dependencies in relational databases. ACM Trans. Database Syst. 5(3), 241–259 (1980)
4. Berthé, V., Rigo, M. (eds.): Combinatorics, Words and Symbolic Dynamics, vol. 135. Cambridge University Press, Cambridge (2016)
5. Bloem, R., Chatterjee, K., Henzinger, T.A., Jobstmann, B.: Better quality in synthesis through quantitative objectives. In: Bouajjani, A., Maler, O. (eds.) CAV 2009. LNCS, vol. 5643, pp. 140–156. Springer, Heidelberg (2009). doi:10.1007/978-3-642-02658-4_14

6. Boker, U., Henzinger, T.A., Otop, J.: The target discounted-sum problem. In: LICS Proceedings, pp. 750–761. IEEE Computer Society (2015)
7. Bouyer, P., Brenguier, R., Markey, N., Ummels, M.: Pure Nash equilibria in concurrent deterministic games. Logical Methods Comput. Sci. 11(2) (2015)
8. Brenguier, R.: Robust equilibria in mean-payoff games. In: Jacobs, B., Löding, C. (eds.) FoSSaCS 2016. LNCS, vol. 9634, pp. 217–233. Springer, Heidelberg (2016). doi:10.1007/978-3-662-49630-5_13
9. Brenguier, R., Clemente, L., Hunter, P., Pérez, G.A., Randour, M., Raskin, J.-F., Sankur, O., Sassolas, M.: Non-zero sum games for reactive synthesis. In: Dediu, A.-H., Janoušek, J., Martín-Vide, C., Truthe, B. (eds.) LATA 2016. LNCS, vol. 9618, pp. 3–23. Springer, Cham (2016). doi:10.1007/978-3-319-30000-9_1
10. Brenguier, R., Raskin, J.-F., Sankur, O.: Assume-admissible synthesis. Acta Inf. **54**(1), 41–83 (2017)
11. Brihaye, T., Bruyère, V., Meunier, N., Raskin, J-F.: Weak subgame perfect equilibria and their application to quantitative reachability. In: CSL Proceedings. LIPIcs, vol. 41, pp. 504–518. Schloss Dagstuhl - Leibniz-Zentrum fuer Informatik (2015)
12. Brihaye, T., De Pril, J., Schewe, S.: Multiplayer cost games with simple nash equilibria. In: Artemov, S., Nerode, A. (eds.) LFCS 2013. LNCS, vol. 7734, pp. 59–73. Springer, Heidelberg (2013). doi:10.1007/978-3-642-35722-0_5
13. Bruyère, V., Hautem, Q., Raskin, J-F.: On the complexity of heterogeneous multidimensional games. In: CONCUR Proceedings. LIPIcs, vol. 59, pp. 11:1–11:15. Schloss Dagstuhl - Leibniz-Zentrum fuer Informatik (2016)
14. Bruyère, V., Meunier, N., Raskin, J-F.: Secure equilibria in weighted games. In: CSL-LICS Proceedings, pp. 26:1–26:26. ACM (2014)
15. Bruyère, V., Le Roux, S., Pauly, A., Raskin, J.-F.: On the existence of weak subgame perfect equilibria. In: Esparza, J., Murawski, A.S. (eds.) FoSSaCS 2017. LNCS, vol. 10203, pp. 145–161. Springer, Heidelberg (2017). doi:10.1007/978-3-662-54458-7_9
16. Buhrke, N., Lescow, H., Vöge, J.: Strategy construction in infinite games with Streett and Rabin chain winning conditions. In: Margaria, T., Steffen, B. (eds.) TACAS 1996. LNCS, vol. 1055, pp. 207–224. Springer, Heidelberg (1996). doi:10.1007/3-540-61042-1_46
17. Calude, C., Jain, S., Khoussainov, B., Li, W., Stephan, F.: Deciding parity games in quasipolynomial time. In: STOC Proceedings. ACM (2017, to appear)
18. Chatterjee, K., Doyen, L., Filiot, E., Raskin, J.-F.: Doomsday equilibria for omega-regular games. Inf. Comput. **254**, 296–315 (2017)
19. Chatterjee, K., Doyen, L., Henzinger, T.A.: Quantitative languages. ACM Trans. Comput. Logic **11**, 23 (2010)
20. Chatterjee, K., Doyen, L., Randour, M., Raskin, J.-F.: Looking at mean-payoff and total-payoff through windows. Inf. Comput. **242**, 25–52 (2015)
21. Chatterjee, K., Dvorák, W., Henzinger, M., Loitzenbauer, V.: Conditionally optimal algorithms for generalized Büchi games. In: MFCS Proceedings. LIPIcs, vol. 58, pp. 25:1–25:15. Schloss Dagstuhl - Leibniz-Zentrum fuer Informatik (2016)
22. Chatterjee, K., Henzinger, M.: Efficient and dynamic algorithms for alternating Büchi games and maximal end-component decomposition. J. ACM **61**(3), 15:1–15:40 (2014)
23. Chatterjee, K., Henzinger, T.A., Jurdzinski, M.: Games with secure equilibria. Theor. Comput. Sci. **365**, 67–82 (2006)
24. Chatterjee, K., Henzinger, T.A., Piterman, N.: Generalized parity games. In: Seidl, H. (ed.) FoSSaCS 2007. LNCS, vol. 4423, pp. 153–167. Springer, Heidelberg (2007). doi:10.1007/978-3-540-71389-0_12

25. Chatterjee, K., Majumdar, R., Jurdziński, M.: On nash equilibria in stochastic games. In: Marcinkowski, J., Tarlecki, A. (eds.) CSL 2004. LNCS, vol. 3210, pp. 26–40. Springer, Heidelberg (2004). doi:10.1007/978-3-540-30124-0_6
26. Condurache, R., Filiot, E., Gentilini, R., Raskin, J-F.: The complexity of rational synthesis. In: Proceedings of ICALP. LIPIcs, vol. 55, pp. 121:1–121:15. Schloss Dagstuhl - Leibniz-Zentrum fuer Informatik (2016)
27. De Pril, J.: Equilibria in Multiplayer Cost Games. Ph. D. thesis, University UMONS (2013)
28. De Pril, J., Flesch, J., Kuipers, J., Schoenmakers, G., Vrieze, K.: Existence of secure equilibrium in multi-player games with perfect information. In: Csuhaj-Varjú, E., Dietzfelbinger, M., Ésik, Z. (eds.) MFCS 2014. LNCS, vol. 8635, pp. 213–225. Springer, Heidelberg (2014). doi:10.1007/978-3-662-44465-8_19
29. Emerson, E.A.: Automata, tableaux, and temporal logics (extended abstract). In: Parikh, R. (ed.) Logic of Programs 1985. LNCS, vol. 193, pp. 79–88. Springer, Heidelberg (1985). doi:10.1007/3-540-15648-8_7
30. Emerson, E.A., Jutla, C.S.: Tree automata, Mu-calculus and determinacy. In: FOCS Proceedings, pp. 368–377. IEEE Computer Society (1991)
31. Emerson, E.A., Lei, C.-L.: Modalities for model checking: branching time logic strikes back. Sci. Comput. Program. 8(3), 275–306 (1987)
32. Fijalkow, N., Horn, F.: Les jeux d'accessibilité généralisée. Tech. Sci. Informatiques 32(9–10), 931–949 (2013)
33. Filar, J., Vrieze, K.: Competitive Markov Decision Processes. Springer, New York (1997)
34. Flesch, J., Kuipers, J., Mashiah-Yaakovi, A., Schoenmakers, G., Solan, E., Vrieze, K.: Perfect-information games with lower-semicontinuous payoffs. Math. Oper. Res. 35, 742–755 (2010)
35. Fudenberg, D., Levine, D.: Subgame-perfect equilibria of finite-and infinite-horizon games. J. Econ. Theor. 31, 251–268 (1983)
36. Gimbert, H., Zielonka, W.: When can you play positionally? In: Fiala, J., Koubek, V., Kratochvíl, J. (eds.) MFCS 2004. LNCS, vol. 3153, pp. 686–697. Springer, Heidelberg (2004). doi:10.1007/978-3-540-28629-5_53
37. Gimbert, H., Zielonka, W.: Games where you can play optimally without any memory. In: Abadi, M., Alfaro, L. (eds.) CONCUR 2005. LNCS, vol. 3653, pp. 428–442. Springer, Heidelberg (2005). doi:10.1007/11539452_33
38. Grädel, E., Thomas, W., Wilke, T. (eds.): Automata, Logics, and Infinite Games: A Guide to Current Research. LNCS, vol. 2500. Springer, Heidelberg (2002)
39. Grädel, E., Ummels, M.: Solution concepts and algorithms for infinite multiplayer games. In: Apt, K., van Rooij, R. (eds.) New Perspectives on Games and Interaction, vol. 4, pp. 151–178. Amsterdam University Press, Amsterdam (2008)
40. Harris, C.: Existence and characterization of perfect equilibrium in games of perfect information. Econometrica 53, 613–628 (1985)
41. Horn, F.: Explicit muller games are PTIME. In: FSTTCS Proceedings. LIPIcs, vol. 2, pp. 235–243. Schloss Dagstuhl - Leibniz-Zentrum fuer Informatik (2008)
42. Hunter, P., Dawar, A.: Complexity bounds for regular games. In: Jędrzejowicz, J., Szepietowski, A. (eds.) MFCS 2005. LNCS, vol. 3618, pp. 495–506. Springer, Heidelberg (2005). doi:10.1007/11549345_43
43. Hunter, P., Raskin, J-F.: Quantitative games with interval objectives. In: FSTTCS Proceedings. LIPIcs, vol. 29, pp. 365–377. Schloss Dagstuhl - Leibniz-Zentrum fuer Informatik (2014)
44. Immerman, N.: Number of quantifiers is better than number of tape cells. J. Comput. Syst. Sci. 22, 384–406 (1981)

45. Jones, N.D.: Space-bounded reducibility among combinatorial problems. J. Comput. Syst. Sci. **11**, 68–75 (1975)
46. Jurdzinski, M.: Deciding the winner in parity games is in UP ∩ co-UP. Inf. Process. Lett. **68**(3), 119–124 (1998)
47. Karp, R.M.: A characterization of the minimum cycle mean in a digraph. Discrete Math. **23**, 309–311 (1978)
48. Kopczyński, E.: Half-positional determinacy of infinite games. In: Bugliesi, M., Preneel, B., Sassone, V., Wegener, I. (eds.) ICALP 2006. LNCS, vol. 4052, pp. 336–347. Springer, Heidelberg (2006). doi:10.1007/11787006_29
49. Kuhn, H.W.: Extensive games and the problem of information. In: Classics in Game Theory, pp. 46–68 (1953)
50. Lothaire, M.: Algebraic Combinatorics on Words, vol. 90. Cambridge University Press, Cambridge (2002)
51. Martin, D.A.: Borel determinacy. Ann. Math. **102**, 363–371 (1975)
52. McNaughton, R.: Infinite games played on finite graphs. Ann. Pure Appl. Logic **65**(2), 149–184 (1993)
53. Nash, J.F.: Equilibrium points in n-person games. In: PNAS, vol. 36, pp. 48–49. National Academy of Sciences (1950)
54. Osborne, M.J., Rubinstein, A.: A Course in Game Theory. MIT Press, Cambridge (1994)
55. Perrin, D., Pin, J.-E.: Infinite Words, Automata, Semigroups, Logic and Games, vol. 141. Elsevier, Amsterdam (2004)
56. Piterman, N., Pnueli, A.: Faster solutions of Rabin and Streett games. In: LICS Proceedings, pp. 275–284. IEEE Computer Society (2006)
57. Purves, R.A., Sudderth, W.D.: Perfect information games with upper semicontinuous payoffs. Math. Oper. Res. **36**(3), 468–473 (2011)
58. Rényi, A.: Representations of real numbers and their ergodic properties. Acta Math. Acad. Scientiarum Hung. **8**(3–4), 477–493 (1957)
59. Le Roux, S., Pauly, A.: Extending finite memory determinacy to multiplayer games. In: Proceedings of SR. EPTCS, vol. 218, pp. 27–40 (2016)
60. Safra, S., Vardi, M.Y.: On omega-automata and temporal logic (preliminary report). In: Proceedings of STOC, pp. 127–137. ACM (1989)
61. Selten, R.: Spieltheoretische Behandlung eines Oligopolmodells mit Nachfrageträgheit. Z. die gesamte Staatswissenschaft **121**, 301–324 (1965). pp. 667–689
62. Solan, E., Vieille, N.: Deterministic multi-player Dynkin games. J. Math. Econ. **39**, 911–929 (2003)
63. Ummels, M.: Rational behaviour and strategy construction in infinite multiplayer games. In: Arun-Kumar, S., Garg, N. (eds.) FSTTCS 2006. LNCS, vol. 4337, pp. 212–223. Springer, Heidelberg (2006). doi:10.1007/11944836_21
64. Ummels, M.: The complexity of nash equilibria in infinite multiplayer games. In: Amadio, R. (ed.) FoSSaCS 2008. LNCS, vol. 4962, pp. 20–34. Springer, Heidelberg (2008). doi:10.1007/978-3-540-78499-9_3
65. Ummels, M., Wojtczak, D.: The complexity of nash equilibria in limit-average games. In: Katoen, J.-P., König, B. (eds.) CONCUR 2011. LNCS, vol. 6901, pp. 482–496. Springer, Heidelberg (2011). doi:10.1007/978-3-642-23217-6_32
66. Ummels, M., Wojtczak, D.: The complexity of Nash equilibria in limit-average games. CoRR, abs/1109.6220 (2011)
67. Vardi, M.Y., Wolper, P.: Reasoning about infinite computations. Inf. Comput. **115**(1), 1–37 (1994)

68. Velner, Y.: Robust multidimensional mean-payoff games are undecidable. In: Pitts, A. (ed.) FoSSaCS 2015. LNCS, vol. 9034, pp. 312–327. Springer, Heidelberg (2015). doi:10.1007/978-3-662-46678-0_20

69. Velner, Y., Chatterjee, K., Doyen, L., Henzinger, T.A., Rabinovich, A.M., Raskin, J.-F.: The complexity of multi-mean-payoff and multi-energy games. Inf. Comput. **241**, 177–196 (2015)

70. von Neumann, J., Morgenstern, O.: Theory of Games and Economic Behavior. Princeton University Press, Princeton (1944)

71. Zwick, U., Paterson, M.: The complexity of mean payoff games on graphs. Theor. Comput. Sci. **158**, 343–359 (1996)

A Comprehensive Introduction to the Theory of Word-Representable Graphs

Sergey Kitaev$^{(\boxtimes)}$

Department of Computer and Information Sciences,
University of Strathclyde, 26 Richmond Street, Glasgow G1 1XH, UK
sergey.kitaev@cis.strath.ac.uk
https://personal.cis.strath.ac.uk/sergey.kitaev/

Abstract. Letters x and y alternate in a word w if after deleting in w all letters but the copies of x and y we either obtain a word $xyxy\cdots$ (of even or odd length) or a word $yxyx\cdots$ (of even or odd length). A graph $G = (V, E)$ is word-representable if and only if there exists a word w over the alphabet V such that letters x and y alternate in w if and only if $xy \in E$.

Word-representable graphs generalize several important classes of graphs such as circle graphs, 3-colorable graphs and comparability graphs. This paper offers a comprehensive introduction to the theory of word-represent-able graphs including the most recent developments in the area.

1 Introduction

The theory of word-representable graphs is a young but very promising research area. It was introduced by the author in 2004 based on the joint research with Steven Seif [20] on the celebrated *Perkins semigroup*, which has played a central role in semigroup theory since 1960, particularly as a source of examples and counterexamples. However, the first systematic study of word-representable graphs was not undertaken until the appearance in 2008 of the paper [18] by the author and Artem Pyatkin, which started the development of the theory. One of the most significant contributors to the area is Magnús M. Halldórsson.

Up to date, nearly 20 papers have been written on the subject, and the core of the book [17] by the author and Vadim Lozin is devoted to the theory of word-representable graphs. It should also be mentioned that the software produced by Marc Glen [7] is often of great help in dealing with word-representation of graphs.

We refer the Reader to [17], where relevance of word-representable graphs to various fields is explained, thus providing a motivation to study the graphs. These fields are algebra, graph theory, computer science, combinatorics on words, and scheduling. In particular, word-representable graphs are important from graph-theoretical point of view, since they generalize several fundamental classes of graphs (e.g. *circle graphs*, *3-colorable graphs* and *comparability graphs*).

© Springer International Publishing AG 2017
É. Charlier et al. (Eds.): DLT 2017, LNCS 10396, pp. 36–67, 2017.
DOI: 10.1007/978-3-319-62809-7_2

A graph $G = (V, E)$ is *word-representable* if and only if there exists a word w over the alphabet V such that letters x and y, $x \neq y$, alternate in w if and only if $xy \in E$ (see Sect. 2 for the definition of alternating letters). Natural questions to ask about word-representable graphs are:

- Are all graphs word-representable?
- If not, how do we characterize word-representable graphs?
- How many word-representable graphs are there?
- What is graph's representation number for a given graph? Essentially, what is the minimal length of a word-representant?
- How hard is it to decide whether a graph is word-representable or not? (complexity)
- Which graph operations preserve (non-)word-representability?
- Which graphs are word-representable in your favourite class of graphs?

This paper offers a comprehensive introduction to the theory of word-representable graphs. Even though the paper is based on the book [17] following some of its structure, our exposition goes far beyond book's content and it reflects the most recent developments in the area. Having said that, there is a relevant topic on a generalization of the theory of word-representable graphs [12, 16] that is discussed in [17, Chap. 6], but we do not discuss it at all.

In this paper we do not include the majority of proofs due to space limitations (while still giving some proofs, or ideas of proofs whenever possible). Also, all graphs we deal with are simple (no loops or multiple edges are allowed), and unless otherwise specified, our graphs are unoriented.

2   Word-Representable Graphs. The Basics

Suppose that w is a word over some alphabet and x and y are two distinct letters in w. We say that x and y *alternate* in w if after deleting in w *all* letters *but* the copies of x and y we either obtain a word $xyxy \cdots$ (of even or odd length) or a word $yxyx \cdots$ (of even or odd length). For example, in the word 23125413241362, the letters 2 and 3 alternate. So do the letters 5 and 6, while the letters 1 and 3 do *not* alternate.

Definition 1. *A graph* $G = (V, E)$ *is* word-representable *if and only if there exists a word* w *over the alphabet* V *such that letters* x *and* y, $x \neq y$, *alternate in* w *if and only if* $xy \in E$. *(By definition,* w *must contain each letter in* V.*)* *We say that* w *represents* G, *and that* w *is a* word-representant.

Definition 1 works for both vertex-labeled and unlabeled graphs because any labeling of a graph G is equivalent to any other labeling of G with respect to word-representability (indeed, the letters of a word w representing G can always be renamed). For example, the graph to the left in Fig. 1 is word-representable because its labeled version to the right in Fig. 1 can be represented by 1213423. For another example, each *complete graph* K_n can be represented by any permutation π of $\{1, 2, \ldots, n\}$, or by π concatenated any number of times. Also,

the *empty graph E_n* (also known as *edgeless graph*, or *null graph*) on vertices
$\{1, 2, \ldots, n\}$ can be represented by $12 \cdots (n-1)nn(n-1) \cdots 21$, or by any other
permutation concatenated with the same permutation written in the reverse
order.

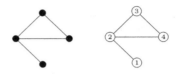

Fig. 1. An example of a word-representable graph

Remark 1. The class of word-representable graphs is *hereditary*. That is, removing a vertex v in a word-representable graph G results in a word-representable
graph G'. Indeed, if w represents G then w with v removed represents G'.
This observation is crucial, e.g. in finding asymptotics for the number of word-
representable graphs [4], which is the only known enumerative result on word-
representable graphs to be stated next.

Theorem 1 ([4]). *The number of non-isomorphic word-representable graphs on
n vertices is given by $2^{\frac{n^2}{3}+o(n^2)}$.*

2.1　k-Representability and Graph's Representation Number

A word w is *k-uniform* if each letter in w occurs k times. For example, the word
243321442311 is 3-uniform, while 23154 is a 1-uniform word (a permutation).

Definition 2. *A graph G is k-word-representable, or k-representable for
brevity, if there exists a k-uniform word w representing it. We say that w k-
represents G.*

The following result establishes equivalence of Definitions 1 and 2.

Theorem 2 ([18]). *A graph is word-representable if and only if it is k-represent-
able for some k.*

Proof. Clearly, k-representability implies word-representability. For the other
direction, we demonstrate on an example how to extend a word-representant
to a uniform word representing the same graph. We refer to [18] for a precise
description of the extending algorithm, and an argument justifying it.

　　Consider the word $w = 3412132154$ representing a graph G on five vertices.
Ignore the letter 1 occurring the maximum number of times (in general, there
could be several such letters all of which need to be ignored) and consider the *initial permutation* $p(w)$ of w formed by the remaining letters, that is, $p(w)$ records
the order of occurrences of the leftmost copies of the letters. For our example,

$p(w) = 3425$. Then the word $p(w)w = 34253412132154$ also represents G, but it contains more occurrences of the letters occurring not maximum number of time in w. This process can be repeated a number of times until each letter occurs the same number of times. In our example, we need to apply the process one more time by appending 5, the initial permutation of $p(w)w$, to the left of $p(w)w$ to obtain a uniform representation of G: 534253412132154.

Following the same arguments as in Theorem 2, one can prove the following result showing that there are *infinitely many* representations for any word-representable graph.

Theorem 3 ([18]). *If a graph is k-representable then it is also $(k + 1)$-representable.*

By Theorem 2, the following notion is well-defined.

Definition 3. Graph's representation number *is the* least k *such that the graph is k-representable. For non-word-representable graphs (whose existence will be discussed below), we let $k = \infty$. Also, we let $\mathcal{R}(G)$ denote G's representation number and $\mathcal{R}_k = \{G : \mathcal{R}(G) = k\}$.*

Clearly, $\mathcal{R}_1 = \{G : G$ is a complete graph$\}$. Next, we discuss $\mathcal{R}_2$.

2.2   Graphs with Representation Number 2

We begin with discussing five particular classes of graphs having representation number 2, namely, *empty graphs*, *trees*, *forests*, *cycle graphs* and *ladder graphs*. Then we state a result saying that graphs with representation number 2 are exactly the class of *circle graphs*.

Empty Graphs. No empty graph E_n for $n \geq 2$ can be represented by a single copy of each letter, so $\mathcal{R}(E_n) \geq 2$. On the other hand, as discussed above, E_n can be represented by concatenation of two permutations, and thus $\mathcal{R}(E_n) = 2$.

Trees and Forests. A simple inductive argument shows that *any tree T can be represented using two copies of each letter*, and thus, if the number of vertices in T is at least 3, $\mathcal{R}(T) = 2$. Indeed, as the base case we have the edge labeled by 1 and 2 that can be 2-represented by 1212. Now, suppose that any tree on at most $n - 1$ vertices can be 2-represented for $n \geq 3$, and consider a tree T with n vertices and with a leaf x connected to a vertex y. Removing the leaf x, we obtain a tree T' that can be 2-represented by a word $w_1 y w_2 y w_3$ where w_1, w_2 and w_3 are possibly empty words not containing y. It is now easy to see that the word $w_1 y w_2 x y x w_3$ 2-represents T (obtained from T' by inserting back the leaf x). Note that the word $w_1 x y x w_2 y w_3$ also represents T.

Representing each tree in a forest by using two letters (trees on one vertex x and two vertices x and y can be represented by xx and $xyxy$, respectively) and concatenating the obtained word-representants, we see that for any forest F having at least two trees, $\mathcal{R}(F) = 2$. Indeed, having two letters in a word-represent for each tree guarantees that no pair of trees will be connected by an edge.

Cycle Graphs. Another class of 2-representable graphs is *cycle graphs*. Note that a cyclic shift of a word-representant may not represent the same graph, as is the case with, say, the word 112. However, if a word-representant is uniform, a cyclic shift does represent the same graph, which is recorded in the following proposition.

Proposition 1 ([18]). *Let $w = uv$ be a k-uniform word representing a graph G, where u and v are two, possibly empty, words. Then the word $w' = vu$ also represents G.*

Now, to represent a cycle graph C_n on n vertices, one can first represent the *path graph* P_n on n vertices using the technique to represent trees, then make a 1-letter cyclic shift still representing P_n by Proposition 1, and swap the first two letters. This idea is demonstrated for the graph in Fig. 2 as follows. The steps in representing the path graph P_6 obtained by removing the edge 16 from C_6 are

$$1212 \rightarrow 121323 \rightarrow 12132434 \rightarrow 1213243545 \rightarrow 121324354656.$$

The 1-letter cyclic shift gives the word 612132435465 still representing P_6 by Proposition 1, and swapping the first two letters gives the sought representation of C_6: 162132435465.

Fig. 2. Cycle graph C_6

Ladder Graphs. The ladder graph L_n with $2n$ vertices, labeled $1, \ldots, n$, $1', \ldots, n'$, and $3n - 2$ edges is constructed following the pattern for $n = 4$ presented in Fig. 3. An inductive argument given in [15] shows that for $n \geq 2$, $\mathcal{R}(L_n) = 2$. Table 1 gives 2-representations of L_n for $n = 1, 2, 3, 4$.

Circle Graphs. A *circle graph* is the *intersection graph* of a set of chords of a circle. That is, it is an unoriented graph whose vertices can be associated with chords of a circle such that two vertices are adjacent if and only if the corresponding chords cross each other. See Fig. 4 for an example of a circle graph on four vertices and its associated chords.

The following theorem provides a complete characterization of $\mathcal{R}_2$.

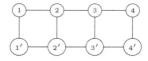

Fig. 3. The ladder graph L_4

Table 1. 2-representations of the ladder graph L_n for $n = 1, 2, 3, 4$

n	2-representation of the ladder graph L_n
1	$11'11'$
2	$1'212'21'2'1$
3	$12'1'323'32'3'121'$
4	$1'213'2'434'43'4'231'2'1$

Theorem 4. *We have*

$$\mathcal{R}_2 = \{G : G \text{ is a circle graph different from a complete graph}\}.$$

Proof. Given a circle graph G, consider its representation on a circle by intersecting chords. Starting from any chord's endpoint, go through all the endpoints in clock-wise direction recording chords' labels. The obtained word w is 2-uniform and it has the property that a pair of letter x and y alternate in w if and only if the pair of chords labeled by x and y intersect, which happens if and only if the vertex x is connected to the vertex y in G. For the graph in Fig. 4, the chords' labels can be read starting from the lower 1 as 13441232, which is a 2-uniform word representing the graph. Thus, G is a circle graph if and only if $G \in \mathcal{R}_2$ with the only exception if G is a complete graph, in which case $G \in \mathcal{R}_1$.

2.3   Graphs with Representation Number 3

Unlike the case of graphs with representation number 2, no characterization of graphs with representation number 3 is know. However, there is a number of interesting results on this class of graphs to be discussed next.

The Petersen Graph. In 2010, Alexander Konovalov and Steven Linton not only showed that the Petersen graph in Fig. 5 is not 2-representable, but also provided two *non-equivalent* (up to renaming letters or a cyclic shift) 3-representations of it:

- 1387296(10)7493541283(10)7685(10)194562 and
- 134(10)58679(10)273412835(10)6819726495.

The fact that the Petersen graph does not belong to $\mathcal{R}_2$ is also justified by the following theorem.

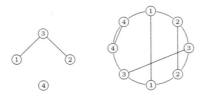

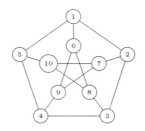

Fig. 4. A circle graph on four vertices and its associated chords

Fig. 5. The Petersen graph

Theorem 5 ([9]). *Petersen's graph is* not *2-representable.*

Proof. Suppose that the graph is 2-representable and w is a 2-uniform word representing it. Let x be a letter in w such that there is a minimal number of letters between the two occurrences of x. Since Petersen's graph is regular of degree 3, it is not difficult to see that there must be exactly three letters, which are all different, between the xs (having more letters between xs would lead to having two equal letters there, contradicting the choice of x).

By symmetry, we can assume that $x = 1$, and by Proposition 1 we can assume that w starts with 1. So, the letters 2, 5 and 6 are between the two 1s, and because of symmetry, the fact that Petersen's graph is *edge-transitive* (that is, each of its edges can be made "internal"), and taking into account that the vertices 2, 5 and 6 are pairwise non-adjacent, we can assume that $w = 12561w_1 6w_2 5w_3 2w_4$ where the w_is are some, possibly empty words for $i \in \{1, 2, 3, 4\}$. To alternate with 6 but not to alternate with 5, the letter 8 must occur in w_1 and w_2. Also, to alternate with 2 but not to alternate with 5, the letter 3 must occur in w_3 and w_4. But then 8833 is a subsequence in w, and thus 8 and 3 must be non-adjacent in the graph, a contradiction.

Prisms. A *prism* Pr_n is a graph consisting of two cycles $12 \cdots n$ and $1'2' \cdots n'$, where $n \geq 3$, connected by the edges ii' for $i = 1, \ldots, n$. In particular, the 3-dimensional cube to the right in Fig. 6 is the prism Pr_4.

Theorem 6 ([18]). *Every prism Pr_n is 3-representable.*

The fact that the triangular prism Pr_3 is not 2-representable was shown in [18]. The following more general result holds.

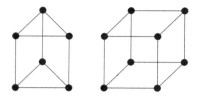

Fig. 6. Prisms Pr_3 and Pr_4

Theorem 7 ([15]). None *of prisms* Pr_n *is 2-representable.*

Theorems 6 and 7 show that $Pr_n \in \mathcal{R}_3$ for any $n \geq 3$.

Colorability of Graphs in $\mathcal{R}_3$. Theorem 11 below shows that $\mathcal{R}_3$ does not even include 2-colorable graphs, and thus any class of c-colorable graphs for $c \geq 3$. Indeed, any c-colorable non-3-representable graph can be extended to a $(c + 1)$-colorable graph by adding an apex (all-adjacent vertex), which is still non-3-representable using the hereditary nature of word-representability (see Remark 1).

A natural question to ask here is: Is $\mathcal{R}_3$ properly included in a class of c-colorable graphs for a constant c? A simple argument of replacing a vertex in the 3-representable triangular prism Pr_3 by a complete graph of certain size led to the following theorem.

Theorem 8 ([15]). *The class* $\mathcal{R}_3$ *is not included in a class of c-colorable graphs for some constant c.*

Subdivisions of Graphs. The following theorem gives a useful tool for constructing 3-representable graphs, that is, graphs with representation number at most 3.

Theorem 9 ([18]). *Let* $G = (V, E)$ *be a 3-representable graph and* $x, y \in V$. *Denote by H the graph obtained from G by adding to it a path of length at least 3 connecting x and y. Then H is also 3-representable.*

Definition 4. *A subdivision of a graph G is a graph obtained from G by replacing each edge xy in G by a simple path (that is, a path without self-intersection) from x to y. A subdivision is called a k-subdivision if each of these paths is of length at least k.*

Definition 5. *An* edge contraction *is an operation which removes an edge from a graph while gluing the two vertices it used to connect. An unoriented graph G is a* minor *of another unoriented graph H if a graph isomorphic to G can be obtained from H by contracting some edges, deleting some edges, and deleting some isolated vertices.*

Theorem 10 ([18]). *For every graph G there are infinitely many 3-representable graphs H that contain G as a minor. Such a graph H can be obtained from G by subdividing each edge into any number of, but at least three edges.*

Note that H in Theorem 10 does not have to be a k-subdivision for some k, that it, edges of G can be subdivided into different number (at least 3) of edges. In either case, the 3-subdivision of any graph G is always 3-representable. Also, it follows from Theorem 9 and the proof of Theorem 10 in [18] that a graph obtained from an edgeless graph by inserting simple paths of length at least 3 between (some) pairs of vertices of the graph is 3-representable.

Finally, note that subdividing *each* edge in any graph into *exactly* two edges gives a bipartite graph, which is word-representable by Theorem 13 (see the discussion in Sect. 2.4 on why a bipartite graph is word-representable).

2.4   Graphs with High Representation Number

In Theorem 28 below we will see that the upper bound on a shortest word-representant for a graph G on n vertices is essentially $2n^2$, that is, one needs at most $2n$ copies of each letter to represent G. Next, we consider two classes of graphs that require essentially $n/2$ copies of each letter to be represented, and these are the longest known shortest word-representants.

Crown Graphs

Definition 6. *A* crown graph *(also known as a* cocktail party graph$)$ $H_{n,n}$ is obtained from the complete bipartite graph $K_{n,n}$ by removing a perfect matching. That is, $H_{n,n}$ is obtained from $K_{n,n}$ by removing n edges such that each vertex was incident to exactly one removed edge.*

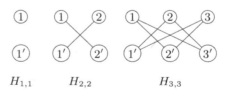

Fig. 7. Crown graphs

See Fig. 7 for examples of crown graphs.

By Theorem 13 below, $H_{n,n}$ can be represented by a concatenation of permutations, because $H_{n,n}$ is a comparability graph defined in Sect. 3 (to see this, just orient all edges from one part to the other). In fact, $H_{n,n}$ is known to require n permutations to be represented. However, can we provide a shorter representation for $H_{n,n}$? It turns out that we can, to be discussed next, but such representations are still long (linear in n).

Note that $H_{1,1} \in \mathcal{R}_2$ by Sect. 2.2. Further, $H_{2,2} \neq K_4$, the complete graph on 4 vertices, and thus $H_{2,2} \in \mathcal{R}_2$ because it can be 2-represented by $121'2'212'1'$. Also, $H_{3,3} = C_6 \in \mathcal{R}_2$ by Sect. 2.2. Finally, $H_{4,4} = \mathrm{Pr}_4 \in \mathcal{R}_3$ by Sect. 2.3. The following theorem gives the representation number $\mathcal{R}(H_{n,n})$ in the remaining cases.

Theorem 11 ([6]). *If $n \geq 5$ then the representation number of $H_{n,n}$ is $\lceil n/2 \rceil$ (that is, one needs $\lceil n/2 \rceil$ copies of each letter to represent $H_{n,n}$, but* not *fewer).*

Crown Graphs with an Apex. The graph G_n is obtained from a crown graph $H_{n,n}$ by adding an apex (all-adjacent vertex). See Fig. 8 for the graph G_3.

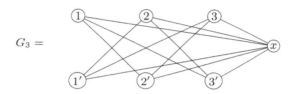

$G_3 =$

Fig. 8. Graph G_3

It turns out that G_n is the *worst* known word-representable graph in the sense that it requires the maximum number of copies of each letter to be represented, as recorded in the following theorem.

Theorem 12 ([18]). *The representation number of G_n is $\lfloor n/2 \rfloor$.*

It is unknown whether there exist graphs on n vertices with representation number between $\lfloor n/2 \rfloor$ and essentially $2n$ (given by Theorem 29).

3   Permutationally Representable Graphs and Their Significance

An orientation of a graph is *transitive* if presence of edges $u \rightarrow v$ and $v \rightarrow z$ implies presence of the edge $u \rightarrow z$. An unoriented graph is a *comparability graph* if it admits a transitive orientation. It is well known [17, Sect. 3.5.1], and is not difficult to show that the smallest non-comparability graph is the cycle graph C_5.

Definition 7. *A graph $G = (V, E)$ is* permutationally representable *if it can be represented by a word of the form $p_1 \cdots p_k$ where p_i is a permutation. We say that G is* permutationally k-representable.

For example, the graph in Fig. 1 is permutationally representable, which is justified by the concatenation of two permutations 21342341.

The following theorem is an easy corollary of the fact that any partially ordered set can be represented as intersection of linear orders.

Theorem 13 ([20]). *A graph is permutationally representable if and only if it is a comparability graph.*

Next, consider a schematic representation of the graph G in Fig. 9 obtained from a graph H by adding an all-adjacent vertex (apex). The following theorem holds.

Theorem 14 ([18]). *The graph G is word-representable if and only if the graph H is permutationally representable.*

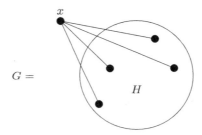

Fig. 9. G is obtained from H by adding an apex

A *wheel graph* W_n is the graph obtained from a cycle graph C_n by adding an apex. It is easy to see that none of cycle graphs C_{2n+1}, for $n \geq 2$, is a comparability graph, and thus none of wheel graphs W_{2n+1}, for $n \geq 2$ is word-representable. In fact, W_5 is the smallest example of a non-word-representable graph (the only one on 6 vertices). Section 6 discusses other examples of non-word-representable graphs.

As a direct corollary to Theorem 14, we have the following important result revealing the structure of neighbourhoods of vertices in a word-representable graph.

Theorem 15 ([18]). *If a graph G is word-representable then the neighbourhood of each vertex in G is permutationally representable (is a comparability graph by Theorem 13).*

The converse to Theorem 15 is *not* true as demonstrated by the counterexamples in Fig. 10 taken from [4,9], respectively.

A *clique* in an unoriented graph is a subset of pairwise adjacent vertices. A *maximum clique* is a clique of the maximum size. Given a graph G, the *Maximum Clique problem* is to find a maximum clique in G. It is well known that the

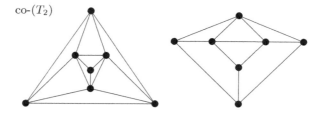

co-(T_2)

Fig. 10. Non-word-representable graphs in which each neighbourhood is permutationally representable

Maximum Clique problem is NP-complete. However, this problem is polynomially solvable for word-representable graphs, which is a corollary of Theorem 15 and is discussed next.

Theorem 16 ([10,11]). *The Maximum Clique problem is polynomially solvable on word-representable graphs.*

Proof. Each neighbourhood of a word-representable graph G is a comparability graph by Theorem 15. It is known that the Maximum Clique problem is solvable on comparability graphs in polynomial time. Thus the problem is solvable on G in polynomial time, since any maximum clique belongs to the neighbourhood of a vertex including the vertex itself.

4   Graphs Representable by Pattern Avoiding Words

It is a very popular area of research to study patterns in words and permutations[1]. The book [14] provides a comprehensive introduction to the field. A *pattern* is a word containing each letter in $\{1, 2, \ldots, k\}$ at least once for some k. A pattern $\tau = \tau_1 \tau_2 \cdots \tau_m$ *occurs* in a word $w = w_1 w_2 \cdots w_n$ if there exist $1 \leq i_1 < i_2 < \cdots < i_m \leq n$ such that $\tau_1 \tau_2 \cdots \tau_m$ is *order-isomorphic* to $w_{i_1} w_{i_2} \cdots w_{i_m}$. We say that w *avoids* τ if w contains no occurrences of τ. For example, the word 42316 contains several occurrences of the pattern 213 (all ending with 6), e.g. the subsequences 426, 416 and 316.

As a particular case of a more general program of research suggested by the author during his plenary talk at the international Permutation Patterns Conference at the East Tennessee State University, Johnson City in 2014, one can consider the following direction (see [17, Sect. 7.8]). Given a set of words avoiding a pattern, or a set of patterns, which class of graphs do these words represent?

---

[1] The patterns considered in this section are ordered, and their study comes from Algebraic Combinatorics. There are a few results on word-representable graphs and (unordered) patterns studied in Combinatorics on Words, namely on squares and cubes in words, that are not presented in this paper, but can be found in [17, Sect. 7.1.3]. One of the results says that for any word-representable graph, there exists a cube-free word representing it.

As a trivial example, consider the class of graphs defined by words avoiding the pattern 21. Clearly, any 21-avoiding word is of the form

$$w = 11 \cdots 122 \cdots 2 \cdots nn \cdots n.$$

If a letter x occurs *at least twice* in w then the respective vertex is isolated. The letters occurring *exactly once* form a *clique* (are connected to each other). Thus, 21-avoiding words describe graphs formed by a clique and an independent set.

Two papers, [5,21], are dedicated to this research direction and will be summarised in this section. So far, apart from Theorem 17 and Corollary 1 below, only 132-avoiding and 123-avoiding words were studied from word-representability point of view. The results of these studies are summarized in Fig. 11, which is taken from [21]. In that figure, and more generally in this section, we slightly abuse the notation and call graphs representable by τ-avoiding words τ-*representable*.

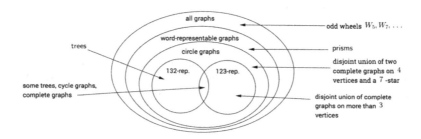

Fig. 11. Relations between graph classes taken from [21]

We note that unlike the case of word-representability without extra restrictions, labeling of graphs *does matter* in the case of pattern avoiding representations. For example, the 132-avoiding word 543212345 represents the graph to the left in Fig. 12, while *no* 132-avoiding word represents the other graph in that figure. Indeed, *no two* letters out of 1, 2, 3 and 4 can occur *once* in a word-representant w or else the respective vertices would *not* form an *independent set*. Say, w.l.o.g. that 1, 2 and 3 occur at *least twice* in w. But then, because 5 is an apex, we can find a copy of 5 in w having at least two letters $x, y \in \{1, 2, 3\}$ both to the left and to the right of the 5, so that the patten 132 in w is *inevitable*. Note that the presence of the vertex 4 in our argument is essential, because if we remove this vertex from the graph to the right in Fig. 12, then the 132-avoiding word 3532151 would represent the obtained graph.

The following theorem has a great potential to be applicable to the study of τ-representable graphs for τ of length 4 or more.

Theorem 17 ([21])**.** *Let G be a word-representable graph, which can be represented by a word avoiding a pattern τ of length $k + 1$. Let x be a vertex in G such that its degree $d(x) \geq k$. Then, any word w representing G that avoids τ must contain no more than k copies of x.*

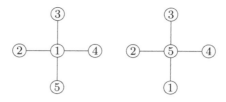

Fig. 12. 132-representable (left) and non-132-representable (right) labelings of the same graph

Proof. If there are *at least* $k + 1$ occurrences of x in w, we obtain a factor (i.e. consecutive subword) $xw_1x \cdots w_kx$, where k neighbours of x in G occur in each w_i. But then w contains *all* patterns of length $k + 1$, in particular, τ Contradiction.

Corollary 1 ([21]). *Let w be a word-representant for a graph which avoids a pattern of length $k + 1$. If some vertex y adjacent to x has degree at least k, then x occurs at most $k + 1$ times in w.*

4.1 132-Representable Graphs

It was shown in [5] that the minimum (with respect to the number of vertices) non-word-representable graph, the wheel graph W_5, is actually a minimum non-132-representable graph (we do not know if there exit other non-132-representable graphs on 6 vertices).

Theorem 18 ([5]). *If a graph G is 132-representable, then there exists a 132-avoiding word w representing G such that any letter in w occurs at most twice.*

Theorems 4 and 18 give the following result.

Theorem 19 ([5]). *Every 132-representable graph is a circle graph.*

Thus, by Theorems 4, 7 and 19, none of prisms Pr_n, $n \geq 3$, is 132-representable. A natural question is if there are circle graphs that are not 132-representable.

Theorem 20 ([21]). *Not all circle graphs are 132-representable. E.g. disjoint union of two complete graphs K_4 is a circle graph, but it is not 132-representable.*

Theorem 21 ([5]). *Any tree is 132-representable.*

Note that in the case of pattern avoiding representations of graphs, Theorem 2 does not necessarily work, because extending a representation to a uniform representation may introduce an occurrences of the pattern(s) in question. For example, while any complete graph K_n can be represented by the 132-avoiding word $n(n-1)\cdots 1$, it was shown in [21] that for $n \geq 3$ no 2-uniform 132-avoiding representation of K_n exists. In either case, [21] shows that any tree can actually be represented by a 2-uniform word thus refining the statement of Theorem 21. For another result on uniform 132-representation see Theorem 27 below.

Theorem 22 ([5]). *Any cycle graph is 132-representable.*

Proof. The cycle graph C_n labeled by $1, 2, \ldots, n$ in clockwise direction can be represented by the 132-avoiding word

$$(n-1)n(n-2)(n-1)(n-3)(n-2)\cdots 45342312.$$

Theorem 23 ([5]). *For $n \geq 1$, a complete graph K_n is 132-representable. Moreover, for $n \geq 3$, there are*

$$2 + C_{n-2} + \sum_{i=0}^{n} C_i$$

different 132-representants for K_n, where $C_n = \frac{1}{n+1}\binom{2n}{n}$ is the n-th Catalan number. Finally, K_1 can be represented by a word of the form $11\cdots 1$ and K_2 by a word of the form $1212\cdots$ (of even or odd length) or $2121\cdots$ (of even or odd length).

As a corollary to the proof of Theorem 23, [5] shows that for $n \geq 3$, the length of any 132-representant of K_n is either n, or $n+1$, or $n+2$, or $n+3$.

4.2   123-Representable Graphs

An analogue of Theorem 19 holds for 123-representable graphs.

Theorem 24 ([21]). *Any 123-representable graph is a circle graph.*

Fig. 13. Star graph $K_{1,6}$

Theorem 25 ([21]). *Any cycle graph is 123-representable.*

Proof. The cycle graph C_n labeled by $1, 2, \ldots, n$ in clockwise direction can be represented by the 123-avoiding word

$$n(n-1)n(n-2)(n-1)(n-3)(n-1)\cdots 23121.$$

Theorem 26 ([21]). *The star graph $K_{1,6}$ in Fig. 13 is not 123-representable.*

It is easy to see that $K_{1,6}$ is a circle graph, and thus not all circle graphs are 123-representable by Theorem 26. Also, by Theorem 26, not all trees are 123-representable.

Based on Theorems 20 and 26, it is easy to come up with a circle graph on 14 vertices that is neither 123- nor 132-representable (see [21]).

As opposed to the situation with 132-representation discussed in Sect. 4.1, any complete graph K_n can be represented by the 123-avoiding 2-uniform word $n(n-1)\cdots 1n(n-1)\cdots 1$ as observed in [21]. Also, it was shown in [21] that any path graph P_n can be 123-represented by a 2-uniform word. We conclude with a general type theorem on uniform representation applicable to both 123- and 132-representations.

Theorem 27 ([21]). *Let a pattern $\tau \in \{123, 132\}$ and $G_1, G_2, \ldots, G_k$ be τ-representable connected components of a graph G. Then G is τ-representable if and only if at most one of the connected components cannot be τ-represented by a 2-uniform word.*

5 Semi-transitive Orientations as a Key Tool in the Theory of Word-Representable Graphs

Recall the definition of a transitive orientation at the beginning of Sect. 3.

A *shortcut* is an *acyclic non-transitively oriented* graph obtained from a directed cycle graph forming a directed cycle on at least four vertices by changing the orientation of one of the edges, and possibly by adding more directed edges connecting some of the vertices (while keeping the graph be acyclic and non-transitive). Thus, any shortcut

- is *acyclic* (that it, there are *no directed cycles*);
- has *at least* 4 vertices;
- has *exactly one* source (the vertex with no edges coming in), *exactly one* sink (the vertex with no edges coming out), and a *directed path* from the source to the sink that goes through *every* vertex in the graph;
- has an edge connecting the source to the sink that we refer to as the *short-cutting edge*;
- is *not* transitive (that it, there exist vertices u, v and z such that $u \to v$ and $v \to z$ are edges, but there is *no* edge $u \to z$).

Definition 8. *An orientation of a graph is* semi-transitive *if it is* acyclic *and* shortcut-free.

It is easy to see from definitions that *any* transitive orientation is necessary semi-transitive. The converse is *not* true, e.g. the following schematic semi-transitively oriented graph is *not* transitively oriented:

Thus semi-transitive orientations generalize transitive orientations.

A way to check if a given oriented graph G is semi-transitively oriented is as follows. First check that G is acyclic; if not, the orientation is not semi-transitive. Next, for a directed edge from a vertex x to a vertex y, consider *each* directed path P having at least three edges without repeated vertices from x to y, and check that the subgraph of G induced by P is transitive. If such non-transitive subgraph is found, the orientation is not semi-transitive. This procedure needs to be applied to each edge in G, and if no non-transitivity is discovered, G's orientation is semi-transitive.

As we will see in Theorem 28, finding a semi-transitive orientation is equivalent to recognising whether a given graph is word-representable, and this is an NP-hard problem (see Theorem 39). Thus, there is no efficient way to construct a semi-transitive orientation in general, and such a construction would rely on an exhaustive search orienting edges one by one, and thus branching the process. Having said that, there are several situations in which branching is not required. For example, the orientation of the partially oriented triangle below can be completed uniquely to avoid a cycle:

For another example, the branching process can normally be shorten e.g. by completing the orientation of quadrilaterals as shown in Fig. 14, which is *unique* to avoid cycles and shortcuts (the diagonal in the last case may require branching).

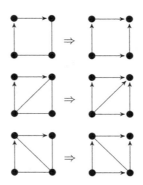

Fig. 14. Completing orientations of quadrilaterals

The main characterization theorem to date for word-representable graphs is the following result.

Theorem 28 ([11]). *A graph G is word-representable if and only if G admits a semi-transitive orientation.*

Proof. The backwards direction is rather complicated and is omitted. An algorithm was created in [11] to turn a semi-transitive orientation of a graph into a word-representant.

The idea of the proof for the forward direction is as follows (see [11] for details). Given a word, say, $w = 2421341$, orient the graph represented by w by letting $x \to y$ be an edge if the *leftmost* x is *to the left* of the *leftmost* y in w, to obtain a semi-transitive orientation:

Any *complete graph* is 1-representable. The algorithm in [11] to turn semi-transitive orientations into word-representants gave the following result.

Theorem 29 ([11]). *Each* non-complete *word-representable graph G is $2(n - \kappa(G))$- representable, where $\kappa(G)$ is the size of the* maximum clique *in G.*

As an immediate corollary of Theorem 29, we have that the *recognition problem* of word-representability is in *NP*. Indeed, any word-representant is of length at most $O(n^2)$, and we need $O(n^2)$ passes through such a word to check alternation properties of all pairs of letters. There is an alternative proof of this complexity observation by Magnús M. Halldórsson in terms of *semi-transitive orientations*. In presenting his proof, we follow [17, Remark 4.2.3].

Checking that a given directed graph G is acyclic is a polynomially solvable problem. Indeed, it is well known that the entry (i, j) of the kth power of the adjacency matrix of G records the number of walks of length k in G from the vertex i to the vertex j. Thus, if G has n vertices, then we need to make sure that the diagonal entries are all 0 in all powers, up to the nth power, of the adjacency matrix of G. Therefore, it remains to show that it is polynomially solvable to check that G is shortcut-free. Let $u \to v$ be an edge in G. Consider the induced subgraph $H_{u \to v}$ consisting of vertices "in between" u and v, that is, the vertex set of $H_{u \to v}$ is

$$\{x \mid \text{there exist directed paths from } u \text{ to } x \text{ and from } x \text{ to } v\}.$$

It is not so difficult to prove that $u \to v$ is not a shortcut (that is, is not a shortcutting edge) if and only if $H_{u \to v}$ is transitive. Now, we can use the well known fact that finding out whether there exists a directed path from one vertex to another in a directed graph is polynomially solvable, and thus it is polynomially solvable to determine $H_{u \to v}$ (one needs to go through n vertices and check the existence of two paths for each vertex). Finally, checking transitivity is also polynomially solvable, which is not difficult to see.

The following theorem shows that word-representable graphs generalize the class of 3-colorable graphs.

Theorem 30 ([11]). *Any 3-colorable graph is word-representable.*

Proof. Coloring a 3-colorable graph in three colors, say, colors 1, 2 and 3, and orienting the edges based on the colors of their endpoints as $1 \to 2 \to 3$, we obtain a semi-transitive orientation. Indeed, obviously there are no cycles, and because the longest directed path involves only three vertices, there are no shortcuts. Theorem 28 can now be applied to complete the proof.

Theorem 30 can be applied to see, for example, that the Petersen graph is word-representable, which we already know from Sect. 2.3. More corollaries to Theorem 30 can be found below.

6   Non-word-representable Graphs

From the discussion in Sect. 3 we already know that the wheel graphs W_{2n+1}, for $n \geq 2$, are not word-representable, and that W_5 is the minimum (by the number of vertices) non-word-representable graph. But then, taking into account the hereditary nature of word-representability (see Remark 1), we have a family $\mathcal{W}$ of non-word-representable graphs characterised by containment of W_{2n+1} ($n \geq 2$) as an induced subgraph.

Note that *each* graph in $\mathcal{W}$ necessarily contains a vertex of degree 5 or more, and also a triangle as an induced subgraph. Natural questions are if there are non-word-representable graphs of *maximum degree* 4, and also if there are *triangle-free* non-word-representable graphs. Both questions were answered in affirmative. The graph to the right in Fig. 10, which was found in [4], addresses the first question, while the second question is addressed by the following construction presented in [10].

Let M be a 4-chromatic graph with girth at least 10 (such graphs exist by a result of Paul Erdős; see [17, Sect. 4.4] for details). The *girth* of a graph is the length of a shortest cycle contained in the graph. If the graph does not contain any cycles (that is, it is an acyclic graph), its girth is defined to be infinity. Now, for every path of length 3 in M add to M an edge connecting path's end vertices. Then the obtained graph is triangle-free and non-word-representable [10].

6.1   Enumeration of Non-word-representable Graphs

According to experiments run by Herman Z.Q. Chen, there are 1, 25 and 929 non-isomorphic non-word-representable connected graphs on six, seven and eight vertices, respectively. These numbers were confirmed and extended to 68,545 for nine vertices, and 4,880,093 for 10 vertices, using a constraint programming (CP)-based method by Özgür Akgün, Ian Gent and Christopher Jefferson.

Figure 15 created by Chen presents the 25 non-isomorphic non-word-represent-able graphs on seven vertices. Note that the only non-word-representable graph on six vertices is the wheel W_5. Further note that the case of seven vertices gives just 10 minimal non-isomorphic non-word-representable graphs, since 15 of the graphs in Fig. 15 contain W_5 as an induced subgraphs (these graphs are the first 11 graphs, plus the 15th, 16th, 18th and 19th graphs).

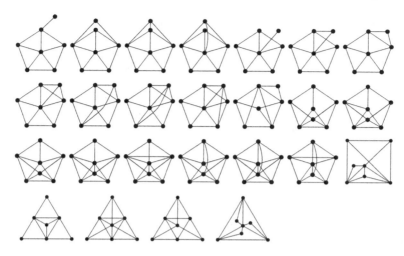

Fig. 15. 25 non-isomorphic non-word-representable graphs on seven vertices

6.2 Non-word-representable Line Graphs

The *line graph* of a graph $G = (V, E)$ is the graph with vertex set E in which two vertices are adjacent if and only if the corresponding edges of G share a vertex. The line graph of G is denoted $L(G)$. Line graphs give a tool to construct non-word-representable graphs as follows from the theorems below.

Theorem 31 ([19]). *Let $n \geq 4$. For any wheel graph W_n, the line graph $L(W_n)$ is non-word-representable.*

Theorem 32 ([19]). *Let $n \geq 5$. For any complete graph K_n, the line graph $L(K_n)$ is non-word-representable.*

$$K_{1,3} = \qquad C_4 = \qquad P_4 =$$

Fig. 16. The claw graph $K_{1,3}$, the cycle graph C_4, and the path graph P_4

The following theorem is especially interesting as it shows how to turn essentially any graph into non-word-representable graph.

Theorem 33 ([19]). *If a connected graph G is not a path graph, a cycle graph, or the claw graph $K_{1,3}$ (see Fig. 16), then the line graph $L^n(G)$ obtained by application of the procedure of taking the line graph to G n times, is non-word-representable for $n \geq 4$.*

7   Word-Representability and Operations on Graphs

In this section we consider some of the most basic operations on graphs, namely, taking the complement, edge subdivision, edge contraction, connecting two graphs by an edge and gluing two graphs in a clique, replacing a vertex with a module, Cartesian product, rooted product and taking line graph.

We do not consider edge-addition/deletion trivially not preserving (non)-word-representability, although there are situations when these operations may preserve word-representability. For example, it is shown in [13] that edge-deletion preserves word-representability on K_4-free word-representable graphs.

Finally, we do not discuss the operations $Y \to \Delta$ (replacing an induced subgraph $K_{1,3}$, the claw, on vertices v_0, v_1, v_2, v_3, where v_0 is the apex, by the triangle on vertices v_1, v_2, v_3 and removing v_0) and $\Delta \to Y$ (removing the edges of a triangle on vertices v_1, v_2, v_3 and adding a vertex v_0 connected to v_1, v_2, v_3) recently studied in [13] in the context of word-representability of graphs.

7.1   Taking the Complement

Starting with a word-representable graph and taking its complement, we may either obtain a word-representable graph or not. Indeed, for example, both any graph on at most five vertices and its complement are word-representable. On the other hand, let G be the graph formed by the 5-cycle (2,4,6,3,5) and an isolated vertex 1. The 5-cycle can be represented by the word 2542643653 (see Sect. 2.2 for a technique to represent cycle graphs) and thus the graph G can be represented by the word 112542643653. However, taking the complement of G, we obtain the wheel graph W_5, which is not word-representable.

Similarly, starting with a non-word-representable graph and taking its complement, we can either obtain a word-representable graph or not. Indeed, the complement of the non-word-representable wheel W_5 is word-representable, as is discussed above. On the other hand, the graph G having two connected components, one W_5 and the other one the 5-cycle C_5, is non-word-representable because of the induced subgraph W_5, while the complement of G also contains an induced subgraph W_5 (formed by the vertices of C_5 in G and any of the remaining vertices) and thus is also non-word-representable.

7.2   Edge Subdivision and Edge Contraction

Subdivision of graphs (see Definition 4) is based on subdivision of individual edges, and it is considered in Sect. 2.3 from 3-representability point of view.

If we change "3-representable" by "word-representable" in Theorem 10 we would obtain a weaker, but clearly still true statement, which is not hard to prove directly via semi-transitive orientations. Indeed, each path of length at least 3 added instead of an edge e can be oriented in a "blocking" way, so that there would be no directed path between e's endpoints. Thus, edge subdivision does not preserve the property of being non-word-representable. The following theorem shows that edge subdivision may be preserved on some subclasses of word-representable graphs, but not on the others.

Theorem 34 ([13]). *Edge subdivision preserves word-representability on K_4-free word-representable graphs, and it does not necessarily preserve word-represent-ability on K_5-free word-representable graphs.*

Recall the definition of edge contraction in Definition 5. By Theorem 10, contracting an edge in a word-representable graph may result in a non-word-representable graph, while in many cases, e.g. in the case of path graphs, word-representability is preserved under this operation.

On the other hand, when starting from a non-word-representable graph, a graph obtained from it by edge contraction can also be either word-representable or non-word-representable. For example, contracting any edge incident with the bottommost vertex in the non-word-representable graph to the right in Fig. 10, we obtain a graph on six vertices that is different from W_5 and is thus word-representable. Finally, any non-word-representable graph can be glued in a vertex with a path graph P (the resulting graph will be non-word-representable), so that contracting any edge in the subgraph formed by P results in a non-word-representable graph.

7.3   Connecting Two Graphs by an Edge and Gluing Two Graphs in a Clique

In what follows, by *gluing* two graphs in a clique we mean the following operation. Suppose $a_1, \ldots, a_k$ and $b_1, \ldots, b_k$ are cliques of size k in graphs G_1 and G_2, respectively. Then gluing G_1 and G_2 in a clique of size k means identifying each a_i with one b_j, for $i, j \in \{1, \ldots, k\}$ so that the neighbourhood of the obtained vertex $c_{i,j}$ is the union of the neighbourhoods of a_i and b_j.

By the hereditary nature of word-representability (see Remark 1), if at least one of two graphs is non-word-representable, then gluing the graphs in a clique, or connecting two graphs by an edge (with the endpoints belonging to different graphs) will result in a non-word-representable graph.

On the other hand, suppose that graphs G_1 and G_2 are word-representable. Then gluing the graphs in a vertex, or connecting the graphs by an edge will result in a word-representable graph. The latter statement is easy to see using the notion of semi-transitive orientation. Indeed, by Theorem 28 both G_1 and G_2 can be oriented semi-transitively, and gluing the oriented graphs in a vertex, or connecting the graphs by an edge oriented arbitrarily, will not result in any cycles or shortcuts created. In fact, it was shown in [15] that if G_1 is k_1-representable (such a k_1 must exist by Theorem 2) and G_2 is k_2-representable, then essentially always the graph obtained either by gluing G_1 and G_2 in a vertex or by connecting the graphs by an edge is $\max(k_1, k_2)$-representable.

Even though gluing two word-representable graphs in a vertex (clique of size 1) always results in a word-representable graph, this is not necessarily true for Gluing graphs in an edge (clique of size 2) or in a triangle (clique of size 3).

We refer to [17, Sect. 5.4.3] for the respective examples. Glueing two graphs in cliques of size 4 or more in the context of word-representability remains an unexplored direction.

7.4   Replacing a Vertex with a Module

A subset X of the set of vertices V of a graph G is a *module* if all members of X have the same set of neighbours among vertices not in X (that is, among vertices in $V \setminus X$). For example, Fig. 17 shows replacement of the vertex 1 in the triangular prism by the module K_3 formed by the vertices a, b and c. Thus, $\{a, b, c\}$ is a module of the graph on the right in Fig. 17.

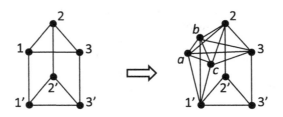

Fig. 17. Replacing a vertex by a module

Theorem 35 ([15]). *Suppose that $G = (V, E)$ is a word-representable graph and $x \in V$. Let G' be obtained from G by replacing x with a module M, where M is any comparability graph (in particular, any clique). Then G' is also word-representable. Moreover, if $\mathcal{R}(G) = k_1$ and $\mathcal{R}(M) = k_2$ then $\mathcal{R}(G') = k$, where $k = \max\{k_1, k_2\}$.*

7.5   Cartesian Product of Two Graphs

The *Cartesian product* $G \square H$ of graphs $G = (V(G), E(G))$ and $H = (V(H), E(H))$ is a graph such that

- the vertex set of $G \square H$ is the Cartesian product $V(G) \times V(H)$; and
- any two vertices (u, u') and (v, v') are adjacent in $G \square H$ if and only if either
  - $u = v$ and u' is adjacent to v' in H, or
  - $u' = v'$ and u is adjacent to v in G.

See Fig. 18 for an example of the Cartesian product of two graphs.

A proof of the following theorem was given by Bruce Sagan in 2014. The proof relies on semi-transitive orientations and it can be found in [17, Sect. 5.4.5].

Theorem 36 (Sagan). *Let G and H be two word-representable graphs. Then the Cartesian product $G \square H$ is also word-representable.*

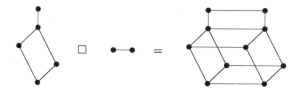

Fig. 18. Cartesian product of two graphs

7.6   Rooted Product of Graphs

The *rooted product* of a graph G and a rooted graph H (i.e. one vertex of H is distinguished), $G \circ H$, is defined as follows: take $|V(G)|$ copies of H, and for every vertex v_i of G, identify v_i with the root vertex of the ith copy of H. See Fig. 19 for an example of the rooted product of two graphs.

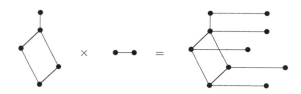

Fig. 19. Rooted product of two graphs

The next theorem is an analogue of Theorem 36 for the rooted product of two graphs.

Theorem 37 ([17]). *Let G and H be two word-representable graphs. Then the rooted product $G \circ H$ is also word-representable.*

Proof. Identifying a vertex v_i in G with the root vertex of the ith copy of H in the definition of the rooted product gives a word-representable graph by the discussion in Sect. 7.3. Thus, identifying the root vertices, one by one, we will keep obtaining word-representable graphs, which gives us at the end word-representability of $G \circ H$.

7.7   Taking the Line Graph Operation

Taking the line graph operation has already been considered in Sect. 6.2. Based on the results presented in that section, we can see that this operation can turn a word-representable graph into either a word-representable graph or non-word-representable graph. Also, there are examples of when the line graph of a non-word-representable graph is non-word-representable. However, it remains an open problem whether a non-word-representable graph can be turned into a word-representable graph by applying the line graph operation.

8   Computational Complexity Results and Word-Representability of Planar Graphs

In this section we will present known complexity results and also discuss word-representability of planar graphs.

8.1   A Summary of Known Complexity Results

Even though the Maximum Clique problem is polynomially solvable on word-representable graphs (see Theorem 16), many classical optimization problems are NP-hard on these graphs. The latter follows from the problems being NP-hard on 3-colorable graphs and Theorem 30.

The justification of the known complexity results presented in Table 2, as well as the definitions of the problems can be found in [17, Section 4.2]. However, below we discuss a proof of the fact that recognizing word-representability is an NP-complete problem. We refer to [17, Sect. 4.2] for any missed references to the results we use.

Table 2. Known complexities for problems on word-representable graphs

Problem	Complexity
Deciding whether a given graph is word-representable	NP-complete
Approximating the graph representation number within a factor of $n^{1-\epsilon}$ for any $\epsilon > 0$	NP-hard
Clique covering	NP-hard
Deciding whether a given graph is k-word-representable for any fixed k, $3 \leq k \leq \lceil n/2 \rceil$	NP-complete
Dominating set	NP-hard
Vertex colouring	NP-hard
Maximum clique	in P
Maximum independent set	NP-hard

Suppose that P is a poset and x and y are two of its elements. We say that x covers y if $x > y$ and there is no element z in P such that $x > z > y$.

The *cover graph* G_P of a poset P has P's elements as its vertices, and $\{x, y\}$ is an edge in G_P if and only if either x covers y, or vice versa. The *diagram* of P, sometimes called a *Hasse diagram* or *order diagram*, is a drawing of the cover graph of G in the plane with x being higher than y whenever x covers y in P. The three-dimensional cube in Fig. 6 is an example of a cover graph.

Vincent Limouzy observed in 2014 that semi-transitive orientations of triangle-free graphs are exactly the 2-*good orientations* considered in [22] by Pretzel (we refer to that paper for the definition of a k-good orientation). Thus, by Proposition 1 in [22] we have the following reformulation of Pretzel's result in our language.

Theorem 38 (Limouzy). *The class of triangle-free word-representable graphs is exactly the class of cover graphs of posets.*

It was further observed by Limouzy, that it is an NP-complete problem to recognize the class of cover graphs of posets. This implies the following theorem, which is a key complexity result on word-representable graphs.

Theorem 39 (Limouzy). *It is an NP-complete problem to recognize whether a given graph is word-representable.*

8.2 Word-Representability of Planar Graphs

Recall that not all planar graphs are word-representable. Indeed, for example, wheel graphs W_{2n+1}, or graphs in Fig. 10, are not word-representable.

Theorem 40 ([10]). *Triangle-free planar graphs are word-representable.*

Proof. By Grötzch's theorem [23], *every* triangle-free planar graph is 3-colorable, and Theorem 30 can be applied.

It remains a challenging open problem to classify word-representable planar graphs. Towards solving the problem, various *triangulations* and certain *subdivisions* of planar graphs were considered to be discussed next. Key tools to study word-representability of planar graphs are *3-colorability* and *semi-transitive orientations*.

Word-Representability of Polyomino Triangulations. A *polyomino* is a plane geometric figure formed by joining one or more equal squares edge to edge. Letting corners of squares in a polyomino be vertices, we can treat polyominoes as graphs. In particular, well known *grid graphs* are obtained from polyominoes in this way. Of particular interest to us are *convex polyominoes*. A polyomino is said to be *column convex* if its intersection with any vertical line is convex (in other words, each column has no holes). Similarly, a polyomino is said to be *row convex* if its intersection with any horizontal line is convex. A polyomino is said to be *convex* if it is row and column convex.

When dealing with word-representability of triangulations of convex polyominoes (such as in Fig. 20), one should watch for *odd wheel graphs* as induced subgraphs (such as the part of the graph in bold in Fig. 20). Absence of such subgraphs will imply 3-colorability and thus word-representability, which is the basis of the proof of the following theorem.

Theorem 41 ([1]). *A triangulation of a convex polyomino is word-representable if and only if it is 3-colorable. There are non-3-colorable word-representable nonconvex polyomino triangulations.*

The case of rectangular polyomino triangulations with a single domino tile (such as in Fig. 21) is considered in the next theorem.

Theorem 42 ([8]). *A triangulation of a rectangular polyomino with a single domino tile is word-representable if and only if it is 3-colorable.*

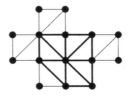

Fig. 20. A triangulation of a poyomino

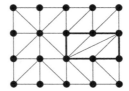

Fig. 21. A triangulation of a rectangular polyomino with a single domino tile

Word-Representability of Near-Triangulations. A *near-triangulation* is a planar graph in which each *inner bounded face* is a triangle (where the *outer face* may possibly not be a triangle).

The following theorem is a far-reaching generalization of Theorems 41 and 42.

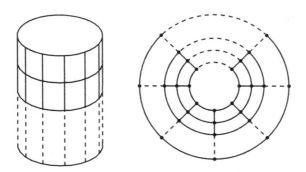

Fig. 22. Grid-covered cylinder

Theorem 43 ([6])**.** *A K_4-free near-triangulation is 3-colorable if and only if it is word-representable.*

Characterization of word-representable near-triangulations (containing K_4) is still an open problem.

Triangulations of Grid-Covered Cylinder Graphs A *grid-covered cylinder*, *GCC* for brevity, is a 3-dimensional figure formed by drawing vertical lines and horizontal circles on the surface of a cylinder, each of which are parallel to the generating line and the upper face of the cylinder, respectively. A GCC can be thought of as the object obtained by gluing the left and right sides of a rectangular grid. See the left picture in Fig. 22 for a schematic way to draw a GCC. The vertical lines and horizontal circles are called the *grid lines*. The part of a GCC between two consecutive vertical lines defines a *sector*.

Any GCC defines a graph, called *grid-covered cylinder graph*, or *GCCG*, whose set of vertices is given by intersection of the grid lines, and whose edges are parts of grid lines between the respective vertices. A typical triangulation of a GCCG is presented schematically in Fig. 23.

Word-representability of triangulations of any GCCG is completely characterized by the following two theorems, which take into consideration the number of sectors in a GCCG.

Theorem 44 ([3]). *A triangulation of a GCCG with* more than three *sectors is word-representable if and only if it contains no W_5 or W_7 as an induced subgraph.*

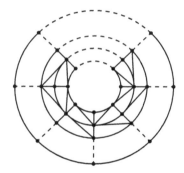

Fig. 23. A triangulation of a GCCG

Theorem 45 ([3]). *A triangulation of a GCCG with three sectors is word-representable if and only if it contains no graph in Fig. 24 as an induced subgraph.*

Subdivisions of Triangular Grid Graphs. The *triangular tiling graph T^∞* is the *Archimedean tiling 3^6* (see Fig. 25). By a *triangular grid graph G* we mean a graph obtained from T^∞ as follows. Specify a finite number of triangles, called *cells*, in T^∞. The edges of G are then all the edges surrounding the specified cells, while the vertices of G are the endpoints of the edges (defined by intersecting lines in T^∞). We say that the specified cells, along with any other cell whose all edges are from G, *belong* to G.

The operation of *face subdivision of a cell* is putting a new vertex inside the cell and making it to be adjacent to every vertex of the cell. Equivalently, face

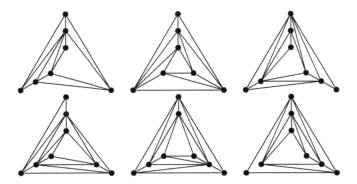

Fig. 24. All minimal non-word-representable induced subgraphs in triangulations of GCCG's with three sectors

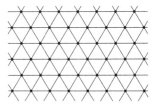

Fig. 25. A fragment of the graph T^∞

subdivision of a cell is replacing the cell (which is the complete graph K_3) by a plane version of the complete graph K_4. A *face subdivision of a set S of cells* of a triangular grid graph G is a graph obtained from G by subdividing each cell in S. The set S of subdivided cells is called a *subdivided set*. For example, Fig. 26 shows K_4, the face subdivision of a cell, and A', a face subdivision of A.

If a face subdivision of G results in a word-representable graph, then the face subdivision is called a *word-representable face subdivision*. Also, we say that a word-representable face subdivision of a triangular grid graph G is *maximal* if subdividing any other cell results in a non-word-representable graph.

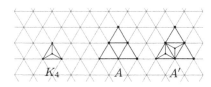

Fig. 26. Examples of face subdivisions: K_4 is the face subdivision of a cell, and A' is a face subdivision of A

An edge of a triangular grid graph G shared with a cell in T^∞ that does not belong to G is called a *boundary edge*. A cell in G that is incident to at least one

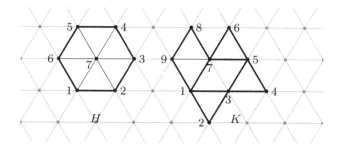

Fig. 27. Graphs H and K, where boundary edges are in bold

boundary edge is called a *boundary cell*. A non-boundary cell in G is called an *interior cell*. For example, the boundary edges in the graphs H and K in Fig. 27 are in bold.

A face subdivision of a triangular grid graph that involves face subdivision of just boundary cells is called a *boundary face subdivision*. The following theorem was proved using the notion of a *smart* orientation (see [2] for details).

Theorem 46 ([2]). *A face subdivision of a triangular grid graph G is word-representable if and only if it has no induced subgraph isomorphic to A'' in Fig. 28, that is, G has no subdivided interior cell.*

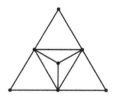

Fig. 28. The graph A''

Theorem 46 can be applied to the *two-dimensional Sierpiński gasket graph* $SG(n)$ to find its maximum word-representable subdivision (see [2] for details).

9 Directions for Further Research

In this section we list some of open problems and directions for further research related to word-representable graphs. The first question though the Reader should ask himself/herself is "Which graphs in their favourite class of graphs are word-representable?".

- Characterize (non-)word-representable planar graphs.
- Characterize word-representable near-triangulations (containing K_4).
- Describe graphs representable by words avoiding a pattern τ, where the notion of a "pattern" can be specified in any suitable way, e.g. it could be a *classical pattern*, a *vincular pattern*, or a *bivincular pattern* (see [14] for definitions).
- Is it true that out of all bipartite graphs on the same number of vertices, *crown graphs* require the *longest* word-representants?
- Are there any graphs on n vertices whose representation requires *more* than $\lfloor n/2 \rfloor$ copies of each letter?
- Is the *line graph* of a non-word-representable graph *always* non-word-represent-able?
- Characterize word-representable graphs in terms of *forbidden subgraphs*.
- Translate a known to you problem on *graphs* to *words* representing these graphs (assuming such words exist), and find an *efficient algorithm* to solve the *obtained problem*, and thus the *original problem*.

The last two problems are of fundamental importance.

References

1. Akrobotu, P., Kitaev, S., Masárová, Z.: On word-representability of polyomino triangulations. Siberian Adv. Math. **25**(1), 1–10 (2015)
2. Chen, T.Z.Q., Kitaev, S., Sun, B.Y.: Word-representability of face subdivisions of triangular grid graphs. Graphs Comb. **32**(5), 1749–1761 (2016)
3. Chen, T.Z.Q., Kitaev, S., Sun, B.Y.: Word-representability of triangulations of grid-covered cylinder graphs. Discr. Appl. Math. **213**(C), 60–70 (2016)
4. Collins, A., Kitaev, S., Lozin, V.: New results on word-representable graphs. Discr. Appl. Math. **216**, 136–141 (2017)
5. Gao, A., Kitaev, S., Zhang, P.: On 132-representable graphs. arXiv:1602.08965 (2016)
6. Glen, M.: Colourability and word-representability of near-triangulations. arXiv:1605.01688 (2016)
7. Glen, M.: Software. http://personal.cis.strath.ac.uk/sergey.kitaev/word-representable-graphs.html
8. Glen, M., Kitaev, S.: Word-representability of triangulations of rectangular polyomino with a single domino tile. J. Comb. Math. Comb. Comput. (to appear)
9. Halldórsson, M.M., Kitaev, S., Pyatkin, A.: Graphs capturing alternations in words. In: Gao, Y., Lu, H., Seki, S., Yu, S. (eds.) DLT 2010. LNCS, vol. 6224, pp. 436–437. Springer, Heidelberg (2010). doi:10.1007/978-3-642-14455-4_41
10. Halldórsson, M.M., Kitaev, S., Pyatkin, A.: Alternation graphs. In: Kolman, P., Kratochvíl, J. (eds.) WG 2011. LNCS, vol. 6986, pp. 191–202. Springer, Heidelberg (2011). doi:10.1007/978-3-642-25870-1_18
11. Halldórsson, M., Kitaev, S., Pyatkin, A.: Semi-transitive orientations and word-representable graphs. Discrete Appl. Math. **201**, 164–171 (2016)
12. Jones, M., Kitaev, S., Pyatkin, A., Remmel, J.: Representing graphs via pattern avoiding words. Electron. J. Comb. **22**(2), 2.53 (2015). 20 pp
13. Kim, J., Kim, M.: Graph orientations on word-representable graphs (in preparation)

14. Kitaev, S.: Patterns in Permutations and Words. Springer, Heidelberg (2011)
15. Kitaev, S.: On graphs with representation number 3. J. Autom. Lang. Comb. **18**(2), 97–112 (2013)
16. Kitaev, S.: Existence of u-representation of graphs. J. Graph Theor. **85**(3), 661–668 (2017)
17. Kitaev, S., Lozin, V.: Words and Graphs. Springer, Heidelberg (2015)
18. Kitaev, S., Pyatkin, A.: On representable graphs. J. Autom. Lang. Comb. **13**(1), 45–54 (2008)
19. Kitaev, S., Salimov, P., Severs, C., Úlfarsson, H.: On the representability of line graphs. In: Mauri, G., Leporati, A. (eds.) DLT 2011. LNCS, vol. 6795, pp. 478–479. Springer, Heidelberg (2011). doi:10.1007/978-3-642-22321-1_46
20. Kitaev, S., Seif, S.: Word problem of the Perkins semigroup via directed acyclic graphs. Order **25**(3), 177–194 (2008)
21. Mandelshtam, Y.: On graphs representable by pattern-avoiding words. arXiv:1608.07614 (2016)
22. Pretzel, O.: On graphs that can be oriented as diagrams of ordered sets. Order **2**(1), 25–40 (1985)
23. Thomassen, C.: A short list color proof of Grötzsch's theorem. J. Comb. Theor. Ser. B **88**(1), 189–192 (2003)

Binomial Coefficients, Valuations, and Words

Eric Rowland$^{(\boxtimes)}$

Department of Mathematics, Hofstra University,
Hempstead, NY 11549, USA
eric.rowland@hofstra.edu

Abstract. The study of arithmetic properties of binomial coefficients has a rich history. A recurring theme is that p-adic statistics reflect the base-p representations of integers. We discuss many results expressing the number of binomial coefficients $\binom{n}{m}$ with a given p-adic valuation in terms of the number of occurrences of a given word in the base-p representation of n, beginning with a result of Glaisher from 1899, up through recent results by Spiegelhofer–Wallner and Rowland.

Keywords: Binomial coefficients $\cdot$ p-adic valuation $\cdot$ Regular sequences

1 Valuations of Binomial Coefficients

In 1852, Kummer [10, pages 115–116] determined the exact power of a prime p that divides a binomial coefficient $\binom{n}{m}$. To state this result, let $\nu_p(n)$ denote the p-adic valuation of n, that is, the exponent of the highest power of p dividing n.

Theorem 1 (Kummer). *Let p be a prime, and let n and m be integers with $0 \le m \le n$. Then $\nu_p(\binom{n}{m})$ is the number of carries involved in adding m to $n - m$ in base p.*

Kummer's theorem is the first of many results to express arithmetic information about binomial coefficients in terms of base-p representations of integers.

Glaisher [6, Sect. 14] seems to have been the first to count binomial coefficients satisfying a given congruence condition. He showed that the number of integers m in the range $0 \le m \le n$ such that $\binom{n}{m}$ is odd is $2^{|n|_1}$. Here $|n|_1$ is the number of 1s in the standard base-2 representation of n.

Half a century later, Glaisher's result was generalized to an arbitrary prime by Fine [5, Theorem 2]. For a prime p and an integer $n \ge 0$, let $\theta_{p,0}(n)$ be the number of integers m in the range $0 \le m \le n$ such that $\binom{n}{m} \not\equiv 0 \mod p$. Let $|n|_w$ be the number of occurrences of the word w in the base-p representation of n. Fine showed that

$$\theta_{p,0}(n) = \prod_{d=0}^{p-1} (d+1)^{|n|_d}.$$

Since the publication of Fine's result, many authors have been interested in generalizations to higher powers of p. A natural quantity to study is the number $\theta_{p,\alpha}(n)$ of binomial coefficients $\binom{n}{m}$, for $0 \le m \le n$, with $\nu_p(\binom{n}{m}) = \alpha$.

© Springer International Publishing AG 2017
É. Charlier et al. (Eds.): DLT 2017, LNCS 10396, pp. 68–74, 2017.
DOI: 10.1007/978-3-319-62809-7_3

Carlitz [3, Eqs. (1.7)–(1.9)] gave a recurrence involving $\theta_{p,\alpha}(n)$ and a secondary quantity $\psi_{p,\alpha}(n)$ defined as the number of integers m in the range $0 \le m \le n$ such that $\nu_p((m+1)\binom{n}{m}) = \alpha$. Namely,

$$\theta_{p,\alpha}(pn+d) = (d+1)\theta_{p,\alpha}(n) + (p-d-1)\psi_{p,\alpha-1}(n-1)$$

$$\psi_{p,\alpha}(pn+d) = \begin{cases} (d+1)\theta_{p,\alpha}(n) + (p-d-1)\psi_{p,\alpha-1}(n-1) & \text{if } 0 \le d \le p-2 \\ p\psi_{p,\alpha-1}(n) & \text{if } d = p-1. \end{cases}$$

As we will see in Sect. 3, this recurrence comes close to giving a matrix generalization of Fine's theorem, but it is not of the right form.

Nonetheless, Carlitz's recurrence can be used to obtain formulas for $\theta_{p,\alpha}(n)$. Let $n_\ell \cdots n_1 n_0$ be the standard base-p representation of n. For $\alpha = 1$, Carlitz [3, Eq. (2.5)] showed

$$\theta_{p,1}(n) = \sum_{i=0}^{\ell-1} (n_\ell+1)(n_{\ell-1}+1)\cdots(n_{i+2}+1)n_{i+1}(p-n_i-1)(n_{i-1}+1)\cdots(n_0+1).$$

Dividing by $\theta_{p,0}(n) = (n_\ell + 1)\cdots(n_0 + 1)$ gives

$$\frac{\theta_{p,1}(n)}{\theta_{p,0}(n)} = \sum_{i=0}^{\ell-1} \frac{n_{i+1}}{n_{i+1}+1} \cdot \frac{p-n_i-1}{n_i+1}.$$

This equation is our first indication that expressions for $\frac{\theta_{p,\alpha}(n)}{\theta_{p,0}(n)}$ can be simpler than expressions for $\theta_{p,\alpha}(n)$ alone. In particular, $\frac{\theta_{p,1}(n)}{\theta_{p,0}(n)}$ is a polynomial in the variables $|n|_w$ for words $w \in \{0, 1, \ldots, p-1\}^*$ of length 2. For $p = 2$ we obtain

$$\theta_{2,1}(n) = 2^{|n|_1} \cdot \frac{1}{2}|n|_{10},$$

which was obtained by Howard [7, Eq. (2.4)] and by Davis and Webb [4, Theorem 7]. For $p = 3$ we have

$$\theta_{3,1}(n) = 2^{|n|_1} 3^{|n|_2} \left(|n|_{10} + \frac{1}{4}|n|_{11} + \frac{4}{3}|n|_{20} + \frac{1}{3}|n|_{21} \right),$$

which also follows from the work of Huard, Spearman, and Williams [8]. For $p = 5$ we have

$$\theta_{5,1}(n) = 2^{|n|_1} 3^{|n|_2} 4^{|n|_3} 5^{|n|_4} \left(2|n|_{10} + \frac{3}{4}|n|_{11} + \frac{1}{3}|n|_{12} + \frac{1}{8}|n|_{13} \right.$$

$$+ \frac{8}{3}|n|_{20} + |n|_{21} + \frac{4}{9}|n|_{22} + \frac{1}{6}|n|_{23}$$

$$+ 3|n|_{30} + \frac{9}{8}|n|_{31} + \frac{1}{2}|n|_{32} + \frac{3}{16}|n|_{33}$$

$$\left. + \frac{16}{5}|n|_{40} + \frac{6}{5}|n|_{41} + \frac{8}{15}|n|_{42} + \frac{1}{5}|n|_{43} \right),$$

and so on.

2   Formulas for Arbitrary Prime Powers

We have seen that $\frac{\theta_{p,1}(n)}{\theta_{p,0}(n)}$ is a polynomial in $|n|_w$. In fact this is true for $\frac{\theta_{p,\alpha}(n)}{\theta_{p,0}(n)}$ in general. Barat and Grabner [2, Sect. 3] showed this implicitly while studying the asymptotic behavior of $\sum_{n=0}^{N} \theta_{p,\alpha}(n)$.

Rowland [12] gave an algorithm for computing a polynomial expression for $\frac{\theta_{p,\alpha}(n)}{\theta_{p,0}(n)}$. For $p = 2$ one computes

$$\theta_{2,2}(n) = 2^{|n|_1} \left(-\frac{1}{8}|n|_{10} + |n|_{100} + \frac{1}{4}|n|_{110} + \frac{1}{8}|n|_{10}^2 \right)$$

(which was obtained by Howard [7, Equation (2.5)] and by Huard, Spearman, and Williams [9, Theorem C]),

$$\theta_{2,3}(n) = 2^{|n|_1} \left(\frac{1}{24}|n|_{10} - \frac{1}{2}|n|_{100} - \frac{1}{8}|n|_{110} + 2|n|_{1000} + \frac{1}{2}|n|_{1010} + \frac{1}{2}|n|_{1100} \right.$$
$$\left. + \frac{1}{8}|n|_{1110} - \frac{1}{16}|n|_{10}^2 + \frac{1}{2}|n|_{10}|n|_{100} + \frac{1}{8}|n|_{10}|n|_{110} + \frac{1}{48}|n|_{10}^3 \right),$$

and so on. The number of nonzero terms in the polynomial $\frac{\theta_{2,\alpha}(n)}{2^{|n|_1}}$ for $\alpha = 0, 1, 2, \ldots$ is sequence A275012 [11]:

$$1, 1, 4, 11, 29, 69, 174, 413, 995, 2364, 5581, 13082, 30600, 71111, 164660, 379682, \ldots$$

The algorithm for computing these polynomials also establishes bounds on their total degree and on the length of words that appear.

Theorem 2 (Rowland [12]). *Let p be a prime, and let $\alpha \geq 0$. Then $\frac{\theta_{p,\alpha}(n)}{\theta_{p,0}(n)}$ is given by a polynomial of degree α in $|n|_w$ for words w satisfying $|w| \leq \alpha + 1$.*

However, the algorithm is not particularly fast, as it constructs a polynomial by summing over certain sets of integer partitions. Spiegelhofer and Wallner [14] produced a faster algorithm by developing a better understanding of the structure of this polynomial. In particular, they showed that the polynomial representation of $\frac{\theta_{p,\alpha}(n)}{\theta_{p,0}(n)}$ is unique, as long as words of the form $0w$ and $w(p-1)$ are not used. Therefore one can talk about *the* coefficient of a given monomial. Moreover, this coefficient can be read off from a certain power series. One can see evidence of this by looking at the coefficient of $|n|_{10}$ in the expressions computed for $\frac{\theta_{2,\alpha}(n)}{2^{|n|_1}}$. The sequence of coefficients is

$$0, \frac{1}{2}, -\frac{1}{8}, \frac{1}{24}, -\frac{1}{64}, \frac{1}{160}, -\frac{1}{384}, \ldots$$

These are the coefficients in the power series for $\log(1 + \frac{x}{2})$ at $x = 0$.

The polynomial

$$T_p(n, x) := \sum_{m=0}^{n} x^{\nu_p(\binom{n}{m})} = \sum_{\alpha \geq 0} \theta_{p,\alpha}(n) x^\alpha$$

is a central object in Spiegelhofer and Wallner's result. Let

$$\overline{T}_p(w, x) := \frac{T_p(\mathrm{val}_p(w), x)}{\theta_{p,0}(\mathrm{val}_p(w))},$$

where $\mathrm{val}_p(w)$ is the integer obtained by reading w in base p. Define the rational function

$$r_p(w, x) := \frac{\overline{T}_p(w, x)\overline{T}_p(w_{LR}, x)}{\overline{T}_p(w_R, x)\overline{T}_p(w_L, x)},$$

where the left and right truncations of a word are defined for $\ell \geq 0$, $c \in \{1, \ldots, p-1\}$, and $d \in \{0, 1, \ldots, p-1\}$ by

$$\begin{array}{lll} \epsilon_L = \epsilon & (c0^\ell)_L = \epsilon & (c0^\ell w)_L = w \\ \epsilon_R = \epsilon & c_R = \epsilon & (wd)_R = w. \end{array}$$

Theorem 3 (Spiegelhofer–Wallner [14]**).** *Let p be a prime, and let $\alpha \geq 0$. Let $w_1, \ldots, w_m$ be words of length ≥ 2 on the alphabet $\{0, 1, \ldots, p-1\}$ that do not begin with 0 or end with $p-1$. Then the coefficient of $|n|_{w_1}^{k_1} \cdots |n|_{w_m}^{k_m}$ in the polynomial $\frac{\theta_{p,\alpha}(n)}{\theta_{p,0}(n)}$ is the coefficient of x^α in the power series expansion for*

$$\frac{1}{k_1!}(\log r_p(w_1, x))^{k_1} \cdots \frac{1}{k_m!}(\log r_p(w_m, x))^{k_m}.$$

3 Matrix Generalizations of Fine's Theorem

Spiegelhofer and Wallner's polynomial $T_p(n, x)$ turns out to have a product formula which generalizes Fine's theorem. Note that the first equation in Carlitz's recurrence,

$$\theta_{p,\alpha}(pn + d) = (d+1)\theta_{p,\alpha}(n) + (p - d - 1)\psi_{p,\alpha-1}(n - 1),$$

can be rewritten in terms of $T_p(n, x)$ as

$$T_p(pn + d, x) = (d+1)T_p(n, x) + \begin{cases} 0 & \text{if } n = 0 \\ (p - d - 1)x^{\nu_p(n)+1}T_p(n - 1, x) & \text{if } n \geq 1. \end{cases}$$

To simplify this equation, let us introduce a secondary polynomial

$$T_p'(n, x) := \begin{cases} 0 & \text{if } n = 0 \\ x^{\nu_p(n)+1}T_p(n - 1, x) & \text{if } n \geq 1, \end{cases}$$

so that $\psi_{p,\alpha-1}(n - 1)$ is the coefficient of x^α in $T_p'(n, x)$. Then we have

$$T_p(pn + d, x) = (d+1)T_p(n, x) + (p - d - 1)T_p'(n, x).$$

We are close to being able to write

$$\begin{bmatrix} T_p(pn+d,x) \\ T'_p(pn+d,x) \end{bmatrix} = M_p(d) \begin{bmatrix} T_p(n,x) \\ T'_p(n,x) \end{bmatrix} \tag{1}$$

for some 2×2 matrix $M_p(d)$, thereby expressing $T_p(pn+d,x)$ and $T'_p(pn+d,x)$ in terms of $T_p(n,x)$ and $T'_p(n,x)$. Carlitz's second equation,

$$\psi_{p,\alpha}(pn+d) = \begin{cases} (d+1)\theta_{p,\alpha}(n) + (p-d-1)\psi_{p,\alpha-1}(n-1) & \text{if } 0 \le d \le p-2 \\ p\psi_{p,\alpha-1}(n) & \text{if } d = p-1, \end{cases}$$

expresses $\psi_{p,\alpha}(pn+d)$ in terms of θ and ψ. But because the coefficient of x^α in $T'_p(n,x)$ is $\psi_{p,\alpha-1}(n-1)$ we instead need to express $\psi_{p,\alpha}(pn+d-1)$ in terms of θ and ψ. The desired equation is

$$\psi_{p,\alpha}(pn+d-1) = d\theta_{p,\alpha}(n) + (p-d)\psi_{p,\alpha-1}(n-1),$$

which is equivalent to

$$T'_p(pn+d,x) = d\,x\,T_p(n,x) + (p-d)\,x\,T'_p(n,x).$$

Therefore, the matrix we seek is

$$M_p(d) = \begin{bmatrix} d+1 & p-d-1 \\ d\,x & (p-d)\,x \end{bmatrix}.$$

Theorem 4 (Rowland [13]). *Let p be a prime, and let $n \ge 0$. Let $n_\ell \cdots n_1 n_0$ be the standard base-p representation of n. Then*

$$T_p(n,x) = \begin{bmatrix} 1 & 0 \end{bmatrix} M_p(n_0) M_p(n_1) \cdots M_p(n_\ell) \begin{bmatrix} 1 \\ 0 \end{bmatrix}.$$

Setting $x = 0$ gives Fine's result as a special case. Namely, the definition of $T'_p(n,x)$ implies $T'_p(n,0) = 0$, so Eq. (1) becomes

$$\begin{bmatrix} \theta_{p,0}(pn+d) \\ 0 \end{bmatrix} = \begin{bmatrix} d+1 & p-d-1 \\ 0 & 0 \end{bmatrix} \begin{bmatrix} \theta_{p,0}(n) \\ 0 \end{bmatrix},$$

or simply

$$\theta_{p,0}(pn+d) = (d+1)\,\theta_{p,0}(n).$$

Moreover, Theorem 4 generalizes naturally to multinomial coefficients. For a k-tuple $\mathbf{m} = (m_1, m_2, \ldots, m_k)$ of non-negative integers, define

$$\text{total}\,\mathbf{m} := m_1 + m_2 + \cdots + m_k$$

and

$$\text{mult}\,\mathbf{m} := \frac{(\text{total}\,\mathbf{m})!}{m_1!\,m_2!\,\cdots\,m_k!}.$$

Let $c_{p,k}(n)$ be the coefficient of x^n in $(1 + x + x^2 + \cdots + x^{p-1})^k$. For each $d \in \{0, 1, \ldots, p - 1\}$, let $M_{p,k}(d)$ be the $k \times k$ matrix whose (i, j) entry is $c_{p,k}(p(j - 1) + d - (i - 1)) x^{i-1}$. For example, let $p = 5$ and $k = 3$; the matrices $M_{5,3}(0), \ldots, M_{5,3}(4)$ are

$$
\begin{bmatrix} 1 & 18 & 6 \\ 0 & 15x & 10x \\ 0 & 10x^2 & 15x^2 \end{bmatrix}, \quad
\begin{bmatrix} 3 & 19 & 3 \\ x & 18x & 6x \\ 0 & 15x^2 & 10x^2 \end{bmatrix}, \quad
\begin{bmatrix} 6 & 18 & 1 \\ 3x & 19x & 3x \\ x^2 & 18x^2 & 6x^2 \end{bmatrix},
$$

$$
\begin{bmatrix} 10 & 15 & 0 \\ 6x & 18x & x \\ 3x^2 & 19x^2 & 3x^2 \end{bmatrix}, \quad
\begin{bmatrix} 15 & 10 & 0 \\ 10x & 15x & 0 \\ 6x^2 & 18x^2 & x^2 \end{bmatrix}.
$$

Theorem 5 (Rowland [13]). *Let p be a prime, let $k \geq 1$, and let $n \geq 0$. Let $e = \begin{bmatrix} 1 & 0 & 0 \cdots 0 \end{bmatrix}$ be the first standard basis vector in $\mathbb{Z}^k$. Let $n_\ell \cdots n_1 n_0$ be the standard base-p representation of n. Then*

$$
\sum_{\substack{\mathbf{m} \in \mathbb{N}^k \\ \text{total } \mathbf{m} = n}} x^{\nu_p(\text{mult } \mathbf{m})} = e \, M_{p,k}(n_0) \, M_{p,k}(n_1) \cdots M_{p,k}(n_\ell) \, e^\top.
$$

The proof essentially amounts to showing that, for $d \in \{0, 1, \ldots, p - 1\}$, $0 \leq i \leq k - 1$, and $\alpha \geq 0$, the map β defined by

$$
\beta(\mathbf{m}) := (\lfloor \mathbf{m}/p \rfloor, \mathbf{m} \bmod p)
$$

is a bijection from the set

$$
A = \left\{ \mathbf{m} \in \mathbb{N}^k : \text{total } \mathbf{m} = pn + d - i \text{ and } \nu_p(\text{mult } \mathbf{m}) = \alpha - \nu_p \left(\frac{(pn + d)!}{(pn + d - i)!} \right) \right\}
$$

to the set

$$
B = \bigcup_{j=0}^{k-1} \left(\left\{ \mathbf{c} \in \mathbb{N}^k : \text{total } \mathbf{c} = n - j \text{ and } \nu_p(\text{mult } \mathbf{c}) = \alpha - \nu_p \left(\frac{n!}{(n - j)!} \right) - j \right\} \right.
$$

$$
\left. \times \left\{ \mathbf{d} \in \{0, 1, \ldots, p - 1\}^k : \text{total } \mathbf{d} = pj + d - i \right\} \right).
$$

The following lemma implies that if $\mathbf{m} \in A$ then $\beta(\mathbf{m}) \in B$.

Lemma 6. *Let p be a prime, $k \geq 1$, $n \geq 0$, $d \in \{0, 1, \ldots, p - 1\}$, and $0 \leq i \leq k - 1$. Let $\mathbf{m} \in \mathbb{N}^k$ with total $\mathbf{m} = pn + d - i$. Let $j = n - \text{total} \lfloor \mathbf{m}/p \rfloor$. Then total$(\mathbf{m} \bmod p) = pj + d - i$, $0 \leq j \leq k - 1$, and*

$$
\nu_p \left(\frac{(pn + d)!}{(pn + d - i)!} \right) + \nu_p(\text{mult } \mathbf{m}) = \nu_p \left(\frac{n!}{(n - j)!} \right) + \nu_p(\text{mult} \lfloor \mathbf{m}/p \rfloor) + j.
$$

We conclude by mentioning a connection to regular sequences. A sequence $s(n)_{n \geq 0}$, with entries in some field, is *p-regular* if the vector space generated by the set of subsequences $\{s(p^e n + i)_{n \geq 0} : e \geq 0 \text{ and } 0 \leq i \leq p^e - 1\}$ is finite-dimensional. For example, the sequence $(\theta_{p,0}(n))_{n \geq 0}$ is a *p*-regular sequence of integers [1, Example 14]. It follows from Theorem 5 and [1, Theorem 2.2] that the sequence of polynomials

$$\left(\sum_{\substack{\mathbf{m} \in \mathbb{N}^k \\ \text{total } \mathbf{m} = n}} x^{\nu_p(\text{mult } \mathbf{m})} \right)_{n \geq 0}$$

is *p*-regular for each k.

References

1. Allouche, J.-P., Shallit, J.: The ring of *k*-regular sequences. Theor. Comput. Sci. **98**, 163–197 (1992)
2. Barat, G., Grabner, P.J.: Distribution of binomial coefficients and digital functions. J. Lond. Math. Soc. **64**, 523–547 (2001)
3. Carlitz, L.: The number of binomial coefficients divisible by a fixed power of a prime. Rend. del Circ. Mat. di Palermo **16**, 299–320 (1967)
4. Davis, K., Webb, W.: Pascal's triangle modulo 4. Fibonacci Q. **29**, 79–83 (1989)
5. Fine, N.: Binomial coefficients modulo a prime. Am. Math. Mon. **54**, 589–592 (1947)
6. Glaisher, J.W.L.: On the residue of a binomial-theorem coefficient with respect to a prime modulus. Q. J. Pure Appl. Math. **30**, 150–156 (1899)
7. Howard, F.T.: The number of binomial coefficients divisible by a fixed power of 2. Proc. Am. Math. Soc. **29**, 236–242 (1971)
8. Huard, J.G., Spearman, B.K., Williams, K.S.: Pascal's triangle (mod 9). Acta Arith. **78**, 331–349 (1997)
9. Huard, J.G., Spearman, B.K., Williams, K.S.: Pascal's triangle (mod 8). Eur. J. Comb. **19**, 45–62 (1998)
10. Kummer, E.: Über die Ergänzungssätze zu den allgemeinen Reciprocitätsgesetzen. J. für die rein und angewandte Math. **44**, 93–146 (1852)
11. The OEIS Foundation, The On-Line Encyclopedia of Integer Sequences. http://oeis.org
12. Rowland, E.: The number of nonzero binomial coefficients modulo p^α. J. Comb. Num. Theor. **3**, 15–25 (2011)
13. Rowland, E.: A matrix generalization of a theorem of Fine. https://arxiv.org/abs/1704.05872
14. Spiegelhofer, L., Wallner, M.: An explicit generating function arising in counting binomial coefficients divisible by powers of primes. https://arxiv.org/abs/1604.07089

Connecting Decidability and Complexity
for MSO Logic

Michał Skrzypczak[✉]

University of Warsaw, Banacha 2, Warsaw, Poland
mskrzypczak@mimuw.edu.pl

Abstract. This work is about studying reasons for (un)decidability of variants of Monadic Second-order (MSO) logic over infinite structures. Thus, it focuses on connecting the fact that a given theory is (un)decidable with certain measures of complexity of that theory.

The first of the measures is the topological complexity. In that case, it turns out that there are strong connections between high topological complexity of languages available in a given logic, and its undecidability. One of the milestone results in this context is the Shelah's proof of undecidability of MSO over reals.

The second complexity measure focuses on the axiomatic strength needed to actually prove decidability of the given theory. The idea is to apply techniques of reverse mathematics to the classical decidability results from automata theory. Recently, both crucial theorems of the area (the results of Büchi and Rabin) have been characterised in these terms. In both cases the proof gives strong relations between decidability of the MSO theory with concepts of classical mathematics: determinacy, Ramsey theorems, weak Konig's lemma, etc.

Keywords: Monadic second-order logic · Decidability · Descriptive set theory · Reverse mathematics

1 Introduction

Monadic Second-order (MSO) logic is one of the fundamental logics used in the areas of verification and model checking. It is a very expressive formalism, comprising most of the other logics used for specifications, like LTL, CTL*, modal μ-calculus, etc. Thus, the decision methods for MSO over infinite words [9]; trees [24]; and certain linear orders [26] are commonly used to easily derive more decidability results. Because of its strength, it seems that MSO lies at the borderline of decidability, a prominent example is the theorem of Shelah [26] stating that the MSO theory of reals is undecidable. Similar undecidability results hold in the case of infinite words, when adding new monadic predicates [25]; or

The author was supported by the Polish National Science Centre grant no. UMO-2016/21/D/ST6/00491.

E. Charlier et al. (Eds.): DLT 2017, LNCS 10396, pp. 75–79, 2017.
DOI: 10.1007/978-3-319-62809-7_4

infinite trees when adding a well-order on the nodes [10]. A separate branch of studies asks which asymptotic extensions of MSO are decidable [1,4,7].

Since variants of MSO seem to occur on both sides of the decidability borderline, one can ask the following question: what makes a given theory (un)decidable? The aim of this work is to survey results answering the above question using two notions of *complexity* of the theory.

2   Descriptive Set Theory

The first notion of complexity that we discuss is the topological complexity of sets available in the given logic. To measure this complexity, one uses Borel and Projective Hierarchies, see [18]. From this perspective, MSO theory of infinite words is very simple, as all the languages definable there are Boolean combinations of Π^0_2-sets [28]. Similarly, the topological complexity of MSO over infinite trees is also under control, the languages definable there occupy exactly the first ω levels of the hierarchy of $\mathcal{R}$-sets (introduced by Kolmogorov in [19]); all of them belong to Δ^1_2; and all of them are measurable [14].

On the other hand [15,26], the MSO theory of reals $(\mathbb{R}, \leq)$ is undecidable, the proofs rely on a construction of a very complex[1] set $Q \subseteq \mathbb{R}$. Shelah [26] conjectured, that when we restrict set quantification to Borel subsets of $\mathbb{R}$ then the theory becomes decidable. This conjecture is still open, the best known result going in that direction follows from Rabin's theorem [24]: MSO of $\mathbb{R}$ with set quantifiers ranging over Π^0_2-sets is decidable.

Bojańczyk in [2] proposed an asymptotic extension of MSO (denoted MSO+U) by a quantifier U that informally allows to express that the delays between consecutive events are unbounded. Although some fragments of MSO+U were proved to be decidable [3–5,8]; the question of decidability of the full logic MSO+U was left open. The first witness that the logic might turn out to be undecidable was given by a result [16] showing that MSO+U over infinite words defines languages lying arbitrarily high in Projective Hierarchy (i.e. Π^1_n-complete). By incorporating the methods of Shelah [26], this led to the following results.

Theorem 1 ([6]). *Assume a set-theoretic axiom that* V = L *(called Axiom of Constructibility [13,17]). Let* $\mathcal{L}$ *be an extension of* MSO *logic that defines a* Π^1_6-*complete set of infinite words. Then the* $\mathcal{L}$-*theory of the infinite tree is undecidable.*

Corollary 2 ([6]). *It is consistent with* ZFC *that the* MSO+U *theory of the infinite tree is undecidable.*

Although the above corollary does not prove undecidability of MSO+U, it implies that there is no concrete algorithm solving that theory which correctness can be proven in ZFC. Thus, there was no hope that the theory may be decidable in the standard sense. This line of research was later surmounted by a direct proof of undecidability of MSO+U over infinite words [7].

---

[1] A posteriori, the considered set Q needs to violate Baire Property.

3   Reverse Mathematics

We will now focus on decidable theories, like in the classical theorems of Büchi [9] and Rabin [24]. We would like to understand how *logically difficult* these theorems are. The standard notion of complexity used for measuring *logical difficulty* of theorems is given by reverse mathematics [12,27]. The recipe is as follows:

1. We choose some logical system to work in, usually it is second-order arithmetic.
2. We formalise the given theorem as a sentence Φ in that logical system.
3. We choose some very weak basic theory (usually it is a theory called RCA_0).
4. Then we prove that over RCA_0, the sentence Φ is equivalent to some known axiom χ.

The last step consists of two implications: one of them boils down to proving that the theorem follows from $RCA_0 + \chi$ (i.e. one can prove it using only χ). The second implication is more tricky, it says that χ is necessary for the theorem to hold, i.e. over RCA_0 the pure fact that Φ holds implies χ.

Over the years, many standard mathematical results were characterised in terms of their axiomatic strength. As it turned out, the set of additional axioms χ that typically appear in this context is very limited. Actually most of the every-day mathematics turns out to be equivalent to one of five standard logical systems, called Big Five [27, page 42]. Among them is Weak König's Lemma (denoted WKL_0), used in many arguments using compactness. An example of a commonly used axiom outside Big Five is Ramsey's Theorem for Pairs and arbitrarily many colours ($RT^2_{<\infty}$), known to be incomparable to WKL_0 [22].

The famous theorem of Büchi states that MSO is decidable over infinite words. From the modern perspective, there are at least two different ways of proving that result: either by determinisation à la McNaughton [23]; or by direct complementation as done by Büchi [9]. The combinatorial core of the former approach is based on applications of König's Lemma, while the latter relies on Ramsey's Theorem. Thus, from the reverse mathematical point of view these seem to be two orthogonal proofs. However, in both cases the respective principles are used in a very limited way: König's Lemma is applied to trees generated by automata; while Ramsey's Theorem is applied to colourings that bear additional algebraic structure.

As it turned out [21], Büchi's decidability theorem is equivalent over RCA_0 to the principle of induction for Σ^0_2-formulae (denoted Σ^0_2-IND). Also, Σ^0_2-IND was shown [21] to be equivalent to an *additive* version of Ramsey's Theorem and to imply the automata-related version of König's Lemma. These results provide a rather complete picture of the logical strengths of principles involved in Büchi's decidability result.

Rabin's Theorem [24] proving decidability of MSO over infinite trees was always believed to be more demanding than Büchi's result. In particular, to the author's best knowledge there is no known proof of Rabin's result that would avoid using automata and determinacy of related games. This observation has been formalised in [20], where the authors proved that (in a strong sense) Rabin's complementation result is equivalent to the statement of determinacy of games with winning conditions being Boolean combinations of Π_2^0-sets.

In both above cases, the respective theorem about MSO turned out to be equivalent to other standard mathematical statements. This suggests that the decidability results about a logic that is robust enough must convey certain knowledge about the whole mathematical universe. Following this idea it seems interesting and promising to study other decidability theorems for robust logics, for instance the result of decidability of MSO over countable linear orders [11,26]. From the point of view of direct implications, this result is somewhere in-between the theorems of Büchi and Rabin.

4   Conclusions

The aim of this survey was to present certain perspectives in which decidability of a given variant of MSO is related to a certain measure of complexity for that logic. The presented examples advocate that, when one faces the question whether a given logic is decidable, it may be useful to analyse the complexity of the logic itself (instead of looking directly for an algorithm or a reduction).

For instance, high topological complexity of languages definable in the logic may indicate (or even prove, see Corollary 2) that the logic cannot be decidable. Similarly, if decidability of the logic has very strong axiomatic consequences, there is no hope to prove it without any strong tools.

On the other hand, if the logic seems to have very limited access to the mathematical universe, one may expect a direct proof of decidability, for instance by a reduction to an appropriately chosen logic that is stronger but still decidable.

References

1. Blumensath, A., Colcombet, T., Parys, P.: On a fragment of AMSO and tiling systems. In: STACS, pp. 19:1–19:14 (2016)
2. Bojańczyk, M.: A bounding quantifier. In: Marcinkowski, J., Tarlecki, A. (eds.) CSL 2004. LNCS, vol. 3210, pp. 41–55. Springer, Heidelberg (2004). doi:10.1007/978-3-540-30124-0_7
3. Bojańczyk, M.: Weak MSO with the unbounding quantifier. Theor. Comput. Syst. **48**(3), 554–576 (2011)
4. Bojańczyk, M.: Weak MSO+U with path quantifiers over infinite trees. In: Esparza, J., Fraigniaud, P., Husfeldt, T., Koutsoupias, E. (eds.) ICALP 2014. LNCS, vol. 8573, pp. 38–49. Springer, Heidelberg (2014). doi:10.1007/978-3-662-43951-7_4
5. Bojańczyk, M., Colcombet, T.: Bounds in ω-regularity. In: LICS, pp. 285–296 (2006)
6. Bojańczyk, M., Gogacz, T., Michalewski, H., Skrzypczak, M.: On the decidability of MSO+U on infinite trees. In: Esparza, J., Fraigniaud, P., Husfeldt, T., Koutsoupias, E. (eds.) ICALP 2014. LNCS, vol. 8573, pp. 50–61. Springer, Heidelberg (2014). doi:10.1007/978-3-662-43951-7_5
7. Bojańczyk, M., Parys, P., Toruńczyk, S.: The MSO+U theory of (n, <) is undecidable. In: STACS, pp. 1–8 (2016)

8. Bojańczyk, M., Toruńczyk, S.: Deterministic automata and extensions of weak MSO. In: FSTTCS, pp. 73–84 (2009)
9. Büchi, J.R.: On a decision method in restricted second-order arithmetic. In: Lane, S.M., Siefkes, D. (eds.) The Collected Works of J. Richard Büchi, pp. 1–11. Springer, New York (1962)
10. Carayol, A., Löding, C., Niwiński, D., Walukiewicz, I.: Choice functions and well-orderings over the infinite binary tree. Cent. Europ. J. of Math. **8**, 662–682 (2010)
11. Carton, O., Colcombet, T., Puppis, G.: Regular languages of words over countable linear orderings. In: Aceto, L., Henzinger, M., Sgall, J. (eds.) ICALP 2011. LNCS, vol. 6756, pp. 125–136. Springer, Heidelberg (2011). doi:10.1007/978-3-642-22012-8_9
12. Friedman, H.: Some systems of second order arithmetic and their use. pp. 235–242 (1975)
13. Gödel, K.: The Consistency of the Axiom of Choice and of the Generalized Continuum Hypothesis with the Axioms of Set Theory. Princeton University Press, New Jersy (1940)
14. Gogacz, T., Michalewski, H., Mio, M., Skrzypczak, M.: Measure properties of game tree languages. In: Csuhaj-Varjú, E., Dietzfelbinger, M., Ésik, Z. (eds.) MFCS 2014. LNCS, vol. 8634, pp. 303–314. Springer, Heidelberg (2014). doi:10.1007/978-3-662-44522-8_26
15. Gurevich, Y., Shelah, S.: Monadic theory of order and topology in ZFC. Annal. Math. Logic **23**(2–3), 179–198 (1982)
16. Hummel, S., Skrzypczak, M.: The topological complexity of MSO+U and related automata models. Fundam. Inf. **119**(1), 87–111 (2012)
17. Jech, T.: Set Theory. Springer, Hiedelberg (2002)
18. Kechris, A.: Classical descriptive set theory. Springer, New York (1995)
19. Kolmogorov, A.N.: Operations sur des ensembles. Mat. Sb. **35**, 415–422 (1928). (in Russian, summary in French)
20. Kołodziejczyk, L.A., Michalewski, H.: How unprovable is Rabin's decidability theorem? In: LICS, pp. 788–797 (2016)
21. Kołodziejczyk, L.A., Michalewski, H., Pradic, P., Skrzypczak, M.: The logical strength of Büchi's decidability theorem. In: CSL, pp. 36:1–36:16 (2016)
22. Liu, J.: RT_2^2 does not imply WKL_0. J. Symbol. Logic **77**(2), 609–620 (2012)
23. McNaughton, R.: Testing and generating infinite sequences by a finite automaton. Inf. Control **9**(5), 521–530 (1966)
24. Rabin, M.O.: Decidability of second-order theories and automata on infinite trees. Trans. Am. Math. Soc. **141**, 1–35 (1969)
25. Rabinovich, A.: On decidability of monadic logic of order over the naturals extended by monadic predicates. Inf. Comput. **205**(6), 870–889 (2007)
26. Shelah, S.: The monadic theory of order. Annal. Math. **102**(3), 379–419 (1975)
27. Simpson, S.G.: Subsystems of Second Order Arithmetic. Perspectives in Logic, 2nd edn. Cambridge University Press, Association for Symbolic Logic, Cambridge, Poughkeepsie (2009)
28. Thomas, W., Lescow, H.: Logical specifications of infinite computations. In: Bakker, J.W., Roever, W.-P., Rozenberg, G. (eds.) REX 1993. LNCS, vol. 803, pp. 583–621. Springer, Heidelberg (1994). doi:10.1007/3-540-58043-3_29

Regular Papers

On Regular Expression Proof Complexity

Simon Beier and Markus Holzer[(⊠)]

Institut für Informatik, Universität Giessen,
Arndtstr. 2, 35392 Giessen, Germany
{simon.beier,holzer}@informatik.uni-giessen.de

Abstract. We investigate the proof complexity of Salomaa's axiom system F_1 for regular expression equivalence. We show that for two regular expression E and F over the alphabet Σ with $L(E) = L(F)$ an equivalence proof of length $O\left(|\Sigma|^4 \cdot \text{TOWER}(\max\{h(E), h(F)\} + 4)\right)$ can be derived within F_1, where $h(E)$ ($h(F)$, respectively) refers to the height of E (F, respectively) and the tower function is defined as $\text{TOWER}(1) = 2$ and $\text{TOWER}(k + 1) = 2^{\text{TOWER}(k)}$, for $k \geq 1$. In other words

$$\text{TOWER}(k) = 2^{2^{2^{\cdot^{\cdot^{\cdot^2}}}}} \Big\} k.$$

This is in sharp contrast to the fact, that regular expression equivalence admisses exponential proof length if *not* restricted to the axiom system F_1. From the theoretical point of view the exponential proof length seems to be best possible, because we show that regular expression equivalence admits a polynomial bounded proof if and only if NP = PSPACE.

1 Introduction

Regular expressions were introduced in the seminal paper of Kleene [6] on nerve nets and finite automata. They allow a beautiful set-theoretic characterization of languages accepted by finite automata. Early result concerning regular expressions can be found in [1,7]. These papers were very influential in shaping automata theory. Compared to automata, regular expressions are better suited for human users and therefore are often used as interfaces to specify certain pattern or languages. On the other hand, automata are regularly used for its manipulation, since the methods develop during the years turned out to be usually more efficient compare to those for regular expressions. For instance, if one wants to check regular expressions for equivalence, they are converted to equivalent nondeterministic finite automata, followed by determinizing them to equivalent deterministic finite state devices, which are finally minimized and checked for equivalence up to isomorphism. It is worth mentioning that a relatively simple decision procedure for the equivalence of regular expressions was given in [3]. In general the equivalence of regular expression is a costly task, because it was classified to be PSPACE-complete in [9]—see also [5]. A completely other method for regular expression equivalence, which is entirely based on regular expressions, is to give a proof in the complete and

© Springer International Publishing AG 2017
É. Charlier et al. (Eds.): DLT 2017, LNCS 10396, pp. 83–95, 2017.
DOI: 10.1007/978-3-319-62809-7_5

sound axiom system F_1 for regular expression equivalence developed in [8]. It is clear that with this approach one cannot overcome the PSPACE barrier, but to our knowledge the proof length complexity of regular expressions and problems related to this question is not studied in the literature in question up to now. In general proof complexity asks the question how difficult it is to prove a theorem—here we view regular expression equivalence as a "theorem" to prove. One natural measure on the complexity of a theorem is its proof length within a certain proof system. Thus, the question on the proof length complexity of Salomaa's axiom system F_1 arises. By a careful analysis of Salomaa's proof on the completeness of the axiom system F_1 we obtain an upper bound on the proof length complexity of regular expression equivalence for expressions E and F over the alphabet Σ, if $L(E) = L(F)$, which is enormous, namely bounded by

$$O(|\Sigma|^4 \cdot \text{Tower}(\max\{h(E), h(F)\} + 4)),$$

where $h(E)$ ($h(F)$, respectively) refers to the height of expression E (F, respectively) and the tower function is defined as $\text{Tower}(1) = 2$ and $\text{Tower}(k+1) = 2^{\text{Tower}(k)}$, for $k \geq 1$. In other words

$$\text{Tower}(k) = \left. 2^{2^{2^{\cdot^{\cdot^{\cdot^2}}}}} \right\} k.$$

On the other hand, what happens, if we do not restrict ourselves to regular expression equivalence proofs in the axiom system F_1? This immediately leads us to proof complexity in general as developed in [2]. The following result is well known in propositional proof complexity: NP=coNP if and only if the set TAUT of all propositional tautologies admits a polynomial bounded proof system. The ultimate goal of propositional proof complexity is to show that there is no propositional proof system allowing for efficient proofs of tautology. But what is the relation of the aforementioned result to regular expression equivalence? We show that within the proof system of [2] that NP=PSPACE if and only if the set

$$\text{EQUIV} = \{ (E, F) \mid E \text{ and } F \text{ are regular expressions with } L(E) = L(F) \}$$

gives rise to a polynomial bounded proof. This is in perfect line with the PSPACE-completeness of EQUIV of [9]. This already shows that we cannot hope for "short" proofs on regular expression equivalence because this is equivalent to a quite unrealistic assumption NP=PSPACE from a computational complexity perspective.

The paper is organized as follows: in the next section we introduce the necessary notations on regular expressions and the axiom system F_1. Then in Sect. 3 we analyse Salomaa's completeness proof of the axiom system F_1 for regular expression equivalence from [8]. This section is the main part of this paper, because first we have to show how to obtain an equational characterization of the involved regular expressions only using the power of the axiom system F_1. Then this characterization is used to develop a proof on equivalence, if both given expressions E and F describe the same set, that is $L(E) = L(F)$. Finally, in Sect. 4, we study the proof complexity of regular expression equivalence in general. Due to space limitations almost all proofs are omitted.

2   Preliminaries

We assume the reader to be familiar with the notations in automata and formal language theory as contained in [4]. Let Σ be an alphabet and Σ^* the set of all words over the alphabet Σ, including the empty word λ. A set $L \subseteq \Sigma^*$ is called a *language*. Operations on languages we are interested in are union, concatenation, and Kleene star.

The *regular expressions* over an alphabet Σ and the languages that they denote are inductively defined as follows:[1] 0 and every letter a with $a \in \Sigma$ are regular expressions, and when E and F are regular expressions, then $E + F$, $E \cdot F$, and E^* are also regular expressions. The language defined by a regular expression is defined as follows: $L(0) = \emptyset$, $L(a) = \{a\}$, $L(E + F) = L(E) \cup L(F)$, $L(E \cdot F) = L(E) \cdot L(F)$, and $L(E^*) = L(E)^*$. Observe, that the empty word must be represented by 0^*, since we have not introduced an extra regular expression denoting λ. We write $E \equiv F$, if the regular expressions E and F are syntactically the same. Of course $E \equiv F$ implies $L(E) = L(F)$, but not the other way around. We assume that bracketing of expressions $E_1 + E_2 + \cdots + E_n$ is done from left-to-right, that is $((\ldots ((E_1 + E_2) + E_3) + \ldots) + E_{n-1}) + E_n$. We use the same convention for concatenation.

In [8] a sound and complete axiom system, called F_1, for regular expression equivalence was given. The axioms of F_1 are

(A_1) $E + (F + G) = (E + F) + G$ (A_7) $0^* \cdot E = E$

(A_2) $E \cdot (F \cdot G) = (E \cdot F) \cdot G$ (A_8) $0 \cdot E = 0$

(A_3) $E + F = F + E$ (A_9) $E + 0 = E$

(A_4) $(E + F) \cdot G = E \cdot G + F \cdot G$ (A_{10}) $E^* = 0^* + E^* \cdot E$

(A_5) $E \cdot (F + G) = E \cdot F + E \cdot G$ (A_{11}) $E^* = (0^* + E)^*,$

(A_6) $E + E = E$

where E, F, and G are regular expressions. The inference rules of F_1 are substitution (R_1)

$$\frac{E = F \qquad C[E] = G}{C[F] = C[E], \quad C[F] = G}$$

and solution of equations (R_2)

$$\frac{E = E \cdot F + G}{E = G \cdot F^*} \text{ if } o(F) = 0$$

---

[1] For convenience, parentheses in regular expressions are sometimes omitted and the concatenation is simply written as juxtaposition. The priority of operators is specified in the usual fashion: concatenation is performed before union, and star before both product and union.

where again E, F, and G are regular expressions and $C[E]$ refers to a regular expression C that contains E as a subexpression. Here

$$o(F) = \begin{cases} 1 & F \text{ possesses the e.w.p.} \\ 0 & \text{otherwise,} \end{cases}$$

where e.w.p. is an abbreviation for *empty word property*. A regular expression E possesses e.w.p. if $\lambda \in L(E)$.

A proof in the axiom system F_1 is a finite sequence of applications of the rules R_1 and R_2 where each equation at the top of the rules is an axiom or appears at the bottom of an earlier rule in the sentence. An equation $E = F$ is derivable within the system F_1 if there is a proof where the equation $E = F$ stands at the bottom of the last rule. A set of equations T is derivable from a finite set of equations S if there is a finite sequence of applications of the rules R_1 and R_2 where each equation at the top of the rules is an axiom or an equation out of S or appears at the bottom of an earlier rule in the sentence and all of the equations of T are contained somewhere in the sequence.

We give a small example of a derivation in the axiom system F_1.

Example 1. Obviously $L(0 \cdot a) = L(0)$ and $L(a \cdot 0) = L(0)$, too. While the first equation $0 \cdot a = 0$ is an axiom in the system F_1, the latter equation $a \cdot 0 = 0$ has to be proven explicitly. The axiom A_2 gives us $a \cdot (0 \cdot 0) = (a \cdot 0) \cdot 0$ and A_8 gives us $0 \cdot 0 = 0$. With R_1 we get

$$\frac{0 \cdot 0 = 0 \qquad a \cdot (0 \cdot 0) = (a \cdot 0) \cdot 0}{a \cdot 0 = a \cdot (0 \cdot 0), \ a \cdot 0 = (a \cdot 0) \cdot 0}$$

Axiom $A6$ tells us $(a \cdot 0 + 0) + (a \cdot 0 + 0) = a \cdot 0 + 0$ and one use of $R1$ gives

$$\frac{(a \cdot 0 + 0) + (a \cdot 0 + 0) = a \cdot 0 + 0 \qquad (a \cdot 0 + 0) + (a \cdot 0 + 0) = a \cdot 0 + 0}{a \cdot 0 + 0 = (a \cdot 0 + 0) + (a \cdot 0 + 0), \quad a \cdot 0 + 0 = a \cdot 0 + 0}$$

Using R_1 again leads to

$$\frac{a \cdot 0 = (a \cdot 0) \cdot 0 \qquad a \cdot 0 + 0 = a \cdot 0 + 0}{(a \cdot 0) \cdot 0 + 0 = a \cdot 0 + 0, \ (a \cdot 0) \cdot 0 + 0 = a \cdot 0 + 0}$$

With R_1 we get $a \cdot 0 + 0 = (a \cdot 0) \cdot 0 + 0$ and Axiom A_9 implies $a \cdot 0 + 0 = a \cdot 0$. Another use of R_1 tells us

$$\frac{a \cdot 0 + 0 = a \cdot 0 \qquad a \cdot 0 + 0 = (a \cdot 0) \cdot 0 + 0}{a \cdot 0 = a \cdot 0 + 0, \ a \cdot 0 = (a \cdot 0) \cdot 0 + 0}$$

Now we use R_2:

$$\frac{a \cdot 0 = (a \cdot 0) \cdot 0 + 0}{a \cdot 0 = 0 \cdot 0^*}$$

The rule R_1 gives $0 \cdot 0^* = a \cdot 0$ and Axiom A_8 leads us to $0 \cdot 0^* = 0$. We use R_1 one last time and obtain

$$\frac{0 \cdot 0^* = a \cdot 0 \qquad 0 \cdot 0^* = 0}{a \cdot 0 = 0 \cdot 0^*, \ a \cdot 0 = 0}$$

So we have found a proof for $a \cdot 0 = 0$ using R_1 seven times and R_2 once.   □

Next we define some descriptional complexity measures for regular expressions. For a regular expression E, we define $|E|.$ and $|E|_*$ to be the numbers of appearances of the symbols "$\cdot$" and "$*$" in E. For a regular expression E over the alphabet Σ the *height* is inductively defined by

$$h(E) = \begin{cases} 0 & \text{if } E \equiv 0 \text{ or } E \equiv a, \text{ for } a \in \Sigma \\ 1 + \max\{h(F), h(G)\} & \text{if } E \equiv F + G \text{ or } E \equiv F \cdot G \\ 1 + h(F) & \text{if } E \equiv F^*. \end{cases}$$

Finally, we introduce equational characterizations of regular expressions, see, e.g., [8]. Assume $\Sigma = \{a_1, a_2, \ldots, a_r\}$. Then for a regular expression E over the alphabet Σ an *equational characterization* is a system of equations

$$E_i = \sum_{j=1}^{r} E_{i,j} a_j + \delta_i, \quad \text{for } i = 1, 2, \ldots, n,$$

for some $n \geq 1$ and regular expressions $E_1, E_2, \ldots, E_n$ over Σ with $E_1 \equiv E$. Furthermore, for every i we have $\delta_i \equiv 0$ or $\delta_i \equiv 0^*$. For each i and j there is a $k \in \{1, 2, \ldots, n\}$ such that $E_{i,j} \equiv E_k$. From [8, Lemma 4] we know that for every regular expression there exists a derivable equational characterization in the axiom system F_1.

3   Regular Expression Proof Complexity Within Salomaa's Axiom System F_1

Our goal is, for regular expressions E and F with $L(E) = L(F)$, to give an upper bound for a proof of the equation $E = F$ in the axiom system F_1. It suffices to determine how often the inference rules R_1 and R_2 are applied. From this one can deduce a trivial upper bound on the proof length. To do this, (i) we will first prove some equations that hold for every regular expression and are often needed in the following (ii) Then, we will give an upper bound for the number of equations in a derivable equational characterization for a regular expression and for the proof of the equational characterization. (iii) Finally, we will show how to derive the equation $E = F$ from the equational characterizations for E and F. We start with the observation that $=$ is an equivalence relation:

Lemma 2. *Let E, F, and G be regular expressions. The equation $E = E$ is derivable with one use of R_1 and no uses of R_2. The equation $F = E$ is derivable from $E = F$ with one use of R_1 and no uses of R_2. The equation $E = G$ is derivable from $E = F$ and $F = G$ with two uses of R_1 and no uses of R_2.*

Next we give upper bounds for the proofs of $E \cdot 0 = 0$ and $E \cdot 0^* = E$.

Lemma 3. *For a regular expression E the equation $E \cdot 0 = 0$ is derivable with seven uses of R_1 and one use of R_2. From the equation $E \cdot 0 = 0$ the equation $E \cdot 0^* = E$ is derivable with four uses of R_1 and one use of R_2.*

3.1   Equational Characterizations for Regular Expressions

We analyze the proof of [8, Lemma 4], which uses the inductive definition of regular expressions, to give an upper bound for the number of equations in a derivable equational characterization for a regular expression and for the proof of the equational characterization. Our results are summarized in Table 1. Here, we only show the result for the equational characterization of $E + F$. The other results from Table 1 can be shown by a similar argumentation. All our regular expressions will be over the alphabet $\{a_1, a_2, \ldots, a_r\}$. In order to prove the results from Table 1 we need the following lemma.

Table 1. Equational characterizations of regular expressions over the alphabet Σ of size r. Here the number of equations derivable with the use of the inference rules R_1 and R_2 in the axiom system F_1 are given. It is assumed that E (F, respectively) has an equational characterization with n (m, respectively) equations.

Regular expression	Axiom system F_1		
	Equational characterization	No. of inference rules used	
	No. of equations	R_1	R_2
0	1	$O(r)$	0
0^*	2	$O(r)$	0
a, for $a \in \Sigma$	3	$O(r)$	0
$E + F$	$n \cdot m$	$O(r^4 nm)$	0
$E \cdot F$	$m \cdot 2^n$	$O((r \cdot n)^4 m \cdot 2^n)$	2
E^*	2^n	$O((r \cdot n)^4 \cdot 2^n)$	2

Lemma 4. *Let $n \geq 1$ and $E_1, E_2, \ldots, E_n$ be regular expressions. Let F_1 be the same expression as E_1 and F_i be the expression $F_{i-1} + E_i$, where the symbols F_{i-1} and E_i are replaced by the expressions they stand for without adding parentheses, for $i = 2, 3, \ldots, n$. Let G and H be regular expressions that both can be obtained from the expression F_n by adding parentheses to give the $+$ symbols that were added to the E_i a priority. Then we can replace an occurrence of G by H in the left-hand side of an equation with $O(n^2)$ uses of R_1 and no use of R_2. Obviously we can deal with $\cdot$ instead of $+$ analogously.* □

Now we are ready to give an upper bound for the number of equations in a derivable equational characterization for a regular expression of the form $E + F$ and for the proof of the equational characterization.

Lemma 5. *Let E be a regular expression with an equational characterization with n equations and F be a regular expression with an equational characterization with m equations. Then from these two equational characterizations an equational characterization for the regular expression $E + F$ with $n \cdot m$ equations is derivable with $O(r^4 nm)$ uses of R_1 and no use of R_2.*

Proof. Let $E_i = \sum_{j=1}^{r} E_{i,j} a_j + \delta_i$, for $i = 1, 2, \ldots, n$, be an equational characterization for E and $F_k = \sum_{j=1}^{r} F_{k,j} a_j + \gamma_k$, for $k = 1, 2, \ldots, m$, be an equational characterization for F. For $(i, k) \in \{1, 2, \ldots, n\} \times \{1, 2, \ldots, m\}$, we get $E_i + F_k = E_i + F_k$ with one use of R_1. Two more uses of R_1 give

$$\sum_{j=1}^{r} E_{i,j} a_j + \delta_i + \left(\sum_{j=1}^{r} F_{k,j} a_j + \gamma_k \right) = E_i + F_k.$$

There are $2r + 2$ summands on the left-hand side. Because of axiom A_3 we can switch positions of two adjacent summands. However it may be necessary to change the positions of parentheses first. So we can switch two adjacent summands with $O(r^2)$ uses of R_1 due to Lemma 4. With switching adjacent summands $O(r^2)$ times we can get the summands in any order we like. Thus we get $\sum_{j=1}^{r}(E_{i,j} a_j + F_{k,j} a_j) + (\delta_i + \gamma_k) = E_i + F_k$ with $O(r^4)$ uses of R_1. Because of Axiom A_4 we get $\sum_{j=1}^{r}((E_{i,j} + F_{k,j}) a_j) + (\delta_i + \gamma_k) = E_i + F_k$ with $2r$ uses of R_1. Let $\epsilon_{i,k}$ be the regular expression 0 if $\delta_i \equiv \gamma_k \equiv 0$, and 0^* otherwise. Axioms A_3, A_6, and A_9 give us $\sum_{j=1}^{r}((E_{i,j} + F_{k,j}) a_j) + \epsilon_{i,k} = E_i + F_k$ with at most two uses of R_1. Another use of R_1 leads to $E_i + F_k = \sum_{j=1}^{r}((E_{i,j} + F_{k,j}) a_j) + \epsilon_{i,k}$. These equations, for $(i, k) \in \{1, 2, \ldots, n\} \times \{1, 2, \ldots, m\}$, are an equational characterization for the regular expression $E + F$.   □

Having the results from Table 1, we are almost ready, given a regular expression, to give an explicit formula for an upper bound for the number of equations in a derivable equational characterization and for the proof of the equational characterization. To this end we define the *tower function* Tow by

$$\mathrm{Tow}(b_1) = 2^{b_1}$$

and

$$\mathrm{Tow}(b_1, b_2, \ldots, b_k) = 2^{b_1 \cdot \mathrm{Tow}(b_2, \ldots, b_k)},$$

for $b_1, b_2, \ldots, b_k > 0$. For $k \geq 0$, the convention $b^{\otimes k} = (b, b, \ldots, b)$, where there are k values b, is used. Then, we define $\mathrm{TOWER}(k) = \mathrm{Tow}\left(1^{\otimes k}\right)$, for $k > 0$. Thus, $\mathrm{TOWER}(k)$ is an exponential tower of height k which has just the number 2 on each level, that is,

$$\mathrm{TOWER}(k) = 2^{2^{2^{\cdot^{\cdot^{\cdot^{2}}}}}} \Big\} k$$

We need some properties of the Tow function, which can easily be seen:

Lemma 6. *For $k > 0$ and $b_1, b_2, \ldots, b_k > 0$, we have:*

1. $\mathrm{Tow}(1, b_1, b_2, \ldots, b_k) = 2^{\mathrm{Tow}(b_1, b_2, \ldots, b_k)}$.
2. $(\mathrm{Tow}(b_1, b_2, \ldots, b_k))^a = \mathrm{Tow}(ab_1, b_2, b_3, \ldots, b_k)$, for $a > 0$.
3. $\mathrm{Tow}(b_1, b_2, \ldots, b_k) = \mathrm{Tow}(b_1, b_2, \ldots, b_{i-1}, b_i \cdot \mathrm{Tow}(b_{i+1}, b_{i+2}, \ldots, b_k))$, for $0 < i < k$.
4. $\mathrm{Tow}(b_{i+1}, b_{i+2}, \ldots, b_k) < \mathrm{Tow}(b_1, b_2, \ldots, b_i, b_{i+1}, \ldots, b_k)$, for $0 < i < k$.   □

To get our upper bound for an equational characterization we use the following estimation.

Lemma 7. *For $k \geq 0$, we have*

$$4 \cdot \mathrm{Tow}\left(2^{\otimes k}, 1, 1\right) \leq \left(\mathrm{Tow}\left(2^{\otimes k}, 1, 1\right)\right)^2 \leq \mathrm{Tower}(k + 3).$$

Now we are ready to give an upper bound formula for an equational characterization for a regular expression:

Theorem 8. *For a regular expression E an equational characterization with at most $\mathrm{Tower}(h(E) + 3)/4$ equations is derivable with $O(r^4 \cdot \mathrm{Tower}(h(E) + 3))$ uses of R_1 and $2 \cdot (|E|. + |E|_*)$ uses of R_2.*

Proof. In the results presented in Table 1 the bound for the number of R_1 uses is given in O-notation. Thus, there is a constant $c > 0$ such that in each of the above mentioned lemmata the number of uses of R_1 is at most c times the bound given inside of O-notation. Now we show by induction on the structure of E that for every regular expression E an equational characterization with at most $\mathrm{Tow}\left(2^{\otimes h(E)}, 1, 1\right)$ equations is derivable with $c \cdot r^4 \cdot \left(\mathrm{Tow}\left(2^{\otimes h(E)}, 1, 1\right)\right)^2$ uses of R_1 and $2 \cdot (|E|. + |E|_*)$ uses of R_2. Then, the result follows by Lemma 7.

It remains to prove the above statement by induction. For the base case let $E \equiv 0$ or $E \equiv a_i$, for $i \in \{1, 2, \ldots, r\}$. Then $h(E) = 0$ and, by the the bounds stated in Table 1 an equational characterization for E with at most $4 = \mathrm{Tow}(1, 1)$ equations is derivable with $c \cdot r$ uses of R_1 and no use of R_2. Now, for the inductive step let $E \equiv F + G$, or $E \equiv F \cdot G$, or $E \equiv F^*$, for regular expressions F and G of height at most $h(E) - 1$, where we set $G \equiv 0$ in the last case. Then, by the induction hypothesis equational characterizations for F and G, both with at most $\mathrm{Tow}\left(2^{\otimes(h(E)-1)}, 1, 1\right)$ equations, are derivable with $2c \cdot r^4 \cdot \left(\mathrm{Tow}\left(2^{\otimes(h(E)-1)}, 1, 1\right)\right)^2$ uses of R_1 and $2 \cdot (|F + G|. + |F + G|_*)$ uses of R_2. By the results stated in Table 1 an equational characterization for E with at most

$$\left(2^{\mathrm{Tow}\left(2^{\otimes(h(E)-1)}, 1, 1\right)}\right)^2 = \left(\mathrm{Tow}\left(1, 2^{\otimes(h(E)-1)}, 1, 1\right)\right)^2 = \mathrm{Tow}\left(2^{\otimes h(E)}, 1, 1\right)$$

equations is derivable with

$$2c \cdot r^4 \cdot \left(\mathrm{Tow}\left(2^{\otimes(h(E)-1)}, 1, 1\right)\right)^2$$
$$+ c \cdot r^4 \left(\mathrm{Tow}\left(2^{\otimes(h(E)-1)}, 1, 1\right)\right)^5 \cdot 2^{\mathrm{Tow}\left(2^{\otimes(h(E)-1)}, 1, 1\right)} \tag{1}$$

uses of R_1 and $2 \cdot (|E|_. + |E|_*)$ uses of R_2. Simplifying Expression (1) using the fact that $2n^2 + n^5 2^n < 2^{4n}$, for $n \geq 0$, which can easily be seen *via* induction, we obtain an upper bound of

$$c \cdot r^4 \cdot 2^{4 \cdot \text{Tow}(2^{\otimes(h(E)-1)},1,1)} = c \cdot r^4 \cdot \text{Tow}\left(4, 2^{\otimes(h(E)-1)}, 1, 1\right)$$

$$= c \cdot r^4 \cdot \left(\text{Tow}\left(2^{\otimes h(E)}, 1, 1\right)\right)^2$$

for the uses of R_1, which proves the stated result.   □

3.2   From Equational Characterizations to the Equality of Regular Expressions

Given regular expressions E and F with $L(E) = L(F)$, we will now show how to derive the equation $E = F$ from their equational characterizations. First, we give [8, Lemma 4], which shows that the coefficients in equations of the form $E = \sum_{i=1}^{r} E_i a_i + \delta$ are in some sense determined by $L(E)$.

Lemma 9. *Let E and F be regular expressions with $L(E) = L(F)$. Furthermore let $E_1, E_2, \ldots, E_r$ and $F_1, F_2, \ldots, F_r$ be regular expressions and $\delta, \gamma \in \{0, 0^*\}$ such that $E = \sum_{i=1}^{r} E_i a_i + \delta$ and $F = \sum_{i=1}^{r} F_i a_i + \gamma$. Then $\delta \equiv \gamma$ and $L(E_i) = L(F_i)$, for all $i \in \{1, 2, \ldots, r\}$.*

In the proof of [8, Theorem 2] it is shown how equational characterizations for regular expressions E and F with $L(E) = L(F)$ can be transformed into a special kind of system of equations, which is later used to derive a proof for the equation $E = F$. We analyze how often the rules R_1 and R_2 are used along the way:

Lemma 10. *Let E be a regular expression with an equational characterization with n equations and F be a regular expression with an equational characterization with m equations and $L(E) = L(F)$. Then, there exists $1 \leq k \leq mn$, for each $i \in \{1, 2, \ldots, k\}$ a pair of regular expressions[2] (G_i, H_i) with $(G_1, H_1) \equiv (E, F)$ and a regular expression ϵ_i, and for each $(i, h) \in \{1, 2, \ldots, k\}^2$ a regular expression $D_{i,h}$ with $\lambda \notin L(D_{i,h})$ such that the system of equations*

$$(G_i, H_i) = \sum_{h=1}^{k} (G_h, H_h) \cdot D_{i,h} + (\epsilon_i, \epsilon_i), \quad \text{for } i = 1, 2, \ldots, k,$$

is derivable from the two given equational characterizations with $O((mn+r^4)mn)$ uses of R_1 and $m + n$ uses of R_2.

---

[2] The notation $(E_1, E_2) \equiv (F_1, F_2)$, for regular expressions E_1, E_2, F_1, and F_2, stands for $E_1 \equiv F_1$ and $E_2 \equiv F_2$. The equation $(E_1, E_2) = (F_1, F_2)$ is a shorthand notation for the system of the two equations $E_1 = F_1$ and $E_2 = F_2$. Furthermore, the expressions $(E_1, E_2) + (F_1, F_2)$ and $(E_1, E_2) \cdot F_1$ define $(E_1 + F_1, E_2 + F_2)$ and $(E_1 \cdot F_1, E_2 \cdot F_1)$, respectively.

Next, we analyze the proof of [8, Lemma 2]. There it is shown how one can get from the system of equations that we derived in the previous lemma to the equation $E = F$. First we make the system of equations smaller by reducing the parameter k:

Lemma 11. *For $k \geq 2$, let (G_i, H_i) be a pair of regular expressions and ϵ_i be a regular expression for each $i \in \{1, 2, \ldots, k\}$ and let $D_{i,h}$ be a regular expression with $\lambda \notin L(D_{i,h})$ for each $(i, h) \in \{1, 2, \ldots, k\}^2$. Then, from the system of equations*

$$(G_i, H_i) = \sum_{h=1}^{k} (G_h, H_h) \cdot D_{i,h} + (\epsilon_i, \epsilon_i), \quad \text{for } i = 1, 2, \ldots, k,$$

we can derive a system of equations of the form

$$(G_i, H_i) = \sum_{h=1}^{k-1} (G_h, H_h) \cdot D'_{i,h} + (\epsilon'_i, \epsilon'_i), \quad \text{for } i = 1, 2, \ldots, k-1,$$

where all the ϵ'_i and $D'_{i,h}$ are regular expressions with $\lambda \notin L(D'_{i,h})$, with $O(k^5)$ uses of R_1 and two uses of R_2.

We can reduce k repeatedly with the previous lemma until $k = 1$. Then, we can show the equation $G_1 = H_1$ with the help of rule R_2:

Corollary 12. *For $k \geq 1$, let (G_i, H_i) be a pair of regular expressions and ϵ_i be a regular expression for each $i \in \{1, 2, \ldots, k\}$ and let $D_{i,h}$ be a regular expression with $\lambda \notin L(D_{i,h})$ for each $(i, h) \in \{1, 2, \ldots, k\}^2$. Then, from the system of equations*

$$(G_i, H_i) = \sum_{h=1}^{k} (G_h, H_h) \cdot D_{i,h} + (\epsilon_i, \epsilon_i), \quad \text{for } i = 1, 2, \ldots, k,$$

one can derive the equation $G_1 = H_1$ with $O(k^6)$ uses of R_1 and $2k$ uses of R_2.

Lemma 10 and Corollary 12 show us how to derive the equation $E = F$ from the equational characterizations of E and F:

Corollary 13. *Let E be a regular expression with an equational characterization with n equations and F be a regular expression with an equational characterization with m equations and $L(E) = L(F)$. Then, the equation $E = F$ is derivable from the two given equational characterizations with $O((mn)^6 + r^4 mn)$ uses of R_1 and $2mn + m + n$ uses of R_2.*   □

With Theorem 8 and Corollary 13 we can derive the equation $E = F$, for regular expressions E and F with $L(E) = L(F)$:

Theorem 14. *Let E and F be regular expressions with $L(E) = L(F)$. Then, the equation $E = F$ is derivable with $O(r^4 \cdot \text{TOWER}(h + 4))$ uses of R_1 and $\text{TOWER}(h + 4)/1024$ uses of R_2, where $h = \max\{h(E), h(F)\}$.*

Proof. By Theorem 8 equational characterizations for E and F, both with at most $\text{TOWER}(h+3)/4$ equations, are derivable with $O(r^4 \cdot \text{TOWER}(h+3))$ uses of R_1 and

$$2 \cdot (|E+F|. + |E+F|_*) < 2 \cdot 2^{h(E+F)} = 2^{h+2}$$

uses of R_2. By Corollary 13 the equation $E = F$ is derivable with

$$O\left(r^4 \cdot (\text{TOWER}(h+3))^{12}\right) \subseteq O\left(r^4 \cdot \text{TOWER}(h+4)\right)$$

uses of R_1 and

$$2^{h+2} + (\text{TOWER}(h+3))^2/8 + \text{TOWER}(h+3)/2 \tag{2}$$

uses of R_2. The term in (2) is bounded from above by

$$(\text{TOWER}(h+3))^2/8 + \text{TOWER}(h+3) < 2^{\text{TOWER}(h+3)-10}$$
$$= \text{TOWER}(h+4) \cdot 2^{-10},$$

where we have used $n^2/8 + n < 2^{n-10}$, for $n \geq 16$. □

4 Proof Complexity of Regular Expressions in General

The study of the efficiency of propositional proof systems dates back to the seminal paper of Cook and Reckhow [2]. There the notion of a proof system in general was introduced, which reads as follows—we literally take the definition from there: if $L \subseteq \Sigma^*$, a *proof system* for L is a deterministic polynomial time computable function $f : \Sigma^* \to L$ such that f is onto. A proof system is *polynomially bounded* if there is a polynomial p such that for all $y \in L$ there is an $x \in \Sigma^*$ such that $y = f(x)$ and $|x| \leq p(|y|)$, where $|z|$ denotes the length of z. If $y = f(x)$, then we will say that x is a proof of y, and x is a *short* proof of y if in addition $|x| \leq p(|y|)$. The ultimate goal of propositional proof complexity is to show that there is no propositional proof system allowing for efficient proofs of tautology.

Up to our knowledge the relation between proof complexity and regular expression equivalence is not investigated so far. In the line of the proof of NP=coNP if and only if the set TAUT of all propositional tautologies admits a polynomial bounded proof system [2], we show a similar relation for the set

$$\text{EQUIV} = \{\,(E, F) \mid E \text{ and } F \text{ are regular expressions with } L(E) = L(F)\,\}$$

and the complexity classes NP and PSPACE—we assume the reader to be familiar with the basics in complexity theory as contained in [4]. Note that it is well known that EQUIV is PSPACE-complete [9].

Theorem 15. NP $=$ PSPACE *if and only if* EQUIV *admits a polynomial bounded proof system.*

Proof. Observe, that EQUIV is in PSPACE. If NP = PSPACE, then EQUIV belongs to NP. Since a language L is in NP if and only if $L = \emptyset$ or L has a polynomially bounded proof system as shown in [2] it follows that the set EQUIV obeys a polynomially bounded proof system as well. Conversely we argue as follows: assume that EQUIV has a polynomially bounded proof system. Then by the aforementioned statement of [2] it implies that the PSPACE-complete set EQUIV is in NP, which in turn gives us PSPACE $\subseteq$ NP due to the closure of both complexity classes under deterministic polynomial many-one reductions. Hence NP = PSPACE.   □

Can we say more on connection between regular expression and proof systems? One possibility is to restrict the equivalence problem for regular expressions. If we consider expressions that use union and concatenation only, the complexity of the equivalence problem drops to coNP-completeness [5]. This is a dramatic change in complexity compared to the general regular expression equivalence problem. In similar veins as in the proof above, we can show the next result, where $\text{EQUIV}_{\text{fin}}$ refers to the equivalence problem of regular expressions with the operations union and concatenation only—the subscript "fin" refers to the fact that the involved expressions can only describe finite languages. The proof is straight forward and thus left to the reader.

Theorem 16. NP=coNP *if and only if* $\text{EQUIV}_{\text{fin}}$ *admits a polynomial bounded proof system.*   □

Let us turn back to the equivalence of regular expressions in general. The most efficient proof that we can come up for EQUIV is simply to convert both expressions into equivalent nondeterministic finite automata, then to determinize these automata in order to obtain equivalent deterministic finite state devices, followed by a minimization, and finally check for isomorphism in order to verify the equivalence. This strategy leads to a proof system f that is in fact deterministic polynomial time computable on the size of the *whole* proof x, which is of exponential length, because the conversion of a regular expression into an equivalent deterministic finite automaton increases its size at most exponential. This is far from a polynomially bounded proof system, but it is the best possible we can come up with at the moment.

References

1. Brzozowski, J.A.: Canonical regular expressions and minimal state graphs for definite events. In: Mathematical Theory of Automata. MRI Symposia Series, vol. 12, pp. 529–561. Polytechnic Press, New York (1962)
2. Cook, S.A., Reckhow, R.A.: The relative efficiency of propositional proof systems. J. Symb. Logic **44**(1), 36–50 (1979)
3. Ginzburg, A.: A procedure of checking equality of regular expressions. J. ACM **14**(2), 355–362 (1967)
4. Hopcroft, J.E., Ullman, J.D.: Introduction to Automata Theory, Languages and Computation. Addison-Wesley, Boston (1979)

5. Hunt, H.B., Rosenkrantz, D.J., Szymanski, T.G.: On the equivalence, containment, and covering problems for the regular and context-free languages. J. Comput. System Sci. **12**, 222–268 (1976)
6. Kleene, S.C.: Representation of events in nerve nets and finite automata. In: Shannon, C.E., McCarthy, J. (eds.) Automata Studies, Annals of Mathematics Studies, vol. 34, pp. 2–42. Princeton University Press, Princeton (1956)
7. McNaughton, R., Yamada, H.: Regular expressions and state graphs for automata. IRE Trans. Electron. Comput. **EC–9**(1), 39–47 (1960)
8. Salomaa, A.: Two complete axiom systems for the algebra of regular events. J. ACM **13**(1), 158–169 (1966)
9. Stockmeyer, L.J., Meyer, A.R.: Word problems requiring exponential time. In: Proceedings of the 5th Symposium on Theory of Computing, pp. 1–9 (1973)

Operational State Complexity and Decidability of Jumping Finite Automata

Simon Beier$^{(\boxtimes)}$, Markus Holzer, and Martin Kutrib

Institut für Informatik, Universität Giessen, Arndtstr. 2, 35392 Giessen, Germany
{simon.beier,holzer,kutrib}@informatik.uni-giessen.de

Abstract. We consider jumping finite automata and their operational state complexity and decidability status. Roughly speaking, a jumping automaton is a finite automaton with a non-continuous input. This device has nice relations to semilinear sets and thus to Parikh images of regular sets, which will be exhaustively used in our proofs. In particular, we prove upper bounds on the intersection and complementation. The latter result on the complementation upper bound answers an open problem from G.J. LAVADO, G. PIGHIZZINI, S. SEKI: Operational State Complexity of Parikh Equivalence [2014]. Moreover, we correct an erroneous result on the inverse homomorphism closure. Finally, we also consider the decidability status of standard problems as regularity, disjointness, universality, inclusion, etc. for jumping finite automata.

1 Introduction

Jumping finite automata were recently introduced in [16] as a machine model for non-local information processing, which is motivated from modern information processing, see, for example, [1]. This non-locality is modeled by a non-continuous input. Roughly speaking, a jumping finite automaton is an ordinary finite automaton, which is allowed to read letters from anywhere in the input string, not necessarily only from the left of the remaining input. Already in [16] quite a large number of questions regarding jumping automata were studied and answered: inclusion relations to well-known formal language families, closure and non-closure results under standard formal language operations, decision problems on jumping finite automata languages, etc. Since then, a series of papers [4,5,18,19] pushed the investigation on jumping finite automata further and obtained results on normalforms for jumping finite automata languages by shuffle expressions, computational complexity of jumping finite automata problems, etc. Nevertheless, still several problems for this new device remain open, such as, for example, questions on the descriptional complexity of operations on jumping finite automata languages.

Since a jumping finite automata reads the input in a non-continuous fashion, obviously, the order of the input letters does not matter. Thus, only the number of symbols in the input is important. Hence, the behavior of jumping automata is somehow related to the notions of Parikh image and Parikh equivalence. It is well known that regular and context-free languages cannot be distinguished *via*

© Springer International Publishing AG 2017
É. Charlier et al. (Eds.): DLT 2017, LNCS 10396, pp. 96–108, 2017.
DOI: 10.1007/978-3-319-62809-7_6

Parikh equivalence, since for both languages families the set of Parikh images coincides with the family of semilinear sets. Recently, several classical results on automata conversions and operations where studied subject to the notion of Parikh equivalence. For instance, in [14] it was shown that the cost of the conversion of an n-state nondeterministic finite automaton into a Parikh equivalent deterministic finite state device is of order $e^{\Theta(\sqrt{n \ln n})}$. Yet another example is the intersection of the Parikh images of an n- and an m-state deterministic finite automaton with an input alphabet of size k. In [15] it was shown that there is a deterministic finite automaton with $O(p(\max\{n, m\})^{q(k)})$ states, where p and q are polynomials, whose Parikh image equals the intersection of the Parikh images of the languages accepted by the automata one started from. A close inspection of these results reveal that there is a nice relation between Parikh images of regular languages and jumping finite automata via semilinear sets. This connection makes it possible to transfer results from Parikh images of finite automata to results on jumping automata and $vice\ versa$. Thus one can read the above mentioned results as results on jumping finite automata, too.

Here we investigate the operational state complexity of jumping finite automata. In particular, we prove upper bounds on the intersection, complementation, and inverse homomorphism. For two jumping finite automata with n and m states and an input alphabet of size k the upper bound on the intersection is $(k \cdot \max\{n, m\})^{O(k^2)}$ in the nondeterministic case and $(k \cdot \max\{n, m\})^{O(k^5)}$ in the deterministic case. Due to the above mentioned close relation between jumping finite automata and Parikh images of regular languages, this bound significantly improves the corresponding result on Parikh images of finite automata languages mentioned above. Also in [15] the bound on complementation is stated as an open problem. Here we show that complementation of an n state jumping finite automaton with an input alphabet of size k can be accepted by a $2^{k^{O(k \cdot \log(k))}} n^{O(k^2 \cdot \log(k))}$ -state jumping automaton. Thus, our result answers the aforementioned open question, too. Moreover, we correct an erroneous result on the inverse homomorphism closure from [16], where we also obtain an upper bound result. All proofs exhaustively use the relation between jumping automata and semilinear sets. A summary of our results on the operational state complexity of jumping automata can be found in Table 1. Finally, we also consider the decidability status of standard problems as regularity, disjointness, universality, inclusion, etc. for jumping finite automata. There we concentrate on the relation between jumping finite automata languages and regular sets. In particular, we study the descriptional complexity of jumping finite automata accepting regular languages compared to ordinary finite automata. We are able to provide an exponential lower bound on the conversion from jumping automata to finite automata, when accepting regular languages only. The question on an upper bound is stated as an open problem.

This paper is organized as follows. In Sect. 2 we give some basic definitions concerning jumping automata and semilinear sets. Then we investigate the operational state complexity of jumping automata in Sect. 3. The operations intersection, complementation, inverse homomorphism, and intersection with regular

Table 1. Operational state complexity results for deterministic jumping finite automata (DJFAs) and nondeterministic jumping finite automata (NJFAs) over an input alphabet Σ of size k (k_2 for the operation of inverse homomorphism). For every operation the parameter n is the maximum of the numbers of states of the operand automata. For the operation of inverse homomorphism we have another alphabet Γ of size k_1, a homomorphism $h : \Gamma^* \to \Sigma^*$, and the parameter $m = \max(\{|h(a)|_b \mid a \in \Gamma, b \in \Sigma\} \cup \{1\})$, where $|h(a)|_b$ stands for the number of appearances of the symbol b in the word $h(a)$. For the last operation (intersection with regular languages) the given bounds are valid whenever the resulting language is accepted by a jumping automaton.

Operation	Number of states of the resulting automaton	
	$X = D$	$X = N$
XJFA $\cap$ XJFA $\to$ XJFA	$(k \cdot n)^{O(k^5)}$	$(k \cdot n)^{O(k^2)}$
$\Sigma^* \backslash$ XJFA $\to$ DJFA	$2^{k^{O(k \cdot \log(k))} n^{O(k^2 \cdot \log(k))}}$	
h^{-1}(NJFA) $\to$ XJFA	$2^{(k_1 k_2 mn)^{5k_1 k_2 + k_2^2 + O(k_1 + k_2)}}$	$(k_1 k_2 mn)^{5k_1 k_2 + k_2^2 + O(k_1 + k_2)}$
XJFA $\cap$ XFA $\to$ XJFA	$(k \cdot n)^{O(k^5)}$	$(k \cdot n)^{O(k^2)}$

languages are considered. Finally in Sect. 4 we deal with the decidability of some problems on jumping automata. Due to space limitations all proofs are omitted.

2   Preliminaries

We assume the reader to be familiar with the basics in automata and formal language theory as contained, for example, in [7]. Let $\mathbb{Z}$ be the set of integers and $\mathbb{N} = \{0, 1, 2, \ldots\}$ be the set of non-negative integers.

For the notion of semilinear sets we follow the notation of Ginsburg and Spanier [6]. For a natural number $k \geq 1$ and finite $C, P \subseteq \mathbb{N}^k$ let $L(C, P)$ denote the subset

$$L(C, P) = \left\{ \boldsymbol{x}_0 + \sum_{\boldsymbol{x}_i \in P} \lambda_i \cdot \boldsymbol{x}_i \;\middle|\; \boldsymbol{x}_0 \in C \text{ and } \lambda_i \in \mathbb{N} \right\}$$

of $\mathbb{N}^k$. Here the $\boldsymbol{x}_0 \in C$ are called the *constants* and the $\boldsymbol{x}_i \in P$ the *periods*. If C is a singleton set we call $L(C, P)$ a *linear* subset of $\mathbb{N}^k$. In this case we simply write $L(\boldsymbol{c}, P)$ instead of $L(\{\boldsymbol{c}\}, P)$. A subset of $\mathbb{N}^k$ is said to be *semilinear* if it is a finite union of linear subsets. We further use $|P|$ to denote the size of a finite subset $P \subseteq \mathbb{N}^k$ and $||P||$ to refer to the value $\max\{||\boldsymbol{x}|| \mid \boldsymbol{x} \in P\}$, where $||\boldsymbol{x}||$ is the maximum norm of $\boldsymbol{x}$, that is, $||(x_1, x_2, \ldots, x_k)|| = \max\{|x_i| \mid 1 \leq i \leq k\}$. Analogously we write $||A||$ for the maximum norm of a matrix A with entries in $\mathbb{Z}$, i.e. the maximum of the absolute values of all entries of A. The elements of $\mathbb{N}^k$ can be partially ordered by the $\leq$-relation on vectors. For vectors $\boldsymbol{x}, \boldsymbol{y} \in \mathbb{N}^k$ we write $\boldsymbol{x} \leq \boldsymbol{y}$ if all components of $\boldsymbol{x}$ are less or equal to the corresponding components of $\boldsymbol{y}$. In this way we especially can speak of *minimal elements* of subsets of $\mathbb{N}^k$. In fact, due to [3] every subset of $\mathbb{N}^k$ has only a finite number of minimal elements.

Let Σ be an alphabet. Then Σ^* is the set of all words over Σ, including the empty word λ. For a language $L \subseteq \Sigma^*$ define the set $\mathsf{perm}(L) = \cup_{w \in L} \mathsf{perm}(w)$, where $\mathsf{perm}(w) = \{v \in \Sigma^* \mid v$ is a permutation of $w\}$. Then a language L is *permutation closed* if $L = \mathsf{perm}(L)$. The length of a word $w \in \Sigma^*$ is denoted by $|w|$. For the number of occurrences of a symbol a in w we use the notation $|w|_a$. We denote the powerset of a set S by 2^S. We use $\subseteq$ for inclusion, and $\subset$ for proper inclusion. The *Parikh-mapping* ψ is the function $\psi(w) = (|w|_{a_1}, |w|_{a_2}, \ldots, |w|_{a_k})$, if $w \in \Sigma^*$ with $\Sigma = \{a_1, a_2, \ldots, a_k\}$. This mapping is extended to languages $L \subseteq \Sigma^*$ by $\psi(L) = \{\psi(w) \mid w \in L\}$. A language L is *semilinear* if its Parikh-mapping is a semilinear set. Furthermore, a language family is said to be *semilinear* if all languages in that family are semilinear. Furthermore, we say that two languages $L, M \subseteq \Sigma^*$ are *Parikh equivalent* if $\psi(L) = \psi(M)$.

Next, we define jumping finite automata where the notion from [16] is slightly adapted. A *nondeterministic general finite automaton* is a quintuple $A = (Q, \Sigma, \delta, q_0, F)$, where Q is the finite set of *states*, Σ is the finite set of *input symbols*, $q_0 \in Q$ is the *initial state*, $F \subseteq Q$ is the set of *accepting states*, and $\delta \colon Q \times (\Sigma \cup \{\lambda\}) \to 2^Q$ is the *transition function*. A nondeterministic general finite automaton is *deterministic* if $\delta(q, a)$ is a singleton set, for every state $q \in Q$ and letter $a \in \Sigma$, and $\delta(q, \lambda) = \emptyset$ for every state $q \in Q$. In this case we simply write $\delta(q, a) = p$ instead of $\delta(q, a) = \{p\}$, assuming that δ is a function from $Q \times \Sigma$ to Q.

One can interpret the general finite automaton A on *configurations* of the form $Q\Sigma^*$ in two ways:

1. As ordinary *finite automaton*: the move relation $\vdash_A$ of A is defined as

$$qw \vdash_A pv \quad \text{iff} \quad w = av, \text{ for } v \in \Sigma^*, a \in \Sigma \cup \{\lambda\}, \text{ and } p \in \delta(q, a).$$

As usual $\vdash_A^*$ refers to the reflexive transitive closure of $\vdash_A$.

2. As *jumping finite automaton*: the jumping relation $\curvearrowright_A$ of A is defined as

$$qw \curvearrowright_A puv \quad \text{iff} \quad w = uav, \text{ for } u, v \in \Sigma^*, a \in \Sigma \cup \{\lambda\}, \text{ and } p \in \delta(q, a).$$

As usual $\curvearrowright_A^*$ refers to the reflexive transitive closure of $\curvearrowright_A$.

In case there is no danger of confusion we simply write $\vdash^*$ and $\vdash$ instead of $\vdash_A^*$ and $\vdash_A$, and $\curvearrowright^*$ and $\curvearrowright$ instead of $\curvearrowright_A^*$ and $\curvearrowright_A$. Thus, we obtain the following languages from a general finite automaton:

1. The language accepted by an ordinary finite automaton $A = (Q, \Sigma, \delta, q_0, F)$ is defined as $L_F(A) = \{w \in \Sigma^* \mid q_0 w \vdash_A^* q_f, \text{ for some } q_f \in F\}$.
2. The language accepted by a jumping finite automaton $A = (Q, \Sigma, \delta, q_0, F)$ is defined as $L_J(A) = \{w \in \Sigma^* \mid q_0 w \curvearrowright_A^* q_f, \text{ for some } q_f \in F\}$.

If there is no danger of confusion we simply write $L(A)$ instead of $L_F(A)$, for a finite automaton A, and $L_J(B)$, for a jumping finite automaton B. If for instance, the NFA A is interpreted as a NJFA A', then one can simply refer by $L_J(A)$ to the language $L(A')$ accepted by A'. This is used to simplify the presentation. For instance, it holds $\psi(L_F(A)) = \psi(L_J(A))$, which will be used later on.

As usual we write NFA (DFA) for nondeterministic (deterministic) finite automata. Moreover, NJFA (DJFA) is an abbreviation for nondeterministic (deterministic) finite jumping automata. The family of languages accepted by a device of type X is denoted by $\mathscr{L}(X)$. Clearly, $\mathscr{L}(NFA) = \mathscr{L}(DFA)$ is the family of regular languages. It is well known that every regular language has a semilinear Parikh mapping; the converse is not true in general. Similarly, for jumping finite automata we have $\mathscr{L}(NJFA) = \mathscr{L}(DJFA)$ [16]. Moreover, in [4] (see also [16]) it is argued that a language is accepted by an NJFA if and only if the language is permutation closed and its Parikh mapping is semilinear. Thus, the family of languages accepted by NJFAs or DJFAs is equal to the family of permutation closed semilinear languages. The closure and non-closure results of the family of all languages accepted by jumping finite automata have been obtained in [5,16,18].

3   Operational State Complexity of Jumping Automata

In this section we consider the operational state complexity of jumping finite automata languages in more detail. Most closure and non-closure results were originally shown in [16]. For some of these results the operational state complexity is easy to determine such as, for example, for union. Clearly, the upper bound is $n + m + 1$ for an NJFA to accept $L(A) \cup L(B)$, for an n-state NJFA A and an m-state NJFA B. On the other hand with the cross-product construction we get the upper bound nm for an DJFA to accept $L(A) \cup L(B)$, for an n-state DJFA A and an m-state DJFA B.

For other language operations on jumping finite automata languages, the operational state complexity is not that obvious such as, for example, intersection and complementation. Using a previous result on Parikh equivalent finite automata from [15], a non-trivial upper bound of order $O(p(\max\{n, m\})^{q(k)})$, where p and q are polynomials, can be derived for a DJFA accepting the intersection of an n- and m-state DJFA language over an alphabet of size k. To our knowledge a bound for Parikh equivalent NFAs is not known yet. Moreover, the state complexity of the complement operation on Parikh equivalent automata is stated as an open problem in [15]. Although, we are working with jumping finite automata, in passing, we improve on the intersection result and solve the open problem on the state complexity of the complementation problem on Parikh equivalent finite state devices. This is due to the close relation between jumping finite automata, semilinear sets, and Parikh equivalent finite automata.

We will prove upper bounds for the language operations intersection, complementation, and inverse homomorphism on NJFA languages. To do this we use the following strategy: first we interpret the NJFAs that we start with as ordinary NFAs and construct a semilinear presentation of the languages under consideration. Here the following result shown in [17, Theorem 4.1] will be used. In fact, it does not explicitly state that the constructed P_i's are linearly independent, but by a careful inspection one observes that they are.

Theorem 1. *Let A be an n-state NFA with input alphabet of size k. Then there exists an index set I with $|I| \in O(k^{4k+6}n^{k^2+3k+5})$ such that, for each $i \in I$, there is a* linearly independent *subset $P_i \subseteq \mathbb{N}^k$ with $||P_i|| \leq n$ and an $\boldsymbol{x}_i \in \mathbb{N}^k$ with $||\boldsymbol{x}_i|| \in O(k^{4k+6}n^{3k+5})$ such that $\psi(L(A)) = \bigcup_{i \in I} L(\boldsymbol{x}_i, P_i)$.*

Then we use a result on the descriptional complexity of our operation applied to semilinear sets from [2]. Finally we convert the resulting semilinear set back to an NFA which we interpret as an NJFA with the following theorem.

Theorem 2. *Let Σ be an alphabet of cardinality k, let I be a finite index set, and for each $i \in I$ let $P_i, C_i \subset \mathbb{N}^k$ be finite subsets. We set $n = \max_{i \in I} ||P_i||$, $m = \max_{i \in I} |P_i|$, $\ell = \max_{i \in I} ||C_i||$, and $L = \max_{i \in I} |C_i|$. Then there exists an NFA A with input alphabet Σ and $k \cdot |I| \cdot (L\ell + mn) + 1$ states such that $\psi(L(A)) = \bigcup_{i \in I} L(C_i, P_i)$.*

When ever it is reasonable we also give upper bounds for the operational state complexity of DJFAs. In order to prove such results we exploit the results for NJFAs and the following two results from [15] that can immediately be adapted to jumping automata.

Lemma 3. *Let A be an NFA with n states and input alphabet $\{a_1, a_2, \ldots, a_k\}$. Then for every $i \in \{1, 2, \ldots, k\}$ there exists an NFA A_i with n states such that $L(A_i) = L(A) \cap \{a_i\}^*$. Furthermore there exists an NFA A_0 with $(k+1) \cdot n + 1$ states such that $L(A_0) = L(A) \backslash \bigcup_{i=1}^{k} L(A_i)$. If A is deterministic, then so are the automata $A_0, A_1, \ldots, A_k$.*

For the construction of A_i in the previous lemma it is sufficient to remove all transitions from A that are not labeled with a_i. Automaton A_0 is then constructed by taking copies of all A_i and a new initial state that leads to the copy of A_i with input symbol a_i. The same constructions can be applied to NJFAs.

The second result from [15] can be adapted to jumping finite automata as well. It gives us an upper bound of order $(k \cdot n)^{O(k^3)}$ for the determinization of NFAs that accept no unary words.

Theorem 4. *Let A be an NFA with n states and an input alphabet of size k such that every word in $L(A)$ contains at least two different symbols. Then there exists a Parikh equivalent DFA with $O(k^{k^3/2+k^2+2k+5}n^{3k^3+6k^2})$ states.*

For the adaption, a given NJFA is interpreted as NFA for which the Parikh equivalent DFA exists. Finally, both are interpreted as jumping automata again. For every word w over the input alphabet, the NFA accepts some permutation of w if and only if the DFA accepts some permutation of w. Since the reinterpretation as jumping automata yields that the NJFA accepts all permutations of some input w if and only if the DJFA accepts all permutations of w, we conclude that both accept the same language. So, on passing we have the following corollary on the determinization of NJFAs that accept no unary words. The bound is the same as in the previous theorem.

Corollary 5. *Let A be an NJFA with n states and an input alphabet of size k such that every word in $L(A)$ contains at least two different symbols. Then there exists an equivalent DJFA with $O(k^{k^3/2+k^2+2k+5}n^{3k^3+6k^2})$ states.* □

3.1   Operational State Complexity of Intersection

In order to deal with the state complexity of the intersection of two NJFA languages we use the following result from [2] on the operational complexity of the intersection of two semilinear sets.

Theorem 6. *Let $\bigcup_{i \in I} L(\mathbf{c}_i, P_i)$ and $\bigcup_{j \in J} L(\mathbf{c}_j, P_j)$ be semilinear subsets of $\mathbb{N}^k$, for some $k \geq 1$. Without loss of generality we may assume that I and J are disjoint finite index sets. We set $n = \max_{i \in I \cup J} \|P_i\|$, $m = \max_{i \in I \cup J} |P_i|$, and $\ell = \max_{i \in I \cup J} \|\mathbf{c}_i\|$. Then for every $(i, j) \in I \times J$ there exist $P_{i,j}, C_{i,j} \subseteq \mathbb{N}^k$ with*

$$\|P_{i,j}\| \leq 3m^2 k^{k/2} n^{k+1},$$
$$\|C_{i,j}\| \leq (3m^2 k^{k/2} n^{k+1} + 1)\ell,$$

and $\left(\bigcup_{i \in I} L(\mathbf{c}_i, P_i)\right) \cap \left(\bigcup_{j \in J} L(\mathbf{c}_j, P_j)\right) = \bigcup_{(i,j) \in I \times J} L(C_{i,j}, P_{i,j}).$

Now we are ready to state our result on the state complexity of intersection of two jumping finite automata. The upper bound for the intersection turns out to be of order $(k \cdot n)^{O(k^2)}$.

Theorem 7. *Let A be an m_1-state NJFA and B be an m_2-state NJFA with an input alphabet of size k. Then $O(k^{9k^2/2+43k/2+21}n^{6k^2+16k+16})$ states, where $n = \max\{m_1, m_2\}$, are sufficient for an NJFA to accept the language $L(A) \cap L(B)$.*

As an immediate byproduct we have shown the following result on ordinary finite automata and Parikh equivalent languages (cf. [15]).

Corollary 8. *Let A be an m_1-state NFA and B be an m_2-state NFA with an input alphabet of size k. Then there is an $O(k^{9k^2/2+43k/2+21}n^{6k^2+16k+16})$-state NFA, where $n = \max\{m_1, m_2\}$, whose Parikh image equals $\psi(L(A)) \cap \psi(L(B))$.* □

Now we turn to the state complexity of the intersection of DJFAs. It will turn out that the upper bound for the intersection in the deterministic case is of order $(k \cdot n)^{O(k^5)}$.

Theorem 9. *Let A be an m_1-state DJFA and B be an m_2-state DJFA with an input alphabet of size k. Then $O(k^{27k^5/2+183k^4/2+196k^3}n^{18k^5+84k^4+144k^3+96k^2+2})$ states, where $n = \max\{m_1, m_2\}$, are sufficient for a DJFA to accept the language $L(A) \cap L(B)$.*

As an immediate consequence from the previous theorem we obtain the following result on the Parikh equivalence of the intersection of DFAs. This improves a previous result on that subject from [15].

Corollary 10. *Let A be an m_1-state DFA and B be an m_2-state DFA with an input alphabet of size k. Then $O(k^{27k^5/2 + 183k^4/2 + 196k^3} n^{18k^5 + 84k^4 + 144k^3 + 96k^2 + 2})$ states, where $n = \max\{m_1, m_2\}$, are sufficient for a DFA, whose Parikh image is equal to the intersection $\psi(L(A)) \cap \psi(L(B))$.* □

3.2   Operational State Complexity of Complementation

The next formal language operation that we consider is the complementation. As stated in [5] the family of jumping automata languages is closed under this operation. To get an upper bound for the state complexity of the complement of an NJFA we use the following result from [2] on the operational complexity of the complement of a semilinear set.

Theorem 11. *Let $k \geq 1$ and $\bigcup_{i \in I} L(\boldsymbol{x}_i, P_i)$ be a semilinear subset of $\mathbb{N}^k$ with $I \neq \emptyset$ and linearly independent sets P_i. We set $n = \max_{i \in I} \|P_i\|$ and $\ell = \max_{i \in I} \|\boldsymbol{x}_i\|$. Define $q = \lceil \log_2 |I| \rceil$. Then there exists an index set J with*

$$|J| \leq (4k(n+1))^{5(k+2)(3k+2)^{q+1}} (\ell+1)^{k \cdot 2^{q+1}}$$

such that, for each $j \in J$, there are $P_j, C_j \subseteq \mathbb{N}^k$ with

$$\|P_j\| \leq (4k(n+1))^{(k+2)(3k+1)^q},$$
$$\|C_j\| \leq (4k(n+1))^{(k+2)(3k+2)^q + k}(\ell+1),$$

and $\mathbb{N}^k \backslash \bigcup_{i \in I} L(\boldsymbol{x}_i, P_i) = \bigcup_{j \in J} L(C_j, P_j)$.

It will turn out that the upper bound for the complementation of an NJFA is of order $2^{k^{O(k \cdot \log(k))}} n^{O(k^2 \cdot \log(k))}$, even if we want the result to be a DJFA.

Theorem 12. *Let A be an n-state NJFA with an input alphabet Σ of size k. Then $2^{k^{(4k+6) \log_2(3k+2) + O(1)}} n^{(k^2 + 3k+5) \log_2(3k+2)} \log_2(n+1)$ states are sufficient for an NJFA or even a DJFA to accept the language $\Sigma^* \backslash L(A)$.*

The upper bound on the complementation of jumping finite automata transfers to finite automata, when considering the Parikh image:

Corollary 13. *Let A be an n-state NFA with an input alphabet of size k. Then there exists a $2^{k^{(4k+6) \log_2(3k+2) + O(1)}} n^{(k^2+3k+5) \log_2(3k+2)} \log_2(n+1)$-state NFA or even a DFA, whose Parikh image is equal to $\mathbb{N}^k \backslash \psi(L(A))$.* □

This answers an open problem stated in [15].

3.3 Operational State Complexity of Inverse Homomorphism

Next we consider the operation of inverse homomorphism. In [16, Theorem 42] it is claimed that the family $\mathscr{L}(\mathrm{NJFA})$ is closed under inverse homomorphism, where the proof relies on an analogous construction as for ordinary finite automata, which reads as follows: Let $A = (Q, \Gamma, \delta, q_0, F)$ be an NFA, Σ be an alphabet, and $h : \Sigma^* \to \Gamma^*$ be a homomorphism. Then the automaton $A' = (Q, \Sigma, \delta', q_0, F)$ accepts the language $h^{-1}(L(A))$, where

$$q \in \delta'(p, a) \quad \text{if and only if} \quad pw_a \vdash_A^* q \quad \text{with } w_a = h(a).$$

The same construction is used in [16, Theorem 42] to show the closure of $\mathscr{L}(\mathrm{NJFA})$ under inverse homomorphism. However, this construction does not work in general as shown by the following counterexample.

Example 14. Consider the NJFA A with the input alphabet $\Sigma = \{a, b\}$ depicted on the left of Fig. 1. It is easy to see that $L(A) = \{w \in \Sigma^* \mid |w|_a = |w|_b\}$. Set

Fig. 1. (Left): NJFA A accepting the set $\{w \in \Sigma^* \mid |w|_a = |w|_b\}$. (Right): NJFA A' induced from A and the homomorphism $h : \{a\}^* \to \{a, b\}^*$ defined by $h(a) = ba$ by the standard construction on NFAs for the inverse homomorphism closure, accepting the set $\{\lambda\}$.

$\Gamma = \{a\}$ and define the homomorphism $h : \Gamma^* \to \Sigma^*$ *via* $h(a) = ba$. Constructing the automaton A' as described above results in the jumping automaton drawn on the right of Fig. 1. But then

$$L(A') = \{\lambda\} \neq a^* = h^{-1}(L(A)),$$

which shows the argument to be invalid. Even a more sophisticated construction taking permutation of words $h(a)$, for $a \in \Sigma$, into account is not properly working.

In the forthcoming we present an explicit construction based on semilinear sets for the claim on the inverse homomorphism closure of $\mathscr{L}(\mathrm{NJFA})$. For this we use the following result from [2] on the operational complexity of the operation of inverse homomorphism on semilinear sets.

Theorem 15. *Let k_1, $k_2 \geq 1$ and $\bigcup_{i \in I} L(c_i, P_i)$ be a semilinear subset of $\mathbb{N}^{k_2}$. We set $n = \max_{i \in I} \|P_i\|$, $m = \max_{i \in I} |P_i|$, and $\ell = \max_{i \in I} \|c_i\|$. Moreover let*

$H \in \mathbb{N}^{k_2 \times k_1}$ be a matrix and $h : \mathbb{N}^{k_1} \to \mathbb{N}^{k_2}$ be the corresponding linear function $\boldsymbol{x} \mapsto H\boldsymbol{x}$. Then for every $i \in I$ there exist Q_i, $C_i \subseteq \mathbb{N}^{k_1}$ with

$$||Q_i|| \leq (k_1 + m + 1)k_2^{\min(k_1+m,k_2)/2} \cdot (||H|| + 1)^{\min(k_1,k_2)}(n+1)^{\min(m,k_2)},$$

$$||C_i|| \leq (k_1 + m + 1)k_2^{\min(k_1+m,k_2)/2} \cdot (||H|| + 1)^{\min(k_1,k_2)}(n+1)^{\min(m,k_2)}\ell,$$

and $h^{-1}\left(\bigcup_{i \in I} L(\boldsymbol{c}_i, P_i)\right) = \bigcup_{i \in I} L(C_i, Q_i)$.

We get an upper bound of order $(k_1 k_2 mn)^{5k_1 k_2 + k_2^2 + O(k_1+k_2)}$ for the operation of inverse homomorphism on NJFAs:

Theorem 16. *Let Γ and Σ be two alphabets of size k_1 and k_2, respectively, and $h : \Gamma^* \to \Sigma^*$ be a homomorphism. Let A be an n-state NJFA with input alphabet Σ and set $m = \max(\{|h(a)|_b \mid a \in \Gamma, b \in \Sigma\} \cup \{1\})$. Then there exists an NJFA with $O(k_1^{3k_1/2+2} k_2^{5k_1 k_2 + 7k_1 + 9k_2 + 13} m^{(k_1+1)\min(k_1,k_2)} n^{4k_1 k_2 + 5k_1 + k_2^2 + 7k_2 + 10})$ states accepting the language $h^{-1}(L(A))$.*

The next corollary is an immediate consequence of the previous theorem.

Corollary 17. *The language family $\mathscr{L}(NJFA)$ is closed under inverse homomorphism.*

Determinizing the resulting NJFA causes an exponential blow up:

Corollary 18. *Let Γ and Σ be two alphabets of size k_1 and k_2, respectively, and $h : \Gamma^* \to \Sigma^*$ be a homomorphism. Let A be an n-state NJFA with input alphabet Σ and set $m = \max(\{|h(a)|_b \mid a \in \Gamma, b \in \Sigma\} \cup \{1\})$. Then there exists a DJFA with $2^{(k_1 k_2 mn)^{5k_1 k_2 + k_2^2 + O(k_1+k_2)}}$ states accepting the language $h^{-1}(L(A))$.*

3.4 Operational State Complexity of Intersection with Regular Languages

Now we consider jumping finite automata languages and their intersection with regular sets in more detail. The language family $\mathscr{L}(NJFA)$ is not closed under intersection with regular languages. However, we have the following result, which will be utilized later.

Lemma 19. *Let A be an NJFA with input alphabet Σ and $R \subseteq \Sigma^*$ a regular language. Then the following statements are equivalent:*

1. *The language $L(A) \cap R$ is in $\mathscr{L}(NJFA)$.*
2. *$L(A) \cap R = L(A) \cap \mathsf{perm}(R)$.*
3. *$L(A) \cap \mathsf{perm}(R) \cap \mathsf{perm}(\Sigma^* \backslash R) = \emptyset$.*

Next we derive an upper bound of order $(k \cdot n)^{O(k^2)}$ on the state complexity of the intersection with regular languages on NJFAs, if the result remains a language accepted by a jumping automaton.

Theorem 20. *Let A be an m_1-state NJFA and B be an m_2-state NFA with an input alphabet of size k. Assume that $L(A) \cap L(B)$ is in $\mathscr{L}(NJFA)$. Then there is an NJFA with $O(k^{9k^2/2+43k/2+21} n^{6k^2+16k+16})$ states, where $n = \max\{m_1, m_2\}$, accepting $L(A) \cap L(B)$.*

For deterministic jumping automata we deduce the following upper bound of order $(k \cdot n)^{O(k^5)}$, which can be proven analogously to Theorem 20, but using Theorem 9 instead of Theorem 7.

Theorem 21. *Let A be an m_1-state DJFA and B be an m_2-state DFA with an input alphabet of size k. Assume that $L(A) \cap L(B)$ is in $\mathscr{L}(NJFA)$. Then there is a DJFA with $O(k^{27k^5/2+183k^4/2+196k^3} n^{18k^5+84k^4+144k^3+96k^2+2})$ states, where $n = \max\{m_1, m_2\}$, accepting $L(A) \cap L(B)$.* □

4 Decidability of Problems Involving Jumping Automata

Already in [16] decidability of some problems on jumping automata were considered. There it was shown that finiteness, infiniteness, membership, and emptiness of NJFA languages is decidable. The status of other decision problems for NJFAs is explicitly stated as open problem. Later in [4] the computational complexity of parsing problems for NJFAs and alternative representations were considered. Decidability and computational complexity results for semilinear sets and Parikh images of regular and/or context-free languages can be found in [8–11]. More recent results on that subject were presented in [12,13,17]. From these papers, and in particular from the latter one, we can deduce the following decidability results on jumping finite state devices, which we state without proof.

Theorem 22. *Let A and B be two NJFAs with input alphabet Σ. Then the following problems are decidable:*

1. *Disjointness: is $L(A) \cap L(B) \neq \emptyset$?*
2. *Universality: is $L(A) = \Sigma^*$?*
3. *Inclusion: is $L(A) \subseteq L(B)$?* □

Instead of relying on results from the literature, some of the above presented decidability statements can be shown by using some results presented earlier in this paper. For instance, the decidability status of the universality problem is seen as follows: Let A be an NJFA with input alphabet Σ. Then by Theorem 12 we can construct an NJFA $\bar{A}$ such that $L(\bar{A}) = \Sigma^* \backslash L(A)$. So, we have $L(A) = \Sigma^*$ if and only if $L(\bar{A}) = \emptyset$. Since emptiness is decidable for NJFAs in the same way as for NFAs the result follows.

In the remainder of this section we again investigate the relation between jumping finite automata and regular languages. As mentioned earlier, the language family $\mathscr{L}(NJFA)$ is not closed under intersection with regular languages. But it is decidable if such an intersection belongs to $\mathscr{L}(NJFA)$:

Theorem 23. *Let A be an NJFA and B be an NFA. Then it is decidable whether the language $L(A) \cap L(B)$ belongs to the language family $\mathscr{L}(NJFA)$ or not.*

As a consequence we get the following result.

Corollary 24. *Let A be an NFA. Then it is decidable whether $L(A)$ is closed under permutation.*

An NP lower bound on the permutation problem was obtained earlier in [5], if an NFA is given. Next an NP upper bound is given for DFAs.

Theorem 25. *Let A be a DFA. Then the problem to decide whether $L(A)$ is not closed under permutation is in NP. If the size of the input alphabet is fixed, then the problem to decide whether $L(A)$ is closed under permutation is in P.*

Finally, the regularity problem for NJFAs is decidable, which can be seen by a result of [6] on semilinear sets. The lower bound on this problem is NP-hardness as recently shown in [5].

Theorem 26. *Let A be an NJFA. Then regularity of $L(A)$ is decidable.*   □

Now the question on the descriptional complexity of NJFAs or DJFAs accepting a regular language compared to ordinary finite automata may arise. The previously presented results show that the increase in the number of states for the conversion of a jumping finite automaton, even a nondeterministic one, accepting a regular language to an ordinary finite automaton is bounded by a recursive function. This substantially depends on (i) the decidability of the regularity for jumping finite automata, (ii) the decidability of the permutation closure for finite automata, and (iii) the decidability of the equivalence for jumping finite automata. Unfortunately, we have no explicit upper bound for the conversion. Nevertheless, we can give an exponential lower bound, which is presented in the next theorem.

Theorem 27. *For any integer $n \geq 1$, there exists an $(n(n+1)/2)$-state DJFA A with input alphabet $\Sigma_n = \{a_1, a_2, \ldots, a_n\}$ accepting a regular language such that any DFA accepting $L(A)$ needs at least $n!$ states.*

References

1. Baeza-Yates, R., Ribeiro-Neto, B.: Modern Information Retrieval: The Concepts and Technology Behind Search. Addison-Wesley, New York (2011)
2. Beier, S., Holzer, M., Kutrib, M.: On the descriptional complexity of operations on semilinear sets. IFIG Research Report 1701, Institut für Informatik, Universität Giessen (2017). http://www.informatik.uni-giessen.de/reports/Report1701.pdf
3. Dickson, L.E.: Finiteness of the odd perfect and primitive abundant numbers with n distinct prime factors. Am. J. Math. **35**, 413–422 (1913)
4. Fernau, H., Paramasivan, M., Schmid, M.L.: Jumping finite automata: characterizations and complexity. In: Drewes, F. (ed.) CIAA 2015. LNCS, vol. 9223, pp. 89–101. Springer, Cham (2015). doi:10.1007/978-3-319-22360-5_8
5. Fernau, H., Paramasivan, M., Schmid, M.L., Vorel, V.: Characterization and complexity results on jumping finite automata. Theor. Comput. Sci. **679**, 31–52 (2017)

6. Ginsburg, S., Spanier, E.H.: Bounded ALGOL-like languages. Trans. AMS **113**, 333–368 (1964)
7. Hopcroft, J.E., Ullman, J.D.: Introduction to Automata Theory, Languages and Computation. Addison-Wesley, Reading (1979)
8. Huynh, D.T.: Deciding the inequivalence of context-free grammars with 1-letter terminal alphabet is Σ_2^p-complete. Theor. Comput. Sci. **33**, 305–326 (1984)
9. Huynh, D.T.: The complexity of equivalence problems for commutative grammars. Inf. Control **66**, 103–121 (1985)
10. Huynh, D.T.: A simple proof for the Σ_2^P upper bound of the inequivalence problem for semilinear sets. Elektr. Informationsverarb. Kybernet. **22**, 147–156 (1986)
11. Huynh, T.-D.: The complexity of semilinear sets. In: Bakker, J., Leeuwen, J. (eds.) ICALP 1980. LNCS, vol. 85, pp. 324–337. Springer, Heidelberg (1980). doi:10.1007/3-540-10003-2_81
12. Kopczyńki, E., To, A.W.: Parikh images of grammar: Complexity and applications. In: Joiannaud, J.P. (ed.) Proceedings of the 25th Annual IEEE Symposium on Logic in Computer Science, pp. 80–89. IEEE (2010)
13. Kopczyński, E.: Complexity of problems of commutative grammars. Log. Methods Comput. Sci. **11**(1) (2015). Paper 9
14. Lavado, G.J., Pighizzini, G., Seki, S.: Converting nondeterministic automata and context-free grammars into Parikh equivalent one-way and two-way deterministic automata. Inf. Comput. **228–229**, 1–15 (2013)
15. Lavado, G.J., Pighizzini, G., Seki, S.: Operational state complexity under parikh equivalence. In: Jürgensen, H., Karhumäki, J., Okhotin, A. (eds.) DCFS 2014. LNCS, vol. 8614, pp. 294–305. Springer, Cham (2014). doi:10.1007/978-3-319-09704-6_26
16. Meduna, A., Zemek, P.: Jumping finite automata. Int. J. Found. Comput. Sci. **23**, 1555–1578 (2012)
17. To, A.W.: Parikh images of regular languages: complexity and applications (2010). http://arxiv.org/abs/1002.1464v2
18. Vorel, V.: On basic properties of jumping finite automata (2015). http://arxiv.org/abs/1511.08396v2
19. Vorel, V.: Two results on discontinuous input processing. In: Câmpeanu, C., Manea, F., Shallit, J. (eds.) DCFS 2016. LNCS, vol. 9777, pp. 205–216. Springer, Cham (2016). doi:10.1007/978-3-319-41114-9_16

Equivalence of Symbolic Tree Transducers

Vincent Hugot[2,3]([✉]), Adrien Boiret[2,3], and Joachim Niehren[1,3]

[1] Inria, Lille, France
[2] University of Lille, Lille, France
[3] Links Project (Inria & Cristal Lab - UMR CNRS 9189), Lille, France
vincent.hugot@inria.fr

Abstract. Symbolic tree transducers are programs that transform data trees with an infinite signature. In this paper, we show that the equivalence problem of deterministic symbolic top-down tree transducers (DTOP) can be reduced to that of classical DTOP. As a consequence the equivalence of two symbolic DTOP can be decided in NEXPTIME, when assuming that all operations related to the processing of data values are in PTIME. This result can be extended to symbolic DTOP with lookahead and thus to deterministic symbolic bottom-up tree transducers.

1 Introduction

Data trees are widely used in various domains of computer science. They represent programs in compiler construction or program analysis, syntactic sentence structure in computational linguistics, all or part of the database instances in semi-structured databases, and structured documents in document processing. The most widely used current formats for data trees are JSON (the Java Script Object Notation) and XML (the eXtensible Markup Language).

We are interested in deciding the equivalence of programs that define transformations on data trees. For instance, we may consider XSLT programs defining XML transformations or Linux installation scripts written in bash that change the file system tree. Our approach is to compile a subclass of such programs into classes of tree transducers for which equivalence is decidable. Here we present a partial landscape of classical classes of tree transducers without data values [7,8,12], where inclusion is read from left to right:

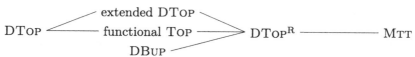

The class DTOP[R] of deterministic top-down tree transducers with regular lookahead by a deterministic bottom-up tree automaton is particularly well

---

This work has been partially supported by CPER Nord-Pas de Calais/FEDER DATA Advanced data science and technologies 2015–2020 and the ANR project Colis, contract number ANR-15-CE25-0001-01.

ⓒ Springer International Publishing AG 2017
É. Charlier et al. (Eds.): DLT 2017, LNCS 10396, pp. 109–121, 2017.
DOI: 10.1007/978-3-319-62809-7_7

behaved [8]. It is closed under composition, which makes it suitable for compilation of programs, and its equivalence problem is decidable in NExpTime, by PTime reduction to the equivalence problem of the class DTop, for which equivalence is decidable in NExpTime [11,14,17]. Furthermore, the class DTop^R subsumes three other classes of tree transducers with pairwise incomparable expressiveness, which capture different aspects of programs: extended DTop with nested pattern matching, functional top-down tree transducers (functional Top) that relax the determinism requirement of DTop, and deterministic bottom-up tree transducers DBup that operate the other way around. The more general class of macro tree transducers (MTT) is much more expressive (and includes lookaheads), but has a long-standing open equivalence problem [9,10,18] and fails to be closed under composition (though its linear size increase subclass has better properties).

In contrast to classical machines that operate on ranked trees over finite signatures, what we need for program verification are generalised machines that operate on data trees with infinite signatures. Most typically the data values which label the nodes of data trees may be strings over some finite alphabet or natural numbers. For dealing with data trees, the classical classes of tree transducers were extended to symbolic classes [15,19,20], and similarly for other kinds of finite state machines. The general idea is to use patterns for describing infinitely many data values in a finite manner, and to allow the transducers to apply transformations on the data values themselves.

We first illustrate by an example that the class of symbolic extended DTop is relevant in practice. For this, we consider the following extended DTop, which performs a routine cleanup and statistics task on a list of log files in a file system, as illustrated in the example of Fig. 1.

$$q\langle \text{NIL} \rangle \rightarrow \text{CONS}\langle \text{ FILE}\langle \text{``log''}, \text{``''} \rangle, \text{ NIL} \rangle \tag{1}$$

$$q\langle \text{CONS}\langle x_1, x_2 \rangle \rangle \rightarrow \text{CONS}\langle q_{\text{name}}\langle x_1 \rangle, q\langle x_2 \rangle \rangle \tag{2}$$

$$q_{\text{name}}\langle \text{FILE}\langle \text{``log''}, x_2 \rangle \rangle \rightarrow \text{FILE}\langle \text{``stats.1''}, \ f_{\text{stats}} @ x_2 \rangle \tag{3}$$

$$q_{\text{name}}\langle \text{FILE}\langle x_1 : \text{``stats.''}(\text{``0''}..\text{``9''})^+, \ x_2 \rangle \rangle \rightarrow \text{FILE}\langle f_{\text{incr}} @ x_1, \ f_{\text{id}} @ x_2 \rangle \tag{4}$$

Such transducers have nested patterns with variables $x_1, x_2, \ldots$ for matching subtrees, expressions for matching data values such as "stats."("0".."9")$^+$ or CONS, and applications of externally defined functions, such as $f_{\text{stats}} @ x_2$, where f_{stats} is a string transformation that produces statistics from its input string (log contents). It should be noted that symbolic functional Top are insufficient for this example since rules such as (3) and (4), with nested patterns, cannot be expressed in a top-down manner. In contrast, symbolic DBup offer an alternative solution for this concrete example.

Veanes and Bjørner [19,20] started the study of symbolic transducers. They showed that equivalence is decidable for symbolic functional Top, if the corresponding problems on data pattern and transformations are. In this paper, we notice that the landscape of tree transducers above remains unchanged when turning classes of classical tree transducers symbolic. Therefore, we can show that the equivalence problem is decidable for the symbolic counterparts of all classes in the landscape except for MTT. To see this, note that any symbolic DTOP is a symbolic functional TOP, so equivalence for symbolic DTOP is decidable. Furthermore, equivalence for symbolic DTOP^{R}s can be reduced to equivalence of symbolic DTOP as in the classical case.

We then start studying the complexity of equivalence for classes of symbolic tree transducers. Our main result is that equivalence for symbolic DTOP^{R}s is in NEXPTIME, under the assumption that operations on patterns and data transformations can be performed in PTIME. If not, one needs to multiply the worst case exponential time with the maximal time needed for such operations. We obtain this result from a novel reduction from the equivalence of symbolic DTOP to the equivalence of classical DTOP, using a weakened version of the *origin-equivalence* in [3]. This reduction allows us to conclude that the equivalence problem of symbolic DTOP is indeed in NEXPTIME as for classical DTOP (and not in 2NEXPTIME as a naive analysis would lead to believe). Due to the modularity of the construction, the equivalence testers obtained for DTOP^{R}s are easy to prove correct, to analyse, and to implement.

2 Tree Automata and Transducers

Some familiarity with formal languages and automata theory, as covered for instance in [5], is assumed.

Given a set S, we denote its cardinality by $|S|$ and its powerset by 2^S. The set of Boolean values is written $\mathbb{B} = \{0, 1\}$. $\mathbb{N}$ is the set of natural integers, including zero. We write $m..n$ the integer interval $[m, n] \cap \mathbb{N}$. We will denote tuples $(a_0, a_1, \ldots, a_n)$ by $a_0\langle a_1, \ldots, a_n \rangle$ or simply as a_0 if $n = 0$.

Let $X = \{x_1, x_2, x_3, \ldots\}$ be a set of **variables**. For $K > 0$, we shall often use the subsets $X_K = \{x_1, \ldots, x_K\}$. A **ranked alphabet** Σ is a (potentially infinite) set disjoint from X paired with a function $ar_\Sigma : \Sigma \to \mathbb{N}$ (or just ar where Σ is clear from context). The set of **ranked trees over Σ with variables in X**, denoted by $\mathcal{T}_\Sigma(X)$, is the least set that contains X and all $a\langle t_1, \ldots, t_n \rangle$ where $a \in \Sigma$, $n = ar_\Sigma(a)$, $t_1, \ldots, t_n \in \mathcal{T}_\Sigma(X)$. $\mathcal{T}_\Sigma(\varnothing)$ is the set of **ground** Σ-trees, also written $\mathcal{T}_\Sigma$. Notions of position, substitution, etc., are all defined as usual.

We next recall the definitions of deterministic top-down tree automata (DTTA) and deterministic top-down tree transducers (DTOP).

Definition 1. A **quasi** DTTA is a tuple $A = (\Sigma, Q, q_{\text{ini}}, rhs)$ such that Σ is a ranked alphabet, Q a finite set, $q_{\text{ini}} \in Q$, and rhs is a partial function that maps pairs $(q, a) \in Q \times \Sigma$ to tuples of $Q^{ar(a)}$. A DTTA is a quasi DTTAfor which Σ is finite, and thus so is rhs.

The elements of Q are called the states of A, q_{ini} its initial state. The rules of A have the form $q\langle a\langle x_1, \ldots, x_n\rangle\rangle \to rhs(q, a)$, where $rhs(q, a)$ is defined and $n = ar(a)$. Each state q of a quasi DTTA A recognises a tree language $[\![q]\!]_A \subseteq \mathcal{T}(\Sigma)$, (or just $[\![q]\!]$ when A is clear from the context) defined by induction on the trees:

we have $a \langle t_1, \ldots, t_n \rangle \in [\![q]\!]$ iff there is a rule $q \langle a \langle x_1, \ldots, x_n \rangle \rangle \to q_1, \ldots, q_n$ and $t_k \in [\![q_k]\!]$ for all $k \in 1..n$. The semantics of the automaton is $[\![A]\!] = [\![q_{\text{ini}}]\!]$.

Bottom-up tree automata (BUTA) are defined similarly and as usual, with rules of the form $a \langle q_1, \ldots, q_n \rangle \to q$.

Definition 2. A **quasi** DTOP is a tuple $M = (\Sigma, \Delta, Q, \text{ax}, \text{rhs})$ such that Σ and Δ are ranked alphabets, Q is a finite set, $q_{\text{ini}} \in Q$, and rhs is a partial function that maps pairs $(q, a) \in Q \times \Sigma$ to $\mathcal{T}_\Delta(Q \times X_{\text{ar}(a)})$. A DTOP is a quasi DTOP for which Σ and Δ are finite, and thus rhs as well.

Σ and Δ are called input and output alphabets; the other components are as in automata. Each state $q \in Q$ has as semantics a partial function $[\![q]\!]$ from $\mathcal{T}_\Sigma$ to $\mathcal{T}_\Delta$, defined by induction on terms $t = a \langle t_1, \ldots, t_n \rangle \in \mathcal{T}_\Sigma$ such that:

$$[\![q]\!](t) \quad = \quad \text{rhs}(q, a) \left[q' \langle x_k \rangle \leftarrow [\![q']\!](t_k) \mid q' \in Q, \ k \in 1..\,\text{ar}(a) \right]. \tag{5}$$

The transformation defined by M is the partial function $[\![M]\!] = [\![q_{\text{ini}}]\!]$.

3   Symbolic Tree Automata and Transducers

In this section, we recall the definitions of symbolic DTTA and symbolic DTOP as in [15]. Symbolic machines are finite representations of potentially infinite quasi DTTA and quasi DTOP. They use **descriptors** to stand for the potentially infinite sets and functions. Given a set S, we call a set D paired with a function $[\![.]\!] : D \to S$ as set of descriptors of elements of S. For instance, we can use the set E of regular expressions $e \in E$ over an alphabet A as descriptors of regular languages $[\![e]\!] \subseteq A^*$. Outside of the definitions, we shall often assimilate the descriptors and their semantics.

Definition 3. A **symbolic** DTTA is a tuple $A = (\Sigma, \Phi, Q, q_{\text{ini}}, \text{rhs})$ such that $(\Phi, Q, q_{\text{ini}}, \text{rhs})$ is a quasi DTTA with a finite set of rules, Φ is an alphabet of descriptors for subsets of the alphabet Σ, with $\text{ar}(a) = \text{ar}(\varphi)$ for any $a \in [\![\varphi]\!]$ and $\varphi \in \Phi$. For all $a \in \Sigma$ and $q \in Q$, there exits at most one $\varphi \in \Phi$ such that $\text{rhs}(q, \varphi)$ is defined and $a \in [\![\varphi]\!]$.

The elements of $\varphi \in \Phi$ are called (descriptors for) guards. A symbolic DTTA A is a finite representation of a (potentially) infinite quasi DTTA A' such that for every rule r of form, $q \langle \varphi \langle x_1, \ldots, x_n \rangle \rangle \to q_1, \ldots, q_n$ of A, and for every $a \in [\![\varphi]\!]$, there is a rule $q \langle a \langle x_1, \ldots, x_n \rangle \rangle \to q_1, \ldots, q_n$ in A'. The semantics of A is defined as that of A': for all $q \in Q$, $[\![q]\!]_A = [\![q]\!]_{A'}$. Symbolic BUTA are defined similarly.

Definition 4. A symbolic DTTA is **effective** if it satisfies the following conditions, which we always assume: (**1**) The set of guards Φ is closed under conjunction and negation, i.e., there exists an algorithm computing some function $\wedge : \Phi \times \Phi \to \Phi$ such that $[\![\varphi \wedge \varphi']\!] = [\![\varphi]\!] \cap [\![\varphi']\!]$ for all $\varphi, \varphi' \in \Phi$, and an algorithm computing some function $\neg : \Phi \to \Phi$ such that $[\![\neg\varphi]\!] = \Sigma \setminus [\![\varphi]\!]$. (**2**) There exists an algorithm deciding membership $a \in [\![\varphi]\!]$ given a guard $\varphi \in \Phi$ and a label $a \in \Sigma$.

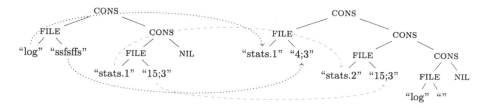

Fig. 1. Log cleanup and statistics: input tree on the left, output on the right.

Definition 5. A **symbolic** DTOP is a tuple $M = (\Sigma, \Delta, \Phi, \mathcal{F}, Q, q_{\text{ini}}, rhs)$ such that $(\Phi, \mathcal{F}, Q, q_{\text{ini}}, rhs)$ is a quasi DTOP with a finite set of rules, $\mathcal{F}$ is an alphabet of descriptors for partial functions from the input alphabet Σ to the output alphabet Δ, with $ar(\llbracket f \rrbracket(a)) = ar(f)$ for every $a \in \text{dom}(f)$ and $f \in \mathcal{F}$. The same conditions on Σ, Φ and rhs apply as for symbolic DTTA above.

The elements of $f \in \mathcal{F}$ are called (descriptors for) data transformations. A symbolic DTOP M is a finite representation of a (potentially) infinite quasi DTOP $M' = (\Sigma, \Delta, Q, ax, rhs')$, such that for every rule $q\langle \varphi\langle x_1, \ldots, x_n \rangle \rangle \rightarrow rhs(q, \varphi)$ of M, and for every $a \in \llbracket \varphi \rrbracket$, there is a rule $q\langle a\langle x_1, \ldots, x_n \rangle \rangle \rightarrow rhs(q, \varphi)[f \leftarrow \llbracket f \rrbracket(a) \mid f \in \mathcal{F}]$ in M'. The semantics of M is defined as that of M': for all $q \in Q$, $\llbracket q \rrbracket_M = \llbracket q \rrbracket_{M'}$.

Definition 6. A symbolic DTOP is **effective** (which is assumed in the remainder) if the underlying DTTA is, and (**1**) There is an algorithm that computes the value of the data transformation $\llbracket f \rrbracket(a)$ for a given $f \in \mathcal{F}$ and $a \in \Sigma$ and returns $\bot$ if it is not defined. (**2**) There is an algorithm that decides whether the image of a data transformation $\llbracket f \rrbracket(\Sigma)$ is empty for a given $f \in \mathcal{F}$.

In **symbolic** DTOP$^{\mathbf{R}}$, a symbolic BUTA on the transducer's input signature Σ first annotates the tree with its states P, and then the symbolic DTOP transforms the annotated tree on $\Sigma \times P$.

Consider the logs of an application on a Unix-flavoured system. The log is a text file named "log", here containing "s" for every successful login, and "f" for every failure. Every week, the old log is discarded and replaced by statistics: the number of successful and failed logins (Fig. 1). For this, we denote by f_{stats} the function counting the numbers n, m of occurrences of "s" and "f" in a string (here a log's contents), and outputting the string n";"m. A fresh "log" is then created. The older log's statistics are named "stats.1", "stats.2", etc. so that higher numbers indicate older stats. To model this with a symbolic DTOP$^{\mathbf{R}}$, we represent the contents of the logs folder by a list (CONS and NIL being the usual constructors) of files, each file being a tree of the form FILE$\langle$"filename", "contents"$\rangle$. The input and output alphabets are thus strings, along with FILE, CONS and NIL. The guards will be a small subset of regular expressions on strings plus descriptors matching CONS and NIL. The lookahead's purpose is to check whether the filename matches a stat file or not (which cannot be done in a top-down transducer's rule, in contrast to the extended rules (3) and (4)) and to annotate the FILE nodes with its findings. Guards are regular expressions. The descriptor matching a specific string is the string itself; $*$ matches everything; "stats."("0".."9")$^+$ is the descriptor matching stats filenames. The lookahead (LA) rules are

$$\text{"stats."("0".."9")}^+\langle\rangle \to p_{\text{stats}} \qquad \text{FILE}\langle p_{\text{stats}}, p\rangle \to p_{\text{stats file}}$$
$$\text{"log"}\langle\rangle \to p \qquad \text{FILE}\langle p, p\rangle \to p$$
$$(\text{"s"} \mid \text{"f"})^*\langle\rangle \to p \qquad \text{CONS}\langle\_, p\rangle \to p \qquad \text{NIL}\langle\rangle \to p$$

The label functions $\mathcal{F}$ are taken as, for instance, the class of rational functions, which we can implement with word transducers with lookahead [6], satisfying all requisite properties. We represent a constant function by the string it produces, the identity by f_{id}, and $f_{\text{increment}}$ for the function taking strings of the form "stats." k, where k is the decimal representation of an integer, and yielding "stats." $(k+1)$. We start in state q:

$$q\langle\text{NIL}:p\rangle \to \text{CONS}\langle \text{FILE}\langle\text{"log"},\text{""}\rangle, \text{NIL}\rangle \tag{1'}$$

$$q\langle\text{CONS}:p\langle x_1,x_2\rangle\rangle \to \text{CONS}\langle q_{\text{name}}\langle x_1\rangle, q\langle x_2\rangle\rangle \tag{2'}$$

$$q_{\text{name}}\langle\text{FILE}:p\langle x_1,x_2\rangle\rangle \to \text{FILE}\langle q_{\text{log}}\langle x_1\rangle, q_{\text{stats}}\langle x_2\rangle\rangle \tag{3a}$$

$$q_{\text{name}}\langle\text{FILE}:p_{\text{stats file}}\langle x_1,x_2\rangle\rangle \to \text{FILE}\langle q_{\text{incr}}\langle x_1\rangle, q_{\text{id}}\langle x_2\rangle\rangle \tag{4a}$$

$$q_{\text{incr}}\langle\text{"stats."("0".."9")}^+:p_{\text{stats}}\rangle \to f_{\text{increment}} \tag{3b}$$

$$q_{\text{id}}\langle*:p\rangle \to f_{\text{id}} \quad q_{\text{stats}}\langle*:p\rangle \to f_{\text{stats}} \quad q_{\text{log}}\langle\text{"log"}:p\rangle \to \text{"stats.1"} . \tag{4b}$$

4   Domain and Composition

To study problems like computation or equivalence on symbolic DTop, it is worth considering those problems as extensions of their counterparts for DTop. Indeed, most difficulties that previous papers [15,19,20] encountered are already relevant for the composition, normalization, or equivalence problems in the finite-labelled case [11,16]. Most of those difficulties come from dealing with the domain of a transducer's transformation. Since these problems have been solved in DTop and proofs in symbolic DTop are essentially identical, we shall only present the results, a reference for the proofs in DTop, and the additional conditions required of Φ and $\mathcal{F}$ for them to carry over to the symbolic case.

The first important results concern automata and their expressive power. Symbolic DTTA which, as said before, we always assume to be effective, and have the classical properties of DTTA (e.g. in [5]).

Lemma 7. (1) *The class of languages described by symbolic DTTA is closed under Boolean set operations.* (2) *If equivalence is decidable on* Φ, *then equivalence is decidable on symbolic DTTA.*

The second important result concerns the domains of symbolic DTop. Several or no states can explore a particular subtree. We use previous results on DTop [11] to see that this domain can be recognized by an automaton.

Lemma 8. *Let M be a symbolic DTop. Then we can build a symbolic DTTA A such that $[\![A]\!] = \mathrm{dom}([\![M]\!])$.*

As pointed out in [15], and contrary to a claim in [20], symbolic DTop are not closed under composition. To find a class closed under composition, a solution presented in the DTop case [11] is to consider transducers *with domain inspection*: a **symbolic DTop with inspection** is a pair $N = (M, A)$ of a symbolic DTop M and a symbolic DTTA A. Its semantic is $[\![N]\!] = [\![M]\!]_{|[\![A]\!]}$, the function of M restricted to the language of A. We know that DTop with inspection are closed under composition [11]. This result extends to symbolic DTop if the set of functions of $\mathcal{F}$ is itself closed under composition, and the images of guards through functions of $\mathcal{F}$ form a suitable set of guards (i.e. they satisfy the requirements for effectiveness).

Lemma 9. *Let $\mathcal{F}$ be closed under composition, and N, N' be two effective symbolic DTop with inspection using functions of $\mathcal{F}$. Then we can build a symbolic DTop with inspection N'' such that $[\![N'']\!] = [\![N]\!] \circ [\![N']\!]$.*

The main intuition behind the generalisation of the classical results [1] is presented in several papers [15,19,20]; roughly, a rule in N'' is the image of the right-hand side of a rule of N' by a state of N. To obtain closure by composition for symbolic DTop – or indeed for DTop, as the problem is fundamentally unchanged by the alphabets – necessitates the use of either very strong restrictions, as in [15], or the use of domain inspection, which we prefer here.

5   Deciding Equivalence

In this section, we show that, given a few basic properties on label transformations (mostly that equivalence is decidable for label transformations) the equivalence problem for symbolic DTop is decidable, regardless of linearity, by reducing that problem to equivalence for DTop, which is known to be decidable. This method does not involve any external SMT solver, unlike [19,20].

There are two basic observations behind this reduction. First, although Σ, Φ and $\mathcal{F}$ may well be infinite, only a finite number of predicates and transformations are actually used in a symbolic DTop (or indeed, any pair of symbolic DTop). These finite subsets of Φ and $\mathcal{F}$ will serve as finite input and output signatures in our reduction. Second, for two symbolic DTop to be equivalent, they generally need to use the same input label to produce some output label: two symbolic DTop using the same function on different input nodes will, in general, produce a different output (see Fig. 2).

To be more specific, we introduce a notion of *origin* similar to the *syntactic alignments* of [2] and the *origins* of [3], and a weakened version of the *origin equivalence* in [3]. We assimilate, in a given tree, the nodes to their addresses according to the Dewey notation. For instance, in Fig. 1, the node 1 is labelled by FILE, and the node 12 in the input is labelled by "ssfsffs". Let us consider a symbolic DTop M, as well as a tree t in its domain. For each node π of the tree $[\![M]\!](t)$, the node at π is created by examining a node μ of t using a rule of M. This input node is unique (see for example Proposition 52 of [2]) and we call it the **origin node** of π for $[\![M]\!](t)$. In the symbolic case, we can also track the function $f|_\varphi$ used to transform the label of μ into the label of π through a rule of guard φ, and call it the **origin function** of π. In Fig. 1, the origin of 12 in the output is 12 in the input, and the origin function is f_{stats}, via rule (4b).

Definition 10. Two equivalent symbolic DTop M, N are *weak-origin-equivalent* iff for all $t \in \text{dom}[\![M]\!]$, for any node π of $[\![M]\!](t)$, the origin nodes of π for M and N are identical, or its origin functions for M and N are constant of same value.

Lemma 11. *If two symbolic* DTop *are equivalent, they are weak-origin-equivalent.*

Intuitively, if M and N are not weak-origin equivalent then, for some tree t, an output node π comes from two different nodes μ and λ in the input using non-constant functions. By changing the label of μ without changing the label of λ, we change $[\![M]\!](t)$ and not $[\![N]\!](t)$, thus proving that M and N are not equivalent.

We now present the reduction from equivalence of symbolic DTop to that of DTop. Let $M = (\Phi, \Sigma, \mathcal{F}, \Delta, P, p_{\text{ini}}, R)$ and $N = (\Phi, \Sigma, \mathcal{F}, \Delta, Q, q_{\text{ini}}, S)$ be two symbolic DTop. We build their DTop **representations**, the DTop $\underline{M}$ and $\underline{N}$. Strictly speaking we should write $\underline{M}^{M,N}$ and $\underline{N}^{M,N}$, as the construction is specific to the pair of transducers under consideration, and the same applies to the representation of each component of the transducers – $\underline{\Phi}$, $\underline{\Sigma}$, etc. – which we define below. In this section we assimilate descriptors φ, f and their semantics $[\![\varphi]\!], [\![f]\!]$ to lighten the notations.

We make the following additional **equivalence-testing assumptions**: for all $\varphi, \psi \in \Phi$, it is decidable whether $\varphi = \psi$. For all $f, g \in \mathcal{F}$ and all $\varphi \in \Phi$, it is decidable whether there exists some $c \in \Delta$ such that $f(\varphi) = \{c\}$, and this c is computable; and it is decidable whether $f|_\varphi = g|_\varphi$.

The finite information relevant to the behaviour of symbolic DTop is which guards are satisfied. Thus we let $\Pi = \{\text{gd}(r) \mid r \in R \cup S\} \subseteq \Phi$ be the subset of guards actually used by either of the two transducers. The **finite alphabet $\underline{\Sigma}$ representing Σ** is defined as $\underline{\Sigma} = 2^\Pi$. The **representation of $a \in \Sigma$** is

$$\underline{a} = \{\pi \in \Pi \mid a \in \pi\} \in \underline{\Sigma} . \tag{6}$$

The **representation of a guard $\varphi \in \Pi$** is $\underline{\varphi} = \{\Pi' \subseteq \Pi \mid \varphi \in \Pi'\} \subseteq \underline{\Sigma}$. The **representation of an input tree $t \in \mathcal{T}(\underline{\Sigma})$** is defined inductively as

$$\underline{a(t_1, \ldots, t_n)} = \{(\underline{a}, b)(u_1, \ldots, u_n) \mid b \in \mathbb{B}, u_i \in \underline{t_i}, \forall i\} \in \mathcal{T}(\underline{\Sigma} \times \mathbb{B}) , \tag{7}$$

where the addition of the bit b, called **obit** (origin bit), will be used to store just enough information about origins to ensure weak origin equivalence between M and N. Accordingly, the **representation**

of a label transformation f restricted to φ, with obit b is defined as

$$\underline{f|_{\varphi,b}} = \begin{cases} c & \text{if } f(\varphi) = \{c\}, \text{ and} \\ (f|_\varphi, b) & \text{otherwise.} \end{cases} \tag{8}$$

The **representation of a rule** $r \in \mathbf{R} \cup \mathbf{S}$, of the form $r = q\langle\varphi\langle x_1, \ldots, x_n\rangle\rangle \to t$, is given by the set $\underline{r}$ of all classical rules

$$q\langle(\rho, b)\langle x_1, \ldots, x_n\rangle\rangle \to t\left[f \leftarrow (f|_\varphi, b) \mid f \in \mathcal{F}\right], \tag{9}$$

for all $b \in \mathbb{B}$, $\rho \in \varphi$. Letting $\underline{R} = \bigcup_{r \in R} \underline{r}$ and $\underline{S} = \bigcup_{s \in S} \underline{s}$, we finally have $\underline{M} = (\underline{\Sigma}, \underline{\Delta}, P, p_{\mathrm{ini}}, \underline{R})$ and $\underline{N} = (\underline{\Sigma}, \underline{\Delta}, Q, q_{\mathrm{ini}}, \underline{S})$.

This representation is built so that the following holds: let t be a tree of $\mathrm{dom}(\llbracket M \rrbracket)$, a node π in $\llbracket M \rrbracket(t)$, its origin node μ and its origin function $f|_\varphi$. Then for any tree $u \in \underline{t}$, the node π in $\llbracket \underline{M} \rrbracket(u)$ is c if $f(\varphi) = \{c\}$, $(f|_\varphi, b)$ otherwise, with b the obit under μ in u. This leads to the following result.

Theorem 12. *Let M, N be two symbolic* DTOP, *as above. Then* $\llbracket M \rrbracket = \llbracket N \rrbracket$ *if and only if* $\llbracket \underline{M} \rrbracket = \llbracket \underline{N} \rrbracket$.

Proof. First, we consider domain equality: the domain of $\underline{M}$ is the set of all the representations of trees of $\mathrm{dom}(\llbracket M \rrbracket)$. Hence $\underline{M}$ and $\underline{N}$ are of same domain if and only if M and N are of same domain. Let us now assume that the domains are the same. Suppose M and N are not equivalent; let t be an input tree such that $u = \llbracket M \rrbracket(t) \neq \llbracket N \rrbracket(t) = v$. We consider an address π that exists both in u and v but where the label at π differs in u and v. In M the origin node of π is μ; the origin function is $f|_\varphi$. In N the origin node of π is λ; the origin function is $g|_\psi$. Since the label at π differs in u and v, there must be a difference of origin node or functions.

If $f|_\varphi \neq g|_\psi$, they can't be constants of same value, otherwise the label at π would be the same in u and v. This means that their representations will differ in $\underline{M}$ and $\underline{N}$. Hence for any $t' \in \underline{t}$, the label at π would differ in $\llbracket \underline{M} \rrbracket(t')$ and $\llbracket \underline{N} \rrbracket(t')$. Hence $\llbracket \underline{M} \rrbracket \neq \llbracket \underline{N} \rrbracket$.

If $f|_\varphi = g|_\psi$, then $\mu \neq \lambda$, and $f|_\varphi$ can't be a constant, otherwise the label at π would be the same in u and v. We pick t' a representation of t where the obit under μ is 1, and the obit under λ is 0. The label at π in $\llbracket \underline{M} \rrbracket(t')$ would be $(f|_\varphi, 1)$, but the label at π in $\llbracket \underline{N} \rrbracket(t')$ would be $(f|_\varphi, 0)$. Hence $\llbracket \underline{M} \rrbracket \neq \llbracket \underline{N} \rrbracket$.

Conversely, suppose $\underline{M}$ and $\underline{N}$ are not equivalent; let t' be an input tree such that $u' = \llbracket \underline{M} \rrbracket(t') \neq \llbracket \underline{N} \rrbracket(t') = v'$. We consider an address π' that exists both in u' and v' but where the label at π' differs in u' and v'. In $\underline{M}$ the origin node of π' is μ', and in $\underline{N}$ the origin node of π' is λ'. Thus, μ' and λ' are also the origin node of π' in M and N for all t such that $t' \in \underline{t}$. If $\mu' \neq \lambda'$, we remark that the label of the nodes at π' cannot be the same constant c. This combined with the disparity of node origins means that M and N are not weak-origin-equivalent, and thus not equivalent. If $\mu' = \lambda'$, then the label at π' in u' and v' are representations of different functions (or constants) $f|_\varphi$ and $g|_\varphi$. There is at least one value $a \in \varphi$ such that $f(a) \neq g(a)$. We pick a tree t such that $t' \in \underline{t}$ and the label at μ' in t

Fig. 2. Using obits to deduce origins. We represent the obits with colours: ○ are nodes of obit 0, ● are nodes of obit 1. We apply two transformations τ_1 and τ_2 on an input tree (here in the middle): τ_1 replaces its right leaf by a copy of the left one, while τ_2 replaces its left leaf by a copy of the right one. Without the obits, these two transformations would look identical. However, as seen above, the obits allow a distinction between τ_1 and τ_2 for some input tree.

is labeled a. The node π' in $[\![M]\!](t)$ is labeled $f(a)$, while the node π' in $[\![N]\!](t)$ is labeled $g(a)$. Hence $[\![M]\!] \neq [\![N]\!]$.   □

Corollary 13. *Under the equivalence-testing assumptions above, the equivalence problem for symbolic* DTOP *is reducible to the equivalence problem on* DTOP, *in* EXPTIME *in the worst case, plus, at worst, an exponential number of operations in* Φ *and* $\mathcal{F}$.

Proof. Given that the construction of DTOP representations in Theorem 12 is effective, we can build them and decide the equivalence of the representations. □

The exact complexity of this algorithm relies on the complexity of the various operations related to the equivalence-testing assumptions (computing intersections, deciding function equivalence) in these sets, but also on the precise number of intersections and negations we have to perform in Φ. To compute these representations, we build all guards φ such that for some label a, $\varphi = (\bigcap_{\pi \in \underline{a}} \pi) \setminus (\bigcup_{\pi' \notin \underline{a}} \pi')$, and decide function equivalence on these φ for the functions used in M and N. In the case where guards are all disjoint, the reduction to DTOP is actually polynomial. In practice, it can be expected that few intersections actually need to be computed. The representations $\underline{a}$ can also be made more parsimonious by taking into account only tests that actually apply during the run, which can be done as in the construction of Lemma 8.

In any case, we can express the number of states and rules in $\underline{M}$ and $\underline{N}$ independently of Φ and $\mathcal{F}$: the states are unchanged, and the number of rules increases, at worst, exponentially.

Lemma 14. *For* M, N *two symbolic* DTOP, *their* DTOP *representations* $\underline{M}$ *and* $\underline{N}$ *are* DTOP *with an exponential number of rules and the same number of states.*

Since the problem of DTOP equivalence is NEXPTIME [13], a naïve approach to calculating the complexity of symbolic DTOP equivalence would yield a 2NEXPTIME algorithm, plus an exponential number of operations in Φ and $\mathcal{F}$. However, upon finer analysis, the complexity of DTOP equivalence is tied to the height of a counter-example between two non-equivalent transducers. This height is, in the worst case scenario, exponential in the number of states in the studied DTOP. Since representations do not create new states, the height of the counter-examples is unchanged, and the exponentials do not compound.

Theorem 15. *The equivalence problem for symbolic* DTOP *is in* NEXPTIME, *plus, at worst, an exponential number of operations in Φ and $\mathcal{F}$.*

5.1 Extension to Symbolic DTOP^R

We want to extend our results from symbolic DTOP to the wider class of symbolic DTOP^R. This class is relevant for several reasons. The first is that it is more expressive than the class of symbolic DTOP with inspection, and subsumes other relevant classes, such as single-valued symbolic DTOP [19,20]. It also possesses interesting properties. Notably, just as it was the case for DTOP^R [8], the class of symbolic DTOP^R is closed under composition.

We want to study the equivalence problem of symbolic DTOP^R. For DTOP, the addition of a regular lookahead does not prevent the equivalence problem from being decidable, as it is polynomially reduced to equivalence on plain DTOP (see [17]). This result can be carried over to the symbolic case, with the same method: annotating with both lookaheads.

Lemma 16. *One can polynomially reduce the equivalence problem of symbolic* DTOP^R *to the equivalence problem of symbolic* DTOP *with inspection.*

This reduction in polynomial time can be combined with our previous results (Corollary 13 and Theorem 15) to provide the following complexity result:

Theorem 17. *Under the assumptions of Corollary 13, the equivalence problem for symbolic* DTOP^R *is decidable in* NEXPTIME, *plus, at worst, an exponential number of operations in Φ and $\mathcal{F}$.*

This result is quite useful, as several DTOP classes are fragments of the class of DTOP^R, and their decidability and complexity results can thus be transposed to the symbolic case. Notably, DBUP and nondeterministic functional TOP can be expressed as DTOP^R.

Corollary 18. *Under the assumptions of Theorem 13, the equivalence problem for deterministic symbolic bottom-up tree transducers and nondeterministic symbolic functional top-down tree transducers is decidable.*

6 Conclusion

The algorithm presented here provides a novel approach to deciding equivalence for symbolic DTOP, and supports non-linear symbolic DTOP, by reduction to DTOP equivalence. Note that decidability of equivalence for DTOP^R [17] works in a comparable way: rather than finding a normal form, the two regular lookaheads are "harmonized" into one, then the problem is reduced to DTOP equivalence. The methods presented in this paper also apply to symbolic DTOP^R without a critical jump in complexity.

Our method does not involve the computation of a normal form, which is a rather classical technique to decide transducer equivalence [4,11], with applications to learning. It is interesting to see if normal forms could be defined for symbolic DTop. This looks challenging, however, as it seems more general than finding normal forms for DTop^R, which remains an open problem.

A first possible extension of our model would be to allow the lookahead to have registers, i.e. to memorize some data from the bottom of the tree to annotate the upper part of the tree with it. Under reasonable restrictions, it is likely that we might adapt our methods to reduce the equivalence problem for these objects to the same problem on DTop^R, thus providing a decidability result.

Furthermore, we would like to find out whether this kind of reduction can be applied to more general classes of transducers such as macro tree transducers (with linear size increase), for which equivalence is decidable [10]. If so, then the decidability results can fairly easily be lifted to symbolic generalisations of the class, at the cost of a few exponential blowups in complexity.

As a final mention, the inversion problem is interesting for symbolic transformations on words and trees, and is relevant to the applications we consider.

References

1. Baker, B.S.: Composition of top-down and bottom-up tree transductions. Inf. Control **41**(2), 186–213 (1979)
2. Boiret, A.: Normalization and learning of transducers on trees and words. Ph.D. thesis, Lille University, France (2016)
3. Bojańczyk, M.: Transducers with origin information. In: Esparza, J., Fraigniaud, P., Husfeldt, T., Koutsoupias, E. (eds.) ICALP 2014. LNCS, vol. 8573, pp. 26–37. Springer, Heidelberg (2014). doi:10.1007/978-3-662-43951-7_3
4. Choffrut, C.: Minimizing subsequential transducers: a survey. Theor. Comput. Sci. **292**(1), 131–143 (2003)
5. Comon, H., Dauchet, M., Gilleron, R., Löding, C., Jacquemard, F., Lugiez, D., Tison, S., Tommasi, M.: Tree automata techniques and applications (2007). http://www.grappa.univ-lille3.fr/tata. Accessed 12th October 2007
6. Elgot, C.C., Mezei, J.E.: On relations defined by generalized finite automata. IBM J. Res. Dev. **9**(1), 47–68 (1965)
7. Engelfriet, J.: Bottom-up and top-down tree transformations - a comparison. Math. Syst. Theor. **9**(3), 198–231 (1975)
8. Engelfriet, J.: Top-down tree transducers with regular look-ahead. Math. Syst. Theor. **10**, 289–303 (1977)
9. Engelfriet, J.: Some Open Questions and Recent Results on Tree Transducers and Tree Languages, pp. 241–286. Academic Press, Cambridge (1980)
10. Engelfriet, J., Maneth, S.: Macro tree translations of linear size increase are MSO definable. SIAM J. Comput. **32**(4), 950–1006 (2003)
11. Engelfriet, J., Maneth, S., Seidl, H.: Deciding equivalence of top-down XML transformations in polynomial time. J. Comput. Syst. Sci. **75**(5), 271–286 (2009)
12. Engelfriet, J., Vogler, H.: Macro tree transducers. J. Comput. Syst. Sci. **31**(1), 71–146 (1985)
13. Ésik, Z.: On functional tree transducers. In: FCT, pp. 121–127 (1979)

14. Ésik, Z.: Decidability results concerning tree transducers II. Acta Cybern. **6**(3), 303–314 (1983)
15. Fülöp, Z., Vogler, H.: Forward and backward application of symbolic tree transducers. Acta Inf. **51**(5), 297–325 (2014)
16. Lemay, A., Maneth, S., Niehren, J.: A learning algorithm for top-down XML transformations. In: PODS, pp. 285–296. ACM (2010)
17. Maneth, S.: Equivalence problems for tree transducers: a brief survey. In: Ésik, Z., Fülöp, Z. (eds.) AFL. EPTCS, vol. 151, pp. 74–93 (2014)
18. Seidl, H., Maneth, S., Kemper, G.: Equivalence of deterministic top-down tree-to-string transducers is decidable. In: FOCS, pp. 943–962 (2015)
19. Veanes, M., Bjørner, N.: Foundations of finite symbolic tree transducers. EATCS **105**, 141–173 (2011)
20. Veanes, M., Bjørner, N.: Symbolic tree transducers. In: Clarke, E., Virbitskaite, I., Voronkov, A. (eds.) PSI 2011. LNCS, vol. 7162, pp. 377–393. Springer, Heidelberg (2012). doi:10.1007/978-3-642-29709-0_32

DFAs and PFAs with Long Shortest Synchronizing Word Length

Michiel de Bondt[1], Henk Don[2], and Hans Zantema[1,3(✉)]

[1] Department of Computer Science, Radboud University Nijmegen,
Nijmegen, The Netherlands
m.debondt@math.ru.nl
[2] Department of Mathematics, Free University, Amsterdam, The Netherlands
h.don@vu.nl
[3] Department of Computer Science, TU Eindhoven, Eindhoven, The Netherlands
h.zantema@tue.nl

Abstract. It was conjectured by Černý in 1964, that a synchronizing DFA on n states always has a synchronizing word of length at most $(n-1)^2$, and he gave a sequence of DFAs for which this bound is reached. Until now a full analysis of all DFAs reaching this bound was only given for $n \leq 4$, and with bounds on the number of symbols for $n \leq 10$. Here we give the full analysis for $n \leq 6$, without bounds on the number of symbols.

For PFAs on $n \leq 6$ states we do a similar analysis as for DFAs and find the maximal shortest synchronizing word lengths, exceeding $(n-1)^2$ for $n = 4, 5, 6$. For arbitrary n we use rewrite systems to construct a PFA on three symbols with exponential shortest synchronizing word length, giving significantly better bounds than earlier exponential constructions. We give a transformation of this PFA to a PFA on two symbols keeping exponential shortest synchronizing word length, yielding a better bound than applying a similar known transformation.

1 Introduction and Preliminaries

A *deterministic finite automaton (DFA)* over a finite alphabet Σ is called *synchronizing*, if it admits a *synchronizing word*. A word $w \in \Sigma^*$ is called *synchronizing* (or directed, or reset), if, starting in any state q, after reading w, one always ends in one particular state q_s. So reading w acts as a reset button: no matter in which state the system is, it always moves to the particular state q_s. Now Černý's conjecture [2] states:

> Every synchronizing DFA on n states admits a synchronizing word of length $\leq (n-1)^2$.

Surprisingly, despite extensive effort, this conjecture is still open, and even the best known upper bounds are still cubic in n. In 1983 Pin [8] established the bound $\frac{1}{6}(n^3 - n)$. Only very recently a slight improvement was claimed by Szykuła [11]. For a survey on synchronizing automata and Černý's conjecture, we refer to [13].

© Springer International Publishing AG 2017
É. Charlier et al. (Eds.): DLT 2017, LNCS 10396, pp. 122–133, 2017.
DOI: 10.1007/978-3-319-62809-7_8

Formally, a *deterministic finite automaton (DFA)* over a finite alphabet Σ consists of a finite set Q of states and a map $\delta : Q \times \Sigma \to Q$.[1] For $w \in \Sigma^*$ and $q \in Q$, we define qw inductively by $q\lambda = q$ and $qwa = \delta(qw, a)$ for $a \in \Sigma$, where λ is the empty word. So qw is the state where one ends, when starting in q and reading the symbols in w consecutively, and qa is a short hand notation for $\delta(q, a)$. A word $w \in \Sigma^*$ is called *synchronizing*, if a state $q_s \in Q$ exists such that $qw = q_s$ for all $q \in Q$.

In [2], Černý already gave DFAs for which the bound of the conjecture is attained: for $n \geq 2$ the DFA C_n is defined to consist of n states $1, 2, \ldots, n$, and two symbols a, b, acting by $qa = q + 1$ for $q = 1, \ldots, n - 1$, $\delta(n, a) = 1$, and $qb = q$ for $q = 2, \ldots, n$, $1b = 2$.

For $n = 4$, this is depicted on the right. For C_n, the string $w = b(a^{n-1}b)^{n-2}$ of length $|w| = (n-1)^2$ satisfies $qw = 2$ for all $q \in Q$, so w is synchronizing. No shorter synchronizing word exists for C_n, as is shown in [2], showing that the bound in Černý's conjecture is sharp.

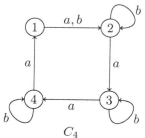

C_4

A DFA on n states is *critical*, if its shortest synchronizing word has length $(n-1)^2$. One goal of this paper is to investigate all critical DFAs up to some size. To exclude infinitely many trivial extensions, we only consider *basic* DFAs: no two distinct symbols act in the same way in the automaton, and no symbol acts as the identity. Obviously, adding the identity or copies of existing symbols has no influence on synchronization.

An extensive investigation was already done by Trahtman in [12]: by computer support and clever algorithms, all critical DFAs on n states and k symbols were investigated for $3 \leq n \leq 7$ and $k \leq 4$, and for $n = 8, 9, 10$ and $k = 2$. Here, a minimality requirement was added: examples were excluded if criticality may be kept after removing one symbol. Then up to isomorphism there are exactly 8 of them, apart from the basic Černý examples: 3 with 3 states, 3 with 4, one with 5 and one with 6. In [4], the minimality requirement and restrictions on alphabet size were dropped and several more examples were found that Trahtman originally expected not to exist. All these are extensions of known examples: in total there are exactly 15 basic critical DFAs for $n = 3$ and exactly 12 basic critical DFAs for $n = 4$. In this paper, we show that for $n = 5, 6$, no more critical DFAs exist than the four known ones, without any restriction on the number of symbols.

A generalization of a DFA is a *Partial Finite Automaton* (PFA); the only difference is that now the transition function δ is allowed to be partial. In a PFA, qw may be undefined, in fact it is only defined if every step is defined. A word $w \in \Sigma^*$ is called *carefully synchronizing* for a PFA, if a state $q_s \in Q$ exists such that qw is defined and $qw = q_s$ for all $q \in Q$. Stated in words: starting in any state q and reading w, every step is defined and one always ends in

---

[1] For synchronization the initial state and the set of final states in the standard definition may be ignored.

state q_s. As being a generalization of DFAs, the shortest carefully synchronizing word may be longer. For $n = 4, 5, 6$ we show that this is indeed the case by finding the maximal shortest carefully synchronizing word length to be 10, 21 and 37, respectively. The maximal length grows exponentially in n, as was already observed by Rystsov [10]. Martyugin [7] established the lower bound $\Omega(3^{n/3})$ with a construction in which the number of symbols is linear in n. In a recent paper, the upper bound $O((3 + \varepsilon)^{\frac{n}{3}})$ was proved [5].

Until recently it was an open question if exponential lower bounds can be achieved with a constant alphabet size. We answer this question by giving a construction of a PFA on n states and three symbols with exponential shortest synchronizing word length. The key idea is that synchronization is forced to mimic exponentially many string rewrite steps, similar to binary counting. Our three-symbol PFA can be transformed to a two-symbol PFA by a standard construction for which we develop a substantial improvement. Independent of our work, recently in [14] it was shown that exponential bounds exist for every constant alphabet size and for two symbols the bound $\Omega(2^{n/35})$ was given. Our basic construction strongly improves this and gives length $\Omega(\phi^{n/3})$ for the three-symbol PFA and length $\Omega(\phi^{n/5})$ for the two-symbol PFA, where $\phi = \frac{1+\sqrt{5}}{2}$. Some optimizations yield further improvements.

The basic tool to analyze (careful) synchronization is the *power automaton*. For any DFA or PFA (Q, Σ, δ), its power automaton is the DFA $(2^Q, \Sigma, \delta')$ where $\delta' : 2^Q \times \Sigma \to 2^Q$ is defined by $\delta'(V, a) = \{q \in Q \mid \exists p \in V : \delta(p, a) = q\}$, if $\delta(p, a)$ is defined for all $p \in V$, otherwise $\delta'(V, a) = \emptyset$. For any $V \subseteq Q, w \in \Sigma^*$, we define Vw as above, using δ' instead of δ. From this definition, one easily proves that $Vw = \{qw \mid q \in V\}$ if qw is defined for all $q \in V$, otherwise $Vw = \emptyset$, for any $V \subseteq Q, w \in \Sigma^*$. A set of the shape $\{q\}$ for $q \in Q$ is called a *singleton*. So a word w is (carefully) synchronizing, if and only if Qw is a singleton. Hence a DFA (PFA) is (carefully) synchronizing, if and only if its power automaton admits a path from Q to a singleton, and the shortest length of such a path corresponds to the shortest length of a (carefully) synchronizing word.

This paper is organized as follows. In Sect. 2 we describe our exhaustive analysis of DFAs on at most 6 states. In Sect. 3 we give our results for PFAs on at most 6 states. In Sect. 4 we present our construction of PFAs on three symbols with exponential shortest carefully synchronizing word length. In Sect. 5 we improve the transformation used by Martyugin [7] and Vorel [14] to reduce to alphabet size two. Section 6 discusses optimizations. We conclude in Sect. 7.

2   Critical DFAs on at Most 6 States

A natural question when studying Černý's conjecture is: what can be said about automata in which the bound of the conjecture is actually attained, the so-called critical automata? Throughout this section we restrict ourselves to basic DFAs. As has already been noted by several authors [4, 12, 13], critical DFAs are rare. There is only one construction known which gives a critical DFA for each n, namely the well-known sequence C_n, discovered by and named after Černý [2].

Apart from this sequence, all known critical DFAs have at most 6 states. In [4], all critical DFAs on less than 5 states were identified, without restriction on the size of the alphabet. For $n = 5$ and 6 it was still an open question if there exist critical (or even supercritical) DFAs, other than those already discovered by Černý, Roman [9] and Kari [6]. In this paper we verify that this is not the case, so for $n = 5$ only two critical DFAs exist (Černý, Roman) and also for $n = 6$ only two exist (Černý, Kari). In fact our results also prove the following theorem (previously only known for $n \leq 5$, see [3]):

Theorem 1. *Every synchronizing DFA on $n \leq 6$ states admits a synchronizing word of length at most $(n - 1)^2$.*

As the number of DFAs on n states grows like 2^{n^n}, an exhaustive search is a non-trivial affair, even for small values of n. The problem is that the alphabet size in a basic DFA can be as large as $n^n - 1$. Up to now for $n = 5, 6$ only DFAs with at most four symbols were checked by Trahtman [12]. Here we give describe our algorithm to investigate all DFAs on 5 and 6 states, without restriction on the alphabet size.

Before explaining the algorithm, we introduce some terminology. A DFA $\mathcal{B}$ obtained by adding some symbols to a DFA $\mathcal{A}$ will be called an *extension* of $\mathcal{A}$. If $\mathcal{A} = (Q, \Sigma, \delta)$, then $S \subseteq Q$ will be called *reachable* if there exists a word $w \in \Sigma^*$ such that $Qw = S$. We say that S is *reducible* if there exists a word w such that $|Sw| < |S|$, and we call w a *reduction word* for S. Our algorithm is mainly based on the following observation:

Property 2. If a DFA $\mathcal{A}$ is synchronizing, and $\mathcal{B}$ is an extension of $\mathcal{A}$, then $\mathcal{B}$ is synchronizing as well and its shortest synchronizing word is at most as long as the shortest synchronzing word for $\mathcal{A}$.

The algorithm roughly runs as follows. We search for (super)critical DFAs on n states, so a DFA is discarded if it synchronizes faster, or if it does not synchronize at all. For a given DFA $\mathcal{A} = (Q, \Sigma, \delta)$ which is not yet discarded or investigated, the algorithm does the following:

1. If $\mathcal{A}$ is synchronizing and (super)critical, we have identified an example we are searching for.
2. If $\mathcal{A}$ is synchronizing and subcritical, it is discarded, together with all its possible extensions (justified by Property 2).
3. If $\mathcal{A}$ is not synchronizing, then find an upper bound L for how fast any synchronizing extension of $\mathcal{A}$ will synchronize (see below). If $L < (n-1)^2$, then discard $\mathcal{A}$ and all its extensions. Otherwise, discard only $\mathcal{A}$ itself.

The upper bound L for how fast any synchronizing extension of $\mathcal{A}$ will synchronize, is found by analyzing distances in the directed graph of the power automaton of $\mathcal{A}$. For $S, T \subseteq Q$, the distance from S to T in this graph is equal to the length of the shortest word w for which $Sw = T$, if such a word exists. We compute L as follows:

1. Determine the size $|S|$ of a smallest reachable set. Let m be the minimal distance from Q to a set of size $|S|$.
2. For each $k \leq |S|$, partition the collection of irreducible sets of size k into strongly connected components. Let m_k be the number of components plus the sum of their diameters.
3. For each reducible set of size $k \leq |S|$, find the length of its shortest reduction word. Let l_k be the maximum of these lengths.
4. Now note that a synchronizing extension of $\mathcal{A}$ will have a synchronizing word of length at most

$$L = m + \sum_{k=2}^{|S|} (m_k + l_k).$$

The algorithm performs a depth-first search. So after investigating a DFA, first all its extensions (not yet considered) are investigated before moving on. Still, we can choose which extension to pick first. We would like to choose an extension that is likely to be discarded immediately together with all its extensions. Therefore, we apply the following heuristic: for each possible extension $\mathcal{B}$ by one symbol, we count how many pairs of states in $\mathcal{B}$ would be reducible. The extension for which this is maximal is investigated first. The motivation is that a DFA is synchronizing if and only if each pair is reducible [2].

Finally, we note that we have described a primitive version of the algorithm here. The algorithm which has actually been used also takes symmetries into account, making it almost $n!$ times faster. For the source code, we refer to [1].

3   PFAs on at Most 6 States

In the remainder of this paper, we study PFAs and shortest carefully synchronizing word lengths. In this section, we focus on PFAs on at most 6 states. In the next section, we construct PFAs with shortest carefully synchronizing words of exponential lengths for general n.

To find PFAs with small number of states and long shortest carefully synchronizing word, we exploit that Property 2 also holds for PFAs. However, for PFAs it is not true that reducibility of all pairs of states guarantees careful synchronization. Therefore, we apply a different search algorithm. Technical details are discussed in [1].

For $n \leq 6$, our algorithm has identified the maximal length of a shortest carefully synchronizing word in a PFA on n states. The results are:

n	2	3	4	5	6
maximal length	1	4	10	21	37

We observe that PFAs exist for $n = 4, 5, 6$ with shortest carefully synchronizing word lengths exceeding $(n-1)^2$. Note that for $n = 5, 6$ this even exceeds the Pin-Frankl bound $\frac{1}{6}(n^3 - n)$ for DFAs from [8]. Where for $n \geq 5$ no critical DFAs are known with more than three symbols, PFAs with long shortest carefully synchronizing word lengths tend to have more symbols: for $n = 4, 5, 6$ states the minimal numbers of symbols achieving the maximal shortest carefully synchronizing word lengths 10, 21 and 37 are 3, 6, 6, respectively. Below we give examples of PFAs on 4, 5 and 6 states reaching these lengths.

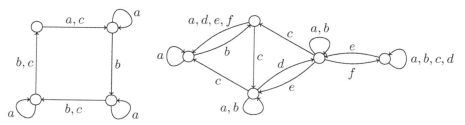

The left one has two synchronizing words of length 10: $abcabab(b+c)ca$. The right one has unique shortest synchronizing word $abcabdbebcabdbfbcdeca$ of length 21.

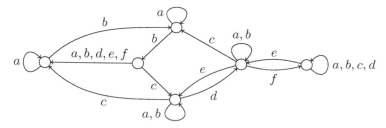

The shortest synchronizing word is $ab^2ab^2cb^2ab^2db^2eb^2cb^2ab^2db^2fb^2cdecb^2a$ for this PFA on 6 states. It is unique and has length 37.

4   Exponential Bounds for PFAs

In this section, we construct for any $k \geq 3$ a strongly connected PFA on $n = 3k$ states and three symbols, for which we show that it is carefully synchronizing, and the shortest carefully synchronizing word has length $\Omega(\phi^{n/3})$ for $\phi = \frac{1+\sqrt{5}}{2} = 1.618\cdots$. The set of states is $Q = \{A_i, B_i, C_i \mid i = 1, \ldots, k\}$. If a set $S \subseteq Q$ contains exactly one element of $\{A_i, B_i, C_i\}$ for every i, it can be represented by a string over $\{A, B, C\}$ of length k. The idea of our construction is that the PFA will mimic rewriting the string C^2A^{k-2} to the string $C^2A^{k-3}B$ with respect to the rewrite system R, which consists of the following three rules

$$BBA \to AAB, \; CBA \to CAB, \; CCA \to CCB.$$

The key argument is that this rewriting is possible, but requires an exponential number of steps. This is elaborated in the following lemma, in which we use $\to_R$ for rewriting with respect to R, that is, $u \to_R v$, if and only if $u = u_1 \ell u_2$ and $v = u_1 r u_2$, for strings u_1, u_2 and a rule $\ell \to r$ in R. Its transitive closure is denoted by $\to_R^+$. We write fib for the standard fibonacci function, defined by $\text{fib}(i) = i$ for $i = 0, 1$, and $\text{fib}(i) = \text{fib}(i-1) + \text{fib}(i-2)$ for $i > 1$. It is well-known that $\text{fib}(n) = \Theta(\phi^n)$.

Lemma 3. *For $k \geq 3$, we have $CCA^{k-2} \rightarrow_R^+ CCA^{k-3}B$. Furthermore, the smallest possible number of steps for rewriting CCA^{k-2} to a string ending in B, is exactly $\mathsf{fib}(k) - 1$.*

Proof. For the first claim we do induction on k. For $k = 3$, we have $CCA \rightarrow_R CCB$. For $k = 4$, we have $CCAA \rightarrow_R CCBA \rightarrow_R CCAB$. For $k > 4$, applying the induction hypothesis twice, we obtain

$$CCA^{k-2} \rightarrow_R^+ CCA^{k-4}BA \rightarrow_R^+ CCA^{k-5}BBA \rightarrow_R CCA^{k-3}B.$$

For the second claim, we define the *weight* $W(u)$ of a string $u = u_1 u_2 \cdots u_k$ over $\{A, B, C\}$ of length k by

$$W(u) = \sum_{i : u_i = B} (\mathsf{fib}(i) - 1).$$

So every B on position i in u contributes $\mathsf{fib}(i) - 1$ to the weight, and the other symbols have no weight.

Now we claim that $W(v) = W(u) + 1$ for all strings u, v with $u \rightarrow_R v$ and u, v only having C's in the first two positions. Since the Cs only occur at positions 1 and 2, by applying $CCA \rightarrow CCB$, the weight increases by $\mathsf{fib}(3) - 1 = 1$ by the creation of B on position 3, and by applying $CBA \rightarrow CAB$, it increases by $\mathsf{fib}(4) - 1 - (\mathsf{fib}(3) - 1) = 1$ since B on position 3 is replaced by B on position 4. By applying $BBA \rightarrow AAB$, the contributions to the weight $\mathsf{fib}(i) - 1$ and $\mathsf{fib}(i+1) - 1$ of the two Bs are replaced by $\mathsf{fib}(i+2) - 1$ of the new B, which is an increase by 1 according to the definition of fib.

So this weight increases by exactly 1 at every rewrite step, hence it requires exactly $\mathsf{fib}(k) - 1$ steps, to go from the initial string CCA^{k-2} of weight 0 to the weight $\mathsf{fib}(k) - 1$ of a B symbol on the last position k, if that is the only B, and more steps if there are more Bs.   □

Now we are ready to define the PFA on $Q = \{A_i, B_i, C_i \mid i = 1, \ldots, k\}$ and three symbols. The three symbols are a start symbol s, a rewrite symbol r and a cyclic shift symbol c. The transitions are defined as follows (writing $\perp$ for undefined):

$A_i s = B_i s = C_i s = C_i,$			for $i = 1, 2,$
$A_i s = B_i s = C_i s = A_i,$			for $i = 3, \ldots, k,$
$A_1 r = \perp,$	$B_1 r = A_1,$	$C_1 r = C_1,$	
$A_2 r = \perp,$	$B_2 r = A_2,$	$C_2 r = C_2,$	
$A_3 r = B_3,$	$B_3 r = \perp,$	$C_3 r = B_2,$	
$A_i r = A_i,$	$B_i r = B_i,$	$C_i r = C_i,$	for $i = 4, \ldots, k,$
$A_i c = A_{i+1},$	$B_i c = B_{i+1},$	$C_i c = C_{i+1},$	for $i = 1, \ldots, k-1,$
$A_k c = A_1,$	$B_k c = B_1,$	$C_k c = C_1.$	

A shortest carefully synchronizing word starts by s, since r is not defined on all states and c permutes all states. After s, the set of reached states is

$S(CCA^{k-2}) = \{C_1, C_2, A_3, \ldots, A_k\}$. Here, for a string $u = a_1 a_2 \cdots a_k$ of length k over $\{A, B, C\}$, we write $S(u)$ for the set of k states, containing A_i if and only if $a_i = A$, containing B_i if and only if $a_i = B$, and containing C_i if and only if $a_i = C$, for $i = 1, 2, \ldots, k$. Note that for $x \in \{A, B, C\}$ and $v \in \{A, B, C\}^{k-1}$, we have $S(vx)c = S(xv)$, so c performs a cyclic shift on strings of length k.

The next lemma states that the symbol r indeed mimicks rewriting: applied on sets of the shape $S(u)$, up to cyclic shift it acts as rewriting on u with respect to R defined above.

Lemma 4. *Let u be a string of the shape CCw, where $w \in \{A, B\}^{k-2}$. If $u \to_R v$ for a string v, then $S(u)c^i r c^{k-i} = S(v)$ for some $i < k$.*

Conversely, if u does not end in B and there exists an i such that r is defined on $S(u)c^i$, then $u \to_R v$ for a string v of the shape CCw, where $w \in \{A, B\}^{k-2}$.

Proof. First assume that $u \to_R v$. If $u = u_1 BBAu_2$ and $v = u_1 AABu_2$, then let $i = |u_2| + 3$, so

$$S(u)c^i r c^{k-i} = S(u_1 BBAu_2)c^i r c^{k-i} = S(BBAu_2 u_1)r c^{k-i}$$
$$= S(AABu_2 u_1)c^{k-i} = S(u_1 AABu_2) = S(v).$$

If $u = u_1 CBAu_2$ and $v = u_1 CABu_2$, then again let $i = |u_2| + 3$, so

$$S(u)c^i r c^{k-i} = S(u_1 CBAu_2)c^i r c^{k-i} = S(CBAu_2 u_1)r c^{k-i}$$
$$= S(CABu_2 u_1)c^{k-i} = S(u_1 CABu_2) = S(v).$$

Finally, if $u = u_1 CCAu_2$ and $v = u_1 CCBu_2$, then $u_1 = \epsilon$ and the result follows for $i = 0$.

Conversely, suppose that $S(u)c^i r$ is defined. Since $S(u)c^k = S(u)$, we may assume that $i < k$ and can write $u = u_1 u_2$, such that $|u_2| = i$. Then $S(u)c^i = S(w)$, where $w = u_2 u_1$. Write $w = a_1 a_2 \cdots a_k$. Since $S(u_2 u_1)r$ is defined, we get $a_1 \neq A$, $a_2 \neq A$ and $a_3 \neq B$. Among these 8 cases, $a_1 = a_2 = a_3 = C$ does not occur since u only contains 2 Cs, and $a_1 a_2 = BC$ or $a_2 a_3 = BC$ does not occur since u does not end in B. The remaining 3 cases are

$$a_1 a_2 a_3 = BBA, \qquad a_1 a_2 a_3 = CBA, \qquad \text{and} \qquad a_1 a_2 a_3 = CCA,$$

where $a_1 a_2 a_3$ is replaced by the corresponding right hand side of the rule by the action of r. Then in $S(u)c^i r c^{k-i}$, the two Cs are on positions 1 and 2 again, and we obtain $S(u)c^i r c^{k-i} = S(v)$ for a string v of the given shape, satisfying $u \to_R v$.   □

Combining Lemmas 3 and 4 and the fact that $\mathsf{fib}(n) = \Omega(\phi^n)$, we obtain the following.

Corollary 5. *There is a word w such that $S(CCA^{k-2})w = S(CCA^{k-3}B)$; the shortest word w for which $S(CCA^{k-2})w$ is of the shape $S(u)c^i$ for u ending in B has length $\Omega(\phi^k)$.*

Now we are ready to prove the lower bound:

Lemma 6. *If w is carefully synchronizing, then $|w| = \Omega(\phi^k)$.*

Proof. Assume that w is a shortest carefully synchronizing word. Then we already observed that the first symbol of w is s, and w yields $S(CCA^{k-2})$ after the first step in the power automaton. By applying only c-steps and r-steps, according to Lemma 4, only sets of the shape $S(u)c^i$ for which $CCA^{k-2} \to_R^+ u$ can be reached, until u ends in B. In this process, each r-step corresponds to a rewrite step. Applying the third symbol s does not make sense, since then we go back to $S(CCA^{k-2})$. According to Corollary 5, in the power automaton at least $\Omega(\phi^k)$ steps are required to reach a set which is not of the shape $S(u)c^i$. So for reaching a singleton, the total number of steps is at least $\Omega(\phi^k)$.   □

Note that for the reasoning until now, the definition of $C_3 r = B_2$ did not play a role, and by s, r all states were replaced by states having the same index. But after the last symbol of u has become B, this $C_3 r = B_2$ will be applied, leading to a subset in which no state of the group A_3, B_3, C_3 occurs any more. Now we arrive at the main theorem.

Theorem 7. *For every n there is a carefully synchronizing PFA on n states and three symbols with shortest carefully synchronizing word length $\Omega(\phi^{n/3})$.*

Proof. If $n = 3k$ we take our automaton, otherwise we add one or two states on which r, c are undefined and s maps to A_1, having no influence on the argument. The bound was proved in Lemma 6; it remains to prove that the automaton is carefully synchronizing, that is, it is possible to end up in a singleton in the power automaton.

Let w be the word from Corollary 5. Since $S(CCA^{k-2})w = S(CCA^{k-3}B)$ and the number of c's in w is divisible by k, we have $C_1 w = C_1$, $C_2 w = C_2$, $A_3 w = A_3, \ldots, A_{k-1} w = A_{k-1}$, $A_k w = B_k$. Hence

$$\{A_1, B_1, C_1\}swcr = \{C_1\}cr = \quad \{C_2\} \subseteq \{A_1, B_1, C_1\}c,$$
$$\{A_2, B_2, C_2\}swcr = \{C_2\}cr = \quad \{B_2\} \subseteq \{A_2, B_2, C_2\},$$
$$\{A_i, B_i, C_i\}swcr = \{A_i\}cr = \{A_{i+1}\} \subseteq \{A_i, B_i, C_i\}c, \quad \text{for } i = 3, 4, \ldots, k-1,$$
$$\{A_k, B_k, C_k\}swcr = \{B_k\}cr = \quad \{A_1\} \subseteq \{A_k, B_k, C_k\}c.$$

So for all $i \neq 2$, $\{A_i, B_i, C_i\}swcr$ is contained in the cyclic successor $\{A_i, B_i, C_i\}c$ of $\{A_i, B_i, C_i\}$. $\{A_2, B_2, C_2\}swcr$ is just contained in $\{A_2, B_2, C_2\}$ itself. Since for any i, one can take the cyclic successor of $\{A_i, B_i, C_i\}$ at most $k-1$ times before ending up in $\{A_2, B_2, C_2\}$, we deduce that

$$\{A_i, B_i, C_i\}(swcr)^{k-1} \subseteq \{A_2, B_2, C_2\} \quad \text{for } i = 1, 2, \ldots, k.$$

As $\{A_2, B_2, C_2\}s = \{C_2\}$, we obtain the carefully synchronizing word $(swcr)^{k-1}s$ of the PFA.   □

5 Reduction to Two Symbols

In this section we construct PFAs with two symbols and exponential shortest carefully synchronizing word length. We do this by a general transformation to two-symbol PFAs, as was done before, e.g. in [14]. There a PFA on n states and m symbols was transformed to a PFA on mn states and two symbols, preserving synchronization length. In the next theorem, we improve this resulting number of states to $(m - 1)n$ or even less, only needing a mild extra condition. Using this result, we reduce our 3-symbol PFA with synchronizing length $\Omega(\phi^{n/3})$ to a 2-symbol PFA with synchronizing length $\Omega(\phi^{n/5})$.

Theorem 8. *Let $P = (Q, \Sigma)$ be a carefully synchronizing PFA with $|Q| = n$, $|\Sigma| = m$, and shortest carefully synchronizing word length $f(n)$. Assume $s \in \Sigma$ and $Q' \subseteq Q$ satisfy the following properties.*

1. *there is some number p such that all symbols are defined on Qs^p for a complete symbol s,*
2. *$qs = q$ for all $q \in Q'$, and*
3. *$qa = qb$ for all $q \in Q'$ and all $a, b \in \Sigma \setminus \{s\}$.*

Let $n' = n - |Q'|$. Then there exists a carefully synchronizing PFA on $n + n' \cdot (m - 2)$ states and 2 symbols, with shortest carefully synchronizing word length at least $f(n)$.

Note that if $Q' = \emptyset$ then only requirement 1 remains, and the resulting number of states is $n + n'(m - 2) = (m - 1)n$.

Proof. Write $Q = \{1, 2, \ldots, n\}$, $Q' = \{n' + 1, \ldots, n\}$, and $\Sigma = \{s, a_1, \ldots, a_{m-1}\}$. Let the states of the new PFA be $P_{1,j}$ for $j = 1, \ldots, n$ and $P_{i,j}$ for $i = 2, \ldots, m-1$, $j = 1, \ldots, n'$. Define the following two symbols a, b on these states:

$$P_{i,j}a = \begin{cases} P_{i+1,j}, & \text{if } i < m - 1, j \leq n', \\ P_{1,js}, & \text{if } i = m - 1, j \leq n', \\ P_{1,j}, & \text{if } i = 1, j > n'. \end{cases}$$

$$\begin{array}{cccccc} P_{1,1} & \cdots & P_{1,n'} & P_{1,n'+1} & \cdots & P_{1,n} \\ P_{2,1} & \cdots & P_{2,n'} & & & \\ \vdots & & \vdots & & & \\ P_{m-1,1} & \cdots & P_{m-1,n'} & & & \end{array}$$

and $P_{i,j}b = P_{1,ja_i}$, for all $i = 1, \ldots, m - 1$ and $j = 1, \ldots, n$ for which $P_{i,j}$ exists and ja_i is defined.

If we arrange the states as indicated above, then on the leftmost n' columns, a moves the states one step downward if possible, and for the bottom row jumps to the top row and acts there as s. For the remainder of the top row a also acts as s (which is the identity). On the leftmost n' columns, the symbol b acts as a_i on row i and then jumps to the top line. For the remainder of the top row, all a_i act in the same way and b acts likewise.

Define $\psi(a_i) = a^{i-1}b$ for $i = 1, \ldots, m-1$, and $\psi(s) = a^{m-1}$. Then on the top line $\psi(a_i)$ acts in the same way as a_i in the original PFA. Similarly, $\psi(s)$ acts as s. On any other row, $\psi(s)$ acts as s, too. Since every symbol a_i is defined on

qs^p for every $q \in Q$, we obtain that $\psi(s)^p b = a^{(m-1)p} b$ is defined on every state and ends up in the top row.

Assume that w is carefully synchronizing in the original PFA. Then by the above observations, $a^{(m-1)p} b\psi(w)$ is carefully synchronizing in the new PFA. Conversely, any carefully synchronizing word of the new PFA can be written as $\psi(w)a^j$, where $0 \leq j \leq m-2$ and $\psi(w)$ is a concatenation of blocks of the form $\psi(l), l \in \Sigma$. Now note that a^j can never synchronize two distinct states in the top row. Therefore, $\psi(w)$ synchronizes the top row and consequently w is synchronizing in the original PFA. Clearly $|\psi(w)a^j| \geq |w| \geq f(n)$.   □

We apply Theorem 8 to our basic construction with $3k$ states and $m = 3$ symbols; note that s, c are defined on all states and r is defined on Qs, so the requirements of Theorem 8 hold for $p = 1$. As r and c act differently on all states, the only option for Q' is $Q' = \emptyset$. Hence we obtain a carefully synchronizing PFA on $(m-1)3k = 6k$ states and two symbols, with shortest carefully synchronizing word length $\Omega(\phi^k)$. For n being the number of states of the new PFA, this is $\Omega(\phi^{n/6})$.

However, instead of our three symbols s, c, r we also get careful synchronization on the three symbols s, c, rc with careful synchronization length of the same order. But then for $i = 4, \ldots, k$ we have $A_i s = A_i$ and $A_i c = A_i rc$, so we may choose $Q' = \{A_4, \ldots, A_k\}$ in Theorem 8, by which $n' = 3k - (k-3) = 2k+3$, yielding a PFA on two symbols and $5k + 3$ states. This results in the following theorem, where for n not of the shape $5k+3$ we add ≤ 4 extra states to achieve this shape, where b is undefined on the new states and a maps the new states to existing states.

Theorem 9. *For every n there is a carefully synchronizing PFA on n states and two symbols with shortest carefully synchronizing word length $\Omega(\phi^{n/5})$.*

6   Further Optimizations

Some further optimizations are possible. For instance, for any $h \geq 2$ we can take $h + 1$ rewrite rules

$$C^i B^{h-i} A \to C^i A^{h-i} B$$

for $i = 0, \ldots, h$, and construct a PFA on the $n = 3k$ states A_i, B_i, C_i for $i = 1, \ldots, k$ with a similar s, c, and r, mimicking the rewrite rules in which the rewriting takes place in the states with indexes $\leq h+1$. For $h = 2$, this coincides with our construction, but for $h > 2$, this gives a better bound $\Omega(a^k)$, where a is the real zero of $x^h - x^{h-1} - \cdots - x - 1$ in between $3/2$ and 2. As this value tends to 2 for increasing h, for every $\epsilon > 0$, we achieve the bound $\Omega((2-\epsilon)^{n/3})$ for three symbols and $\Omega((2-\epsilon)^{n/5})$ for two symbols. For further improvements to $\Omega((4-\epsilon)^{n/5}) = \Omega((2-\epsilon)^{2n/5})$ for three symbols and $\Omega((4-\epsilon)^{n/6}) = \Omega((2-\epsilon)^{n/3})$ for two symbols, we refer to the extended version [1].

7   Conclusions

For every n we constructed a PFA on n states and 3 symbols for which careful synchronization is forced to mimic rewriting with respect to a string rewriting system. This system requires an exponential number of steps to reach a string of a particular shape. The resulting exponential synchronization length is much larger than the cubic upper bound for synchronization length of DFAs. We show that for $n = 4$ the shortest synchronization length for a PFA already can exceed the maximal shortest synchronization length for a DFA. For $n = 4, 5, 6$ we found greatest possible shortest synchronization lengths, both for DFAs and PFAs, where for DFAs until now this was only fully investigated for $n \leq 4$, that is, by not assuming any bound on the number of symbols. Both for DFAs and PFAs better techniques are needed to do the same analysis for $n = 7$ or higher.

References

1. de Bondt, M., Don, H., Zantema, H.: DFAs and PFAs with long shortest synchronizing word length (2017). https://arxiv.org/abs/1703.07618
2. Černy, J.: Poznámka k homogénnym experimentom s konečnými automatmi. Matematicko-fyzikálny časopis, Slovensk. Akad. Vied **14**(3), 208–216 (1964)
3. Černy, J., Piricka, A., Rosenauerova, B.: On directable automata. Kybernetika **7**(4), 289–298 (1971)
4. Don, H., Zantema, H.: Finding DFAs with maximal shortest synchronizing word length. In: Drewes, F., Martín-Vide, C., Truthe, B. (eds.) LATA 2017. LNCS, vol. 10168, pp. 249–260. Springer, Cham (2017). doi:10.1007/978-3-319-53733-7_18
5. Gerencsér, B., Gusev, V.V., Jungers, R.M.: Primitive sets of nonnegative matrices and synchronizing automata (2016). https://arxiv.org/abs/1602.07556
6. Kari, J.: A counterexample to a conjecture concerning synchronizing word in finite automata. EATCS Bull. **73**, 146–147 (2001)
7. Martyugin, P.V.: A lower bound for the length of the shortest carefully synchronizing words. Russ. Math. (Iz. VUZ) **54**(1), 46–54 (2010)
8. Pin, J.E.: On two combinatorial problems arising from automata theory. Ann. Discrete Math. **17**, 535–548 (1983)
9. Roman, A.: A note on Černý conjecture for automata with 3-letter alphabet. J. Autom. Lang. Comb. **13**(2), 141–143 (2008)
10. Rystsov, I.: Asymptotic estimate of the length of a diagnostic word for a finite automaton. Cybernetics **16**(2), 194–198 (1980)
11. Szykuła, M.: Improving the upper bound the length of the shortest reset word (2017). https://arxiv.org/abs/1702.05455
12. Trahtman, A.N.: An efficient algorithm finds noticeable trends and examples concerning the Černy conjecture. In: Královič, R., Urzyczyn, P. (eds.) MFCS 2006. LNCS, vol. 4162, pp. 789–800. Springer, Heidelberg (2006). doi:10.1007/11821069_68
13. Volkov, M.V.: Synchronizing automata and the Černý conjecture. In: Martín-Vide, C., Otto, F., Fernau, H. (eds.) LATA 2008. LNCS, vol. 5196, pp. 11–27. Springer, Heidelberg (2008). doi:10.1007/978-3-540-88282-4_4
14. Vorel, V.: Subset synchronization and careful synchronization of binary finite automata. Int. J. Found. Comput. Sci. **27**(5), 557–578 (2016)

On the Mother of All Automata: The Position Automaton

Sabine Broda[1]([✉]), Markus Holzer[2], Eva Maia[1], Nelma Moreira[1], and Rogério Reis[1]

[1] CMUP and DCC, Faculdade de Ciências da Universidado do Porto,
Rua do Campo Alegre, 1021, 4169-007 Porto, Portugal
{sbb,emaia,nam,rvr}@dcc.fc.up.pt
[2] Institut für Informatik, Universität Giessen, Arndtstr. 2, 35392 Giessen, Germany
holzer@informatik.uni-giessen.de

Abstract. We contribute new relations to the taxonomy of different conversions from regular expressions to equivalent finite automata. In particular, we are interested in ordinary transformations that construct automata such as, the follow automaton, the partial derivative automaton, the prefix automaton, the automata based on pointed expressions recently introduced and studied, and last but not least the position, or Glushkov automaton ($\mathcal{A}_{\mathrm{POS}}$), and their double reversed construction counterparts. We deepen the understanding of these constructions and show that with the artefacts used to construct the Glushkov automaton one is able to capture most of them. As a byproduct we define a *dual* version $\mathcal{A}_{\overline{\mathrm{POS}}}$ of the position automaton which plays a similar role as $\mathcal{A}_{\mathrm{POS}}$ but now for the reverse expression. It turns out that although the conversion of regular expressions and reversal of regular expressions to finite automata seems quite similar, there are significant differences.

1 Introduction

It is well known that regular expressions define exactly the same languages as deterministic or nondeterministic finite automata. The conversion between these representations has been intensively studied for more than half a century—see, e.g., Gruber and Holzer [11] for a recent survey on this subject w.r.t. descriptional complexity. There are a few classical algorithms and variants thereof for converting finite automata into equivalent regular expressions and as shown in [17] all these approaches are more or less reformulations of the same underlying algorithmic idea, and they yield (almost) the same regular expressions. For the converse transformation, that is, the conversion of regular expressions into equivalent finite automata, the situation is much more diverse, since the algorithmic underlying ideas already are different. Nevertheless, for some of the

S. Broda, E. Maia, N. Moreira, R. Reis—Partially supported by CMUP (UID/MAT/00144/2013), which is funded by FCT (Portugal) with national (MEC) and European structural funds through the programs FEDER, under the partnership agreement PT2020.

© Springer International Publishing AG 2017
É. Charlier et al. (Eds.): DLT 2017, LNCS 10396, pp. 134–146, 2017.
DOI: 10.1007/978-3-319-62809-7_9

algorithms the constructed automata can still be related to each other by determinisation and/or quotients w.r.t. equivalence relations. For instance, by Ilie
and Yu [12] it was shown that for a regular expression α the follow automa-
ton $\mathcal{A}_\mathrm{F}(\alpha)$ is isomorphic ($\simeq$) to the quotient of the position or Glushkov [10]
automaton $\mathcal{A}_\mathrm{POS}(\alpha)$ w.r.t. the relation $\equiv_\mathrm{F}$, that is, $\mathcal{A}_\mathrm{F}(\alpha) \simeq \mathcal{A}_\mathrm{POS}(\alpha)/\!\equiv_\mathrm{F}$.
Another relation is that the determinisation of the position automaton $\mathcal{A}_\mathrm{POS}(\alpha)$
is the McNaughton and Yamada [14] automaton $\mathcal{A}_\mathrm{MY}(\alpha)$, or in mathematical
notation $\mathrm{D}(\mathcal{A}_\mathrm{POS}(\alpha)) = \mathcal{A}_\mathrm{MY}(\alpha)$. From the variety of contructions from regular
expressions to equivalent finite automata these are only two examples where the
position automaton plays a central role.

We contribute further relations to the taxonomy of conversions from regular
expressions to finite automata—see Fig. 1 on page 11. Arrows, that are displayed
in bold in that figure correspond to new contributions in this paper. Provenance
of results that are not original is well indicated. Besides the above mentioned
follow automaton $\mathcal{A}_\mathrm{F}$ we also consider the partial derivative automaton $\mathcal{A}_\mathrm{PD}$ of
Mirkin [15] and Antimirov [2], the prefix automaton $\mathcal{A}_\mathrm{Pre}$ of Yamamoto [19], and
contructions based on a recent approach of Asperti et al. [3] and by Nipkow and
Traytel [16] by pointed expressions that lead to the mark after and mark before
automata $\mathcal{A}_\mathrm{MA}$ and $\mathcal{A}_\mathrm{MB}$, respectively. Pointed expressions are an alternative
representation of sets of positions. For the follow automaton $\mathcal{A}_\mathrm{F}$ we show that
it can be directly computed from the expression by labelling states not with
positions but with their Follow sets and their finality, and that the quotient of
the determinised follow automaton w.r.t. a right-invariant relation $\equiv_s$, which is
a generalization of the $\equiv_\mathrm{F}$-relation, leads to the mark before automaton $\mathcal{A}_\mathrm{MB}$. It
is known that $\mathcal{A}_\mathrm{MA}$ is isomorphic to the Yamada-McNaughton automaton $\mathcal{A}_\mathrm{MY}$,
which is proven to be the determinisation of the prefix automaton $\mathcal{A}_\mathrm{Pre}$. From
$\mathcal{A}_\mathrm{MA}$ to $\mathcal{A}_\mathrm{MB}$ we present a homomorphism, showing that $\mathcal{A}_\mathrm{MA}$ cannot be smaller
than $\mathcal{A}_\mathrm{MB}$—compare with [16].

When considering pointed expressions with only one point marking we obtain
the position automaton in case of the mark after interpretation, while the other
interpretation leads us to a *dual* version $\mathcal{A}_{\overleftarrow{\mathrm{POS}}}$ of the position automaton. We
show that the double reverse construction $\mathcal{A}_\mathrm{POS}(\alpha^\mathrm{R})^\mathrm{R}$ is isomorphic to $\mathcal{A}_{\overleftarrow{\mathrm{POS}}}(\alpha)$
and that the determinisation of $\mathcal{A}_{\overleftarrow{\mathrm{POS}}}(\alpha)$ yields $\mathcal{A}_\mathrm{MB}(\alpha)$. Our study provides evidence that $\mathcal{A}_{\overleftarrow{\mathrm{POS}}}$ plays a similar role as $\mathcal{A}_\mathrm{POS}$, but for the reverse expression α^R
instead of the expression α. This is supported by the fact that $\mathrm{D}(\mathcal{A}_\mathrm{POS}(\alpha^\mathrm{R})^\mathrm{R})$
is isomorphic to $\mathrm{D}(\mathcal{A}_\mathrm{F}(\alpha^\mathrm{R})^\mathrm{R})$ and $\mathrm{D}(\mathcal{A}_\mathrm{PD}(\alpha^\mathrm{R})^\mathrm{R})$. It is worth mentioning that
the double reverse automata $\mathcal{A}_\mathrm{Pre}(\alpha^\mathrm{R})^\mathrm{R}$ and its determinisation $\mathrm{D}(\mathcal{A}_\mathrm{Pre}(\alpha^\mathrm{R})^\mathrm{R})$
get out of the line since the latter automaton turns out to be *not* isomorphic to
$\mathrm{D}(\mathcal{A}_\mathrm{POS}(\alpha^\mathrm{R})^\mathrm{R})$. This shows, that although the taxonomy of "ordinary" conversions and double reversal conversions is quite similar, there are subtle differences
that break the symmetry. Most proofs of our results are based on Glushkov's
position concept which turns out to be highly valuable and can be used to
describe automata constructions that look different at first sight, not only for
the implementation use but also from the theoretical perspective. Due to space
limitations, most proofs are omitted.

2   Preliminaries

In this section we review some basic definitions about regular expressions and finite automata and fix notation. Given an alphabet (finite set of *letters* or *alphabet symbols*) Σ, the set RE of *regular expressions*, α, over Σ is defined inductively by: $\emptyset$, ε and every letter σ_i is a regular expression, when α and α' are regular expressions then $(\alpha + \alpha')$, $(\alpha \cdot \alpha')$, and $(\alpha^\star)$ are regular expressions. The *language* associated to α is denoted by $\mathcal{L}(\alpha)$ and defined as usual. The *alphabetic size* $|\alpha|_\Sigma$ is its number of letters. We denote the subset of Σ containing the symbols that occur in α by Σ_α. We define $\varepsilon(\alpha)$ by $\varepsilon(\alpha) = \varepsilon$ if $\varepsilon \in \mathcal{L}(\alpha)$, and $\varepsilon(\alpha) = \emptyset$ otherwise. By abuse of notation, we consider $\varepsilon S = S$ and $\emptyset S = \emptyset$, for any set S. A *nondeterministic finite automaton* (NFA) is a five-tuple $A = \langle Q, \Sigma, \delta, I, F \rangle$ where Q is a finite set of states, Σ is a finite alphabet, $I \subseteq Q$ is the set of initial states, $F \subseteq Q$ is the set of final states, and $\delta : Q \times (\Sigma \cup \{\varepsilon\}) \to 2^Q$ is the transition function. We consider the size of an NFA as its number of states. An NFA that has transitions labelled with ε is an ε-NFA. In this paper, excepted when explicitly mentioned, we will consider NFAs without ε transitions. The transition function can be extended to words and to sets of states in the natural way. When $I = \{q_0\}$, we use $I = q_0$. We define the *finality* function ε on Q by $\varepsilon(q) = \varepsilon$ if $q \in F$ and $\varepsilon(q) = \emptyset$, otherwise. For $S \subseteq Q$ we have $\varepsilon(S) = \varepsilon$ iff there is some state $q \in S$ with $\varepsilon(q) = \varepsilon$, and $\varepsilon(S) = \emptyset$ otherwise. An NFA accepting a non-empty language is *trim* if every state is accessible from an initial state and every state leads to a final state. The language accepted by A is $\mathcal{L}(A) = \{ w \in \Sigma^\star \mid \delta(I, w) \cap F \neq \emptyset \}$. Two automata are *equivalent* if they accept the same language. If two automata A and B are isomorphic, we write $A \simeq B$.

An NFA is *deterministic* (DFA) if $|\delta(q, \sigma)| \leq 1$, for all $(q, \sigma) \in Q \times \Sigma$, and $|I| = 1$. In this case, we simply write $\delta(p, a) = q$ instead of $\delta(p, a) = \{q\}$. We can convert an NFA A into an equivalent DFA $\mathrm{D}(A)$ by the *determinisation* operation D, using the well-known subset construction, where only subsets reachable from the initial subset of $\mathrm{D}(A)$ are used. Formally, $\mathrm{D}(A) = \langle Q_\mathrm{D}, \Sigma, \delta_\mathrm{D}, I_\mathrm{D}, F_\mathrm{D} \rangle$, where $Q_\mathrm{D} \subseteq 2^Q$, $I_\mathrm{D} = I$, $\delta_\mathrm{D}(S, \sigma) = \bigcup_{q \in S} \delta(q, \sigma)$ for $S \subseteq Q$, $\sigma \in \Sigma$, and $F_\mathrm{D} = \{ S \in Q_\mathrm{D} \mid S \cap F \neq \emptyset \}$. Note that $S \in F_\mathrm{D}$ if and only if $\varepsilon(S) = \varepsilon$.

An equivalence relation $\equiv$ on Q is *right invariant* w.r.t. an NFA A if and only if: $\equiv \subseteq (Q - F)^2 \cup F^2$; and $\forall p, q \in Q, \sigma \in \Sigma$, if $p \equiv q$, then $\forall p' \in \delta(p, \sigma) \; \exists q' \in \delta(q, \sigma)$ such that $p' \equiv q'$. Given a set of states $S \subseteq Q$, we denote $S/\equiv \; = \{ [q] \mid q \in S \}$. Note that $p \equiv q$ implies $\delta(p, \sigma)/\equiv \; = \delta(q, \sigma)/\equiv$, for $p, q \in Q$ and $\sigma \in \Sigma$. Furthermore, if A is deterministic, then $p \equiv q$ implies $\delta(p, \sigma) \equiv \delta(q, \sigma)$. If $\equiv$ is a right-invariant relation on Q, the *quotient automaton* $A/\equiv$ is given by $A/\equiv \; = \langle Q/\equiv, \Sigma, \delta/\equiv, [q_0], F/\equiv \rangle$, where $\delta/\equiv ([p], \sigma) = \{ [q] \mid q \in \delta(p, \sigma) \} = \delta(p, \sigma)/\equiv$. It is easy to see that $\mathcal{L}(A/\equiv) = \mathcal{L}(A)$. Given a right-invariant relation $\equiv$ w.r.t. an NFA A, we can consider the natural extension of $\equiv$ w.r.t. $\mathrm{D}(A)$, where for $X, Y \subseteq 2^Q$ we have $X \equiv Y$ if and only if $X/\equiv \; = Y/\equiv$. The following lemma relates determinisation with these right-invariant relations.

Lemma 1. $\mathrm{D}(A/\equiv) = \mathrm{D}(A)/\equiv$, *if $\equiv$ is a right-invariant relation w.r.t. A.*

3   The Position Automaton and *"The Rest"*

In this section we recall the definition of the position automaton and several related automata constructions. In particular, the determinisation of the position automaton, some ε-NFAs, derivative based constructions and the follow automaton are considered. We show that the latter can be obtained directly from the regular expression.

To decide if a word is represented by a regular expression, one can scan the symbols of the regular expression in a specific way. For instance, given $\alpha = a(bb + aba)^\star b$ the word $abbabab$ can be obtained by scanning the first a, the two consecutive bs and then the second a, the third b, the third a, and the last b. This illustrates that uniquely identifying each letter of a regular expression is important for word recognition. Formally, given $\alpha \in RE$, one can mark each occurrence of a letter σ with its position in α, considering reading it from left to right. The resulting regular expression is a *marked* regular expression $\overline{\alpha}$ with all symbols distinct and over the alphabet $\Sigma_{\overline{\alpha}}$. Then, a position $i \in [1, |\alpha|_\Sigma]$ corresponds to the symbol σ_i in $\overline{\alpha}$, and consequently to exactly one occurrence of σ in α. For instance, $\overline{\alpha} = a_1(b_2 b_3 + a_4 b_5 a_6)^\star b_7$. The same notation is used for unmarking, $\overline{\overline{\alpha}} = \alpha$. Let $\mathsf{Pos}(\alpha) = \{1, 2, \ldots, |\alpha|_\Sigma\}$, and $\mathsf{Pos}_0(\alpha) = \mathsf{Pos}(\alpha) \cup \{0\}$.

Positions were used by Glushkov [10] to define an NFA equivalent to α, usually called the *position automaton* or Glushkov automaton ($\mathcal{A}_{\mathrm{POS}}(\alpha)$). Each state of the automaton, except for the initial state, corresponds to a position and there exists a transition from a position i to position j by a letter σ such that $\overline{\sigma_j} = \sigma$, if σ_i can be followed by σ_j in some word represented by $\overline{\alpha}$. More formally this reads as follows: the sets characterising the positions that can begin, end or be followed in words of $\mathcal{L}(\overline{\alpha})$ are, $\mathsf{First}(\alpha) = \{\, i \mid \sigma_i w \in \mathcal{L}(\overline{\alpha}) \,\}$, $\mathsf{Last}(\alpha) = \{\, i \mid w\sigma_i \in \mathcal{L}(\overline{\alpha}) \,\}$, and $\mathsf{Follow}(\alpha) = \{\, (i, j) \mid u\,\sigma_i \sigma_j v \in \mathcal{L}(\overline{\alpha}) \,\}$ respectively. We also define $\mathsf{Last}_0(\alpha) = \mathsf{Last}(\alpha) \cup \varepsilon(\alpha)\{0\}$. Furthermore, given $i \in \mathsf{Pos}(\alpha)$ and $S \in 2^{\mathsf{Pos}_0(\alpha)}$, let $\mathsf{Follow}(\alpha, i) = \{\, j \mid (i, j) \in \mathsf{Follow}(\alpha) \,\}$, $\mathsf{Follow}(\alpha, 0) = \mathsf{First}(\alpha)$ and $\mathsf{Follow}(\alpha, S) = \bigcup_{i \in S} \mathsf{Follow}(\alpha, i)$. The *position automaton* for α is

$$\mathcal{A}_{\mathrm{POS}}(\alpha) = \langle \mathsf{Pos}_0(\alpha), \Sigma, \delta_{\mathrm{POS}}, 0, \mathsf{Last}_0(\alpha) \rangle,$$

where $\delta_{\mathrm{POS}}(i, \sigma) = \{\, j \mid j \in \mathsf{Follow}(\alpha, i) \text{ and } \sigma = \overline{\sigma_j} \,\}$.

Proposition 2 ([10]). *$\mathcal{L}(\mathcal{A}_{\mathrm{POS}}(\alpha)) = \mathcal{L}(\alpha)$.*

The following example gives some intuition on the construction of the position automaton and its behaviour. Note that $\varepsilon(0) = \varepsilon(\alpha)$, and we will use either one or the other as it will be more convenient.

Example 3. Consider $\alpha = (b + ab)^\star + b^\star$ with $\overline{\alpha} = (b_1 + a_2 b_3)^\star + b_4^\star$. Then, $\mathsf{First}(\alpha) = \{1, 2, 4\}$, $\mathsf{Last}_0(\alpha) = \{0, 1, 3, 4\}$ and $\mathsf{Follow}(\alpha) = \{(1, 1), (1, 2), (2, 3), (3, 1), (3, 2), (4, 4)\}$. The position automaton $\mathcal{A}_{\mathrm{POS}}$ for α is depicted below.

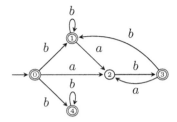

□

Note that each state, different from 0, in the position automaton corresponds to a symbol σ_i in $\overline{\alpha}$, where σ is the symbol just read. Thus, one can define a function Select that selects from a set of positions $S \subseteq \mathsf{Pos}(\alpha)$, those that correspond to a given letter, i.e., $\mathsf{Select} : 2^{\mathsf{Pos}(\alpha)} \times \Sigma \to 2^{\mathsf{Pos}(\alpha)}$ defined is by

$$\mathsf{Select}(S, \sigma) = \{\, i \mid i \in S \text{ and } \overline{\sigma_i} = \sigma \,\}.$$

Then, δ_{POS} can be defined by composing Follow with Select, i.e.,

$$\delta_{\mathsf{POS}}(i, \sigma) = \mathsf{Select}(\mathsf{Follow}(\alpha, i), \sigma). \tag{1}$$

The same notion[1] was used by McNaughton and Yamada [14] to define an automaton which corresponds to the determinisation of the position automaton. With the definition of δ_{POS} in (1) and considering the determinisation algorithm, the McNaughton and Yamada DFA can be defined as

$$\mathcal{A}_{\mathrm{MY}}(\alpha) = \mathrm{D}(\mathcal{A}_{\mathsf{POS}}(\alpha)) = \langle Q_{\mathrm{MY}}, \Sigma, \delta_{\mathrm{MY}}, \{0\}, F_{\mathrm{MY}} \rangle,$$

where $Q_{\mathrm{MY}} \subseteq 2^{\mathsf{Pos}_0(\alpha)}$, $F_{\mathrm{MY}} = \{\, S \in Q_{\mathrm{MY}} \mid \varepsilon(S) = \varepsilon \,\}$ and for $S \in 2^{\mathsf{Pos}(\alpha)}$ and $\sigma \in \Sigma$, $\delta_{\mathrm{MY}}(S, \sigma) = \mathsf{Select}(\mathsf{Follow}(\alpha, S), \sigma)$.

Proposition 4 ([14]). $\mathcal{L}(\mathcal{A}_{\mathrm{MY}}(\alpha)) = \mathcal{L}(\alpha)$.

Example 5. Applying the McNaughton-Yamada construction to α from Example 3, we obtain the following DFA, $\mathcal{A}_{\mathrm{MY}}(\alpha)$:

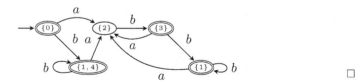

□

---

[1] Some authors use slightly different notions of marking [8, 14], which have in common that each symbol in the marked expression corresponds to exactly one occurrence of a symbol in the original expression.

In the forthcoming we review the Thompson like construction of the follow automaton $\mathcal{A}_F$, which was introduced by Ilie and Yu in [12]. We show that one can directly construct this automaton by an appropriate state labeling inspired by the position automaton $\mathcal{A}_{POS}$. Then we recall some constructions of automata from regular expressions based on derivatives and variants, such as, e.g., Brzozowski's construction by derivatives and the construction of Mirkin [15] and Antimirov [2] by partial derivatives. Here we focus on known results characterizing these automata as quotients of the position automaton.

3.1   The Follow Automaton $\mathcal{A}_F$

The most used conversion from regular expressions to equivalent ε-NFAs is the Thompson conversion [18], $\mathcal{A}_{\varepsilon\text{-}T}$. An improved use of ε-transitions lead to the definition of the ε-follow automaton [12]. From a Thompson automaton, if ε-transitions are eliminated in an adequate manner, the position automaton is obtained [1,9]. Eliminating ε-transitions from the ε-follow automaton, the resulting automaton is the *follow automaton* $\mathcal{A}_F(\alpha)$ which was introduced by Ilie and Yu [12] in 2003.

Proposition 6 ([12]). $\mathcal{L}(\mathcal{A}_F(\alpha)) = \mathcal{L}(\alpha)$.

They also showed that the follow automaton is a quotient of the position automaton, obtained by identifying positions with the same Follow set. For instance, in the position automaton of Example 3, one can see that $\mathsf{Follow}(\alpha, 1) = \mathsf{Follow}(\alpha, 3) = \{1, 2\}$, and that 1 and 3 are both accepting states. Formally, Ilie and Yu considered the right-invariant equivalence relation $\equiv_F$ defined on the set of states $\mathsf{Pos}_0(\alpha)$, w.r.t. $\mathcal{A}_{POS}(\alpha)$, by

$$i \equiv_F j \;\Leftrightarrow\; \mathsf{Follow}(\alpha, i) = \mathsf{Follow}(\alpha, j) \text{ and } \varepsilon(i) = \varepsilon(j),$$

and showed that $\mathcal{A}_F(\alpha) \simeq \mathcal{A}_{POS}(\alpha)\big/_{\equiv_F}$.

We show that $\mathcal{A}_F(\alpha)$ can be directly computed from α, by labelling states not with positions $i \in \mathsf{Pos}_0(\alpha)$, but with their Follow sets and their finality. Let

$$\mathcal{A}_F(\alpha) = \langle \mathsf{F}(\alpha), \Sigma, \delta_F, (\mathsf{Follow}(\alpha, 0), \varepsilon(0)), F_F \rangle,$$

where $\mathsf{F}(\alpha) = \{ (\mathsf{Follow}(\alpha, i), \varepsilon(i)) \mid i \in \mathsf{Pos}_0(\alpha) \} \subseteq 2^{\mathsf{Pos}(\alpha)} \times \{\varepsilon, \emptyset\}$, $F_F = \{ (S, c) \in \mathsf{F}(\alpha) \mid c = \varepsilon \}$ and for $(S, c) \in \mathsf{F}(\alpha)$ and $\sigma \in \Sigma$,

$$\delta_F((S, c), \sigma) = \{ (\mathsf{Follow}(\alpha, j), \varepsilon(j)) \mid j \in \mathsf{Select}(S, \sigma) \}.$$

The transition function δ_F is defined as a composition of Select with Follow, instead of Follow with Select as for δ_{POS} (and δ_{MY}). It is necessary to include the finality of a position in the label of the corresponding state, since there might be positions with the same Follow, but different finalities. This can, for instance, be observed in the automaton for $\alpha = a(b^*c)^*$. With this definition of $\mathcal{A}_F$ we obtain an alternative proof of the result by Ilie and Yu.

Proposition 7. $\mathcal{A}_{\mathrm{F}}(\alpha) \simeq \mathcal{A}_{\mathrm{POS}}(\alpha)/{\equiv_{\mathrm{F}}}$.

Proof. Consider φ_{F} : $\mathsf{Pos}_0(\alpha)/{\equiv_{\mathrm{F}}} \longrightarrow \mathsf{F}(\alpha)$ defined by $\varphi_{\mathrm{F}}([i]) = (\mathsf{Follow}(\alpha, i), \varepsilon(i))$. By definition, φ_{F} is a bijection and preserves initial as well as final states. Furthermore, for $[i] \in \mathsf{Pos}_0(\alpha)/{\equiv_{\mathrm{F}}}$ and $\sigma \in \Sigma$ we have

$$\varphi_{\mathrm{F}}(\delta_{\mathrm{POS}}/{\equiv_{\mathrm{F}}}([i], \sigma)) = \varphi_{\mathrm{F}}(\{\, [j] \mid j \in \mathsf{Select}(\mathsf{Follow}(\alpha, i), \sigma) \,\})$$
$$= \{\, \varphi_{\mathrm{F}}([j]) \mid j \in \mathsf{Select}(\mathsf{Follow}(\alpha, i), \sigma) \,\}$$
$$= \delta_{\mathrm{F}}((\mathsf{Follow}(\alpha, i), \varepsilon(i)), \sigma) = \delta_{\mathrm{F}}(\varphi_{\mathrm{F}}([i]), \sigma).$$

This shows that φ_{F} is an isomorphism.   □

3.2   Derivative Based Constructions

Brzozowski [5] defined a DFA equivalent to a regular expression using the notion of derivative. The derivative of $\alpha \in RE$ w.r.t. $\sigma \in \Sigma$ is $\sigma^{-1}\alpha$, such that $\mathcal{L}(\sigma^{-1}\alpha) = \{\, w \mid \sigma w \in \mathcal{L}(\alpha) \}$. This notion can be extended to words: $\varepsilon^{-1}\alpha = \alpha$ and $(\sigma w)^{-1}\alpha = w^{-1}(\sigma^{-1}\alpha)$. The set of all derivatives of α, $\{\, w^{-1}\alpha \mid w \in \Sigma^{\star} \}$ may not be finite. For finiteness, Brzozowski considered the quotient of that set modulo some regular expressions equivalences.

The partial derivative automaton $\mathcal{A}_{\mathrm{PD}}(\alpha)$ of a regular expression α was defined independently by Mirkin [15] and Antimirov [2]. Champarnaud and Ziadi stated the equivalence of the two formulations [6], and proved that $\mathcal{A}_{\mathrm{PD}}$ is a quotient of the $\mathcal{A}_{\mathrm{POS}}$ by a right-invariant relation $(\equiv_c)$ [7].

The *prefix automaton* $\mathcal{A}_{\mathrm{Pre}}$ was introduced by Yamamoto [19] as a quotient of the $\mathcal{A}_{\varepsilon\text{-T}}$ automaton. Maia et al. [13] characterised the $\mathcal{A}_{\mathrm{Pre}}$ automaton as a solution of a system of left RE equations and express it as a quotient of $\mathcal{A}_{\mathrm{POS}}$ by a left-invariant equivalence relation $(\equiv_{\ell})$, i.e., a right-invariant relation w.r.t. the reversal of $\mathcal{A}_{\mathrm{POS}}$, cf. Sect. 5.

4   Automata Based on Pointed Expressions

Next we review two automata constructions, $\mathcal{A}_{\mathrm{MB}}$ and $\mathcal{A}_{\mathrm{MA}}$, that are based on recent approaches of Asperti et al. [3] and by Nipkow and Traytel [16] using pointed expressions. In a pointed regular expression, several positions are selected, and are graphically marked with a *point* corresponding to a letter. Those automata correspond to two different interpretations of a *pointed* expression, i.e., of a given set of positions S: in the first case, given a letter σ one selects which positions from S correspond to that letter and then determines which possible positions can follow; in the second case the set of positions S corresponds to where one can be *after* reading the letter σ. For instance, the pointed regular expression $a(\bullet bb + \bullet aba)^{\star} \bullet b$ characterises the set of positions $\{2, 4, 7\}$. Intuitively, these are the positions which have been reached after reading some prefix of an input word. Asperti et al. thought that their algorithm "*au point*" computed a DFA isomorphic to $\mathcal{A}_{\mathrm{MY}}(\alpha)$, but Nipkow and Traytel [16] showed that their construction led to a dual automaton and called it *mark before*, $\mathcal{A}_{MB}$, while $\mathcal{A}_{MY}$ was isomorphic to a *mark after*, $\mathcal{A}_{MA}$. Using the notation of the previous section, a transition in $\mathcal{A}_{MA}$ is a composition of Follow with Select similarly as described in (1), while in $\mathcal{A}_{MB}$ it will be a composition of Select with Follow. Because of the behaviour of the transition function δ_{MY} of $\mathcal{A}_{MY}(\alpha)$, Nipkow and Traytel called this construction *mark after* ($\mathcal{A}_{MA}(\alpha)$).

In this section, we show that the $\mathcal{A}_{MB}$ is isomorphic to a quotient of the determinisation of $\mathcal{A}_F$, and as a corollary it follows that $\mathcal{A}_{MA}$ ($\mathcal{A}_{MY}$) cannot be smaller than $\mathcal{A}_{MB}$ (as already stated by Nipkow and Traytel). Moreover, we also consider the case, where one restricts pointed regular expressions with only *one* point marking a position. Obviously, the *mark after* automaton of single pointed expressions is related to the position automaton.

4.1   The Automaton $\mathcal{A}_{MB}$ *Versus* $D(\mathcal{A}_F)$

As mentioned above, Asperti *et al.* introduced the notion of pointed regular expression in order to obtain a compact representation of a set of positions. However, a point was used to mark a position to be visited when reading a letter instead of a position reached after reading the letter, as is the case for $\mathcal{A}_{POS}$ and $\mathcal{A}_{MA}$. The resulting construction was called *mark before*, $\mathcal{A}_{MB}$, by Nipkow and Traytel. In our framework, this means that δ_{MB} is a composition of Follow with Select. Formally, given $\alpha \in RE$, let

$$\mathcal{A}_{MB}(\alpha) = \langle Q_{MB}, \Sigma, \delta_{MB}, (\mathsf{Follow}(\alpha, 0), \varepsilon(0)), F_{MB} \rangle,$$

where $Q_{MB} \subseteq 2^{\mathsf{Pos}(\alpha)} \times \{\emptyset, \varepsilon\}$, and for $(S, c) \in Q_{MB}$ and $\sigma \in \Sigma$,

$$\delta_{MB}((S, c), \sigma) = (\mathsf{Follow}(\alpha, \mathsf{Select}(S, \sigma)), \varepsilon(\mathsf{Select}(S, \sigma))),$$

and $F_{MB} = \{ (S, c) \mid c = \varepsilon \}$. In Q_{MB} we consider only the states that are accessible from the initial state by δ_{MB}.

Proposition 8 ([3,16]). $\mathcal{L}(\mathcal{A}_{MB}(\alpha)) = \mathcal{L}(\alpha)$.

Example 9. Consider again the regular expression α from Example 3. The $\mathcal{A}_{MB}(\alpha)$ DFA is depicted below.

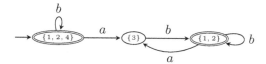

Note that the first state label is the set $\mathsf{First}(\alpha)$, and one can see that two states are saved when comparing with $\mathcal{A}_{MA}$, in Example 5.   □

One could expect that the $\mathcal{A}_{MB}$ construction was isomorphic to the determinisation of $\mathcal{A}_F$. But we will see that in general that is not the case. The determinisation of $\mathcal{A}_F$, $D(\mathcal{A}_F(\alpha)) = \langle Q_{D(\mathcal{A}_F)}, \Sigma, \{(\mathsf{Follow}(\alpha, 0), \varepsilon(0))\}, \delta_{D(\mathcal{A}_F)}, F_{D(\mathcal{A}_F)} \rangle$, can be obtained by the subset construction.

Example 10. Considering again the regular expression α from Example 3, the $\mathcal{A}_F(\alpha)$ and $D(\mathcal{A}_F(\alpha))$ are respectively:

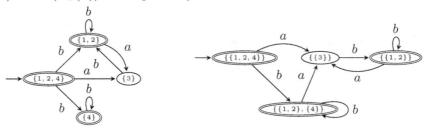

It is clear that $D(\mathcal{A}_F(\alpha))$ is not isomorphic to $\mathcal{A}_{MB}(\alpha)$ (see Example 9). However if one merges the states labeled by $\{(\{1,2,4\},\varepsilon)\}$ and $\{(\{1,2\},\varepsilon),(\{4\},\varepsilon)\}$ in $D(\mathcal{A}_F(\alpha))$, the DFA $\mathcal{A}_{MB}(\alpha)$ is obtained. Next we prove that if certain sets of sets in the determinisation of $\mathcal{A}_F$ are flattened the resulting automaton is isomorphic to $\mathcal{A}_{MB}$.   □

Let $\equiv_s$ be the equivalence relation on $2^{F(\alpha)}$ defined by,

$$I \equiv_s J \Leftrightarrow \bigcup_{(S,\_)\in I} S = \bigcup_{(S,\_)\in J} S \text{ and } \varepsilon(I) = \varepsilon(J).$$

Proposition 11. $\mathcal{L}\left(D(\mathcal{A}_F(\alpha))/\equiv_s\right) = \mathcal{L}(D(\mathcal{A}_F(\alpha)))$.

Proposition 12. $D(\mathcal{A}_F(\alpha))/\equiv_s \simeq \mathcal{A}_{MB}(\alpha)$.

Nipkow and Traytel presented a homomorphism from $\mathcal{A}_{MA}$ to $\mathcal{A}_{MB}$, showing that $\mathcal{A}_{MA}$ cannot be smaller than $\mathcal{A}_{MB}$. The same result is a direct corollary of the above results.

Corollary 13. *Let φ_F be defined as in the proof of Proposition 7. Then we have* $\varphi_F\left(D\left(\mathcal{A}_{POS}(\alpha)/\equiv_F\right)\right)/\equiv_s = \varphi_F\left(D(\mathcal{A}_{POS}(\alpha))/\equiv_F\right)/\equiv_s \simeq \mathcal{A}_{MB}(\alpha)$.

Example 14. Considering $\mathcal{A}_{MY}$ ($\mathcal{A}_{MA}$) from Example 5, we have $\{1\} \equiv_F \{3\}$, since $[1]_{\equiv_F} = [3]_{\equiv_F}$. Furthermore,

$$\varphi_F(\{[0]_{\equiv_F}\}) = \{(\{1,2,4\},\varepsilon)\} \equiv_s \{(\{1,2\},\varepsilon),(\{4\},\varepsilon)\} = \varphi_F(\{[1]_{\equiv_F},[4]_{\equiv_F}\}). \quad □$$

4.2   The Dual Position Automaton

If one considers pointed regular expressions with only one point marking a position to be visited when reading a letter, an NFA, dual of $\mathcal{A}_{POS}$ ($\mathcal{A}_{\overleftarrow{POS}}$), can be defined. We show that its determinisation yields $\mathcal{A}_{MB}$. Given a regular expression α, with $n = |\alpha|_\Sigma = |\mathsf{Pos}(\alpha)|$, the set of states of $\mathcal{A}_{\overleftarrow{POS}}$ is $\mathsf{Pos}(\alpha)$ plus an unique final state $n+1$. The set of initial states is $\mathsf{Follow}(\alpha,0) \cup \varepsilon(\alpha)\{n+1\}$. From a state $i \in \mathsf{Pos}(\alpha)$ reading $\sigma \in \Sigma$ one can move to $\mathsf{Follow}(\alpha,i)$ if $\overline{\sigma_i} = \sigma$. That is, by first selecting $\mathsf{Select}(\{i\},\sigma)$ which is i if $\overline{\sigma_i} = \sigma$, and empty otherwise, and then applying Follow. Moreover, if $\varepsilon(i) = \varepsilon$ there is a transition to $n + 1$. Formally,

$$\mathcal{A}_{\overleftarrow{\text{POS}}}(\alpha) = \langle \text{Pos}(\alpha) \cup \{n+1\}, \Sigma, \delta_{\overleftarrow{\text{POS}}}, \text{Follow}(\alpha, 0) \cup \varepsilon(\alpha)\{n+1\}, \{n+1\} \rangle,$$

with $\delta_{\overleftarrow{\text{POS}}}(i, \sigma) = \text{Follow}(\alpha, \text{Select}(\{i\}, \sigma)) \cup \varepsilon(\text{Select}(\{i\}, \sigma))\{n+1\}$. This means that $\delta_{\overleftarrow{\text{POS}}}(i, \sigma) = \text{Follow}(\alpha, i) \cup \varepsilon(i)\{n+1\}$, only if $i \in \text{Pos}(\alpha)$ and $\overline{\sigma_i} = \sigma$, being the empty set otherwise.

Example 15. Considering again the regular expression α from Example 3, the $\mathcal{A}_{\overleftarrow{\text{POS}}}(\alpha)$ is the following:

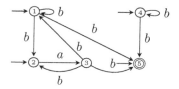

Observe that for each state of $\mathcal{A}_{\overleftarrow{\text{POS}}}$ all transitions *leaving* it have the same label. This is exactly the opposite of the position automaton $\mathcal{A}_{\text{POS}}$, where for each state all transitions *into it* have the same label.

Proposition 16. $\text{D}(\mathcal{A}_{\overleftarrow{\text{POS}}}(\alpha)) \simeq \mathcal{A}_{\text{MB}}(\alpha).$

5   Reversals and Automata Constructions

Given a language L the reversal of L, L^R, is the language obtained by reversing all the words in L. The reversal of a regular expression α is denoted by α^R, and is inductively defined by: $\alpha^R = \alpha$ for $\alpha \in \Sigma \cup \{\varepsilon, \emptyset\}$, $(\alpha + \beta)^R = \beta^R + \alpha^R$, $(\alpha\beta)^R = \beta^R \alpha^R$ and $(\alpha^\star)^R = (\alpha^R)^\star$. The reversal α^R describes $\mathcal{L}(\alpha)^R$. In the same way, given an automaton $\mathcal{A} = \langle Q, \Sigma, \delta, I, F \rangle$ its reversal is $\mathcal{A}^R = \langle Q, \Sigma, \delta^R, F, I \rangle$, where $\delta^R(q, \sigma) = \{p \mid q \in \delta(p, \sigma)\}$ and $\mathcal{L}(\mathcal{A}^R) = \mathcal{L}(\mathcal{A})^R$.

Given α, any of the automata constructions in the previous sections can be applied to α^R. If one reverses the resulting automaton, an alternative automaton construction for $\mathcal{L}(\alpha)$ is obtained. In this section we establish some relations between the direct constructions and the double reversed ones. We show that $\mathcal{A}_{\text{POS}}(\alpha^R)^R \simeq \mathcal{A}_{\overleftarrow{\text{POS}}}$. We also show that determinising any quotient of $\mathcal{A}_{\text{POS}}(\alpha^R)^R$ by a right-invariant relation is the same as determinising $\mathcal{A}_{\text{POS}}(\alpha^R)^R$ and thus, by Proposition 16, the resulting automata are all isomorphic to $\mathcal{A}_{\text{MB}}$. The same does not hold if one considers quotients by a left-invariant relation, and we illustrate that with the $\mathcal{A}_{\text{Pre}}$ construction.

Our first result on reversals of expressions and automata reads as follows:

Proposition 17. $\mathcal{A}_{\text{POS}}(\alpha^R)^R \simeq \mathcal{A}_{\overleftarrow{\text{POS}}}(\alpha).$

For the position automaton several quotients were presented in Sect. 3 for which different DFAs could be obtained by determinisation. For the dual construction, $\mathcal{A}_{\overline{\text{POS}}}$, the determinisation of any quotient by a right-invariant relation is isomorphic to $\mathcal{A}_{\text{MB}}$. The following simple lemma explains the reason.

Lemma 18. *Let A be a trim NFA and consider $\equiv$ a right-invariant relation w.r.t. A. Then, $\mathrm{D}(A^{\mathrm{R}}/\equiv) = \mathrm{D}(A^{\mathrm{R}})$.*

This result has direct consequences for all constructions, that can be obtained as a quotient of the position automaton by some right-invariant relation. In particular, we have

Proposition 19. $\mathrm{D}(\mathcal{A}_{\text{POS}}(\alpha^{\mathrm{R}})^{\mathrm{R}}) \simeq \mathrm{D}(\mathcal{A}_{\text{PD}}(\alpha^{\mathrm{R}})^{\mathrm{R}}) \simeq \mathrm{D}(\mathcal{A}_{\text{F}}(\alpha^{\mathrm{R}})^{\mathrm{R}}) \simeq \mathcal{A}_{\text{MB}}(\alpha)$.

Note that Lemma 18 does not hold for left-invariant relations. In particular, one can consider the $\mathcal{A}_{\text{Pre}}$ construction mentioned in Sect. 3.2.

Proposition 20. *For $\alpha = a^{\star} + (a + b)a^{\star}$, $\mathrm{D}(\mathcal{A}_{\text{Pre}}(\alpha^{\mathrm{R}})^{\mathrm{R}}) \not\simeq \mathrm{D}(\mathcal{A}_{\text{POS}}(\alpha^{\mathrm{R}})^{\mathrm{R}})$.*

However, Lemma 18 also implies that if a relation $\equiv$ is a left-invariant equivalence relation w.r.t an NFA A then $\mathrm{D}(A/\equiv) = \mathrm{D}(A)$. In particular, the determinisation of $\mathcal{A}_{\text{Pre}}$ is isomorphic to the determinisation of $\mathcal{A}_{\text{POS}}$, i.e., $\mathcal{A}_{\text{MY}}$.

Proposition 21. $\mathrm{D}(\mathcal{A}_{\text{Pre}}(\alpha)) \simeq \mathcal{A}_{\text{MY}}(\alpha)$.

6   Taxonomy

In Fig. 1 the relations between the different automata are graphically represented. The two top nodes correspond to regular expressions. Each other node corresponds to a particular automaton, up to isomorphism, and edges between two nodes represent transformation algorithms, such as epsilon elimination ($\overline{\varepsilon}$) determinisation (D), reversal (R), quotient by some equivalence relation, or a specific construction. Edges in bold correspond to contributions in this paper. Different nodes represent objects for which there is some witness that distinguishes them. The relation between $\mathrm{D}(\mathcal{A}_{\text{PD}}(\alpha))$ and $\mathcal{A}_{\text{B}}(\alpha)$ was obtained by Nipkow and Traytel, and the ones between $\mathcal{A}_{\text{B}}(\alpha)$ and $\mathcal{A}_{\text{MB}}(\alpha)$ were obtained by Asperti et al.. The resulting automaton (between $\mathcal{A}_{\text{B}}(\alpha)$ and $\mathcal{A}_{\text{MB}}(\alpha)$ in the diagram) is the only one for which we do not have witnesses distinguishing it from the others. Brzozowski [4] showed that for a trim NFA $\mathcal{A}$, $\mathrm{D}(\mathcal{A})$ is minimal if $\mathcal{A}^{\mathrm{R}}$ is deterministic. Consequently, one obtains the nice property that, whenever $X(\alpha)$ is a deterministic automaton, for instance $\mathcal{A}_{\text{MB}}$ and $\mathcal{A}_{\text{MA}}$, then $\mathrm{D}(X(\alpha^{\mathrm{R}})^{\mathrm{R}})$ is the minimal DFA for $\mathcal{L}(\alpha)$. Experimental results suggest that $\mathcal{A}_{\text{MB}}(\alpha)$ is never larger than $\mathrm{D}(\mathcal{A}_{\text{PD}}(\alpha))$, so that should be investigated in future work.

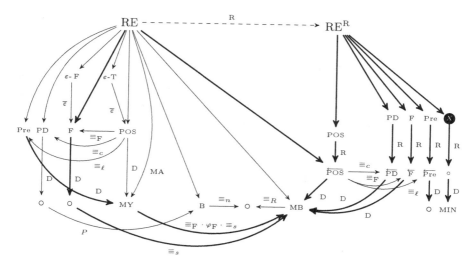

Fig. 1. Taxonomy of conversions from regular expressions to finite automata. Bold arrows correspond to relations studied in this work. Here X may be any construction that yields a DFA.

References

1. Allauzen, C., Mohri, M.: A unified construction of the glushkov, follow, and antimirov automata. In: Královič, R., Urzyczyn, P. (eds.) MFCS 2006. LNCS, vol. 4162, pp. 110–121. Springer, Heidelberg (2006). doi:10.1007/11821069_10
2. Antimirov, V.M.: Partial derivatives of regular expressions and finite automaton constructions. Theor. Comput. Sci. **155**(2), 291–319 (1996)
3. Asperti, A., Coen, C.S., Tassi, E.: Regular expressions, au point. CoRR abs/1010.2604. http://arxiv.org/abs/1010.2604 (2010)
4. Brzozowski, J.A.: Canonical regular expressions and minimal state graphs for definite events. In: Mathematical Theory of Automata, MRI Symposia Series, pp. 529–561. Polytechnic Press, Polytechnic Institute of Brooklyn, NY, 12 (1962)
5. Brzozowski, J.: Derivatives of regular expressions. J. Assoc. Comput. Mach. **11**, 481–494 (1964)
6. Champarnaud, J.M., Ziadi, D.: From Mirkin's prebases to Antimirov's word partial derivatives. Fundam. Inform. **45**(3), 195–205 (2001)
7. Champarnaud, J.M., Ziadi, D.: Canonical derivatives, partial derivatives and finite automaton constructions. Theor. Comput. Sci. **289**, 137–163 (2002)
8. Chen, H., Yu, S.: Derivatives of regular expressions and an application. In: Dinneen, M.J., Khoussainov, B., Nies, A. (eds.) WTCS 2012. LNCS, vol. 7160, pp. 343–356. Springer, Heidelberg (2012). doi:10.1007/978-3-642-27654-5_27
9. Giammarresi, D., Ponty, J.L., Wood, D.: Glushkov and Thompson constructions: A synthesis. HKUST TCSC-98-11, The Department of Science and Engineering, Theoretical Cmputer Science Group, The Hong Kong University of Science and Technology (1998)
10. Glushkov, V.M.: The abstract theory of automata. Russ. Math. Surveys **16**, 1–53 (1961)

11. Gruber, H., Holzer, M.: From finite automata to regular expressions and back–a summary on descriptional complexity. Intern. J. Found. Comput. Sci. **26**(8), 1009–1040 (2015)
12. Ilie, L., Yu, S.: Follow automata. Inf. Comput. **186**(1), 140–162 (2003)
13. Maia, E., Moreira, N., Reis, R.: Prefix and right-partial derivative automata. In: Beckmann, A., Mitrana, V., Soskova, M. (eds.) CiE 2015. LNCS, vol. 9136, pp. 258–267. Springer, Cham (2015). doi:10.1007/978-3-319-20028-6_26
14. McNaughton, R., Yamada, H.: Regular expressions and state graphs for automata. IEEE Trans. Comput. **9**, 39–47 (1960)
15. Mirkin, B.G.: An algorithm for constructing a base in a language of regular expressions. Eng. Cybern. **5**, 51–57 (1966)
16. Nipkow, T., Traytel, D.: Unified decision procedures for regular expression equivalence. In: Klein, G., Gamboa, R. (eds.) ITP 2014. LNCS, vol. 8558, pp. 450–466. Springer, Cham (2014). doi:10.1007/978-3-319-08970-6_29
17. Sakarovitch, J.: Elements of Automata Theory. Cambridge University Press, Cambridge (2009)
18. Thompson, K.: Regular expression search algorithm. Commun. ACM **11**(6), 410–422 (1968)
19. Yamamoto, H.: A new finite automaton construction for regular expressions. In: Bensch, S., Freund, R., Otto, F. (eds.) 6th NCMA. vol. 304, pp. 249–264 (2014)

Two-Way Two-Tape Automata

Olivier Carton[1]([✉]), Léo Exibard[2], and Olivier Serre[1]

[1] IRIF, Université Paris Diderot & CNRS, Paris, France
Olivier.Carton@irif.fr
[2] Département d'Informatique, ENS de Lyon, Lyon, France

Abstract. In this article we consider two-way two-tape (alternating) automata accepting pairs of words and we study some closure properties of this model. Our main result is that such alternating automata are not closed under complementation for non-unary alphabets. This improves a similar result of Kari and Moore for picture languages. We also show that these deterministic, non-deterministic and alternating automata are not closed under composition.

Keywords: Alternating · Multi-tape automata · Complementation

1 Introduction

In this article we consider two-way two-tape (alternating) automata that are designed to recognize binary relations between finite words. Although these automata are quite natural as read-only Turing machines, almost no work has been devoted to this model of computation. We study their properties, with a special focus on their closure properties, in particular under complementation and composition.

Finite states machines with inputs and outputs are widely used in many different areas like coding [11], computer arithmetics [12], natural language processing [13] and program analysis [2]. The simplest model is obtained by adding outputs to a classical finite-state (non)-deterministic one-way automaton to get a machine known as a transducer. In a transducer, the input word is only scanned once by a one-way head and the output is produced by the transitions used along the reading. A transducer can equivalently be seen as a machine with two tapes, one for the input and the other for the output, that are scanned once by two one-way heads. Relations realized by this kind of machines are called rational. They have been intensively studied since the early days of automata theory [3] and they enjoy some nice properties [14]. They are, for instance, closed under composition, but not under complementation.

Rational relations turn out to form a rather small class, hence classes of stronger transducers have been introduced by enriching transducers with extra features like two-wayness and/or alternation. A well studied class is that of two-way transducers in which the input word is scanned by a two-way head and

Funded by the DeLTA project (ANR-16-CE40-0007).

É. Charlier et al. (Eds.): DLT 2017, LNCS 10396, pp. 147–159, 2017.
DOI: 10.1007/978-3-319-62809-7_10

the output is produced (or equivalently scanned) by a one-way head. In that class, deterministic machines are of special interest as they are equivalent to MSO-transductions and turn out to be closed under composition [4].

In this article, we consider machines with two tapes which are both scanned by two-way heads. In this model, it is important to consider the output word as already written, and not produced on some output tape to ensure consistency, as it is scanned several times.

Another key feature, used to increase either the expressive power or the succinctness of finite state machines, is alternation, which allows the machine to spawn several copies of itself. Alternation often provides for free closure under complementation for machines on finite structures like words or trees. Indeed, the dual machine obtained by swapping existential and universal states and complementing the acceptance condition accepts the complement language as long as all computations terminate: if the run of a machine may loop, the closure under complementation of alternating machines may no longer hold.

Picture automata introduced in [1] scan a 2-dimensional array of symbols with moves in the four cardinal directions to either accept or reject it. As in the case of two-way automata, the border of the array is marked by special symbols. A run of such an automaton may loop as it can scan the same position twice with the same control state. These picture automata differ from classical word or tree automata where all variants (deterministic, non-deterministic, alternating) are equivalent. Indeed, it has been shown by Kari and Moore that alternating picture automata are not closed under complementation as soon as the alphabet size is greater than 1 [8]. This means that loops are inherent to the model and that they cannot be removed.

In this article, we show that two-tape two-way alternating automata are not closed under complementation either. Picture automata are actually very close to the model that we consider. In particular they coincide for unary alphabets. Indeed, over a unary alphabet, a pair of words is merely a pair of integers (their lengths) and this is equivalent to a two-dimensional array on a unary alphabet. However, as soon as the alphabets have cardinality at least 2, the models are distinct. Indeed, for an alphabet of size k, the number of $m \times n$-arrays is k^{mn} while the number of pairs of words is only k^{m+n}.

The fact that the two models coincide for unary alphabets allows us to recover immediately some separation and undecidability properties. For instance, it has been shown that such deterministic picture automata are strictly less powerful than non-deterministic ones which are, in turn, less powerful than alternating automata [8]. These results carry over the two-tape two-way automata that we consider.

To prove that alternating automata are not closed under complementation, we use a counter-example that is close to the one in [9] for picture automata. However, since our coding is different, we need some extra arguments to show that it is accepted by an alternating automaton and, to show that its complement cannot, we use a more direct proof.

2 Two-Way Two-Tape Automata

In this article, Σ is a finite alphabet and Σ^* denotes the set of finite words over Σ. For a word $u \in \Sigma^*$, we denote its length by $|u|$, and for each $1 \leq i \leq |u|$ we denote by u_i its i-th letter. From now on, $\vdash$ (begin) and $\dashv$ (end) are reserved characters not belonging to Σ and marking word boundaries. For simplicity of notation, we let $\Sigma_{\vdash}^{\dashv} = \Sigma \cup \{\vdash, \dashv\}$. For $u \in \Sigma^*$, we let $u_{\vdash}^{\dashv} = \vdash u \dashv$, with $u_{\vdash 0}^{\dashv} = \vdash$, $u_{\vdash |u|+1}^{\dashv} = \dashv$ and $u_{\vdash i}^{\dashv} = u_i$ for each $1 \leq i \leq |u|$.

A *(non-deterministic) two-way two-tape finite automaton* is a tuple $\mathcal{A} = (Q, \Sigma, \Delta, I, F)$, where Q is the set of *states*, and $I, F \subseteq Q$ are respectively the sets of *initial* and *final* states. We call $\Delta \subseteq (Q \times \Sigma_{\vdash}^{\dashv} \times \Sigma_{\vdash}^{\dashv}) \times (Q \times \{\triangleleft, \triangledown, \triangleright\} \times \{\triangleleft, \triangledown, \triangleright\})$ the *transition relation*. We use the notation $p \xrightarrow{a_1, a_2 | d_1, d_2} q$ for $(p, a_1, a_2, q, d_1, d_2) \in \Delta$. We require that the reading heads cannot cross the words boundaries, i.e. for every transition $p \xrightarrow{a_1, a_2 | d_1, d_2} q$ and every $i = 1, 2$ if $a_i = \vdash$ (*resp.* $a_i = \dashv$) then $d_i \neq \triangleleft$ (*resp.* $d_i \neq \triangleright$).

An automaton $\mathcal{A}$ is said to be *deterministic* whenever for every state $p \in Q$ and every letters $a_1, a_2 \in \Sigma$, there exists at most one $q \in Q, d_1, d_2 \in \{\triangleleft, \triangledown, \triangleright\}$ such that $p \xrightarrow{a_1, a_2 | d_1, d_2} q$.

Note that extending the model to more than two tapes is straightforward. Note also that if we restrict the model to a single tape we retrieve the classical notion of two-way automata on finite words (Fig. 1).

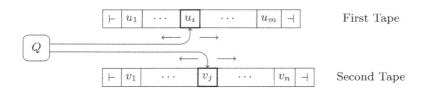

Fig. 1. Schema of a two-way two-tape finite automaton

A *configuration* is a triple $(q, i, j) \in Q \times \mathbb{N} \times \mathbb{N}$, where q is the current state, and i (resp. j) is the position of the reading head on the first (resp. second) tape (recall that the $\vdash$ marker is by convention at position 0).

We say that a configuration (q_2, i_2, j_2) is a *successor* of a configuration (q_1, i_1, j_1) with regard to input (u, v), written $(q_1, i_1, j_1) \xrightarrow{(u,v)} (q_2, i_2, j_2)$, when automaton $\mathcal{A}$ can go from one configuration to the next in a single step, i.e. if $q_1 \xrightarrow{(u_{\vdash}^{\dashv})_{i_1}, (v_{\vdash}^{\dashv})_{j_1} | d, e} q_2$, and if $i_2 = i_1 + \chi(d)$ and $j_2 = j_1 + \chi(e)$, where $\chi(\triangleleft) = -1$, $\chi(\triangledown) = 0$ and $\chi(\triangleright) = 1$.

A *run* of $\mathcal{A}$ on input $(u, v) \in \Sigma^* \times \Sigma^*$ is a (possibly infinite) sequence $(p_k, i_k, j_k)_{1 \leq k < n}$, where $n \in \mathbb{N} \cup \{\infty\}$, of successive configurations: for every $1 \leq k < n$, one has $(p_k, i_k, j_k) \xrightarrow{(u,v)} (p_{k+1}, i_{k+1}, j_{k+1})$.

An *initial run* is a run starting with an initial configuration, i.e. one such that $p_0 \in I$ and $i_0 = j_0 = 0$. It is *accepting* when it is moreover finite and contains a final state $f \in F$, i.e. there exists $k \leq n$ such that $p_k \in F$.

The *relation* $\mathcal{R}(\mathcal{A})$ *accepted by* a two-way two-tape automaton $\mathcal{A}$ is the set of pairs (u, v) such that there exists an accepting run of $\mathcal{A}$ on input (u, v).

We now introduce alternating automata, which generalize non-deterministic automata. An *alternating automaton* is an automaton whose set of control states is partitioned into *existential* ($Q_\exists$) and *universal* ($Q_\forall$) states. A configuration is defined as in the non-deterministic setting and it is existential (*resp.* universal) if the control state is.

Runs of alternating automata are (possibly infinite) trees whose nodes are labeled by configurations, and such that each inner node u labeled by a configuration C satisfies the following conditions:

- If C is existential then u has a single son that is labeled by a successor configuration of C.
- If C is universal and if $\{C_1, \ldots, C_k\}$ denotes all successor configurations of C, then u has k sons each of them labeled by a different C_i for $1 \leq i \leq k$.

A run is *accepting* if it is finite, its root is labeled by an initial configuration and all its leaves are labeled either by accepting configurations, or by universal configurations that have no successor configuration. Again, we define the relation accepted by an alternating automaton as those pairs of words over which there is an accepting run. Note that non-deterministic automata correspond to the case where $Q_\forall = \emptyset$.

First we remark that if we restrict the model by forbidding the reading heads to go to the left, we obtain a 1-way model that recognizes the rational relations, i.e. those realized by transducers [14]. Not surprisingly this is a restriction as two-way two-tape automata can, for instance, recognize deterministically the relation $\{(u, \tilde{u}) \mid u \in \Sigma^*\}$, where $\tilde{u}$ denotes the reverse of u. The corresponding automaton is depicted in Fig. 2. This relation cannot be recognized by classical one-way transducers.

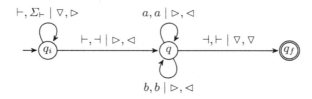

Fig. 2. Automaton recognizing $\{(u, \tilde{u}) \mid u \in \Sigma^*\}$ when $\Sigma = \{a, b\}$

Our model captures much more complex relations. It is shown in the next section that two-way two-tape automata can recognize the relation Coprime $= \{(a^p, a^q) \mid p \wedge q = 1\}$, where $p \wedge q$ denotes the greatest common divisor of p and q, by implementing a variant of the Euclidean algorithm.

3   Picture Languages

For brevity we simply recall here some key notions on picture languages: we refer the reader to [5] for an excellent survey on the topic with complete definitions of all objects discussed below.

A *picture* p of dimensions $m \times n$ is a matrix over a finite alphabet Σ. For every $1 \leq i \leq m$ and $1 \leq j \leq n$, we write $p_{i,j}$ for the content of the cell at position (i,j). To recognize pictures, we add a special marker $\# \notin \Sigma$ all around the picture p, i.e. we adopt the convention that for all $0 \leq i \leq m + 1$ and $0 \leq j \leq n + 1$, $p_{0,j} = p_{m+1,j} = p_{i,0} = p_{i,n+1} = \#$. We write Σ^{**} for the set of all pictures over Σ. A *picture language* is thus a subset of Σ^{**}.

In order to recognize picture languages, 4-way automata were first introduced in [1]. Such an automaton has a single head which is able to move on a two-dimensional array of symbols (surrounded by markers) in the four directions (up, down, left, right) and accepts when reaching a final state. A schema is provided in Fig. 3.

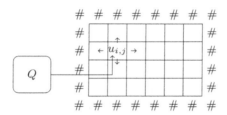

Fig. 3. Schema of a 4-way automaton

The link between two-way two-tape automata and picture automata is provided by special pictures, called *products of words*, that we now define. For $u, v \in \Sigma^*$, we define the picture $u \otimes v = ((u_i, v_j))_{\substack{1 \leq i \leq |u| \\ 1 \leq j \leq |v|}}$ over the product alphabet $\Sigma \times \Sigma$.

Thus, any relation $\mathcal{R} \subseteq \Sigma^* \times \Sigma^*$ is mapped to a picture language $L_{\mathcal{R}}^{\otimes} \subseteq (\Sigma \times \Sigma)^{**}$. Moreover, over unary alphabets, pictures languages and binary relations over words are in one-to-one correspondance as the pair (a^m, a^n) unambiguously represents an image of dimensions $m \times n$ and conversely. Consequently, for $L \subseteq \{a\}^{**}$, L is recognizable by a 4-way deterministic (resp. non-deterministic, alternating) automaton *iff* $\mathcal{R}$ is recognizable by a deterministic (resp. non-deterministic, alternating) two-way two-tape automaton where $\mathcal{R}$ is the unique relation such that $L = L_{\mathcal{R}}^{\otimes}$.

Thus, all the results known for unary picture languages also hold in our model when $|\Sigma| = 1$. In particular, in [8], it is shown that determinism is strictly weaker than non-determinism, the latter being weaker than alternation: DFA $\subsetneq$ NFA $\subsetneq$ AFA where DFA (resp. NFA, AFA) is the class of relations recognized by deterministic (resp. non-deterministic, alternating) two-way two-tape automata.

The relation $\mathcal{R} = \{(a^w, a^h) \mid \exists i, j \in \mathbb{N}, w = ih + j(h+1)\}$ is such that $\mathcal{R} \notin$ DFA, $\mathcal{R} \in$ NFA, $\overline{\mathcal{R}} \notin$ NFA, $\overline{\mathcal{R}} \in$ AFA, where $\overline{\mathcal{R}}$ denotes the complement of $\mathcal{R}$ in $\Sigma^* \times \Sigma^*$. In [10], it is shown that the emptiness problem is undecidable even for a unary alphabet.

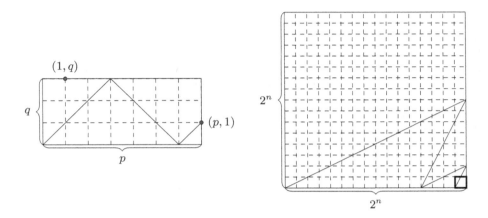

Fig. 4. Euclidean algorithm (left) and squares of size 2^n (right)

The following examples from [9] show that picture automata and two-way two-tape automata are really expressive. The relation Coprime $= \{(a^p, a^q) \mid p \wedge q = 1\}$ can be recognized by a deterministic automaton implementing a variant of the Euclidean algorithm. The automaton follows diagonals until it reaches either one of $(p, 1)$ or $(1, q)$ or one of $(0, 0)$, $(p, 0)$, $(0, q)$ or (p, q). In the former case, it accepts and in the latter case, it rejects.

A more sophisticated deterministic automaton can recognize the relation $\{(2^n, 2^n) \mid n \in \mathbb{N}\}$ following the schema in Fig. 4 (right). The automaton moves with a ratio of $1/2$ as long as it is possible (i.e. while the remaining length is divisible by 2), and it accepts if it reaches the bottom-right square. An even more sophisticated deterministic automaton can accept the relation $\{(2^{2^n}, 2^{2^n}) \mid n \in \mathbb{N}\}$.

4   Alternating Two-Way Two-Tape Are Not Closed Under Complementation

Our main result is the following. Note that the case of unary alphabets is still an open problem.

Theorem 1. *The class of relations recognized by alternating two-way two-tape automata is not closed under complementation as soon as the alphabet is not unary.*

In this proof we denote by AFA the class of relations recognized by alternating two-way two-tape automata, and by co-AFA the class of relations whose complement is recognized by alternating two-way two-tape automata. Hence, we aim to prove that AFA and co-AFA are distinct, and this is achieved by defining a well-chosen relation $\mathcal{P}$ and show that $\mathcal{P} \in$ AFA but $\mathcal{P} \notin$ co-AFA. The first point is proved by giving an explicit alternating two-way two-tape automaton recognizing $\mathcal{P}$ (Lemma 2); the second point is proved by contradiction using a game-theoretic approach combined with a combinatorial argument (Lemma 3).

The proof is inspired by the one in [8] establishing that the set of picture languages recognized by alternating 4-way automata is not closed under complementation. One difference is that the counter-example they exhibit cannot be used in our setting as the images that are considered are not product of words (as defined in Sect. 3). Hence, even if the general idea — coding a permutation to build a counter-example — is similar to the one in [8], our coding is different and therefore new ideas are needed to prove that we can accept this relation with an alternating two-way two-tape automaton. The proof that the complement is not recognizable by an alternating two-way two-tape automaton, even if it shares ideas with the one in [8], is somehow more direct as it does not appeal to an intermediate class.

In the following, if $c_1, \ldots, c_n$ are n words of length n over alphabet $\{0, 1\}$ we identify the tuple $(c_1, \ldots, c_n)$ with the $n \times n$ matrix whose i-th column is c_i. We define the relation $\mathcal{P}$ by

$$\mathcal{P} = \{(a^n, c_1\# \cdots \#c_n\$c_1\# \cdots \#c_n) \mid n \in \mathbb{N}, (c_1, \ldots, c_n) \in \mathfrak{S}_n\}$$

where $\mathfrak{S}_n$ denotes the set of all $(n \times n \ \{0, 1\}$-matrix coding) permutations over $\{1, \ldots, n\}$. See Fig. 5 for an example.

$$
\begin{array}{|ccc|}
0\ 0\ 1\ \$\ 0\ 0\ 1 \\
1\ 0\ 0\ \$\ 1\ 0\ 0 \\
0\ 1\ 0\ \$\ 0\ 1\ 0 \\
\end{array}
\qquad (a^3, 010\#001\#100\$010\#001\#100)
$$

Fig. 5. Two identical permutations separated by \$s and the corresponding encoding

As announced, we start by showing that $\mathcal{P} \in$ AFA.

Lemma 2. *There exists an alternating two-way two-tape automaton recognizing $\mathcal{P}$.*

Proof. We describe below how to check whether a pair $(a^n, c) \in a^* \times \{0, 1, \$, \#\}^*$ is in $\mathcal{P}$. As the a^n word is used only to store position we sometimes refer to the content of the first tape as the *counter*.

To decide if $(a^n, c) \in \mathcal{P}$, we have to check two things. The first one is whether c is of the form $\sigma_1\$\sigma_2$, where σ_1, σ_2 are the encodings of two permutations of dimension n and the second one is whether $\sigma_1 = \sigma_2$.

For the first step, we only explain how to check the property for σ_1 as the case of σ_2 is checked in the same way. Assume $\sigma_1 = c_0 \# c_1 \# \cdots \# c_n$ (checking that there are n block is easy). We first need to check that each c_i (i.e. each column) has length n and contains exactly one 1: this is easy (the length condition being checked thanks to the first tape). We then need to check that for each $k = 1, \ldots, n$ there is exactly one c_i whose k-th letter is a 1 (i.e. each row contains exactly one 1). This property is checked for each $k = 1, \ldots, n$ in increasing order starting from $k = 1$, and at the beginning of step k the head in the counter in the first tape is at position k and the head in the second tape is at the beginning of σ_1: now going left on the counter and right on the second tape at the same speed the k-th symbol of c_1 is reached (just before the left marker is read on the first tape); then the automaton goes back to the beginning of c_1 and the counter is increased back to k; finally the second head goes to the beginning of c_2 (going to the right until reading a $\#$), then c_2 is processed in the same way, and so on until reading a $\$$ meaning that c_n was processed and that one can go back to the beginning of the second tape and start again but now for $k + 1$.

We are now left with checking that $\sigma_1 = \sigma_2$ knowing that both σ_1 and σ_2 encode a permutation. For that we use the same approach as the one in [8, Lemma 2, Condition (*)]. More precisely, we use the following property: two permutations are different *iff* there exists an inversion i.e. there exists $i \neq i', \sigma_1(i) < \sigma_1(i')$ but $\sigma_2(i) > \sigma_2(i')$. This can be checked by trying all possible way of moving in the associated picture (as illustrated in Fig. 5) in the following fashion (see Fig. 6): from a 1 move right to another column but stay on the same side of the column of $\$$s. Find the 1 on that column. Then move to the other 1 that is on the same row, on the opposite side of the column of $\$$s, and repeat: then, the machine enters an infinite loop *iff* it finds an inversion [8].

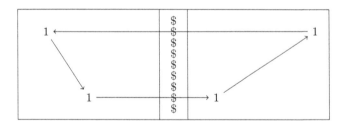

Fig. 6. Loop corresponding to an inversion

The first step (finding a 1) is easy and done using universal states (which ensures that we do the check for all possible 1). Moving to another column on the right is easy: simply use a universal state and keep moving the head to the right passing one or more $\#$ but stopping before reading the $\$$. Finding the 1 on the column is easy (look for it in between the left and right $\#$). Finally, moving to the other 1 on the same row on the opposite side of the $\$$ is achieved by using the counter to keep in memory the row number j, and then go to the other side and check every column until finding the one with a 1 on the j-th row (this is done with the same trick used previously to check that each row contains a single 1). If at some point when moving to a column to the right the automaton hit a \$ then it goes to a final state. The automaton accepts the desired language for the following reason: if an inversion exists, it is found (thanks to universal choices) and leads to a looping hence, rejecting, computation; if not, whatever the universal choices are the automaton is getting closer and closer to the \$ and eventually hit it thus reaching a final state.

Therefore, we built an alternating two-way two-tape automaton recognizing $\mathcal{P}$. More generally, the ideas used in the first part of the proof can be used to prove the following: for any picture language L recognizable by a 4-way automaton, the relation $\{(a^{\max(n,m)}, c_1 \# \cdots \# c_m) \mid (c_1, \ldots, c_m)$ is an $n \times m$ picture in L$\}$ is recognizable by a two-way two-tape automaton.   □

We now show that the complement of $\mathcal{P}$ cannot be recognized by an alternating two-way two-tape automaton.

Lemma 3. *Relation* $\mathcal{P} \notin$ *co-AFA.*

Proof. The proof is by contradiction assuming the existence of an alternating two-way two-tape automaton $\mathcal{A} = (Q, \Sigma, \Delta, I, F)$ such that a pair (u, v) is accepted by $\mathcal{A}$ *iff* $(u, v) \notin \mathcal{P}$. Hence, a pair (a^n, c) is rejected by $\mathcal{A}$ *iff* if belongs to $\mathcal{P}$.

We now rephrase the fact that (a^n, c) is rejected by $\mathcal{A}$, to that end we use a game-theoretic flavoured argument. Indeed, another way of thinking about a run of an alternating automaton is as a two-player games where the existential player is in charge of choosing the transition in existential configurations while the universal player takes care of universal configurations. The winning condition for the existential player is that a final configuration is eventually reached. In particular $\mathcal{A}$ rejects its input *iff* the universal player has a winning strategy, i.e. a playing policy ensuring no final state is ever reached whatever the existential player does. Moreover, this strategy can be chosen to be positional, i.e. only depending on the configuration, not on how it was reached (see *e.g.* [6]).

Fix an input of the form $(a^n, \sigma\$\sigma) \in \mathcal{P}$. A strategy for the universal player and the proof that it is a winning one can be described as follows. For each position $0 \leq j \leq 2n + 2$ on the second tape (including the markers) we associate a $n + 2$ tuple $\tau_j = (\tau_{0,j}, \ldots, \tau_{n+1,j})$ of functions describing what configurations can appear when the heads are at position (i, j) and for universal ones what the strategy of the universal player is. Hence for each $0 \leq i \leq n + 1$, $\tau_{i,j}$ is a map from Q such that:

- if $q \in Q_\exists$, $\tau_{i,j}(q)$ is either $\bot$ (meaning (q, i, j) is not reachable) or $\top$ (reachable)
- if $q \in Q_\forall$, $\tau_{i,j}(q)$ is either $\bot$ (not reachable) or a transition in Δ starting with (q, x, y) where x (*resp.* y) is the letter at position i (*resp.* j) in the first (*resp.* second) tape, including markers. The latter case means that configuration (q, i, j) can appear and gives the corresponding move in the universal player strategy.

Such an object $\tau_0, \ldots, \tau_{2n+2}$ is a *proof of reject* if it satisfies the following four conditions.

1. For each initial state $q \in I$, $\tau_{0,0}(q) \neq \bot$ (all initial configurations are allowed).
2. For each final state $q \in F$ and each i, j, $\tau_{i,j}(q) = \bot$ (no final configuration can be reached).
3. For each existential state q and each i, j such that $\tau_{i,j}(q) = \top$, for each possible successor (q', i', j') of (q, i, j) one has that $\tau_{i',j'}(q') \neq \bot$ (all successors of allowed existential configurations are allowed).
4. For each universal state q and each i, j such that $\tau_{i,j}(q) \neq \bot$, if one denotes by (q', i', j') the configuration reached from (q, i, j) applying transition $\tau_{i,j}(q)$ then $\tau_{i',j'}(q') \neq \bot$ (an allowed universal configurations has its successor described by the strategy allowed).

It is then standard to notice that $\mathcal{A}$ has no accepting run over $(a^n, \sigma\$\sigma)$ *iff* there is a proof of reject for it.

Now, for a fixed n, consider the number of possible values for the central part $(\tau_n, \tau_{n+1}, \tau_{n+2})$ of a proof: it is smaller than $((|\Delta| + 1)^{|Q|})^{3n}$. Hence, for n large enough it is smaller than $n!$. For such n it means that there are two distinct permutations $\sigma \neq \sigma'$ such that the proofs of reject for $\sigma\$\sigma$ and for $\sigma'\$\sigma'$ coincide on their central part: hence, they can be combined (glue the left part of the first proof with the right part of the second), leading a proof of reject for $\sigma\$\sigma'$. But this leads a contradiction as $\sigma\$\sigma' \notin \mathcal{P}$ and therefore is accepted by $\mathcal{A}$.   □

5   Union, Intersection and Composition

In this section, we study the closure under union, intersection and composition of two-way two-tape automata.

5.1   Closure Under Union and Intersection

Concerning closure under union and intersection we have the following picture.

Lemma 4. *Relations recognized by deterministic (resp. non-deterministic, resp. alternating) two-way two-tape automata are closed under union and intersection.*

Proof (sketch). For non-deterministic automata union is for free; intersection is simply obtained by simulating the first automaton, and if it reaches an accepting state, simulating the second automaton. For alternating automata both closures are for free. For deterministic automata, intersection can be obtained as in the non-deterministic case. Concerning union, one can no longer use non-determinism and hence, one needs to simulate successively the two automata. However this does not work in general as a two-way two-tape automaton can reject by entering an infinite loop. Hence, one must first get rid of such phenomenon: this can be achieved thanks to a method due to Sipser [15], which ensures that a deterministic two-way two-tape automaton never rejects by looping.   □

Remark that preventing deterministic two-way two-tape automata from rejecting by looping can be used to prove that the class of relations they recognize is also closed under complementation, and therefore forms a Boolean algebra.

5.2   Non-closure Under Composition

In this section, we prove that the class of relations recognized by two-way two-tape automata is not closed under composition, even in the deterministic case.

Theorem 5. *Relations recognized by deterministic (*resp. *non-deterministic, resp. alternating) two-way two-tape automata are not closed under composition.*

Proof. The proof works the same way for all three models. One first establishes (see Lemma 6 below) an elementary bound on the growth of the functions recognized by our model, and then exhibit (see Lemma 7 below) a recognizable relation breaking this bound when self-composed.   □

We now give the lemmas used in the proof of Theorem 5. The following lemma bounds the growth of the functions, that is functional relations recognized by two-way two-tape automata (a similar statement can easily be obtained for alternating two-way two-tape automata with a doubly exponential bound instead).

Lemma 6. *If f is a function recognized by a two-way two-tape automaton, then there exists $k \in \mathbb{N}$ such that for all $n \in \mathbb{N}$ and $u \in \Sigma^n$, one has $|f(u)| \le \binom{2nk}{nk+1}$, where $\binom{n}{k}$ denotes the binomial coefficient.*

Proof. Let $\mathcal{A}$ be a two-way two-tape automaton recognizing a function f. Let $n \in \mathbb{N}$, and $u \in \Sigma^n$. Finally, let k be the size of $\mathcal{A}$. Fixing u, we easily build from $\mathcal{A}$ a two-way automaton with nk states accepting the singleton language $\{f(u)\}$. Now, thanks to a result by Kapoutsis [7] establishing that any m-state two-way automaton can be transformed into an equivalent equivalent non-deterministic finite automaton with $\binom{2m}{m+1}$ states, we have that $\{f(u)\}$ is accepted by a non-deterministic finite automaton with $\binom{2nk}{nk+1}$ states. As the shortest word accepted by an m-state non-deterministic finite automaton has length at most m, it concludes the proof.   □

For $i \in \mathbb{N}$ and $n \ge \log_2(i)$ let $\langle i \rangle_n \in \{0,1\}^n$ be the writing of i in base 2 on n bits, with the less significant bit on the left (possibly padded with zeros). For example, $\langle 6 \rangle_5 = 01100$. We are now ready to exhibit a function with an exponential growth and recognized by a deterministic two-way two-tape automaton.

Lemma 7. *The function $\mathcal{E} = \{(w, \#w_0\#w_1\#\dots\#w_{2^{|w|}-1}\#) \mid w_i = \langle i \rangle_{|w|}\}$ is recognizable by a deterministic two-way two-tape automaton.*

The proof is based on the concept of so-called synchronous relations. A relation is *synchronous* if it is recognized by a synchronous transducer, which can be simulated by a two-way two-tape automaton whose two heads always move simultaneously to the right and which accepts once the two inputs words are entirely scanned. An example of such a relation is $\mathcal{I} = \{(\langle i \rangle_n, \langle i+1 \rangle_n) \mid n \ge 1$ and $0 \le i \le 2^n - 1\}$ since an integer can deterministically be incremented starting from its less significant digit. Note that a synchronous transducer can always be made deterministic as it is a classical automaton over the product alphabet.

Lemma 8. *Let $\mathcal{R}$ be a synchronous relation. Then $\{(w, \#u\#v\#) \mid (u,v) \in \mathcal{R}, |w| = |u| = |v|\}$ is recognized by a deterministic two-way two-tape automaton.*

The rough idea to prove Lemma 8 is to use the first tape to keep track of the position when alternatively reading the u and v part of the word written on the second tape.

Now, using Lemma 8 with relation $\mathcal{I}$ we conclude that the relation $\mathcal{J} = \{(w, \# \langle i \rangle_{|w|} \# \langle i+1 \rangle_{|w|} \# \mid 0 \leq i \leq 2^{|w|} - 1\}$ is recognizable by a deterministic two-way two-tape automaton.

Finally, $\mathcal{E}$ is recognized by iterating the methods for $\mathcal{J}$ (this is made possible thanks to the presence of $\#$) and stopping once the second word is only composed of 1 s.

References

1. Blum, M., Hewitt, C.: Automata on a 2-dimensional tape. In: 8th Annual Symposium on Switching and Automata Theory (SWAT 1967), pp. 155–160. IEEE (1967)
2. Cohen, A., Collard, J.F.: Instance-wise reaching definition analysis for recursive programs using context-free transductions. In: International Conference on Parallel Architectures and Compilation Techniques, pp. 332–339. IEEE (1998)
3. Elgot, C.C., Mezei, J.E.: On relations defined by generalized finite automata. IBM J. Res. Dev. **9**, 47–68 (1965)
4. Engelfriet, J., Hoogeboom, H.J.: MSO definable string transductions and two-way finite-state transducers. ACM Trans. Comput. Log. **2**(2), 216–254 (2001)
5. Giammarresi, D., Restivo, A.: Recognizable picture languages. Int. J. Pattern Recogn. Artif. Intell. **6**, 241–256 (1992)
6. Grädel, E., Thomas, W., Wilke, T. (eds.): Automata, Logics, and Infinite Games: A Guide to Current Research. LNCS, vol. 2500. Springer, Heidelberg (2002)
7. Kapoutsis, C.: Removing bidirectionality from nondeterministic finite automata. In: Jędrzejowicz, J., Szepietowski, A. (eds.) MFCS 2005. LNCS, vol. 3618, pp. 544–555. Springer, Heidelberg (2005). doi:10.1007/11549345_47
8. Kari, J., Moore, C.: New results on alternating and non-deterministic two-dimensional finite-state automata. In: Ferreira, A., Reichel, H. (eds.) STACS 2001. LNCS, vol. 2010, pp. 396–406. Springer, Heidelberg (2001). doi:10.1007/3-540-44693-1_35
9. Kari, J., Moore, C.: Rectangles and squares recognized by two-dimensional automata. In: Karhumäki, J., Maurer, H., Păun, G., Rozenberg, G. (eds.) Theory Is Forever. LNCS, vol. 3113, pp. 134–144. Springer, Heidelberg (2004). doi:10.1007/978-3-540-27812-2_13
10. Kari, J., Salo, V.: A survey on picture-walking automata. In: Kuich, W., Rahonis, G. (eds.) Algebraic Foundations in Computer Science. LNCS, vol. 7020, pp. 183–213. Springer, Heidelberg (2011). doi:10.1007/978-3-642-24897-9_9
11. Lind, D., Marcus, B.: An Introduction to Symbolic Dynamics and Coding. Cambridge University Press, Cambridge (1995)
12. Lothaire, M.: Algebraic Combinatorics on Words, Chap. 7, pp. 230–268. Cambridge University Press, Cambridge (2002)

13. Roche, E., Schabes, Y.: Finite-State Language Processing, Chap. 7. MIT Press, Cambridge (1997)
14. Sakarovitch, J.: Elements of Automata Theory. Cambridge University Press, Cambridge (2009)
15. Sipser, M.: Halting space-bounded computations. Theor. Comput. Sci. **10**(3), 335–338 (1980)

Undecidability and Finite Automata

Jörg Endrullis[1], Jeffrey Shallit[2(✉)], and Tim Smith[2]

[1] Department of Computer Science, Vrije Universiteit Amsterdam,
De Boelelaan 1081a, 1081 HV Amsterdam, The Netherlands
joerg@endrullis.de
[2] School of Computer Science, University of Waterloo,
Waterloo, ON N2L 3G1, Canada
{shallit,timsmith}@uwaterloo.ca

Abstract. Using a novel rewriting problem, we show that several natural decision problems about finite automata are undecidable (i.e., recursively unsolvable). In contrast, we also prove three related problems are decidable. We apply one result to prove the undecidability of a related problem about k-automatic sets of rational numbers.

Keywords: Finite automata · Undecidability · Conjugate · Power

1 Introduction

Starting with the first result of Turing [16], computer scientists have assembled a large collection of natural decision problems that are undecidable (i.e., recursively unsolvable); see, for example, the book of Rozenberg and Salomaa [12].

Although some of these results deal with relatively weak computing models, such as pushdown automata [2,6], few, if any, are concerned with the very simplest model: the finite automaton. One exception is the following decision problem, due to Engelfriet and Rozenberg [5, Theorem 15]: given a finite automaton M with an input alphabet of both primed and unprimed letters (i.e., an alphabet $\Sigma \cup \Sigma'$, where $\Sigma' = \{a' : a \in \Sigma\}$), decide if M accepts a word w where the primed letters, after the primes have been removed, form a word identical to that formed by the unprimed letters. This problem was also mentioned by Hoogeboom [7]. This problem is easily seen to be undecidable, as it is a disguised version of the classical Post correspondence problem [10].

In this paper we start by proving a novel lemma on rewriting systems. In Sect. 3, this lemma is then applied to give a new example of a natural problem on finite automata that is undecidable. In Sect. 4 we prove that a related problem on the so-called k-automatic sets of rational numbers is also undecidable. In Sect. 5 we prove the undecidability of yet another problem about finite automata. Finally, in Sect. 5 we show that it is decidable if a finite automaton accepts two distinct conjugates.

© Springer International Publishing AG 2017
É. Charlier et al. (Eds.): DLT 2017, LNCS 10396, pp. 160–172, 2017.
DOI: 10.1007/978-3-319-62809-7_11

2   A Lemma on Rewriting Systems

For our purposes, a *rewriting system* S over an alphabet Σ consists of a finite set of context-free rules of the form $\ell \rightarrow r$, where $\ell, r \in \Sigma^*$. Such a rewriting rule applies to the word $\alpha\ell\beta \in \Sigma^*$, and converts it to $\alpha r\beta$. We indicate this by writing $\alpha\ell\beta \Longrightarrow \alpha r\beta$. We use $\Longrightarrow^*$ for the reflexive, transitive closure of $\Longrightarrow$ (so that $\gamma \Longrightarrow^* \zeta$ means that there is a sequence of 0 or more rules taking the word γ to ζ). A rewriting system is said to be *length-preserving* if $|\ell| = |r|$ for all rewriting rules $\ell \rightarrow r$.

Many undecidable decision problems related to rewriting systems are known [3]. However, to the best of our knowledge, the following one is new.

REWRITE-POWER

Instance: An alphabet Σ containing the symbols a and b (and possibly other symbols), and a length-preserving rewriting system S.

Question: Does there exist an integer $n \geq 1$ such that $a^n \Longrightarrow^* b^n$?

Lemma 1. *The decision problem* REWRITE-POWER *is undecidable.*

Proof. The standard approach for showing that a rewriting problem is undecidable is to reduce from the halting problem, by encoding a Turing machine M and simulating its computation using the rewriting rules; for example, see [3].

The difficulty with applying that approach in the present case is the lack of asymmetry, i.e., the fact that the initial word consists of all a's. Because there is no distinguished symbol with which to start the simulation, unwanted parallel simulations of M could occur at different parts of the word.

To deal with this difficulty, we construct a rewriting system that permits multiple simulations of M to arise, but employs a delimiter symbol $\$$ to ensure that they do not interfere with each other. Each simulation works on its own portion of the word, and changes it to b's (ending by changing the delimiter symbol as well) only if M halts.

Here are the details. We use the one-tape model of Turing machine from Hopcroft and Ullman [8], where $M = (Q, \Omega, \Gamma, \delta, q_0, \mathsf{B}, q_f)$. Here Q is the set of states of M, with $q_0 \in Q$ the start state and $q_f \in Q$ the unique final state. Let Ω be the input alphabet and Γ the tape alphabet, with $\mathsf{B} \in \Gamma$ being the distinguished blank symbol. Let δ be the (partial) transition function, with domain $Q \times \Gamma$ and range $Q \times \Gamma \times \{L, R\}$. We assume without loss of generality that M halts, i.e., M has no next move, iff it is in state q_f. We also assume that $a, b, \$ \notin \Gamma$.

We construct our length-preserving rewriting system mimicking the computations of M as follows. Let $\Sigma = \Gamma \cup Q \cup \{a, b, \$\}$. Let S contain the following rewriting rules:

$$aa \rightarrow \$q_0 \tag{1}$$
$$a \rightarrow \mathsf{B} \tag{2}$$
$$q_i c \rightarrow d q_j \quad \text{if } (q_j, d, R) \in \delta(q_i, c) \text{ with } c, d \in \Gamma \tag{3}$$
$$f q_i c \rightarrow q_j f d \quad \text{if } (q_j, d, L) \in \delta(q_i, c) \text{ and } f \in \Gamma \tag{4}$$
$$q_f c \rightarrow c q_f \quad \text{for all } c \in \Gamma \tag{5}$$
$$c q_f \rightarrow q_f b \quad \text{for all } c \in \Gamma \tag{6}$$
$$\$q_f \rightarrow bb \tag{7}$$

Rule (1) starts a new simulation. Each simulation has its own head symbol q_i and left end-marker \$, and never moves the head symbol past an a, b, or \$. This ensures that the simulations are kept separate from each other. Rule (2) converts an a to a blank symbol, available for use in a simulation. Rules (3) and (4) are used to simulate the transitions of M. Once a simulation reaches q_f (meaning that M has halted), its head can be moved to the right using rule (5), past all of the symbols it has read, and then back to the left using rule (6), changing all of those symbols to b's. Finally, the simulation can be stopped using rule (7).

We argue that M halts when run on a blank tape iff $a^n \Longrightarrow^* b^n$ for some $n \geq 1$.

If M halts when run on a blank tape, then there is some number of tape cells k which it uses. Let $n = k + 2$. Then S can use rule (1) to start a simulation with the two a's at the left end of the word, use rule (2) to convert the k remaining a's to blank cells, and run the simulation using rules (3) and (4). Eventually M halts in the final state q_f. At that point, S can move the head to the right end of the word using rule (5), convert all k tape cells to b's using rule (6), and then convert the end-marker \$ and tape head to b's with rule (7). Thus we have $a^n \Longrightarrow^* b^n$.

For the other direction, if the initial word a^n is ever transformed into b^n, it means that one or more simulations were run, each of which operated on a portion of the word without interference from the others. The absence of interference can be deduced from the shape of the rules. There is only one occurrence of the marker \$ in the left-hand sides of the rules, namely as the leftmost symbol in the left-hand side of rule (7). Furthermore, \$ can only be rewritten to b which does not occur in any left-hand side.

Each simulation runs M on a blank tape, and uses a number of tape cells bounded by the length of its portion of the word. Since every portion of the word was transformed into b's, M halted in every one of the simulations (otherwise the word would still contain one or more head symbols and end-markers). The completion of any one of these simulations is enough to show that M halts when run on a blank tape.

Therefore M halts when run on a blank tape iff there exists an $n \geq 1$ such that $a^n \Longrightarrow^* b^n$, completing the reduction from the halting problem. Since the halting problem is undecidable, REWRITE-POWER is also undecidable. □

3   An Undecidable Problem on Finite Automata

Our model of finite automaton is the usual one (e.g., [8]). We now consider a decision problem on finite automata. To state it, we need the notion of the product of two words of the same length. Let Σ, Δ be alphabets, and let $w \in \Sigma^*$, $x \in \Delta^*$, with $|w| = |x|$. Then by $w \times x$ we mean the word over the alphabet $\Sigma \times \Delta$ whose projection π_1 over the first coordinate is w and whose projection π_2 over the second coordinate is x. More precisely, if $w = a_1 a_2 \cdots a_n$ and $x = b_1 b_2 \cdots b_n$, then $w \times x = [a_1, b_1][a_2, b_2] \cdots [a_n, b_n]$. In this case $\pi_1(w \times x) = w$ and $\pi_2(w \times x) = x$. For example, if $y = [\texttt{t}, \texttt{h}][\texttt{e}, \texttt{o}][\texttt{r}, \texttt{e}][\texttt{m}, \texttt{s}]$, then $\pi_1(y) = \texttt{term}$ and $\pi_1(y) = \texttt{hoes}$. To simplify notation, we often write $\left[\begin{smallmatrix} w \\ x \end{smallmatrix}\right]$ in place of $w \times x$. For example,

$$\begin{bmatrix} \texttt{cat} \\ \texttt{dog} \end{bmatrix} \quad \text{means the same thing as} \quad \begin{bmatrix} \texttt{c} \\ \texttt{d} \end{bmatrix}\begin{bmatrix} \texttt{a} \\ \texttt{o} \end{bmatrix}\begin{bmatrix} \texttt{t} \\ \texttt{g} \end{bmatrix} \quad \text{and} \quad [\texttt{c}, \texttt{d}][\texttt{a}, \texttt{o}][\texttt{t}, \texttt{g}].$$

Consider the following decision problem:

ACCEPTS-SHIFT

Instance: An alphabet Γ, a letter $c \notin \Gamma$, and a finite automaton M with input alphabet $(\Gamma \cup \{c\})^2$.

Question: Does M accept a word of the form $xc^n \times c^n x$ for some $x \in \Gamma^*$ and $n \geq 0$?

Theorem 2. *The decision problem* ACCEPTS-SHIFT *is undecidable.*

Proof. We reduce from the problem REWRITE-POWER. An instance of this decision problem is a set S of length-preserving rewriting rules, an alphabet Σ, and letters $a, b \in \Sigma$. Define $\Gamma = \Sigma \cup \{d\}$, where $d \notin \Sigma$ is a new symbol. Now we define the following regular languages:

$$E = \left\{ \begin{bmatrix} e \\ e \end{bmatrix} : e \in \Sigma \right\} \quad \text{and} \quad R = \left\{ \begin{bmatrix} r \\ \ell \end{bmatrix} : \ell \to r \in S \right\}$$

$$L = \begin{bmatrix} d \\ c \end{bmatrix}\begin{bmatrix} a \\ c \end{bmatrix}^+ \begin{bmatrix} d \\ d \end{bmatrix} \left(E^* R E^* \begin{bmatrix} d \\ d \end{bmatrix} \right)^* \begin{bmatrix} c \\ b \end{bmatrix}^+ \begin{bmatrix} c \\ d \end{bmatrix}.$$

Let $M = (Q, \Delta, \delta, q_0, F)$ be a deterministic finite automaton accepting L, with $\Delta = (\Gamma \cup \{c\})^2$. Clearly M can be constructed effectively from the definitions.

We claim that, for all $n \geq 2$, we have $a^{n-1} \Longrightarrow^* b^{n-1}$ iff the language $L = L(M)$ contains a word of the form $xc^n \times c^n x$. The crucial observation is that

$$u \Longrightarrow v \quad \text{iff} \quad \begin{bmatrix} v \\ u \end{bmatrix} \in E^* R E^*. \tag{8}$$

This follows immediately from the definitions of E and R.

$\Longrightarrow$: Suppose

$$u_0 := a^{n-1} \Longrightarrow u_1 \Longrightarrow \cdots \Longrightarrow u_m = b^{n-1}$$

with $m \geq 1$ and $n \geq 2$. Then

$$\begin{bmatrix} u_{i+1} \\ u_i \end{bmatrix} \in E^* R E^*$$

for $0 \leq i < m$. Then

$$\begin{bmatrix} d \\ d \end{bmatrix}\begin{bmatrix} u_1 \\ u_0 \end{bmatrix}\begin{bmatrix} d \\ d \end{bmatrix}\begin{bmatrix} u_2 \\ u_1 \end{bmatrix} \cdots \begin{bmatrix} d \\ d \end{bmatrix}\begin{bmatrix} u_m \\ u_{m-1} \end{bmatrix}\begin{bmatrix} d \\ d \end{bmatrix} \in \begin{bmatrix} d \\ d \end{bmatrix}\left(E^* R E^* \begin{bmatrix} d \\ d \end{bmatrix}\right)^* .$$

Hence

$$\begin{bmatrix} d \\ c \end{bmatrix}\begin{bmatrix} u_0 \\ c^{n-1} \end{bmatrix}\begin{bmatrix} d \\ d \end{bmatrix}\begin{bmatrix} u_1 \\ u_0 \end{bmatrix}\begin{bmatrix} d \\ d \end{bmatrix}\begin{bmatrix} u_2 \\ u_1 \end{bmatrix} \cdots \begin{bmatrix} d \\ d \end{bmatrix}\begin{bmatrix} u_m \\ u_{m-1} \end{bmatrix}\begin{bmatrix} d \\ d \end{bmatrix}\begin{bmatrix} c^{n-1} \\ u_m \end{bmatrix}\begin{bmatrix} c \\ d \end{bmatrix} \in L,$$

as desired. The first component is $du_0 du_1 d \cdots du_m dc^n$, while the second component is $c^n du_0 du_1 \cdots du_{m-1} du_m d$. Taking $x = du_0 du_1 d \cdots du_m d$, we see that $xc^n \times c^n x \in L$.

$\Longleftarrow$: Assume that $xc^n \times c^n x \in L$ for some word x with $n \geq 2$. Now L consists only of words of the form

$$w = \begin{bmatrix} d \\ c \end{bmatrix}\begin{bmatrix} a \\ c \end{bmatrix}^i\begin{bmatrix} d \\ d \end{bmatrix}\begin{bmatrix} v_0 \\ u_0 \end{bmatrix}\begin{bmatrix} d \\ d \end{bmatrix}\begin{bmatrix} v_1 \\ u_1 \end{bmatrix}\begin{bmatrix} d \\ d \end{bmatrix} \cdots \begin{bmatrix} v_m \\ u_m \end{bmatrix}\begin{bmatrix} d \\ d \end{bmatrix}\begin{bmatrix} c \\ b \end{bmatrix}^j\begin{bmatrix} c \\ d \end{bmatrix}$$

where $u_t \Longrightarrow v_t$ for $1 \leq t \leq m$ and $i, j \geq 1$. Observe that

$$\pi_1(w) = da^i dv_0 dv_1 \cdots dv_m dc^{j+1} \quad \text{and} \quad \pi_2(w) = c^{i+1} du_0 du_1 \cdots du_m db^j d,$$

so if $\pi_1(w) = xc^n$ and $\pi_2(w) = c^n x$ we must have $i = j = n - 1$ and $x = da^i dv_0 dv_1 \cdots dv_m d = du_0 du_1 \cdots du_m db^j d$. Since d is a new symbol, not in the alphabet of Σ, it follows that $u_0 = a^i$, $u_1 = v_0$, $u_2 = v_1, \ldots, u_m = v_{m-1}$, and $b^j = v_m$.

But then $u_0 \Longrightarrow v_0 = u_1$, $u_1 \Longrightarrow v_1 = u_2$, and so forth, up to $u_{m-1} \Longrightarrow v_{m-1} = u_m$, and finally $u_m \Longrightarrow v_m = b^j$. So $u_0 \Longrightarrow^* v_m$, and therefore $a^{n-1} \Longrightarrow^* b^{n-1}$. This completes the proof. $\qquad\square$

Remark 3. In the decision problem ACCEPTS-SHIFT, the undecidability of the problem arises, in an essential way, from words of the form $xc^n \times c^n x$ where $n < |x|$, and not from those words with $n \geq |x|$ as one might first suspect. More formally, the related decision problem defined below is actually solvable in cubic time.

ACCEPTS-LONG-SHIFT

Instance: An alphabet Σ, a letter $c \notin \Sigma$, and a finite automaton M with input alphabet $(\Sigma \cup \{c\})^2$.

Question: Does M accept a word of the form $xc^n \times c^n x$ for some $n \geq |x|$?

Theorem 4. *The decision problem* ACCEPTS-LONG-SHIFT *is solvable in cubic time.*

Proof. Suppose $x = a_1 a_2 \cdots a_m$. If $y = x c^n \times c^n x$ and $n \geq |x|$ then

$$y = [a_1, c] \cdots [a_m, c][c, c]^{n-m}[c, a_1][c, a_2] \cdots [c, a_m].$$

Given a DFA $M = (Q, \Sigma, \delta, q_0, F)$, we can create a nondeterministic finite automaton M' that accepts all x for which the corresponding y is accepted by M. The idea is that M' has state set $Q' = Q \times Q \times Q$; on input x the machine M' "guesses" a state $q \in Q$, and stores it in the second component, and then simulates M on input $x \times c^m$ in the first component, starting from q_0 and reaching some state p, and simulates M on input $c^m \times x$ in the third component, starting from q. Finally, M' accepts if the third component is an element of F and if there exists a path from p to q labeled $[c, c]^i$ for some $i \geq 0$. Now we can test whether M' accepts a word by using depth-first or breadth-first search on the transition diagram of M', whose size is at most cubic in terms of the size of M. $\qquad\square$

4 Application to k-automatic Sets of Rational Numbers

Recently the second author and co-authors defined a notation of k-automaticity for sets of non-negative rational numbers [11,13], in analogy with the more well-known concept for sets of non-negative integers [4].

For an integer $k \geq 2$ define $\Sigma_k = \{0, 1, \ldots, k-1\}$. If $w \in \Sigma_k^*$, define $[w]_k$ to be the integer represented by the word w in base k (assuming the most significant digit is at the left). Let M be a finite automaton with input alphabet $\Sigma_k \times \Sigma_k$. We define $\mathrm{quo}_k(M) \subseteq \mathbb{Q}^{\geq 0}$ to be the set

$$\left\{ \frac{[\pi_1(x)]_k}{[\pi_2(x)]_k} : x \in L(M) \right\}.$$

Furthermore, we call a set $T \subseteq \mathbb{Q}^{\geq 0}$ k-automatic if there exists a finite automaton M such that $T = \mathrm{quo}_k(M)$.

We first consider the following decision problem:

ACCEPTS-POWER

Instance: An integer $k \geq 2$, and a finite automaton M with input alphabet $(\Sigma_k)^2$.

Question: Is $\mathrm{quo}_k(L(M)) \cap \{k^i : i \geq 0\}$ nonempty?

Theorem 5. *The problem* ACCEPTS-POWER *is undecidable.*

Proof. The basic idea is to reduce once more from REWRITE-POWER, using the same construction as in the proof of Theorem 2. Our reduction produces an instance of ACCEPTS-SHIFT consisting of an alphabet Γ of cardinality ℓ, a letter

$c \notin \Gamma$, and a finite automaton M. By renaming symbols, if necessary, we can assume the symbols of Γ are the digits $1, 2, \ldots, \ell$ and c is the digit 0. It then suffices to take $k = \ell + 1$. Then $y \in L(M)$ with $\text{quo}_k(y)$ a power of k if and only if $y = x0^n \times 0^n x$ for some x and some $n \geq 0$. Note that, by our construction in the proof of Theorem 2, if M accepts $x0^n \times 0^n x$, then x contains no 0's.   □

Now consider a family of analogous decision problems ACCEPTS-POWER(k), where in each problem k is fixed.

Theorem 6. *For each integer $k \geq 2$, the decision problem* ACCEPTS-POWER(k) *is undecidable.*

Proof. We have to overcome the problem that k can depend on the size of Γ. To do so, we recode all words over the alphabet $\{0, 1\}$. It suffices to use the morphism φ defined by

$$\varphi(c) = 0^{m+1} \qquad\qquad \varphi(a_i) = 1^i 0^{m-i} 1$$

where $\Sigma = \{a_1, a_2, \ldots, a_m\}$. In the proof of Theorem 2, we replace E, R, L by E', R', L', as follows:

$$E' = \left\{ \begin{bmatrix} \varphi(e) \\ \varphi(e) \end{bmatrix} : e \in \Sigma \right\} \quad \text{and} \quad R' = \left\{ \begin{bmatrix} \varphi(r) \\ \varphi(\ell) \end{bmatrix} : \ell \to r \in S \right\}$$

$$L' = \begin{bmatrix} \varphi(d) \\ \varphi(c) \end{bmatrix} \begin{bmatrix} \varphi(a) \\ \varphi(c) \end{bmatrix}^+ \begin{bmatrix} \varphi(d) \\ \varphi(d) \end{bmatrix} \left(E^* R E^* \begin{bmatrix} \varphi(d) \\ \varphi(d) \end{bmatrix} \right)^* \begin{bmatrix} \varphi(c) \\ \varphi(b) \end{bmatrix}^+ \begin{bmatrix} \varphi(c) \\ \varphi(d) \end{bmatrix}.$$

The construction works because the blocks for symbols of Σ begin and end with at least one 1, while the block for c consists of all 0's. Therefore, if the first coordinate of an element of L' has a suffix in 0^+, this can only arise from $\varphi(c)$, and the same for prefixes of the second coordinate.   □

5   Problems About Conjugates

Recall that we say two words x and y are *conjugates* if one is a cyclic shift of the other; that is, if there exist u, v such that $x = uv$ and $y = vu$.

The undecidability result of the previous section suggests studying the following related natural decision problem.

ACCEPTS-GENERAL-SHIFT

Instance: A finite automaton M with input alphabet Σ^2.

Question: Does M accept a word of the form $x \times y$ for conjugates $x, y \in \Sigma^*$?

Theorem 7. *The decision problem* ACCEPTS-GENERAL-SHIFT *is undecidable.*

Proof. We reduce from the problem ACCEPTS-SHIFT. An instance of this problem is an alphabet Γ, a letter $c \notin \Gamma$, and a finite automaton M with input alphabet $(\Gamma \cup \{c\})^2$.

First check whether M accepts a word of the form $x \times x$ for some $x \in \Gamma^*$. (This is decidable because the language $\{x \times x : x \in \Gamma^*\}$ is regular.) If so, ACCEPTS-SHIFT(Γ, c, M) = "yes". Otherwise, construct a finite automaton M' whose language is

$$L(M) \cap \{sc^+ \times c^+t \mid s, t \in \Gamma^*\}.$$

Notice that ACCEPTS-SHIFT(Γ, c, M') = ACCEPTS-SHIFT(Γ, c, M). Clearly we have that if ACCEPTS-GENERAL-SHIFT(M') = "no", then ACCEPTS-SHIFT (Γ, c, M') = "no".

So suppose that ACCEPTS-GENERAL-SHIFT(M') = "yes". Then M' accepts a word $w = x \times y$ for words $x = uv$, $y = vu$ where $u, v \in (\Gamma \cup \{c\})^*$. We now show that $w = zc^n \times c^n z$ for some $z \in \Gamma^*$ and $n \geq 1$.

By the construction of M', uv ends with c, vu begins with c, and any two occurrences of c in uv or vu have only c's between them. Hence if u or v is empty, then $w = c^n \times c^n$ for some $n \geq 1$, and we can take $z = \epsilon$, the empty word. So say neither u nor v is empty. Then v begins and ends with c, and hence v is in c^+. It follows that if u contains c, then u begins and ends with c, so again $w = c^n \times c^n$ for some $n \geq 1$, and we can take $z = \epsilon$. So say u does not contain c. Then $w = uc^n \times c^n u$ with $u \in \Gamma^+$ and $n = |v|$, and we can take $z = u$.

So $w = zc^n \times c^n z$ for some word $z \in \Gamma^*$ and $n \geq 1$. Therefore we have ACCEPTS-SHIFT(Γ, c, M') = "yes". This completes the reduction. Then since ACCEPTS-SHIFT is undecidable by Theorem 2, ACCEPTS-GENERAL-SHIFT is also undecidable. $\qquad\square$

Now we turn to two other decision problems, both inspired by the problem ACCEPTS-GENERAL-SHIFT. The first is

ACCEPTS-DISTINCT-CONJUGATES

Instance: A DFA $M = (Q, \Sigma, \delta, q_0, F)$.

Question: Does M accept two distinct conjugates uv and vu?

We will prove

Theorem 8. ACCEPTS-DISTINCT-CONJUGATES *is decidable.*

To prove this theorem, we need the concept of primitive word and primitive root. A nonempty word x is said to be *primitive* if it cannot be written in the form $x = y^i$ for a word y and an integer $i \geq 2$. The *primitive root* of a word x is the unique primitive word t such that $x = t^j$ for some $j \geq 1$.

Lemma 9. *If a DFA M of n states accepts two distinct conjugates, then it accepts two distinct conjugates uv and vu, with at least one of u and v of length $\leq n^2$.*

Proof. Let $L = L(M)$, the language accepted by $M = (Q, \Sigma, \delta, q_0, F)$, where $|Q| = n$. Suppose that there exist $uv \in L$, $vu \in L$, but $uv \neq vu$. Without loss of generality, assume $|uv|$ is as small as possible. Assume, contrary to what we want to prove, that both $|u|$ and $|v|$ are $> n^2$.

Consider the acceptance path of uv through M: it looks like $\delta(q_0, u) = q_1$ and $\delta(q_1, v) = p_1$ for some $q_1 \in Q$ and $p_1 \in F$. Similarly, consider the acceptance path of vu through M: it looks like $\delta(q_0, v) = q_2$ and $\delta(q_2, u) = p_2$ for some $q_2 \in Q$ and $p_2 \in F$.

Now create a new DFA $M' = (Q \times Q, \Sigma, \delta', q_0', F')$ by the usual product construction, where $\delta'([r, s], a) := [\delta(r, a), \delta(s, a)]$ and $q_0' = [q_0, q_1]$ and $F = \{[q_2, p_1]\}$. Then M' has n^2 states and accepts v.

Since $|v| > n^2$, the acceptance path for v in M' visits $\geq n^2 + 2$ states and hence some state is repeated, giving us a loop of at most n^2 states that can be cut out. Hence we can write $v = v_1 v_2 v_3$, where $v_2 \neq \epsilon$ and $v_1 v_3 \neq \epsilon$, and M' accepts $v_1 v_3$. In M, then, it follows that $\delta(q_1, v_1 v_3) = p_1$ and $\delta(q_0, v_1 v_3) = q_2$, and hence M accepts the conjugates $uv_1 v_3$ and $v_1 v_3 u$. Since $|uv_1 v_3| < |uv|$, the minimality of $|uv|$ implies that these conjugates cannot be distinct, and so we must have

$$uv_1 v_3 = v_1 v_3 u. \tag{9}$$

We can now repeat the argument of the previous paragraph for the word u. We get a decomposition $u = u_1 u_2 u_3$ where $u_2 \neq \epsilon$ and $u_1 u_3 \neq \epsilon$, and we get

$$vu_1 u_3 = u_1 u_3 v. \tag{10}$$

Finally, the acceptance paths in M we have created imply that we can cut out both u_2 and v_2 simultaneously from uv and vu, and still get words accepted by M. So $u_1 u_3 v_1 v_3$ and $v_1 v_3 u_1 u_3$ are both accepted. Again, by minimality, we get that

$$u_1 u_3 v_1 v_3 = v_1 v_3 u_1 u_3. \tag{11}$$

Now, by the Lyndon-Schützenberger theorem (see, e.g., [9,14]), Eq. (11) implies the existence of a nonempty word t and integers i, j such that $u_1 u_3 = t^i$, $v_1 v_3 = t^j$. Without loss of generality, we can assume that t is primitive.

Applying the same theorem to Eq. (10) tells us that there exists k such that $v = t^k$. And applying the same theorem once more to Eq. (9) tells us that there exists ℓ such that $u = t^\ell$. But then $uv = vu$, a contradiction. $\qquad\square$

Remark 10. We observe that the bound of n^2 in the previous result is optimal, up to a constant multiplicative factor. Consider the languages

$$L_t = (a^t)^+ b(a^{t+1})^+ bb \cup (a^t)^+ bb(a^{t+1})^+ bb.$$

Then it is easy to see that L_t can be accepted by a (complete) DFA of $n = 3t + 8$ states. The shortest pair of distinct conjugates in L_n, however, are $a^{t(t+1)}ba^{t(t+1)}bb$ and $a^{t(t+1)}bba^{t(t+1)}b$, corresponding to $u = a^{t(t+1)}b$ of length $t^2 + t + 1$ and $v = a^{t(t+1)}bb$ of length $t^2 + t + 2$. Thus both u and v are of length $n^2/9 + O(n)$.

We can now prove Theorem 8.

Proof. Given $L = L(M)$, for each nonempty word x define the language

$$L_x = \{y \in \Sigma^* : xy \in L, \ yx \in L, \ xy \neq yx\}.$$

We observe that each L_x is a regular language. To see this, note that we can write $L_x = L_1 \cap L_2 \cap L_3$, where

$$L_1 = \{y \in \Sigma^* : xy \in L\}$$
$$L_2 = \{y \in \Sigma^* : yx \in L\}$$
$$L_3 = \{y \in \Sigma^* : xy \neq yx\}.$$

Both L_1 and L_2 are easily seen to be regular, and finite automata accepting them are easily constructed from M. To see that the same holds for L_3, note that if $xy = yx$ with x nonempty, then by the Lyndon-Schützenberger theorem it follows that $y \in t^*$, where t is the primitive root of x. Hence $L_3 = \overline{t^*}$. Therefore we can construct a finite automaton M_x accepting L_x.

Finally, here is the decision procedure. By Lemma 9 we know that if an n-state DFA M accepts a pair of words uv and vu with $uv \neq vu$, then it must accept a pair with either $|u| \leq n^2$ or $|v| \leq n^2$. Thus, it suffices to enumerate all $u \in \Sigma^*$ of lengths $1, 2, \ldots, n^2$, and compute M_u for each u. If at least one M_u has $L(M_u)$ nonempty, then answer "yes"; otherwise answer "no".   □

An alternative approach for proving Theorem 8 was suggested by the referee, as follows:

Proof. We show that ACCEPTS-DISTINCT-CONJUGATES is decidable by reducing it to the functionality problem for nondeterministic streaming string transducers (NSSTs) [1]. An NSST is a one-way nondeterministic automaton equipped with a fixed set of variables in which to store strings. At each step, it reads a symbol from the input, changes state, and updates its string variables in parallel with a "copyless assignment". At the end of the input, it produces an output string based on its string variables and final state. See [1] for details.

Consider the relation R defined as the set of pairs $\{(uv, vu) \in L(M) \times L(M)\}$. An NSST T can implement R as a transduction as follows. T uses two string variables X and Y. Let w be T's input. T updates X with $X := X\sigma$ whenever a symbol σ of w is read, until some nondeterministic transition, after which, as long as symbols σ of w are read, the following updates are performed: $X := X$ and $Y := Y\sigma$. When the end of w is reached, $w = uv$ for $u = X$ and $v = Y$. During the computation, T uses its finite-state control to check that w is in $L(M)$. At the same time, T checks that vu is in $L(M)$ by guessing a state q of M, simulating M on u starting from q, verifying that M ends u in an accepting state, simulating M on v starting from q_0, and verifying that M ends v in q. If all of these checks succeed, T outputs $YX = vu$.

We say that T is *functional* if for every input string, T produces at most one output string. If T is functional, then for every $x \in L(M)$, x has at most one conjugate in $L(M)$. Then since every string is a conjugate of itself, x has no conjugates other than x in $L(M)$, and so M does not accept distinct conjugates. On the other hand, if M does not accept distinct conjugates, then for each input x, T can only produce x as output, and so T is functional. Therefore T is functional iff the answer to ACCEPTS-DISTINCT-CONJUGATES is "no". By Theorem 4.1 of [1], checking if T is functional is decidable. Therefore ACCEPTS-DISTINCT-CONJUGATES is decidable.   □

Our second decision problem is

ACCEPTS-NON-CONJUGATES

Instance: A DFA $M = (Q, \Sigma, \delta, q_0, F)$.

Question: Does M accept two words of the same length that are not conjugates?

We prove

Theorem 11. ACCEPTS-NON-CONJUGATES *is decidable.*

Proof. Given a formal language L over an ordered alphabet Σ, we define lexlt(L) to be the union, over all $n \geq 0$, of the lexicographically least word of length n in L, if it exists. As is well-known (see, e.g., [15, Lemma 1]), if L is regular, then so is lexlt(L). Furthermore, given a DFA for L, we can algorithmically construct a DFA for lexlt(L).

We also define cyc(L) to be the union, over all words $w \in L$, of the conjugates of w. Again, as is well-known (see, e.g., [14, Theorem 3.4.3]), if L is regular, then so is cyc(L). Furthermore, given a DFA for L, we can algorithmically construct a DFA for cyc(L).

We claim that L contains two words x and y of the same length that are non-conjugates if and only if L is not a subset of cyc(lexlt(L)).

Suppose such x, y exist. Let t be the lexicographically least word in L of length $|x|$. If t is a conjugate of x, then y is not a conjugate of t, so $y \notin$ cyc(lexlt(L)). On the other hand, if t is not a conjugate of x, then $x \notin$ cyc(lexlt(L)). In both cases L is not a subset of cyc(lexlt(L)).

Suppose L is not a subset of cyc(lexlt(L)). Then there is some word of some length n in L, say x, that is not a conjugate of the lexicographically least word of length n, say y. Then x and y are the desired two words.

Putting this all together, we get our decision procedure for the decision problem ACCEPTS-NON-CONJUGATES: given the DFA M for L, construct the DFA M' for $L-$cyc(lexlt(L)) using the techniques mentioned above. If M' accepts at least one word, then the answer for ACCEPTS-NON-CONJUGATES is "yes"; otherwise it is "no".   □

6   Final Remarks

We still do not know whether the following problem from [11, p. 363] is decidable:

ACCEPTS-INTEGER

Instance: A finite automaton M with input alphabet $(\Sigma_k)^2$.

Question: Is $\text{quo}_k(L(M)) \cap \mathbb{N}$ nonempty?

Unfortunately our techniques do not seem immediately applicable to this problem.

We mention two other problems about finite automata whose decidability is still open:

1. Given a DFA M with input alphabet $\{0,1\}$, decide if there exists at least one prime number p such that M accepts the base-2 representation of p.

Remark 12. An algorithm for this problem would allow resolution of the existence of a Fermat prime $2^{2^k} + 1$ for $k > 4$.

2. Given a DFA M with input alphabet $\{0,1\}$, decide if there exists at least one integer $n \geq 0$ such that M accepts the base-2 representation of n^2.

Acknowledgments. We thank Hendrik Jan Hoogeboom and the referees for their helpful comments.

References

1. Alur, R., Deshmukh, J.V.: Nondeterministic streaming string transducers. In: Aceto, L., Henzinger, M., Sgall, J. (eds.) ICALP 2011. LNCS, vol. 6756, pp. 1–20. Springer, Heidelberg (2011). doi:10.1007/978-3-642-22012-8_1
2. Bar-Hillel, Y., Perles, M., Shamir, E.: On formal properties of simple phrase structure grammars. Z. Phonetik. Sprachwiss. Kommuniationsforsch. **14**, 143–172 (1961)
3. Book, R.V., Otto, F.: String-Rewriting Systems. Springer, New York (1993). doi:10. 1007/978-1-4613-9771-7
4. Cobham, A.: Uniform tag sequences. Math. Syst. Theor. **6**, 164–192 (1972)
5. Engelfriet, J., Rozenberg, G.: Fixed point languages, equality languages, and representation of recursively enumerable languages. J. Assoc. Comput. Mach. **27**, 499–518 (1980)
6. Ginsburg, S., Rose, G.F.: Some recursively unsolvable problems in ALGOL-like languages. J. Assoc. Comput. Mach. **10**, 29–47 (1963)
7. Hoogeboom, H.J.: Are there undecidable properties of non-turing-complete automata? Posting on stackexchange, 20 October 2012. http://cs.stackexchange.com/questions/1697/are-there-undecidable-properties-of-non-turing-complete-automata

8. Hopcroft, J.E., Ullman, J.D.: Introduction to Automata Theory, Languages, and Computation. Addison-Wesley, Reading (1979)
9. Lyndon, R.C., Schützenberger, M.P.: The equation $a^M = b^N c^P$ in a free group. Mich. Math. J. **9**, 289–298 (1962)
10. Post, E.: Absolutely unsolvable problems and relatively undecidable propositions: account of an anticipation. In: Davis, M. (ed.) The Undecidable, pp. 338–433. Raven Press, Hewlett (1965)
11. Rowland, E., Shallit, J.: Automatic sets of rational numbers. Int. J. Found. Comput. Sci. **26**, 343–365 (2015)
12. Rozenberg, G., Salomaa, A.: Cornerstones of Undecidability. Prentice-Hall, New York (1994)
13. Schaeffer, L., Shallit, J.: The critical exponent is computable for automatic sequences. Int. J. Found. Comput. Sci. **23**, 1611–1626 (2012)
14. Shallit, J.: A Second Course in Formal Languages and Automata Theory. Cambridge University Press, Cambridge (2009)
15. Shallit, J.O.: Numeration systems, linear recurrences, and regular sets. Inf. Comput. **113**, 331–347 (1994)
16. Turing, A.M.: On computable numbers, with an application to the Entscheidungsproblem. Proc. Lond. Math. Soc. **42**, 230–265 (1936)

On the Power of Permitting Semi-conditional Grammars

Zsolt Gazdag[1]($\boxtimes$) and Krisztián Tichler[2]

[1] Department of Foundations of Computer Science,
University of Szeged, Szeged, Hungary
gazdag@inf.u-szeged.hu
[2] Department of Algorithms and Their Applications,
Eötvös Loránd University, Budapest, Hungary
ktichler@inf.elte.hu

Abstract. Permitting semi-conditional grammars are such extensions of context-free grammars where each rule is associated with a word v, and such a rule can be applied to a sentential form u only if v is a subword of u. In this paper we show that the class of languages generated by permitting semi-conditional grammars with no erasing rules is strictly included in the class of context-sensitive languages.

Keywords: Conditional grammars · Permitting context · Generative power

1 Introduction

Context-free (CF) grammars are extensively studied since they serve as formal models in many areas of computer science. One of their good properties is that their membership problem is efficiently solvable. These grammars were invented by Noam Chomsky to describe the structures of words in sentences of natural languages. However, it turned out that certain natural languages contain phenomena such as cross-serial dependencies, that cannot be handled by CF grammars (see e.g. [12]). The more powerful context-sensitive (CS) grammars are able to model cross-dependencies, but the membership problem for them is already PSPACE-complete. In [11] Joshi gave several properties that a grammar should have in order to be able to model natural languages. These properties are the ability to handle *limited cross-serial dependencies*, the *constant growth* of the associated language, and the *polynomial time solvability of the membership problem*. Grammars satisfying these properties are called *mildly context-sensitive grammars*.

This research was supported by a research grant from the Faculty of Informatics, Eötvös Loránd University.

Z. Gazdag—Research of this author was partially supported by the Hungarian National Research, Development and Innovation Office (NKFIH) under grant K 108448.

© Springer International Publishing AG 2017
É. Charlier et al. (Eds.): DLT 2017, LNCS 10396, pp. 173–184, 2017.
DOI: 10.1007/978-3-319-62809-7_12

One way to enrich CF grammars with context sensitivity and, in turn, raise their generative power is to control their derivations by context conditions. For example, in *random context grammars* (RCG's) [18] two sets of nonterminals, a permitting P and a forbidding one Q, are associated to every context-free rule. Then a rule is applicable, if it is applicable in the context-free sense and nonterminals in Q do not occur, while every nonterminal in P does occur in the current sentential form. If in an RCG each rule is associated with an empty forbidding set (resp. permitting set), then the grammar is called a *permitting* (resp. *forbidding*) RCG.

It turned out that RCG's with erasing rules have equal power to that of Turing machines, thus recently a restricted variant of them was introduced and investigated [15]. In these grammars the permitting and forbidding sets are associated to the nonterminals rather than to the rules. Moreover, one of these sets is always a singleton and the other one is empty. We will call these grammars *restricted random context grammars* (rRCG's) in this paper. It turned out that even with this very limited ability of controlling the derivations these grammars are equivalent to random context grammars [2, 15]. Moreover, permitting rRCG's are as powerful as permitting RCG's [2], and this is the case for the forbidding variants too if erasing rules are allowed [8].

In [16] a variant of random context grammars, called *semi-conditional grammars* (SCG's) were introduced. In these grammars every rule r is associated with two words, a permitting word v_1 and a forbidding one v_2, and r is applicable only if v_1 is a subword of the sentential form, but v_2 is not. Moreover, an SCG G is of degree (i, j) if the length of its permitting words is at most i and that of the forbidding words is at most j. It was shown in [16] that SCG's without erasing rules and with degree $(1, 2)$ or $(2, 1)$ have equal power to that of CS grammars. This clearly means that these grammars are too powerful to meet all the conditions of mild context-sensitivity. It turned out in [16], on the other hand, that these grammars with degree $(1, 1)$ cannot generate all languages in CS. The invention of SCG's was motivated by the grammars of Kelemen [13], where only a permitting word was associated to each rule (we call these grammars *permitting semi-conditional grammars* in this paper). In [16] it remained an open question whether permitting SCG's can generate all CS languages. In this paper we show that there is a CS language that cannot be generated by any permitting RCG if the use of erasing rules is not allowed. Some results concerning grammars mentioned in this introduction are given in Fig. 1.

The proof of our result is based on a pumping lemma similar to the one presented in [6], where it was shown that the class of languages generated by permitting RCG's with no erasing rules is strictly included in the class of languages generated by RCG's. In the proof of the pumping lemma in [6] the following property was essential: sufficiently long derivations of a permitting RCG with no erasing rules always contain two sentential forms α and β such that β is derived from α and, for every nonterminal A, $|\alpha|_A \leq |\beta|_A$ (here $|\alpha|_A$ and $|\beta|_A$ denote the number of occurrences of A in α and β, respectively). This property follows from Dickson's lemma [4] which states that any infinite sequence

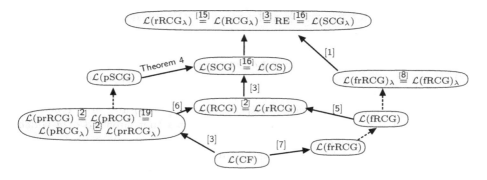

Fig. 1. A comparison of the power of some variants of grammars mentioned in the introduction. Arrows with solid lines represent strict inclusions, while arrows with dashed lines indicate inclusions which are not known to be strict. References to the presented equalities or strict inclusions are also given. Inclusions represented by dashed lines follow from definitions. SCG_λ, RCG_λ, and $rRCG_\lambda$ (resp. SCG, RCG, and rRCG) denote the classes of the corresponding grammars with erasing rules (resp. with no erasing rules). For a class of grammars C, $\mathcal{L}(C)$ denotes the class of languages generated by grammars in C, and pC (resp. fC) denotes that subclass of C, where only permitting (resp. forbidding) context conditions are used.

$v_1, v_2, \ldots$ of n-vectors over the natural numbers contains an infinite sub-sequence $v_{i_1} \le v_{i_2} \le \ldots$, where $\le$ is the componentwise ordering of n-vectors.

In the proof of our pumping lemma (Lemma 2) we need to find such sentential forms α and β in a derivation of a permitting SCG G that satisfy a stronger condition: if u is a permitting word of G, then β should contain at least as many occurrences of u as the number of these is in α. To do so we will use Higman's lemma [10], which ensures that in any infinite sequence $v_1, v_2, \ldots$ of words, there is an infinite subsequence $v_{i_1} \le_s v_{i_2} \le_s \ldots$, where $\le_s$ is the subsequence (or scattered subword) relation. However, to find an appropriate α and β we cannot apply directly Higman's lemma to the sentential forms of a derivation, but rather to certain carefully defined words obtained from these sentential forms.

The paper is organized as follows. First, we introduce the necessary notions and notations. Then, in Sect. 3 we present the main result of the paper. Finally, we give some concluding remarks in Sect. 4. Due to space reasons, some proofs are omitted in the paper. The interested reader can find them in [9].

2   Preliminaries

We define here the necessary notions, however we assume that the reader is familiar with the basic concepts of the theory of formal languages. For a comprehensive guide we refer to [17]. An *alphabet* Σ is a finite, nonempty set of symbols whose elements are also called *letters*. *Words* over Σ are finite sequences of letters in Σ. As usual, Σ^* denotes the set of all words over Σ including the *empty word* ε. For a letter $a \in \Sigma$ and a word $u \in \Sigma^*$, $|u|$ denotes the *length of* u and $|u|_a$ is the *number of occurrences of a in u*. $\mathbb{N}$ denotes the set of natural numbers. For $n, m \in \mathbb{N}$, $n < m$, $[n, m]$ denotes the set $\{n, n+1, \ldots, m\}$. If $n = 1$, then $[n, m]$ is denoted by $[m]$. The *set of positions in* u (pos(u) for short) is $[|u|]$.

Let $u \in \Sigma^*$. A word v is a *scattered subword* of u, if v can be obtained from u by erasing some (possibly zero) letters. Moreover, v is a *subword* of u if there are words $u_1, u_2 \in \Sigma^*$ such that $u = u_1 v u_2$. Let $i \in$ pos(u) and $m \geq 1$ be such that $i + m - 1 \in$ pos(u). Then subw(u, i, m) denotes that subword of u which starts on the ith position and has length m. It will always be clear from the context whether we consider an arbitrary subword of u or that one which starts on a certain position. Those subwords of u that have length m are also called m-*subwords*. The *subsequence relation* $\leq_s$ over Σ^* is a binary relation defined as follows. For $u, v \in \Sigma^*$, $u \leq_s v$, if u is a scattered subword of v. Let $f : [k] \rightarrow [l]$ $(k, l \geq 1)$ be a (partial) function. The *domain* and *range* of f, denoted by $dom(f)$ and $ran(f)$, respectively, are defined as follows: $dom(f) = \{i \in [k] \mid \exists j \in [l] : f(i) = j\}$ and $ran(f) = \{i \in [l] \mid \exists j \in [k] : f(j) = i\}$. If $I \subsetneq [k]$, then $f|_I$ denotes the restriction of f to I. Let $u, v \in \Sigma^*$ and $f : \text{pos}(v) \rightarrow \text{pos}(u)$ be a (partial) function. If, for every $i \in dom(f)$, subw($v, i, 1$) = subw($u, f(i), 1$), then we call f *letter-preserving*. A *well-quasi-ordering* (wqo for short) on a set S is a reflexive, transitive binary relation $\leq$ such that any infinite sequence $a_1, a_2, \ldots$ $(a_i \in S, i \geq 1)$ contains a pair $a_j \leq a_k$ with $j < k$. The following result is due to [10] (see also [14]).

Proposition 1. *Let Σ be an alphabet. Then $\leq_s$ is a wqo on Σ^*. Consequently, for every infinite sequence $u_1, u_2, \ldots$ ($u_i \in \Sigma^*, i \geq 1$), there is an infinite subsequence $u_{i_1} \leq_s u_{i_2} \leq_s \ldots$.*

A *semi-conditional grammar* (SCG for short) is a 4-tuple $G = (V, \Sigma, R, S)$, where V and Σ are alphabets of the *nonterminal* and *terminal* symbols, respectively (it is assumed that $V \cap \Sigma = \emptyset$), $S \in V$ is the *start symbol*, and R is a finite set of *production rules* of the form $(A \rightarrow \alpha, p, q)$, where $A \in V, \alpha \in (V \cup \Sigma)^+$ (that is $A \rightarrow \alpha$ is a usual non-erasing context-free rule), and $p, q \in (V \cup \Sigma)^*$. For such a rule r, the words p and q are called the *permitting* and *forbidding contexts* of r, respectively. The *right-hand side* of r (denoted by rhs(r)) is α. We will often denote $V \cup \Sigma$ by V_G. The *derivation relation* $\Rightarrow_G$ of G is defined as follows. For every word $u_1, u_2, \alpha \in V_G^*$ and $A \in V$, $u_1 A u_2 \Rightarrow_G u_1 \alpha u_2$ if and only if there is a rule $(A \rightarrow \alpha, p, q) \in R$ such that (i) p is a subword of $u_1 A u_2$, and (ii) if $q \neq \varepsilon$, then q is not a subword of $u_1 A u_2$. We will often write $\Rightarrow$ instead of $\Rightarrow_G$ when G is clear from the context. As usual, the *reflexive, transitive closure of* $\Rightarrow$ is denoted by $\Rightarrow^*$ and the *language generated by* G is $L(G) = \{u \in \Sigma^* \mid S \Rightarrow^* u\}$. A word $\alpha \in V_G^*$ is called a *sentential form of* G (or just a *sentential form* if G is clear from the context). A *derivation der* from α to β is a sequence $\alpha_1 \Rightarrow \alpha_2 \Rightarrow \ldots \Rightarrow \alpha_{n+1}$ of sentential forms, for some $n \geq 0$ such that $\alpha_1 = \alpha$ and $\alpha_{n+1} = \beta$. The *length of der* (denoted by $|der|$) is n. Let α and β be sentential forms and *der* a derivation from α to β. The *sentential form vector of der* (denoted by sfv(der)) is $(u_1, \ldots, u_k)$ ($k = |\alpha|, u_i \in V_G^*, i \in [k]$), such that $\beta = u_1 \ldots u_k$ and, for every $i \in [k]$, u_i is derived from subw$(\alpha, i, 1)$. Let $der : \alpha_1 \Rightarrow \ldots \Rightarrow \alpha_n$ $(n \geq 1)$ and $der' : \alpha_n \Rightarrow \ldots \Rightarrow \alpha_k$ $(k \geq n)$. Then $der\ der'$ denotes the derivation $\alpha_1 \Rightarrow \ldots \Rightarrow \alpha_k$.

If, for every rule $(A \to \alpha, p, q)$ in R, $q = \varepsilon$, then G is a *permitting SCG*, or a *pSCG* for short. Let $G = (V, \Sigma, R, S)$ be a pSCG. The *set of permitting contexts of* G is $pw(G) = \{p \mid (r, p, q) \in R\}$ and $\max_{pw(G)} = \max\{|u| \mid u \in pw(G)\}$. We denote by $\mathcal{L}(\text{pSCG})$ and $\mathcal{L}(\text{CS})$ the families of languages generated by pSCG's and context-sensitive grammars, respectively.

3   The Main Result

Here we show that pSCG's are strictly weaker than context sensitive grammars by proving that the language $L = \{a^{2^{2^n}} \mid n \geq 0\}$ cannot be generated by any pSCG (Theorem 4). The proof, roughly, consists of the following main steps. First we define the notion of m-embedding (Definition 1). Intuitively, a word α can be m-embedded to a word β, if there is an injective mapping of the m-subwords of α to the m-subwords of β such that this mapping preserves the order of these words and satisfies certain additional conditions. Then we show that if a pSCG $G = (V, \Sigma, R, S)$ with $m = \max_{pw(G)}$ has a derivation der from α to β $(\alpha, \beta \in V_G^*)$ such that $|\alpha| < |\beta|$ and α can be m-embedded to β, then the der can be "pumped" so that the obtained derivation is a valid derivation of G (cf. Lemma 2, which we will often refer to as our *pumping lemma*). Finally, we show that sufficiently long derivations of G always contain sentential forms α and β such that α can be m-embedded to β (Lemma 3). To this end we will use the fact that $\leq_s$ is a wqo on V_G^*.

Definition 1. Let Σ be an alphabet, $\alpha, \beta \in \Sigma^*$, $k = |\alpha|$, $l = |\beta|$, and $m \geq 1$. An m-embedding of α to β is a strictly increasing function $g : [k - m + 1] \to [l]$ such that the following (partial) mapping $f : \text{pos}(\beta) \to \text{pos}(\alpha)$ is letter-preserving and well defined: for every $i \in [k - m + 1]$ and $\kappa \in [0, m - 1]$, $f(g(i) + \kappa) = i + \kappa$. If g is an m-embedding, then the above f is denoted by $\text{inv}_m(g)$. Moreover, if an m-embedding of α to β exists, then we denote this by $\alpha \rightsquigarrow_m \beta$.

Example 1. *Here we give two examples to demonstrate the notion of m-embedding.*

(1) *Let $\alpha = BAAB$ and $\beta = BAAAB$. Any 3-subword of α is a subword of β, too. Due to the letter-preserving property, only the following g can be a 3-embedding of α to β: $g(1) = 1$ and $g(2) = 3$. The mapping $f = \text{inv}_m(g)$ is letter-preserving, but not well defined. Indeed, with $i = 1$ and $\kappa = 2$ we get $f(g(i) + \kappa) = f(3) = 3$, while with $i = 2$ and $\kappa = 0$, $f(g(i) + \kappa) = f(3) = 2$. This implies that there is no 3-embedding of α to β.*

(2) *Let $\alpha = ABBAC$, $\beta = AABBAABAC$ and g be the following strictly increasing function: $g(1) = 2, g(2) = 3$, and $g(3) = 7$. Then the mapping $f = \text{inv}_m(g)$ is letter-preserving and well defined: $f(i) = i - 1$ $(i \in [2, 5])$ and $f(i) = i - 4$ $(i \in [7, 9])$. Thus g is a 3-embedding of α to β.*

The following properties of m-embeddings will be useful in what follows.

Proposition 2. *Let Σ be an alphabet, $m \geq 1$, and $\alpha, \beta \in \Sigma^*$. Assume that g is an m-embedding of α to β and $f = \mathrm{inv}_m(g)$. Then the following statements hold.*

(i) *For every $i \in \mathrm{pos}(\alpha)$, $|\{t \in \mathrm{pos}(\beta) \mid f(t) = i\}| \leq m$.*
(ii) *If $|\alpha| = |\beta|$, then $\alpha = \beta$.*
(iii) *If $i, j \in \mathrm{pos}(\alpha)$ with $j = i + 1$, then $g(j) - g(i) = 1$ or $g(j) - g(i) \geq m$.*

Proof. See [9].

We will also need the following operation which inserts words into certain positions of a word. Let Σ be an alphabet and $\alpha = X_1 \ldots X_k$ ($k \geq 1, X_i \in \Sigma, i \in [k]$). Let moreover $u_1, \ldots, u_l \in \Sigma^*$ and $f : [k] \to [l]$ be a (partial) function. The *substitution of* $\mathbf{u} = (u_1, \ldots, u_l)$ *into* α *by* f (denoted by $\mathrm{subst}(\mathbf{u}, \alpha, f)$) is the word $\beta = v_1 \ldots v_k$, where v_i ($i \in [k]$) is defined as follows. If $f(i)$ is defined, then let $v_i = u_{f(i)}$, and let $v_i = X_i$ otherwise.

Sometimes we will need to extend a function f used in a substitution. An *extension of* f (with respect to α) is a function $\hat{f}$ defined as follows. For every $i \in dom(f)$, $\hat{f}(i) = f(i)$, and for every $i \in [k] - dom(f)$, $\hat{f}$ is either undefined or defined as follows: if there is a $j \in dom(f)$ such that $\mathrm{subw}(\alpha, i, 1) = \mathrm{subw}(\alpha, j, 1)$, then take such a j and let $\hat{f}(i) = f(j)$. Notice that f is always an extension of itself.

Example 2. *Let $\Sigma = \{A, B, C\}$, $\alpha = ABCBB$ and $\mathbf{u} = (u_1, u_2, u_3)$, where $u_1 = AA, u_2 = ABC, u_3 = CC$. Let furthermore $f : [5] \to [3]$ be the following partial function. $f(2) = f(3) = 3, f(5) = 1$. Then $\mathrm{subst}(\mathbf{u}, \alpha, f) = Au_3u_3Bu_1 = ACCCCBAA$ and f has two possible extensions other than f. $\hat{f}(1)$ is undefined. $\hat{f}(2) = \hat{f}(3) = 3, \hat{f}(5) = 1$ and $\hat{f}(4)$ is either 1 or 3 resulting in $\mathrm{subst}(\mathbf{u}, \alpha, \hat{f})$ equal to AC^4A^4 or AC^6A^2, respectively.*

The following lemma will be crucial in the proof of our pumping lemma.

Lemma 1. *Let $G = (V, \Sigma, R, S)$ be a pSCG, $m = \max_{pw(G)}$, and $\alpha, \alpha', \beta \in V_G^*$. Assume that $\alpha \Rightarrow_G^* \alpha'$ and $\alpha \rightsquigarrow_m \beta$. Let der be a derivation from α to α', g an m-embedding of α to β, and $f = \mathrm{inv}_m(g)$. Then, for every extension $\hat{f}$ of f, $\beta \Rightarrow_G^* \beta_{\hat{f}}$, where $\beta_{\hat{f}} = \mathrm{subst}(\mathrm{sfv}(der), \beta, \hat{f})$.*

Proof. Let $\beta' = \mathrm{subst}(\mathrm{sfv}(der), \beta, f)$. We first show that $\beta \Rightarrow_G^* \beta'$ by induction on $n = |der|$. If $n = 0$, then one can see that $\beta' = \beta$, and thus the statement trivially holds. Assume that it holds for n. We prove it for $n + 1$. In this case der can be written as $der = der_1 der_2$, where der_1 is $\alpha_0 \Rightarrow_G \ldots \Rightarrow_G \alpha_n$, der_2 is $\alpha_n \Rightarrow_G \alpha_{n+1}$, $\alpha_0 = \alpha$, and $\alpha_{n+1} = \alpha'$. Let $\beta_n = \mathrm{subst}(\mathrm{sfv}(der_1), \beta, f)$. By the induction hypothesis, there is a derivation der_1' from β to β_n. Let $(u_1, \ldots, u_k) = \mathrm{sfv}(der_1)$ ($k = |\alpha|$). Assume that G rewrites a nonterminal A during $\alpha_n \Rightarrow_G \alpha_{n+1}$ using a rule $r = (A \to \gamma, p, \varepsilon)$ (see Fig. 2 for an example).

Let $i \in [k]$ and $\kappa \in pos(u_i)$ be such that the rewritten A occurs on the κth position of u_i. Let $i_1 < i_2 \ldots < i_\xi$ be all the positions in $pos(\beta)$ with $f(i_j) = i$ ($j \in [\xi]$). Let $(v_1, \ldots, v_l) = \mathrm{sfv}(der_1')$ ($l = |\beta|$). Then, for every $j \in [\xi]$, $v_{i_j} = u_i$ and thus, for every such j, there is a position $\kappa_j \in pos(\beta_n)$ satisfying that κ_j corresponds to the κth position in v_{i_j}. Clearly $\mathrm{subw}(\beta_n, \kappa_j, 1) = A$ and $\beta' = \beta_{n+1} = \mathrm{subst}((\gamma), \beta_n, h)$ where $h : pos(\beta) \rightarrow \{1\}$ is defined as follows: $h(j) = 1$ if $j \in \{\kappa_1, \ldots, \kappa_\xi\}$, and it is undefined otherwise. Therefore, to prove $\beta_n \Rightarrow_G^* \beta'$ it is enough to show that G can use r to rewrite each nonterminal A that occurs on a position κ_j ($j \in [\xi]$) in β_n.

Since G can apply r at the step $\alpha_n \Rightarrow_G \alpha_{n+1}$, α_n should contain the permitting context p. Then there are $\mu \in [k]$ and $\nu \in [0, m-1]$ such that p occurs in the subword $u_\mu \ldots u_{\mu+\nu}$ of α_n (notice that G has no erasing rules). Since g is an m-embedding of α to β, it is clear that $v_{g(\mu)} \ldots v_{g(\mu)+\nu} = u_\mu \ldots u_{\mu+\nu}$. Thus, there is at most one index $j \in [\xi]$ such that $v_{g(\mu)} \ldots v_{g(\mu)+\nu}$ contains that A which occurs on the κ_jth position in β_n. If no such j exists, then G can rewrite all A's occurring on positions κ_j ($j \in [\xi]$) in β_n, since the permitting context p is always present as a subword of $v_{g(\mu)} \ldots v_{g(\mu)+\nu}$. Otherwise let $j \in [\xi]$ be such that the subword $v_{g(\mu)} \ldots v_{g(\mu)+\nu}$ includes that A which occurs on the κ_jth position of β_n. Then G should rewrite first those A's in β_n that occur on positions other that κ_j and, at the last step, that A which occurs on the κ_jth position. Therefore $\beta_n \Rightarrow_G^* \beta_{n+1} = \beta'$ which implies that $\beta \Rightarrow_G^* \beta'$.

To finish the proof of the lemma consider a derivation der' from β to β'. Looking at the inductive proof of $\beta \Rightarrow^* \beta'$, one can see that, for each derivation step in der', there is a $j \in dom(f)$ such that $[j, j+m-1] \subseteq dom(f)$, and the necessary permitting context is in a sentential form derived from $\mathrm{subw}(\beta, j, m)$. In other words, those letters in β that are on such positions which are not included in $dom(f)$ do not occur in the permitting contexts used during der'. Assume that $i \in pos(\beta) - dom(f)$ such that $\hat{f}(i) = f(j)$, for some $j \in dom(f)$. Let u be the $f(j)$th word in $\mathrm{sfv}(der)$ and $X = \mathrm{subw}(\beta, j, 1)$. Then G derives u during der' from this X. On the other hand, by the definition of $\hat{f}$, $\mathrm{subw}(\beta, i, 1) = X$. Thus, der' can be extended to such a derivation where G, using the appropriate rules simultaneously, derives u also from that X which occurs on the ith position of β. Following this way of thinking one can see that der' can be extended to a derivation of $\beta_{\hat{f}}$ from β which completes the proof of the lemma.

To prove out pumping lemma we need the following preparation. Let $g : \alpha \leadsto_m \beta$ be an m-embedding. The mapping

$$\mathrm{cmp}_g(i) = \begin{cases} g(i) & \text{for } i \in [k-m+1] \\ g(k-m+1)+i-(k-m+1) & \text{for } i \in [k-m+2, k] \end{cases}$$

is called the *completion* of g. Note that $dom(\mathrm{cmp}_g) = pos(\alpha)$ and by the definition of an m-embedding cmp_g is letter-preserving. If $f = \mathrm{inv}_m(g)$ then $f(\mathrm{cmp}_g(i)) = i$ holds for $i \in [k]$.

Proposition 3. *Let $\alpha \in \Sigma^*$, $|\alpha| = k$, $z_i \in \Sigma^*$, $z_i \neq \varepsilon$ ($i \in [k]$) and $\beta = z_1 \cdots z_k$ with $|\beta| = l$. Suppose that $g : \alpha \leadsto_m \beta$ with $f = \mathrm{inv}_m(g)$ and $\bar{g} = \mathrm{cmp}_g$.*

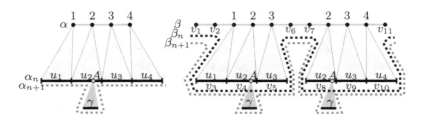

Fig. 2. The inductive proof of Lemma 1 assuming $m = 3$

Let us introduce the notations $x_i = \begin{cases} z_{f(i)} & \text{if } i \in dom(f) \\ \text{subw}(\beta, i, 1) & \text{if } i \notin dom(f) \end{cases}$,

$\zeta(i, r) = \sum_{j=1}^{i-1} |z_j| + r$ $(i \in [k], r \in [|z_i|])$ and $\xi(i, r) = \sum_{j=1}^{i-1} |x_j| + r$ $(i \in [l], r \in [|x_i|])$. Then for the mapping $g'(\zeta(i, r)) := \xi(\bar{g}(i), r)$ $(\zeta(i, r) \in [l - m + 1])$,

$$g' : \beta \leadsto_m \text{subst}(\mathbf{u}, \beta, f)$$

holds, where $\mathbf{u} = (z_1, \ldots, z_k)$.

Proof. See [9].

Lemma 2. *Let* $G = (V, \Sigma, R, S)$ *be a pSCG and* $m = \max_{pw(G)}$. *Suppose that* $\alpha \Rightarrow^* \beta$, $\beta \Rightarrow^* \gamma$, $\alpha \leadsto_m \beta$, *and* $|\alpha| < |\beta|$ *hold for some* $\alpha, \beta, \gamma \in V_G^*$. *Then there is a* $\gamma' \in \Sigma^*$ *such that (i)* $\alpha \Rightarrow^* \gamma'$ *and (ii)* $|\gamma| < |\gamma'| \leq (m + 1)|\gamma|$.

Proof. Let $k = |\alpha|$, $l = |\beta|$, and g be an m-embedding of α to β. Let moreover $f = \text{inv}_m(g)$, and der', der'' be any derivations from α to β and from β to γ, respectively. Applying Proposition 3 with these parameters and $\Sigma = V_G$, $\mathbf{u} = \text{sfv}(der')$ we get $g' : \beta \leadsto_m \beta'$, where $\beta' = \text{subst}(\mathbf{u}, \beta, f)$. Let $f' = \text{inv}_m(g')$. By Lemma 1 $\beta \Rightarrow^* \beta'$.

Let $\hat{f}'$ be the following function. For every $\tau \in pos(\beta')$, if $\tau \in dom(f')$, then let $\hat{f}'(\tau) = f'(\tau)$. Otherwise let $\tau = \xi(i, r)$, for some $i \in [l]$ and $r \in [|x_i|]$, and we define $\hat{f}'(\tau)$ as follows. If $i \in dom(f)$, then let $\hat{f}'(\tau) = \zeta(f(i), r)$, and let $\hat{f}'(\tau) = i$, otherwise. Notice that $\hat{f}'$ is a letter preserving function form β' to β. Indeed, if $i \in dom(f)$, then $x_i = z_{f(i)}$, and $x_i = \text{subw}(\beta, i, 1)$, otherwise. Let $\tau \in pos(\beta') - dom(f')$. Since g' is an m-embedding of β to β', there is a $\tau' \in pos(\beta')$ with $f'(\tau') = \hat{f}'(\tau)$. Then $\text{subw}(\beta', \tau, 1) = \text{subw}(\beta, \hat{f}'(\tau), 1) = \text{subw}(\beta, f'(\tau'), 1) = \text{subw}(\beta', \tau', 1)$. Thus $\hat{f}'$ is an extension of f'.

Now let $\gamma' = \text{subst}((v_1, \ldots, v_l), \beta', \hat{f}')$, where $(v_1, \ldots, v_l) = \text{sfv}(der'')$. By Lemma 1, $\beta' \Rightarrow^*_G \gamma'$. This, together with $\alpha \Rightarrow^*_G \beta$ and $\beta \Rightarrow^*_G \beta'$ implies $\alpha \Rightarrow^*_G \gamma'$, i.e., Statement (i) of the lemma holds. Statement (ii) can be seen as follows. Since g' is an m-embedding of β to β', for every $i \in [l]$, there is a $\tau \in pos(\beta')$ with $f'(\tau) = i$. Thus, each v_i $(i \in [l])$ is substituted for a position in β' by f'. Therefore $|\gamma| = |v_1 \ldots v_l| \leq |\text{subst}((v_1, \ldots, v_l), \beta', f')| \leq |\text{subst}((v_1, \ldots, v_l), \beta', \hat{f}')| = |\gamma'|$. Moreover, since $|\alpha| < |\beta|$, there is an $i \in [k]$ such that $|z_i| \geq 2$. Let $j \in [l]$ with

$f(j) = i$. Then $|x_j| \geq 2$, so $|\beta| = l < l + 1 \leq \sum_{s=1}^{l} |x_s| = |\beta'|$. This implies that $|\gamma| = |\gamma'|$ cannot hold, consequently $|\gamma| < |\gamma'|$.

On the other hand, by (i) of Proposition 2, for every $i \in [k]$, z_i is substituted for at most m different positions in β by f. Moreover, one can see that, for every $i \in dom(f)$, $\hat{f}'$ is an injective function from $[\xi(i, 1), \xi(i, |x_i|)]$ to $[l]$. Furthermore, $\hat{f}'$ is injective on the set $\{\tau \mid \tau = \xi(i, 1), i \in [l] - dom(f)\}$, too. Consequently, for every $i \in [l]$, v_i is substituted for at most $m + 1$ different positions in β' by $\hat{f}'$. Therefore, $|\gamma'| \leq (m + 1)|\gamma|$ should hold finishing the proof of Statement (ii).

Next we demonstrate some of the constructions used in the previous proof.

Example 3. *Let* $G = (V, \Sigma, R, S)$ *be a pSCG,* $\alpha = ABBA$, $\beta = EABBFBACA$ $(A, B, C, E, F \in V \cup \Sigma)$, *and* $m = 2$. *Let* $\gamma \in \Sigma^*$, *and assume that* G *has two derivations* der' *and* der'' *from* α *to* β *and from* β *to* γ, *respectively. Clearly* $|\alpha| < |\beta|$ *and* $\alpha \leadsto_m \beta$ *with the following* m-*embedding* g: $g(1) = 2$, $g(2) = 3$, *and* $g(3) = 6$. *Then, according to Lemma 2, we can give a* $\gamma' \in \Sigma^*$ *with the following properties:* (i) $\alpha \Rightarrow^* \gamma'$ *and* (ii) $|\gamma| < |\gamma'| \leq (m + 1)|\gamma|$.

Assume, for instance, that $\mathrm{sfv}(der') = (z_1, z_2, z_3, z_4)$, *where* $z_1 = E$, $z_2 = AB$, $z_3 = BF$, *and* $z_4 = BACA$. *Let* $f = \mathrm{inv}_m(g)$. *Then* $\beta' = \mathrm{subst}(\mathrm{sfv}(der'), \beta, f) = x_1 \ldots x_9$, *where* $x_1 = E$, $x_2 = E$, $x_3 = AB$, $x_4 = BF$, $x_5 = F$, $x_6 = BF$, $x_7 = BACA$, $x_8 = C$, *and* $x_9 = A$. *Now, if we define* g' *according to the proof of Lemma 2, then we get that* $dom(g') = [1, 8]$, *and* $g'(i) = i + 1$ *if* $i \in [1, 3]$, *and* $g'(i) = i + 4$, *otherwise. It is easy to verify that* g' *is an* m-*embedding of* β *to* β'. *Let* $f' = \mathrm{inv}_m(g')$. *Then* $dom(f') = [2, 5] \cup [8, 13]$. *Let us define now* $\hat{f}'$ *according to the proof of Lemma 2, that is,* $\hat{f}'(1) = 1$, $\hat{f}'(6) = \hat{f}'(7) = 5$, $\hat{f}'(14) = 8$, *and* $\hat{f}'(15) = 9$. *One can check that, for every* $\tau \in [1, 15] - dom(f')$, $\mathrm{subw}(\beta', \tau, 1) = \mathrm{subw}(\beta, \hat{f}'(\tau), 1)$. *Assume that* $\mathrm{sfv}(der'') = v_1 \ldots v_9$, *where* $v_i \in \Sigma^*$ $(i \in \mathrm{pos}(\beta))$. *Then* $\gamma' = \mathrm{subst}(\mathrm{sfv}(der''), \beta', \hat{f}') = v_1 v_1 \ldots v_5 v_5 \ldots v_9 v_8 v_9$. *By* (i) *of Lemma 2,* $\alpha \Rightarrow^* \gamma'$, *and it is easy to check that* γ' *satisfies Statement* (ii) *too.*

The following proposition together with Lemma 3 will be used to show that sufficiently long derivations of pSCG's always contain sentential forms α and β satisfying the conditions of Lemma 2. The statement can be seen using Proposition 1 (see also the proof of Lemma 2 in [6]).

Proposition 4. *Let* Σ *be an alphabet and* $n_1, n_2, \ldots$ *an infinite sequence of numbers in* $\mathbb{N}$. *Then there is* $M \in \mathbb{N}$ *such that, for every sequence* $v_1, v_2, \ldots, v_n$ *where* $n \geq M$ *and* $v_i \in \Sigma^*$ *with* $|v_i| \leq n_i$ $(i \in [n])$, *there are numbers* $i < j$ *in* $[M]$ *satisfying* $v_i \leq_s v_j$.

Let $G = (V, \Sigma, R, S)$ be a pSCG and $m \geq 1$. We will apply the above result to appropriate derivations of G in order to find sentential forms α and β satisfying $\alpha \leadsto_m \beta$. However, Proposition 4 ensures only that we can find such α and β which satisfy $\alpha \leq_s \beta$. Clearly, this does not imply $\alpha \leadsto_m \beta$. Thus we will apply Proposition 4 not directly to the derivations of G but to sequences of words derived from these derivations. To this end we will use two functions wdo and p defined below.

Let Σ be an alphabet and $m \geq 1$. We denote by $\Sigma^{\leq m}$ the set of all words in Σ^* with length at most m. Since $\Sigma^{\leq m}$ is a finite set, we will treat it as an alphabet. Now let wdo : $\Sigma^* \to (\Sigma^{\leq m})^*$ be defined as follows. Let $u \in \Sigma^*$. If $|u| < m$, then let $\mathrm{wdo}(u) = u$ (that is, u on the right-hand side is considered as a letter in $\Sigma^{\leq m}$). If $|u| \geq m$, then let $\mathrm{wdo}(u) = \mathrm{subw}(u, 1, m) \ldots \mathrm{subw}(u, |u| - m + 1, m)$ (again, $\mathrm{subw}(u, i, m)$ ($i \in [1, |u| - m + 1]$) is considered as a letter in $\Sigma^{\leq m}$). The name wdo comes from the word *window*, since for a word u, $\mathrm{wdo}(u)$ is that word whose letters are determined by moving a window of length m on u from left to right. The intuition behind the definition of wdo is the following: if $\mathrm{wdo}(\alpha) \leq_s \mathrm{wdo}(\beta)$, then every m-subword of α has to be an m-subword of β too. On the other hand $\mathrm{wdo}(\alpha) \leq_s \mathrm{wdo}(\beta)$ still does not imply $\alpha \leadsto_m \beta$ (see, for example, the first item in Example 1). Thus we will use the following function p before applying wdo on the sentential forms of G. Let Σ be an alphabet. Then $\hat{\Sigma}$ denotes the alphabet $\{a^{(i)} \mid a \in \Sigma, i \in [m]\}$. Now let $p : \Sigma^* \to \hat{\Sigma}^*$ be defined as follows. For a word $u = a_1 \ldots a_k \in \Sigma^*$ ($a_i \in \Sigma, i \in [k]$), $p(u) = a_1^{(1 \bmod m)} \ldots a_k^{(k \bmod m)}$. Intuitively, p associates the number $i \bmod m$ to the ith letter of u (we put this number in parentheses in order not to confuse it with the usual notation of the iteration of a letter). We will see in the proof of the next lemma that for two sentential forms α and β of G, $\mathrm{wdo}(p(\alpha)) \leq_s \mathrm{wdo}(p(\beta))$ implies $\alpha \leadsto_m \beta$.

Lemma 3. *Let $G = (V, \Sigma, R, S)$ be a pSCG and $m \geq 1$. Then there is $M \in \mathbb{N}$ such that the following holds. For every derivation $\alpha_0 \Rightarrow \alpha_1 \Rightarrow \ldots \Rightarrow \alpha_n$ of G with $n \geq M$, there are $i < j$ in $[M]$ such that $\alpha_i \leadsto_m \alpha_j$.*

Proof. Let $m = \max_{pw(G)}$, $\rho = \max\{|\mathrm{rhs}(r)| \mid r \in R\}$ and consider the sequence $n_1, n_2, \ldots$ where $n_i = i\rho$ ($i \geq 1$). Let moreover M be the number given in Proposition 4 and $\alpha_0 \Rightarrow \alpha_1 \Rightarrow \ldots \Rightarrow \alpha_n$ be a derivation of G with $n \geq M$. Clearly, $|\mathrm{wdo}(p(\alpha_i))| \leq n_i$, for every $i \in [n]$. Then, by Proposition 4, there are numbers $i < j$ in $[M]$ such that $\mathrm{wdo}(p(\alpha_i)) \leq_s \mathrm{wdo}(p(\alpha_j))$. We show that $\alpha_i \leadsto_m \alpha_j$. To simplify the notation, let us denote $\mathrm{wdo}(p(\alpha_i))$ and $\mathrm{wdo}(p(\alpha_j))$ by u and v, respectively. If $|u| = 1$, then $|\alpha_i| \leq m$ and α_i is a subword of α_j. In this case $\alpha_i \leadsto_m \alpha_j$ trivially holds. Assume now that $|u| \geq 2$, and let $k = |u|$ and $l = |\alpha_j|$. Since u is a scattered subword of v, there are $i_1 < \ldots < i_k$ in $\mathrm{pos}(v)$ such that $u = \mathrm{subw}(v, i_1, 1) \ldots \mathrm{subw}(v, i_k, 1)$. Then let $g : [k] \to [l]$ be a strictly increasing function defined as $g(\nu) = i_\nu$ ($\nu \in [k]$). Notice that $k = |\alpha_i| - m + 1$. Let moreover $f : \mathrm{pos}(\alpha_j) \to \mathrm{pos}(\alpha_i)$ be a (partial) function defined as $f(g(\nu) + \kappa) = \nu + \kappa$ ($\nu \in [k], \kappa \in [0, m - 1]$). To see that g is an m-embedding of α_i to α_j it is enough to show that f is letter preserving and well-defined.

Let $\nu \in [k]$. Using the definition of wdo we get that $\mathrm{subw}(p(\alpha_i), \nu, m) = \mathrm{subw}(p(\alpha_j), g(\nu), m)$ and in turn $\mathrm{subw}(\alpha_i, \nu, m) = \mathrm{subw}(\alpha_j, g(\nu), m)$. Thus f is letter preserving. Now, let $\nu \in [k - 1]$. Using again the definition of wdo we get that $\mathrm{subw}(p(\alpha_i), \nu, m) = \mathrm{subw}(p(\alpha_j), g(\nu), m)$ and $\mathrm{subw}(p(\alpha_i), \nu + 1, m) = \mathrm{subw}(p(\alpha_j), g(\nu + 1), m)$. Thus, the upper index added by p to the first letter of

$\mathrm{subw}(\alpha_i, \nu, m)$ should match that of $\mathrm{subw}(\alpha_j, g(\nu), m)$. Similar observation holds for the words $\mathrm{subw}(\alpha_i, \nu + 1, m)$ and $\mathrm{subw}(\alpha_j, g(\nu + 1), m)$. This implies that either $g(\nu+1) - g(\nu) = 1$ or $g(\nu+1) - g(\nu) \geq m$ should hold. It is easy to see that in both cases the definition of f is consistent. Therefore g is an m-embedding of α_i to α_j.

Theorem 4. $\mathcal{L}(\mathrm{pSCG}) \subsetneq \mathcal{L}(\mathrm{CS})$.

Proof. By [16] $\mathcal{L}(\mathrm{pSCG}) \subseteq \mathcal{L}(\mathrm{CS})$. Thus, since $L = \{a^{2^{2^n}} \mid n \geq 0\}$ is clearly included in $\mathcal{L}(\mathrm{CS})$, it is enough to show that $L \notin \mathcal{L}(\mathrm{pSCG})$. Assume on the contrary that $L \in \mathcal{L}(\mathrm{pSCG})$ and let G be a pSCG with $L(G) = L$. Let moreover $m = \max_{pw(G)}$. Since L is not a context-free language, we can assume that $m \geq 1$. Then let M be the number of Lemma 3, and $u = a^{2^{2^{mMN}}}$, where $N = \max\{|\mathrm{rhs}(r)| \mid r \in R\}$. Let moreover $der : S = \alpha_0 \Rightarrow \alpha_1 \Rightarrow \ldots \Rightarrow \alpha_n = u$ be one of the shortest derivations of G from S to u. Clearly $n \geq M$. Thus, by Lemma 3, there are $i < j$ in $[M]$ such that $\alpha_i \leadsto_m \alpha_j$. We can assume that $|\alpha_i| < |\alpha_j|$. Indeed, assume on the contrary that this is not the case. Then, since G has no erasing rules, $|\alpha_i| = |\alpha_j|$. This, using (ii) of Proposition 2, implies that $\alpha_i = \alpha_j$. This yields that $der' : \alpha_0 \Rightarrow \ldots \Rightarrow \alpha_i \Rightarrow \alpha_{j+1} \Rightarrow \ldots \Rightarrow \alpha_n$ is also a derivation of G from S to u with $|der'| < n$. However this contradicts the assumption that der is a shortest derivation from S to u. Applying Lemma 2 we get that there is an $u' \in \Sigma^*$ such that $\alpha_i \Rightarrow^* u'$ and $|u| < |u'| \leq (m+1)|u|$. Since $S \Rightarrow^* \alpha_i$, also $S \Rightarrow^* u'$ holds. Consequently, $u' \in L$.

Clearly, the shortest word $v \in L$ with $|u| < |v|$ is $a^{2^{2^{mMN+1}}}$. On the other hand, $|u'| \leq (m+1)2^{2^{mMN}} < 2^{2^{mMN}} \cdot 2^{2^{mMN}} = 2^{2^{mMN+1}} = |v|$. Thus $|u'| < |v|$ yielding $u' \notin L$ which is a contradiction. Therefore $L \notin \mathcal{L}(\mathrm{pSCG})$.

4   Conclusions

In this paper we have investigated permitting semi-conditional grammars introduced by Kelemen [13]. We showed that these grammars are strictly weaker than context-sensitive grammars when erasing rules are not allowed. However, it is still open whether this remains true if erasing rules are allowed. In [19] it was shown that allowing erasing rules does not increase the generative power of permitting random context grammars. To decide whether this holds also for permitting semi-conditional grammars is a possible topic for future work. It is also an interesting question, for example, whether the inclusion $\mathcal{L}(\mathrm{pRCG}) \subseteq \mathcal{L}(\mathrm{pSCG})$ depicted in Fig. 1 is strict or not.

Acknowledgement. We are grateful to Erzsébet Csuhaj Varjú for introducing us the topic of this paper and also for her many useful comments on it. We thank the anonymous reviewers for their constructive comments, which helped us to improve the manuscript.

References

1. Bordihn, H., Fernau, H.: Accepting grammars and systems: an overview. In: Developments in Language Theory, Magdeburg, Germany (1995)
2. Dassow, J., Masopust, T.: On restricted context-free grammars. J. Comput. Syst. Sci. **78**(1), 293–304 (2012)
3. Dassow, J., Păun, G.: Regulated Rewriting in Formal Language Theory. Springer, New York (1989)
4. Dickson, L.E.: Finiteness of the odd perfect and primitive abundant numbers with n distinct prime factors. Am. J. Math. **35**(4), 413–422 (1913)
5. Ewert, S., van der Walt, A.: A shrinking lemma for random forbidding context languages. Theor. Comput. Sci. **237**(1–2), 149–158 (2000)
6. Ewert, S., van der Walt, A.: A pumping lemma for random permitting context languages. Theor. Comput. Sci. **270**(1–2), 959–967 (2002)
7. Gazdag, Z.: A note on context-free grammars with rewriting restrictions. In: Brodnik, A., Galambos, G. (eds.) Proceedings of the 2010 Mini-Conference on Applied Theoretical Computer Science. University of Primorska Press, Koper (2011)
8. Gazdag, Z.: Remarks on some simple variants of random context grammars. J. Autom. Lang. Comb. **19**(1–4), 81–92 (2014)
9. Gazdag, Z., Tichler, K.: On the power of permitting semi-conditional grammars, extended version. https://www.researchgate.net/publication/312587701_On_the_Power_of_Permitting_Semi-conditional_Grammars
10. Higman, G.: Ordering by divisibility in abstract algebras. Proc. Lond. Math. Soc. **3**(1), 326–336 (1952)
11. Joshi, A.K.: Tree adjoining grammars: how much context-sensitivity is required to provide reasonable structural descriptions? In: Dowty, D.R., Karttunen, L., Zwicky, A.M. (eds.) Natural Language Parsing, pp. 206–250. Cambridge University Press, Cambridge (1985)
12. Jurafsky, D., Martin, J.H.: Speech and Language Processing: An Introduction to Natural Language Processing, Computational Linguistics, and Speech Recognition. Prentice Hall PTR, Upper Saddle River, NJ, USA (2000)
13. Kelemen, J.: Conditional grammars: motivations, definitions, and some properties. In: Proceedings of the Conference on Automata, Languages and Mathematical Sciences, Salgótarján, pp. 110–123 (1984)
14. Kruskal, J.B.: Well-quasi-ordering, the tree theorem, and Vazsonyi's conjecture. Trans. Am. Math. Soc. **95**(2), 210–225 (1960)
15. Masopust, T.: Simple restriction in context-free rewriting. J. Comput. Syst. Sci. **76**(8), 837–846 (2010)
16. Păun, G.: A variant of random context grammars: semi-conditional grammars. Theor. Comput. Sci. **41**, 1–17 (1985)
17. Salomaa, A.: Formal Languages. Academic Press, New York, London (1973)
18. van der Walt, A.: Random context languages. Inf. Process. **71**, 66–68 (1972)
19. Zetzsche, G.: On erasing productions in random context grammars. In: Abramsky, S., Gavoille, C., Kirchner, C., Meyer auf der Heide, F., Spirakis, P.G. (eds.) ICALP 2010. LNCS, vol. 6199, pp. 175–186. Springer, Heidelberg (2010). doi:10.1007/978-3-642-14162-1_15

On the Interplay Between Babai and Černý's Conjectures

François Gonze[1]($\boxtimes$), Vladimir V. Gusev[1,2], Balázs Gerencsér[3],
Raphaël M. Jungers[1], and Mikhail V. Volkov[2]

[1] ICTEAM Institute, Université catholique de Louvain, Louvain-la-Neuve, Belgium
{francois.gonze,vladimir.gusev,raphael.jungers}@uclouvain.be
[2] Ural Federal University, Ekaterinburg, Russia
mikhail.volkov@usu.ru
[3] Alfréd Rényi Institute of Mathematics, Budapest, Hungary
gerencser.balazs@renyi.mta.hu

Abstract. Motivated by the Babai conjecture and the Černý conjecture, we study the reset thresholds of automata with the transition monoid equal to the full monoid of transformations of the state set. For automata with n states in this class, we prove that the reset thresholds are upper-bounded by $2n^2 - 6n + 5$ and can attain the value $\frac{n(n-1)}{2}$. In addition, we study diameters of the pair digraphs of permutation automata and construct n-state permutation automata with diameter $\frac{n^2}{4} + o(n^2)$.

1 Background and Overview

Completely reachable automata, i.e., deterministic finite automata in which every non-empty subset of the state set occurs as the image of the whole state set under the action of a suitable input word, appeared in the study of descriptional complexity of formal languages [26] and in relation to the Černý conjecture [13]. In [6] an emphasis has been made on automata in this class with minimal transition monoid size. In the present paper we focus on automata being in a sense the extreme opposites of those studied in [6], namely, on automata of maximal transition monoid size. In other words, we consider automata *with full transition monoid*, i.e., transition monoid equal to the full monoid of transformations of the state set; clearly, automata with this property are completely reachable. There are several reasons justifying special attention to automata with full transition monoid. First, as observed in [6], the membership problem for this class of automata is decidable in polynomial time (of the size of the input automaton) while the complexity of membership in the class of all completely reachable automata still remains unknown. Second, this class contains automata that

---

V. Gusev and M.V. Volkov were supported by RFBR grant no. 16-01-00795, Russian Ministry of Education and Science project no. 1.3253.2017, and the Competitiveness Enhancement Program of Ural Federal University. Balázs Gerencsér was supported by PD grant no. 121107, National Research, Development and Innovation Office of Hungary. This work was supported by the French Community of Belgium and by the IAP network DYSCO. Raphaël Jungers is a Fulbright Fellow and a FNRS Research Associate.

© Springer International Publishing AG 2017
É. Charlier et al. (Eds.): DLT 2017, LNCS 10396, pp. 185–197, 2017.
DOI: 10.1007/978-3-319-62809-7_13

correspond to Brzozowski's most complex regular languages [7] and to other regular languages that play a distinguished role in descriptive complexity analysis. Finally, and most importantly from our viewpoint, automata with full transition monoid are synchronizing and their synchronization issues constitute a sort of meeting point for two famous open problems—the *Babai conjecture* and the *Černý conjecture*. Next, we recall these conjectures and outline the contribution of the present paper in view of these problems.

1.1   The Babai Conjecture

Let A be a set of generators of a finite group G. The *Cayley graph* $\Gamma(G, A)$ consists of G as the set of vertices and the edges $\{g, ga\}$ for all $g \in G$, $a \in A$. The *diameter* of $\Gamma(G, A)$ is the maximum among the lengths of shortest paths between any two vertices. In group theory terms, the diameter of $\Gamma(G, A)$ is the smallest ℓ such that every $g \in G$ can be represented as $g = a_1^{\varepsilon_1} a_2^{\varepsilon_2} \cdots a_\ell^{\varepsilon_\ell}$, where $\varepsilon_i \in \{1, -1\}$ and $a_i \in A$ for all $i = 1, \ldots, \ell$. The *diameter* $\mathrm{diam}(G)$ of G is the maximal diameter of $\Gamma(G, A)$ among all generating sets A of G. The notion of group diameter is related to the growth rate in groups, expander graphs, random walks on groups and their mixing times, see, e.g., [23,33]. Recently, the following conjecture received significant attention:

Conjecture 1 (Babai [4]). *The diameter of each non-abelian finite simple group G does not exceed $(\log |G|)^{O(1)}$, where the implied constant is absolute.*

Note that for the case of the symmetric group S_n, this conjecture readily implies $\mathrm{diam}(S_n) \leq n^{O(1)}$. (The group S_n is not simple but for $n \geq 5$ it contains a non-abelian simple subgroup of index 2.)

The Babai conjecture was proved for various classes of groups, but despite intensive research effort it remains open, see [22] for an overview. In the case of S_n, a recent breakthrough gives only a quasipolynomial upper bound, namely, $\exp(O((\log n)^4 \log \log n)$, and it relies on the Classification of Finite Simple Groups [22]. It is even more astonishing if we compare it to the best known lower bound in this case: for the classical set of generators consisting of the transposition $(1, 2)$ and the full cycle $(1, 2, \ldots, n)$, every permutation in S_n can be expressed as a product of at most $\sim \frac{3n^2}{4}$ (asymptotically) generators [40].

1.2   The Černý Conjecture

Recall that a deterministic finite state automaton (DFA) is a triple[1] $\langle Q, \Sigma, \delta \rangle$, where Q is a finite set of states, Σ is a finite set of input symbols called the *alphabet*, and δ is a function $\delta \colon Q \times \Sigma \to Q$ called the *transition function*. A *word* is a sequence of letters from the alphabet. The *length* of a word is the number of its letters. We can look at $\delta(q, a)$ as the result of the *action* of the letter $a \in \Sigma$ at the state $q \in Q$. We extend this action to the action of words

---

[1] As initial and final states play no role in our considerations, we omit them.

over Σ on Q denoting, for any word w and any state $q \in Q$, the state resulting in successive applications of the letters of w from left to right by $q \cdot w$. For a subset $P \subseteq Q$, we write $P \cdot w$ for the set $\{p \cdot w \mid p \in P\}$.

A DFA $\mathscr{A} = \langle Q, \Sigma, \delta \rangle$ is called *synchronizing* if there exist a word w and a state f such that $Q \cdot w = \{f\}$. Any such word is called a *synchronizing* or *reset* word. The minimum length of reset words for $\mathscr{A}$ is called the *reset threshold* of $\mathscr{A}$ and is denoted by $\mathrm{rt}(\mathscr{A})$. Synchronizing automata appear in various branches of mathematics and are related to synchronizing codes [5], part orienting problems [27,28], substitution systems [16], primitive sets of matrices [19], synchronizing groups [3], convex optimization [20], and consensus theory [11].

Conjecture 2 (Černý [9,10]). *The reset threshold of an n-state synchronizing automaton is at most $(n-1)^2$.*

If the conjecture holds true, then the value $(n-1)^2$ is optimal, since for every n there exists an n-state automaton $\mathscr{C}_n$ with the reset threshold equal to $(n-1)^2$ [9].

The Černý conjecture has gained a lot of attention in automata theory. It has been shown to hold true in various special classes [14,21,24,31,35,36], but in the general case, it remains open for already half a century. For more than 30 years, the best upper bound was $\frac{n^3-n}{6}$, obtained in [15,30] and independently in [25]. Recently, a small improvement on this bound has been reported in [37]: the new bound is still cubic in n but improves the coefficient $\frac{1}{6}$ at n^3 by $\frac{4}{46875}$. A survey on synchronizing automata and the Černý conjecture can be found in [38].

In order to make the relationship between the Černý and the Babai conjectures more visible, we borrow from [2] the idea of restating the former in terms similar to those used in the formulation of the latter. Let T_n be the full transformation monoid of an n-element set Q. A transformation $t \in T_n$ is a *constant* if there exists $f \in Q$ such that for all $q \in Q$ we have $t(q) = f$. We can state the Černý conjecture as follows: for every set of transformations $A \subseteq T_n$, if the submonoid generated by A contains a constant, then there exists a constant g such that $g = a_1 a_2 \cdots a_\ell$, where $\ell \leq (n-1)^2$ and $a_i \in A$ for all $i = 1, \ldots, \ell$. It is easy to see that this formulation is equivalent to the original one by treating the letters of an automaton as the transformations of its state set since reset words precisely correspond to constant transformations.

1.3   Our Contributions

The first part of our paper is devoted to the following *hybrid Babai–Černý problem*[2]: given a set of generators A of the full transformation monoid T_n, what is the length $\ell(A)$ of the shortest product $a_1 a_2 \cdots a_\ell$ with $a_i \in A$ which is equal to a constant? Namely, we are interested in the bounds on $\ell(A)$ that depend only on n. The hybrid Babai–Černý problem is a special case of the Černý problem.

---

[2] During the preparation of this paper we discovered that the same question was also posed in [34, Conjecture 3], though its connection with Babai's problem was not registered there.

Indeed, it is a restriction to the class of DFAs with the *transition monoid*, i.e., the transformation monoid generated by the actions of letters, equal to T_n. Of course, the general cubic upper bound is valid, but not the lower bound, since the Černý automata $\mathscr{C}_n$ do not belong to this class (even though they are completely reachable, see [6]). In Sect. 2 we establish that the growth rate of $\ell(n)$ is $\Theta(n^2)$, more precisely, we show that $\frac{n(n-1)}{2} \leq \ell(n) \leq 2n^2 - 6n + 5$. We also present the exact values of $\ell(n)$ for small values of n resulting from our computational experiments. Our contribution can be also seen as a progress towards resolution of Conjecture 3 from [34].

The second part of our paper is devoted to a "local" version of the Babai problem where we restrict our attention to the action on the set of (unordered) pairs. Let A be a set of permutations from S_n. The *pair digraph* $P(A)$ consists of pairs $\{i, j\}$ as the set of vertices and the edges $(\{i, j\}, \{ia, ja\})$ for all i, j and $a \in A$. The *diameter* of $P(A)$, denoted $\operatorname{diam} P(A)$, is the maximum among the lengths of shortest (directed) paths between any two vertices. We study the behavior of $\operatorname{diam} P(A)$ in terms of n. The problem comes from analysis of certain aspects of Markov chains and group theory [17], but our interest in it is mainly motivated by its importance for the theory of synchronizing automata. Indeed, every synchronizing automaton $\mathscr{A}$ must have a letter a, say, whose action merges a pair of states. Thus, one can construct a reset word for $\mathscr{A}$ by successively moving pairs of states to a pair merged by a. If $\mathscr{A}$ possesses sufficiently many letters acting as permutations (as automata with the full transition monoid do), one can move pairs by these permutations, and hence, upper bounds on the diameter of the corresponding pair digraph induce upper bounds on $\operatorname{rt}(\mathscr{A})$.

Clearly, $\operatorname{diam} P(A) \leq \frac{n(n-1)}{2}$ for all $A \subseteq S_n$. In Sect. 3 we establish the lower bound $\frac{n^2}{4} + o(n^2)$ on $\operatorname{diam} P(A)$ by presenting a series of examples with only two generators for every odd n.

Due to the space constraints, proofs of Theorem 1, Lemmas 5 and 6 and Theorem 9 have been moved to the extended version and are available on the page https://arxiv.org/abs/1704.04047.

1.4   Related Work

The diameters of groups and semigroups constitute a relatively well studied topic. A general discussion on diameters and growth rates of groups can be found in [23]. Various results about the diameter of T_n and its submonoids are described in [29,34]. The length of the shortest representation of a constant (including the case of partially defined transformations) is typically studied in the framework of synchronizing automata, see [1,38,39].

2   Automata with Full Transition Monoid

2.1   Naïve Construction

Recall that, on the one hand, the Černý automata $\mathscr{C}_n$ from [9] have two letters of which one acts as a cyclic permutation and the other fixes all states, except

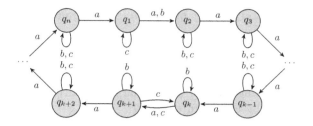

Fig. 1. The automaton $\mathscr{CB}_{n,k}$

one, which is mapped to the next element in the cyclic order defined by the cyclic permutation. On the other hand, the extremal case of the Babai conjecture for S_n is composed of a cyclic permutation and the transformation which fixes all elements except two, which are neighbors in the cyclic order defined by the cyclic permutation. Therefore, one could wonder if a combination of these transformations could result in a DFA with both large reset threshold and full transition monoid.

The construction is defined as follows. There are n states $q_1, \ldots, q_n$ and three letters a, b, and c. The letter a acts as a cyclic permutation on the states, following their indices. The letter b fixes all states, except q_1, which is mapped to q_2 by b. The letter c fixes all states, except q_k and q_{k+1}, for some k, which are swapped by c. The resulting automaton $\mathscr{CB}_{n,k}$ is shown in Fig. 1. We notice that if we remove the letter c, we obtain the automaton $\mathscr{C}_n$ from the Černý family providing the largest currently existing lower bound in the Černý problem, and if we remove the letter b, we obtain a generating set of the group S_n providing the largest currently existing lower bound in the Babai problem for S_n. Also observe that in the case where $k = 2$, our automaton is nothing but Brzozowski's "Universal Witness" [7] recognizing the most complex regular language, i.e., the language witnessing at once practically all tight lower bounds found for the state complexity of various operations with regular languages, see [7, Theorem 6]. The next result shows that, however, the reset threshold of the automaton $\mathscr{CB}_{n,k}$ is upper-bounded by $O(n \log n)$, while, as we show later, among automata with full transition monoid there exist ones whose reset threshold is a quadratic function of their state number.

Theorem 1. *The automaton* $\mathscr{CB}_{n,k}$ *has a reset word of length at most* $4n\lceil \log_2 n \rceil$.

2.2   Random Sampling and Exhaustive Search

Every DFA with the transition monoid T_n necessarily has permutation letters that generate the whole symmetric group S_n and a letter of rank $n-1$ (i.e., a letter whose image has $n-1$ elements). It is a well known fact that the converse is also true, i.e., the transition monoid of any automaton with permutation letters generating S_n and a letter of rank $n - 1$ is equal to T_n, see, e.g., [18, Theorem 3.1.3].

Relying on a group-theoretic result by Dixon [12], Cameron [8] observed that an automaton formed by two permutation letters taken uniformly at random and an arbitrary non-permutation letter is synchronizing with high probability. We give an extension by using another non-trivial group-theoretical result, namely, the following theorem by Friedman *et al.* [17].

Theorem 2. *For every r and $d \geq 2$ there is a constant C such that for d permutations $\pi_1, \pi_2, \ldots, \pi_d$ of S_n taken uniformly at random, the following property F_r holds with probability tending to 1 as $n \to \infty$: for any two r-tuples of distinct elements in $\{1, 2, \ldots, n\}$, there is a product of less than $C \log n$ of the π_i's which maps the first r-tuple to the second.*

Corollary 3. *There is a constant C such that the reset threshold of an n-state automaton with two random permutation letters and an arbitrary non-permutation letter does not exceed $Cn \log n$ with probability that tends to 1 as $n \to \infty$.*

Proof. Let $\mathscr{A} = \langle Q, \Sigma, \delta \rangle$ stand for the automaton in the formulation of the corollary. We let $a \in \Sigma$ be the non-permutation letter and assume that the two permutation letters in Σ have the property F_2 of Theorem 2 for $r = 2$ with some constant C. By Theorem 2 this assumption holds true with probability that tends to 1 as $n \to \infty$.

There exists two different states $q_1, q_2 \in Q$ such that $q_1 \cdot a = q_2 \cdot a$. The set $Q \cdot a$ contains less than n elements. If $|Q \cdot a| = 1$, then a is a reset word for $\mathscr{A}$. If $|Q \cdot a| > 1$, take two different states $p_1, p_2 \in Q \cdot a$. By F_2, there is a product w of less than $C \log n$ of the permutation letters such that $p_i \cdot w = q_i$ for $i = 1, 2$. Now $|Q \cdot awa| < |Q \cdot a|$. If $|Q \cdot awa| = 1$, awa is a reset word for $\mathscr{A}$. If $|Q \cdot awa| > 1$, we apply the same argument to a pair of different states in $Q \cdot awa$. Clearly, the process results in a reset word in at most $n - 1$ steps while the suffix appended at each step is of length at most $C \log n + 1$. Hence the length of the reset word constructed this way is at most $(C + 1)n \log n$.   □

Corollary 3 indicates that one can hardly discover an n-state automaton with the transition monoid equal to T_n and sufficiently large reset threshold by a random sampling. Therefore, we performed an exhaustive search among all automata with two permutation letters generating S_n and one letter of rank $n - 1$. Our computational results are summarized in Table 1.

Table 1. The largest reset thresholds of n-state automata two permutation letters generating S_n and one letter of rank $n - 1$

Number of states	2	3	4	5	6	7
Reset threshold	1	4	8	14	19	27

As n grows, the reset thresholds of the obtained examples become much smaller than $(n - 1)^2$. We were unable to derive a series of n-state three-letter automata with the transition monoid T_n and quadratically growing reset thresholds. We suspect that the reset threshold of automata in this class is $o(n^2)$.

In the case of unbounded alphabet, for every n, we present an n-state automaton $\mathscr{V}_n$ with the transition monoid T_n such that $\mathrm{rt}(\mathscr{V}_n) = \frac{n(n-1)}{2}$. The state set of $\mathscr{V}_n$ is $Q_n = \{q_0, \ldots, q_{n-1}\}$ and the input alphabet consists of n letters $a_1, \ldots, a_n$. The transition function is defined as follows:

$$\begin{cases} q_i \cdot a_j = q_i & \text{for } 0 \leq i < n, \ 1 \leq j < n, \ i \neq j, \ i \neq j+1, \ j \neq n, \\ q_i \cdot a_i = q_{i-1} & \text{for } 0 < i \leq n-1, \\ q_i \cdot a_{i+1} = q_{i+1} & \text{for } 0 \leq i < n-1, \\ q_0 \cdot a_n = q_1 \cdot a_n = q_0, \\ q_i \cdot a_n = q_i & \text{for } 2 \leq i \leq n-1. \end{cases}$$

Simply speaking, every letter a_i for $i \leq n-1$ swaps the states q_i and q_{i-1} and fixes the other states. The letter a_n brings both q_0 and q_1 to q_0 and fixes the other states. The automaton $\mathscr{V}_5$ is depicted in Fig. 2.

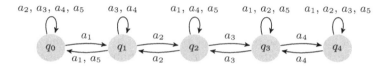

Fig. 2. The automaton $\mathscr{V}_5$

Recall that a state z of an DFA is said to be a *sink state* (or *zero*) if $z \cdot a = z$ for every input letter a. It is known that every n-state synchronizing automaton with zero can be reset by a word of length $\frac{n(n-1)}{2}$, cf. [31]. To show that this upper bound is tight for each n, Rystsov [31] constructed an n-state and $(n-1)$-letter synchronizing automaton $\mathscr{R}_n$ with zero which cannot be reset by any word of length less than $\frac{n(n-1)}{2}$. In fact, our automaton $\mathscr{V}_n$ is a slight modification of $\mathscr{R}_n$ as the latter automaton is nothing but $\mathscr{V}_n$ without the letter a_1.

Theorem 4. *For every n, the automaton $\mathscr{V}_n$ has T_n as its transition monoid and* $\mathrm{rt}(\mathscr{V}_n) = \frac{n(n-1)}{2}$.

Proof. The letters $a_1, \ldots, a_{n-1}$ generate S_n because the product $a_1 \cdots a_{n-1}$ is a full cycle and any full cycle together with any transposition generates S_n. Since the letter a_n has rank $n-1$, it together with $a_1, \ldots, a_{n-1}$ generates T_n.

The automaton $\mathscr{V}_n$ is synchronizing because so is the restricted automaton $\mathscr{R}_n$, and $\mathrm{rt}(\mathscr{V}_n) \leq \frac{n(n-1)}{2}$ because every reset word for $\mathscr{R}_n$ resets $\mathscr{V}_n$ as well. It remains to verify that the length of any reset word for $\mathscr{V}_n$ must be at least $\frac{n(n-1)}{2}$. Let w be a reset word of minimum length for $\mathscr{V}_n$. Since a_n is the only non-permutation letter, we must have $w = w'a_n$ for some w' such that $|Q_n \cdot w'| > 1$. This is only possible when $Q_n \cdot w' = \{q_0, q_1\}$ whence $Q_n \cdot w = \{q_0\}$. Consider the function f from the set of all non-empty subsets of Q_n into the set of non-negative integers defined as follows: if $S = \{q_{s_1}, \ldots, q_{s_t}\}$, then $f(S) = \sum_{i=1}^{t} s_i$. Clearly, $f(\{q_0\}) = 0$ and $f(Q_n) = \frac{n(n-1)}{2}$. For any set S and any letter a_j, we have $f(S \cdot a_j) \geq f(S) - 1$ since each letter only exchanges two adjacent states or maps q_1 and q_0 to q_0. Thus, when we apply the word w letter-by-letter, the value of f after the application of the prefix of w of length i cannot be less than $\frac{n(n-1)}{2} - i$. Hence, to reach the value 0, we need at least $\frac{n(n-1)}{2}$ letters. $\square$

2.3   Upper Bound on the Reset Threshold

We now provide a quadratic upper bound on the reset words of automata with the transition monoid equal to T_n. Our proof is inspired by the method of Rystsov [32] adapted to our case.

Let $\mathscr{A} = \langle Q, \Sigma, \delta \rangle$ be a DFA. Given a proper non-empty subset $R \subset Q$ and a word w over Σ, we say that R *can be extended by w* if the cardinality of the set $Rw^{-1} = \{q \in Q \mid q \cdot w \in R\}$ is greater than $|R|$. Now assume that $|Q| = n$ and the transition monoid of $\mathscr{A}$ coincides with the full transformation monoid T_n. Then there is a letter x of rank $n - 1$. The set $Q \setminus Q \cdot x$ consists of a unique state, which is called the *excluded state* for x and is denoted by $\text{excl}(x)$. Furthermore, the set $Q \cdot x$ contains a unique state p such that $p = q_1 \cdot x = q_2 \cdot x$ for some $q_1 \neq q_2$; this state p is called the *duplicate state* for x and is denoted by $\text{dupl}(x)$. We notice that a non-empty subset $R \subset Q$ can be extended by x if and only if $\text{dupl}(x) \in R$ and $\text{excl}(x) \notin R$. Moreover, if a word w is a product of permutation letters, R can be extended by the word wx if and only if $\text{dupl}(x) \in Rw^{-1}$ and $\text{excl}(x) \notin Rw^{-1}$. To better understand which extensions are possible, we construct a series of directed graphs (digraphs) Γ_i, $i = 0, 1, \ldots$, with the set Q as the vertex set.

The digraph Γ_0 has the set $E_0 = \{(\text{excl}(x), \text{dupl}(x))\}$ as its edge set. Let Π be the set of permutation letters of $\mathscr{A}$. Notice that Π generates the symmetric group S_n. By Π^i we denote the set of words of length at most i over the letters in Π. The digraph Γ_i for $i > 0$ has the edge set $E_i = \{(\text{excl}(x) \cdot w, \text{dupl}(x) \cdot w) \mid w \in \Pi^i\}$. The digraphs Γ_i, $i = 0, 1, \ldots$, form a sort of stratification for the graph Γ_∞ with the edge set $E_\infty = \cup_{i=0}^{\infty} E_i$; the latter digraph has been studied in [6, 32] (in the context of arbitrary completely reachable automata). Observe that none of the digraphs Γ_i, $i = 0, 1, \ldots$, have loops.

Recall that a digraph is said to be *strongly connected* if for every pair of its vertices, there exists a directed path from the first vertex to the second. We need the two following lemmas.

Lemma 5. *If the digraph Γ_k is strongly connected, then every proper non-empty subset in Q can be extended by a word of length at most $k + 1$.*

Lemma 6. *The digraph Γ_{2n-3} is strongly connected.*

Theorem 7. *Let $\mathscr{A}$ be an n-state automaton with the transition monoid equal to T_n. The reset threshold of $\mathscr{A}$ is at most $2n^2 - 6n + 5$.*

Proof. Let x be a letter of rank $n-1$ and $h = \text{dupl}(x)$. We extend the set $\{h\}$ by x, getting a subset R_2 with $|R_2| \geq 2$. Lemmas 5 and 6 imply that proper non-empty subsets in Q can be extended by words of length at most $2n-2$. Starting with R_2, we extend subsets until we reach the full state set. Let u_i be the word of length at most $2n-2$ used for the i-th of these extensions and let m be the number of the extensions. Observe that $m \leq n-2$. Clearly, the word $u_m \cdots u_1 x$ resets $\mathscr{A}$ and has the length at most $1 + (n-2)(2n-2) = 2n^2 - 6n + 5$.   □

Remark 8. Let $\mathscr{A} = \langle Q, \Sigma, \delta \rangle$ be an n-state DFA that has a letter of rank $n-1$, and let P be the subgroup of the symmetric group S_n generated by the permutation letters from Σ. Our proof of Theorem 7 actually works in the case if P is a *2-transitive* group, that is, P acts transitively on the set of ordered pairs of Q.

3   Bounds on the Diameter of the Pair Digraph

In this section we present a lower bound on the largest diameter of the pair digraph $P(A)$ for $A \subseteq S_n$. We proceed by presenting subsets $A \subseteq S_n$ for every odd n whose diameter is $\frac{n^2}{4} + o(n^2)$. In order to simplify the presentation, we mostly use automata terminology and describe the corresponding examples as a family of automata $\mathscr{F}_n = \langle Q_n, A, \delta \rangle$ (the input letters of $\mathscr{F}_n$ form the subset A). We let $Q_n = \{q_1, \ldots, q_n\}$ and denote pairs of states such as $\{q_i, q_j\}$ simply by $q_i q_j$.

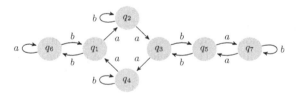

Fig. 3. The automaton $\mathscr{F}_7$

The automaton $\mathscr{F}_7$ shown in Fig. 3 is the first of the family $\mathscr{F}_n$. The digraph of pairs of its states is shown in Fig. 4. One can verify that the shortest word mapping $q_2 q_4$ to $q_4 q_7$ has length 15.

The automata of the family are obtained recursively, starting with $\mathscr{F}_7$. From $\mathscr{F}_n$, we construct $\mathscr{F}_{n+2}$. The effect of the letters is the same for the states $q_1, \ldots, q_{n-2}$ in $\mathscr{F}_n$ and $\mathscr{F}_{n+2}$. The effect of the letters a and b at the states q_{n-1}, q_n, q_{n+1} and q_{n+2} is defined as follows: the letters mapping q_{n-1} and q_n to themselves in $\mathscr{F}_n$ exchange q_{n-1} with q_{n+1} and q_n with q_{n+2} respectively in $\mathscr{F}_{n+2}$. The other letter maps q_{n+1} and q_{n+2} to themselves and q_{n-1}, q_n to q_{n-3} and respectively q_{n-2}. The result is shown in Fig. 5 (for $n \equiv 3 \pmod 4$), in which k stands for $\frac{n-5}{2}$.

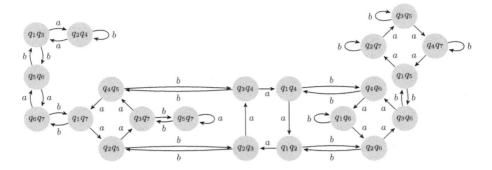

Fig. 4. The pair digraph of $\mathscr{F}_7$

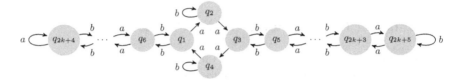

Fig. 5. The automaton $\mathscr{F}_{2k+5}$, with k odd

Theorem 9. *For odd $n \geq 7$, the diameter of the pair digraph of the automaton $\mathscr{F}_n$ is at least $\frac{n^2}{4} + o(n^2)$.*

Proof Sketch. For the automaton $\mathscr{F}_n$ ($n > 7$, $n \equiv 3 \pmod 4$), we claim that any word mapping $q_2 q_4$ to $q_{k+2} q_{k+4}$ with $k = \frac{n-5}{2}$ has length at least $\frac{n^2}{4} + \frac{5n}{4} - 7$. For this, we define a function N which associates a non-negative integer $N(q_i q_j)$ to each pair $q_i q_j$, $i < j$. This function is such that if a pair $q_i q_j$ is mapped by a or b to a pair $q_{i'} q_{j'}$, then $N(q_{i'} q_{j'}) \geq N(q_i q_j) - 1$. This implies that if $(q_i q_j) \cdot w = q_s q_t$ for some word w, then the length of w is at least $N(q_i q_j) - N(q_s q_t)$. The number assigned to $q_{k+2} q_{k+4}$ is 0, while the number given to $q_2 q_4$ is equal to $\frac{n^2}{4} + \frac{5n}{4} - 7$, thus, the claim holds. The values of the function N are provided in the extended version.

In addition, a word of length $\frac{n^2}{4} + \frac{5n}{4} - 7$ that maps $q_2 q_4$ to $q_{k+2} q_{k+4}$ is described in the extended version. Therefore $\frac{n^2}{4} + \frac{5n}{4} - 7$ is the exact value of the "distance" between these two particular pairs.

A similar argument holds for $n \equiv 1 \pmod 4$, with the distance between two particular pairs of states being at least $\frac{n^2}{4} + \frac{5n}{4} - 7.5$.   □

Our numerical experiments confirm that $\frac{n^2}{4} + \frac{5n}{4} - 7$ is indeed the diameter of the pair graph of the automaton $\mathscr{F}_n$ for $n \equiv 3 \pmod 4$ from $n = 11$ to $n = 31$ while $\frac{n^2}{4} + \frac{5n}{4} - 7.5$ is the exact value of the diameter for $n \equiv 1 \pmod 4$ from $n = 13$ to $n = 29$.

We have also computed the largest diameter of the pair digraph $P(A)$ for all $A \subseteq S_n$ with $|A| = 2$ and $n = 5, 7, 9$ and performed a number of random sampling experiments with two permutations for larger values of n. The experimental results suggest that the pair digraph of the automaton $\mathscr{F}_n$ has the largest diameter among all possible pair digraphs. Thus, we formulate the following:

Conjecture 3. *The diameter of the pair digraph for a subset of S_n is bounded above by $\frac{n^2}{4} + o(n^2)$.*

4   Conclusion

We studied the hybrid Babai–Černý problem, where the question is to find tight bounds on the reset threshold for automata with the full transition monoid. We presented a series of n-state automata $\mathscr{V}_n$ in this class with the reset threshold equal to $\frac{n(n-1)}{2}$, thus establishing a lower bound for the problem, and found an upper bound with the same growth rate, namely, $2n^2 + o(n^2)$. We also described a series of n-state automata with diameter of the pair digraph equal to $\frac{n^2}{4} + o(n^2)$.

For follow-up work, one direction is to refine the bounds with respect to the constants that do not match yet. Also, a lower bound for the hybrid problem using only three letters (generators) is of interest, since the number of letters of the presented family $\mathscr{V}_n$ is equal to the number of states.

References

1. Ananichev, D.S., Gusev, V.V., Volkov, M.V.: Primitive digraphs with large exponents and slowly synchronizing automata. J. Math. Sci. **192**(3), 263–278 (2013)
2. Ananichev, D.S., Volkov, M.V.: Some results on Černy type problems for transformation semigroups. In: Araújo, I.M., Branco, M.J.J., Fernandes, V.H., Gomes, G.M.S. (eds.) Semigroups and Languages, pp. 23–42. World Scientific (2004)
3. Araújo, J., Cameron, P.J., Steinberg, B.: Between primitive and 2-transitive: Synchronization and its friends. CoRR abs/1511.03184 (2015)
4. Babai, L., Seress, A.: On the diameter of permutation groups. Eur. J. Combin. **13**(4), 231–243 (1992)
5. Berstel, J., Perrin, D., Reutenauer, C.: Codes and Automata. CUP, Cambridge (2009)
6. Bondar, E.A., Volkov, M.V.: Completely reachable automata. In: Câmpeanu, C., Manea, F., Shallit, J. (eds.) DCFS 2016. LNCS, vol. 9777, pp. 1–17. Springer, Cham (2016). doi:10.1007/978-3-319-41114-9_1
7. Brzozowski, J.A.: In search of most complex regular languages. Int. J. Found. Comput. Sci. **24**(6), 691–708 (2013)
8. Cameron, P.J.: Dixon's theorem and random synchronization. Discrete Math. **313**(11), 1233–1236 (2013)
9. Černý, J.: Poznámka k homogénnym experimentom s konečnými automatami. Mat.-fyz. Časopis Slovenskej Akadémie Vied **14**(3), 208–216 (1964). in Slovak
10. Černý, J., Pirická, A., Rosenauerová, B.: On directable automata. Kybernetica **7**, 289–298 (1971)

11. Chevalier, P.Y., Hendrickx, J.M., Jungers, R.M.: Reachability of consensus and synchronizing automata. In: 54th IEEE Conference on Decision and Control (CDC), pp. 4139–4144. IEEE (2015)
12. Dixon, J.D.: The probability of generating the symmetric group. Math. Z. **110**, 199–205 (1969)
13. Don, H.: The Černý conjecture and 1-contracting automata. Electr. J. Comb. **23**(3), P3.12 (2016)
14. Dubuc, L.: Sur les automates circulaires et la conjecture de Černý. RAIRO Informatique Théorique et Applications **32**, 21–34 (1998). in French
15. Frankl, P.: An extremal problem for two families of sets. Eur. J. Combin. **3**, 125–127 (1982)
16. Frettlöh, D., Sing, B.: Computing modular coincidences for substitution tilings and point sets. Discrete Comput. Geom. **37**, 381–407 (2007)
17. Friedman, J., Joux, A., Roichman, Y., Stern, J., Tillich, J.P.: The action of a few permutations on r-tuples is quickly transitive. Random Struct. Algorithms **12**(4), 335–350 (1998)
18. Ganyushkin, O., Mazorchuk, V.: Classical Finite Transformation Semigroups: An Introduction. Springer, London (2009)
19. Gerencsér, B., Gusev, V.V., Jungers, R.M.: Primitive sets of nonnegative matrices and synchronizing automata. CoRR abs/1602.07556 (2016)
20. Gonze, F., Jungers, R.M.: On the synchronizing probability function and the triple rendezvous time for synchronizing automata. SIAM J. Discrete Math. **30**(2), 995–1014 (2016)
21. Grech, M., Kisielewicz, A.: The Černý conjecture for automata respecting intervals of a directed graph. Discrete Math. Theoret. Comput. Sci. **15**(3), 61–72 (2013)
22. Helfgott, H.A., Seress, A.: On the diameter of permutation groups. Ann. Math. **179**(2), 611–658 (2014)
23. Helfgott, H.A.: Growth in groups: ideas and perspectives. Bull. Amer. Math. Soc. **52**(3), 357–413 (2015)
24. Kari, J.: Synchronizing finite automata on Eulerian digraphs. Theoret. Comput. Sci. **295**, 223–232 (2003)
25. Klyachko, A.A., Rystsov, I.K., Spivak, M.A.: An extremal combinatorial problem associated with the bound on the length of a synchronizing word in an automaton. Cybern. Syst. Anal. **23**(2), 165–171 (1987)
26. Maslennikova, M.: Reset complexity of ideal languages. In: Bieliková, M. (ed.) SOFSEM 2012. Proceedings of the Institute of Computer Science Academy of Sciences of the Czech Republic, vol. II, pp. 33–44 (2012)
27. Natarajan, B.K.: An algorithmic approach to the automated design of parts orienters. In: 27th FOCS, pp. 132–142. IEEE (1986)
28. Natarajan, B.K.: Some paradigms for the automated design of parts feeders. Int. J. Robot. Res. **8**(6), 89–109 (1989)
29. Panteleev, P.: Preset distinguishing sequences and diameter of transformation semigroups. In: Dediu, A.-H., Formenti, E., Martín-Vide, C., Truthe, B. (eds.) LATA 2015. LNCS, vol. 8977, pp. 353–364. Springer, Cham (2015). doi:10.1007/978-3-319-15579-1_27
30. Pin, J.E.: On two combinatorial problems arising from automata theory. Ann. Discrete Math. **17**, 535–548 (1983)
31. Rystsov, I.K.: Reset words for commutative and solvable automata. Theoret. Comput. Sci. **172**(1), 273–279 (1997)
32. Rystsov, I.K.: Estimation of the length of reset words for automata with simple idempotents. Cybern. Syst. Anal. **36**(3), 339–344 (2000)

33. Saloff-Coste, L.: Random walks on finite groups. In: Kesten, H. (ed.) Probability on Discrete Structures, pp. 263–346. Springer, Heidelberg (2004)
34. Salomaa, A.: Composition sequences for functions over a finite domain. Theoret. Comput. Sci. **292**(1), 263–281 (2003)
35. Steinberg, B.: The averaging trick and the Černý conjecture. Int. J. Found. Comput. Sci. **22**(7), 1697–1706 (2011)
36. Steinberg, B.: The Černý conjecture for one-cluster automata with prime length cycle. Theoret. Comput. Sci. **412**(39), 5487–5491 (2011)
37. Szykuła, M.: Improving the upper bound the length of the shortest reset words. CoRR abs/1702.05455 (2017)
38. Volkov, M.V.: Synchronizing automata and the Černý conjecture. In: Martín-Vide, C., Otto, F., Fernau, H. (eds.) LATA 2008. LNCS, vol. 5196, pp. 11–27. Springer, Heidelberg (2008). doi:10.1007/978-3-540-88282-4_4
39. Vorel, V.: Subset synchronization of transitive automata. In: Ésik, Z., Fülöp, Z. (eds.) AFL 2014. EPTCS, vol. 151, pp. 370–381 (2014)
40. Zubov, A.Y.: On the diameter of the group S_N with respect to a system of generators consisting of a complete cycle and a transposition. Tr. Diskretn. Mat. **2**, 112–150 (1998). in Russian

Differences Between 2D Neighborhoods According to Real Time Computation

Anael Grandjean[(✉)]

LIRMM, Université de Montpellier, 161 rue Ada, 34392 Montpellier, France
anael.grandjean@lirmm.fr

Abstract. Cellular automata are a parallel model of computation. This paper presents studies about the impact of the choice of the neighborhood on small complexity classes, mainly the real time class. The main result states that given two neighborhoods $\mathcal{N}$ and $\mathcal{N}'$, if $\mathcal{N}$ has a limiting vertex in some direction and $\mathcal{N}'$ have no vertex in that direction then there is a language recognizable in real time with $\mathcal{N}'$ and not with $\mathcal{N}$. One easy corollary is that real time classes for two neighborhoods may be incomparable (and such neighborhoods are easy to construct).

1 Introduction

Cellular automata (CA) were first introduced in the 1940s (published posthumously in 1966) by J. von Neumann and S. Ulam as a mathematical model to study self-replication [5]. A cellular automaton is made of an infinity of identical cells, disposed on the $\mathbb{Z}^2$ grid, each connected to its four nearest neighbors. Each cell is in a specific state from a finite set common to all cells (for example dead or alive). The evolution of this system is discrete and massively parallel. At each time step, every cell changes its state into a new one determined by its previous state and those of its neighbors.

Although this model was initially designed to be studied as a dynamical system, A.R. Smith III proved that it is possible to embed Turing machines in cellular automata [7], proving this system to be a convenient model for massively parallel computation.

The initial model of Ulam and von Neumann has quickly been modified, mainly regarding two aspects. The first one is the two-dimensionality of the model. Replacing $\mathbb{Z}$ by $\mathbb{Z}^d$ it is easy to adapt cellular automata to any dimension. Another way to modify the model is to change the definition of neighbors, that is changing the way cells are connected to each other.

Our main interest is to understand the differences (in terms of computational properties) induced by the choice of the neighbors. It is well known that every neighborhood gives the same computational speed up to a multiplicative constant. Therefore we are interested in classes of very small complexity, namely the real time class. The real time is in some way the time needed by the automaton to read its input. The real time class is therefore the set of languages recognizable in real time. This class obviously depends on the dimension and on the neighborhood of the automata.

© Springer International Publishing AG 2017
É. Charlier et al. (Eds.): DLT 2017, LNCS 10396, pp. 198–209, 2017.
DOI: 10.1007/978-3-319-62809-7_14

It has been shown by Poupet [6] that one-dimensional neighborhoods can be partitioned into two categories: those which connect cells in both directions and those where cells are only connected in one direction. Moreover, allowing transmission of information in both directions really adds computational power in real time [1]. The two-dimensional case is much more complex. Although some results show the equivalence in real time of some neighborhoods [2,3], Terrier showed in [8] that Moore and von Neumann neighborhoods are not equivalent for real time computation. More precisely there is a language recognizable in real time with von Neumann neighborhood but not with Moore neighborhood. Terrier later expanded the result to every neighborhood with a limiting vertex (instead of just Moore neighborhood) [9].

This language is the first example of a language recognizable in real time with some neighborhoods but not with some others. This is the first step towards a more precise classification of languages. Indeed we could classify languages inside the already small class of real time recognizable languages by looking at the set of neighborhoods which allows to recognize it. Even though the whole classification is a really long term goal, studying the algorithmic properties induced by different neighborhoods seem to be the best way to work towards it.

The following paper presents two results. The first one is a generalization of Terrier's proof in order to expand the result to a whole class of languages instead of the specific one she introduced. The second result, probably the most interesting one, consists in showing that two neighborhoods with different limiting vertices induce incomparable real time classes. An immediate corollary of this theorem is that there is an infinite family of neighborhoods which classes are all incomparable with each other.

2 Definitions

2.1 Cellular Automata

Definition 1 (Cellular Automaton). *A cellular automaton (CA) is a quadruple $\mathcal{A} = (d, Q, \mathcal{N}, \delta)$ where:*

- *$d \in \mathbb{N}$ is the dimension of $\mathcal{A}$;*
- *Q is a finite set whose elements are called* states*;*
- *$\mathcal{N} \subset \mathbb{Z}^d$ is a finite set called* neighborhood *of $\mathcal{A}$ such that $0 \in \mathcal{N}$;*
- *$\delta : Q^{\mathcal{N}} \to Q$ is the local transition function of $\mathcal{A}$.*

A configuration *of the automaton is a mapping $\mathfrak{C} : \mathbb{Z}^d \to Q$. The elements of $\mathbb{Z}^d$ are called* cells *and for a given cell $c \in \mathbb{Z}^d$, we say that $\mathfrak{C}(c)$ is the state of c in the configuration $\mathfrak{C}$. The set of all configurations over Q is denoted $\mathrm{Conf}(Q)$. For a given configuration $\mathfrak{C} \in \mathrm{Conf}(Q)$ and a cell $c \in \mathbb{Z}^d$, define the* neighborhood *of c in $\mathfrak{C}$*

$$\mathcal{N}_{\mathfrak{C}}(c) = \begin{cases} \mathcal{N} \to Q \\ n \mapsto \mathfrak{C}(c+n) \end{cases}$$

Fig. 1. von Neumann neighborhood (left) and Moore neighborhood (right).

From the local transition function δ, we define the global transition function $\Delta_{\mathcal{A}}$ *of the automaton. The image of a configuration $\mathfrak{C}$ by $\Delta_{\mathcal{A}}$ is obtained by replacing the state of each cell c by the image by δ of the neighborhood of c in $\mathfrak{C}$:*

$$\Delta_{\mathcal{A}} : \begin{cases} \mathrm{Conf}(Q) \to \mathrm{Conf}(Q) \\ \mathfrak{C} \mapsto \begin{cases} \mathbb{Z}^d \to Q \\ c \mapsto \delta(\mathcal{N}_{\mathfrak{C}}(c)) \end{cases} \end{cases}$$

In this article, we will only consider 2-dimensional CA with various neighborhoods. From now on every definition is given for 2-dimensional neighborhoods. Two neighborhoods are of particular interest: von Neumann neighborhood and Moore neighborhood (see Fig. 1).

2.2   Neighborhoods

Definition 2 (Convex Hull). *A neighborhood $\mathcal{N}$, as a subset of $\mathbb{N}^2$ can also be seen as a subset of $\mathbb{R}^2$. As such, we define its* continuous convex hull *(denoted* $\mathrm{CH}(\mathcal{N})$*) is the smallest convex set of $\mathbb{R}^2$ that contains $\mathcal{N}$. It is a polygon which vertices are elements of $\mathcal{N}$. Those vertices are called the vertices of $\mathcal{N}$.*

In the paper we will also use the following notations:

$$n_x = (max\{x \in \mathbb{Q} \,|\, (x,0) \in \mathrm{CH}(\mathcal{N})\}, 0).$$

$$n_y = (0, max\{y \in \mathbb{Q} \,|\, (0,y) \in \mathrm{CH}(\mathcal{N})\}).$$

$$\mathcal{N}^{k+1} = \mathcal{N}^k \oplus \mathcal{N}.$$

Definition 3 (Completeness). *We say that a neighborhood $\mathcal{N}$ is* complete, *if $\mathbb{Z}^2 = \bigcup \mathcal{N}^k$.*

Definition 4 (Convexity). *We say that a neighborhood $\mathcal{N}$ is* convex, *if $\mathcal{N} = \mathrm{CH}(\mathcal{N}) \cap \mathbb{Z}^2$.*

2.3   Language Recognition

In this paper, we focus on performing language recognition with 2-dimensional cellular automata. The languages we are interested in are languages of 2-dimensional finite words. One can think of several ways to deal with finite words with infinite cellular automata. Different ways of defining the recognition process can be found for example in [4] or [8]. We choose here to define an initial configuration where a finite number of cells are in states that represent the letters of the input word and all the others are in a specific quiescent state we denote by $\sharp$. The result of the computation will be read on the bottom left corner of the input word. The bottom left cell will enter in an accepting state only for words of the language.

Definition 5 (Language Recognizer). *Given a finite alphabet Σ and a language $L \subseteq \Sigma^{**}$, a 2-dimensional CA $\mathcal{A}$ with states Q is said to recognize L in time $f : \mathbb{N}^2 \to \mathbb{N}$ with accepting states $Q_a \subseteq Q$ and quiescent state $q_0 \in Q$ if, $\Sigma \subseteq Q$ and for any word w of size $n \times m$ in Σ^{**}, starting from the configuration*

$$\mathbb{Z}^2 \to Q$$
$$(x, y) \mapsto \begin{cases} w_{x,y} & \text{if} \quad x \in [\![0, n-1]\!] \quad and \quad y \in [\![0, m-1]\!] \\ q_0 & \text{otherwise} \end{cases}$$

the state of the origin at time $f(n, m)$ is in Q_a if and only if $w \in L$.

Definition 6 (Real Time). *The real time (denoted $RT(\mathcal{N}, n, m)$) corresponds to the minimal time such that the state of the origin cell depend on the whole input word. It is a function of both the neighborhood and the size of the input. Formally:*

$$RT(\mathcal{N}, n, m) = min\{t \in \mathbb{N} \mid [\![0, n-1]\!] \times [\![0, m-1]\!] \subset \mathcal{N}^t\}$$

Definition 7 (Real Time class). *We say that a language L is recognizable in real time with a neighborhood $\mathcal{N}$ if there is a cellular automaton with neighborhood $\mathcal{N}$ which recognizes L in real time. That is each word of L of size $(n \times m)$ is either accepted or rejected in at most $RT(\mathcal{N}, n, m)$ time steps. We denote by $CA_{\mathcal{N}}(RT)$ the class of languages recognizable in real time with $\mathcal{N}$.*

In the whole paper, we are only interested in complete neighborhoods, for convenience we will now omit the word complete, and simply talk about neighborhoods. Moreover, the following theorem, by V. Poupet and M. Delacourt in [2] allows us to consider only convex neighborhoods.

Theorem 8 (Poupet and Delacourt 2007). *Given any complete neighborhood $\mathcal{N}$, we have $CA_{CH(\mathcal{N})}(RT) \subseteq CA_{\mathcal{N}}(RT)$.*

3   Difference Between Real and Linear Time

In this section we recall a language introduced by V. Terrier in [8]. This language cannot be recognized in real time by any neighborhood with some specific vertex, called limiting vertex. Her proof involves some precise counting argument. We will expand her result to a set of languages for which the same kind of counting argument works. We will then, in the next section, show how to build such a language so that it can be recognized by some specific neighborhood.

Definition 9 (Limiting vertex). *Given a neighborhood $\mathcal{N}$ and $v = (a, b) \in \mathcal{N}$, we say that v is a* limiting vertex *if:*

- *v is a vertex of $\mathcal{N}$,*
- *both a and b are positive,*
- *$[|0, a|] \times [|0, b|] \subseteq \mathcal{N}$.*

This definition of a limiting vertex may seem pretty specific but it is in fact rather reasonable to consider neighborhoods with such a vertex. To have a vertex with both coordinates positive means that there is a south west direction in which the information can travel faster by going south and west at the same time than by going south then west. The *limiting* condition means that in order to go south (or west) it is at least as fast to go straight south (or west) than to use this diagonal movement.

0	0	0	0	0	0	0	1	0	1	0	0
1	0	0	1	0	1	0	0	0	0	0	1
0	0	0	0	0	0	0	0	0	1	0	1
0	0	0	0	0	0	0	0	0	0	1	1
0	0	1	0	0	0	0	0	0	0	0	0
0	0	0	0	0	0	0	1	1	0	0	0
0	0	0	0	0	0	0	0	0	0	0	0
1	0	0	0	0	0	1	0	0	0	0	0
0	0	1	0	0	0	0	0	0	0	0	0
0	1	0	0	0	1	0	0	0	0	1	0
0	0	0	0	0	0	0	0	0	0	1	0
1	0	0	0	0	0	0	1	0	0	0	0

0	0	0	0	0	0	0	1	0	1	1	0
1	0	0	1	0	1	0	0	0	0	0	1
0	0	0	0	0	0	0	0	0	1	0	1
0	0	0	0	0	0	0	0	0	0	1	1
0	0	1	0	0	0	0	0	0	0	0	0
0	0	0	0	0	0	0	1	1	0	0	0
0	0	0	0	0	0	0	0	0	0	0	0
1	0	0	0	0	0	1	0	0	0	0	0
0	0	1	0	0	0	0	0	0	0	0	0
0	1	0	0	0	1	0	0	0	0	1	0
0	0	0	0	0	0	0	0	0	0	1	0
1	0	0	0	0	0	0	1	0	0	0	0

Fig. 2. Two words of size 12×12, the one on the left is in L, the one on the right is not. Yellow cells are the one containing the code, and therefore marked. (Color figure online)

The language L is composed of 2-dimensional words (called pictures) over $\{0, 1\}^2$ that is each cell contains a couple of bits. On the second component, the only 1's are on the right of the topmost row and on the top of the rightmost column. Those strings can be seen as integers. We denote those numbers x and y. We say that a picture W of size $n \times m$ is in L if $W(n - x, m - y) = 1$. If $x > n$ or $y > m$ then W is not in L (see Fig. 2).

Theorem 10 (V. Terrier 2004). *Let $\mathcal{N}$ be a neighborhood with at least one limiting vertex $v = (a, b)$, then $L \notin CA_{\mathcal{N}}(RT)$.*

For our purpose, we will need to extend the theorem to a bigger set of languages. We will keep the main idea of the language L, the coordinates of a cell c are given in a part of the input, far from the origin, and the result depends on the bit contained in cell c. We then try to find the minimal conditions such that the counting argument involved in V. Terrier's proof can be adapted.

First of all we consider two-dimensional words with any finite alphabet Σ. We then define the following set of languages:

Definition 11 (Encoded Languages). *An* encoded *language any language* L *such that there exist a function* $C : \Sigma^{**} \mapsto \mathbb{Z}^2$ *with the following properties:*

- $w \in L \Leftrightarrow w(C(w)) = 1$,
- *there is an integer* k *such that* $C(w)$ *only depends on the letters of* w *placed on the* $k \log(n + m)$ *rightmost columns and* $k \log(n + m)$ *topmost lines, and on* n *and* m *where* $(n \times m)$ *is the size of* w,
- *denote by* $I(n, m)$ *the image of* C *restricted to words of size* $(n \times m)$, *then* $|I(n, m) \bigcap [\![0, n]\!] \times [\![0, m]\!]| \geq \frac{nm}{l}$, *for some integer* l.

In order to simplify the construction in Sect. 4 we want to restrict the languages to some inputs. Formally we define a restriction on the size of the input as follows:

Definition 12 (Restriction of a language). *If* L *is a 2-dimensional language over* Σ, *we say that* L' *is a* size-restriction *of* L *if there is a property* P *over couples of integers such that* $w \in L' \Leftrightarrow (w \in L$ & $P(n, m))$ *where* w *is of size* $(n \times m)$.

Theorem 13. *Let* L *be a size-restriction of an encoded language,* $\mathcal{N}$ *a neighborhood with a limiting vertex* (x, y). *If* L *contains arbitrarily large words of size* $(n \times m)$ *with* $\frac{n}{m} = \frac{x}{y}$ *then it is not recognizable in real time with neighborhood* $\mathcal{N}$.

Proof. The main idea of the proof is that the information contained in the furthest cells from the origin have to be brought back to the origin using a nearly optimal path. For some inputs, the optimal path is a straight line to the origin. In such an input, this information can only interact with a restricted number of cells, along the path. We can then show that this number is small enough that it cannot contain all the information contained in the original input, in particular some information about cells that could be designed by the encoding are lost.

We will now suppose that such a language is recognizable and find a contradiction.

Let $\mathcal{N}$ be a neighborhood with a limiting vertex $v = (a, b)$. Let L be a size-restriction of an encoded language with arbitrarily large inputs of ratio $\frac{a}{b}$. Suppose there is an automaton $\mathcal{A}$ with this neighborhood which recognizes L.

Now take n and m such that $\frac{n}{m} = \frac{a}{b}$ and L contains a word of size $(n \times m)$. Therefore, there is a function C such that L contains each word w of size $(n \times m)$ such that $w(C(w)) = 1$. Accordingly to our definition, there is a constant c such that C only depends on cells in a rectangle of size $c. \log(n + m) \times c. \log(m + n)$ in the north east corner of the input.

To formally prove our claims, we introduce the following notations:

Definition 14 (Notations)

- $\mathrm{RT} = \mathrm{RT}(\mathcal{N}, n, m)$
- $U = [\![n - c \log(n + m), n]\!] \times [\![m - c \log(n + m), m]\!]$.
- $V = ([\![0, n]\!] \times [\![0, m]\!]) \backslash U$.
- $K(t) = -\mathcal{N}^t(U) \cap \mathcal{N}^{\mathrm{RT} - t}(0)$.
- $T(t) = \mathcal{N}(K(t)) \backslash K(t)$.

The set U is the set of cells containing relevant information for the coding C. V denotes the other cells of the input. The set $K(t)$ is the set of cells which at time t depends on the input on U and are still relevant for the recognition (meaning that they can affect the state of the origin before real time). The set $T(t)$ contains each cell needed to compute the states of the cells of $K(t+1)$ at time $t+1$, knowing the state of the cells of $K(t)$ at time t. For convenience we will say states of $K(t)$ for states of the cells of $K(t)$ at time t, and states of $T(t)$ for states of cells of $T(t)$ at time t.

First we can remark that the states of $T(t)$ do not depend on the input on U. Indeed, the states of $T(t)$ are relevant for the computation of the states of $K(t+1)$ and therefore relevant for the recognition. If any such state depended on the input in U, it would be in $K(t)$. In other words, the states of $T(t)$ are uniquely determined by the input on V.

Moreover we can see that the knowledge of the states of $T(t)$ for all t allows us to compute $K(\mathrm{RT})$ given $K(0)$. Now notice that $U = K(0)$ and $K(\mathrm{RT})$ contains only the origin. Therefore two inputs on V which lead to the same states in $T(t)$ are equivalent, that is for any input on U the result of the computation is the same with any of the two inputs on V. From this it is easy to see that there is at most $|Q|^T$ equivalence classes where $T = \sum_{t=0}^{\mathrm{RT}} |T(t)|$

The third constraint on the function C (having an image big enough inside V) implies that there is at least $2^{\frac{nm}{T}}$ equivalence classes for inputs on V. Indeed if two inputs differ on a cell inside the image of C, the result of the computation with both inputs should not be the same when U designs this cell.

The rest of the proof consists in bounding the size of $T(t)$ in order to achieve a contradiction regarding this number of equivalence classes.

Lemma 15. *The size of $K(t)$ is bounded by a $O(\log(n) + log(m))$ which does not depend on t.*

Proof. First of all, it is easy to see that $|T(t)|$ is bounded by $|K(t)| \times |\mathcal{N}|$. This is why we are now trying to bound $|K(t)|$.

The part $\mathcal{N}^{\mathrm{RT} - t}(0)$ have the shape of the convex hull of $\mathcal{N}$, centered in the origin cell. One extremum point is $(\mathrm{RT} - t)v$. Consider $W(t)$ the minimal cone anchored in $(\mathrm{RT} - t)v$ which contains $\mathcal{N}^{\mathrm{RT} - t}(0)$ (see Fig. 3). As v is a vertex, this cone is less than a half plane, and because v is limiting, it contains more than the south west quarter of the plane. In particular $W(0)$ contains the whole input.

Now consider $\mathcal{N}^{-t}(U)$. We denote by u_0 the cell $(n - c\log(n + m), m - c log(m + n))$, on the bottom left of U.

The shape of this part is slightly more complicated but one can check that one extremum point is $((n - c\log(n + m), m - c log(m + n)) - tv) = u_0 - tv$. We similarly define $W'(t)$ the minimal cone anchored in $u_0 - tv$, containing $\mathcal{N}^{-t}(U)$. Now remark that those two cones are symmetric up to translation. Indeed, the angles of the cones are the same and only depend on the shape of $\mathcal{N}$.

$K(t)$ is then contained in the intersection of $W(t)$ and $W'(t)$ (see Fig. 3). As the angle of the cones does not depend on time, the size of $K(t)$ only depends on the distance between the two anchors of the cones. Moreover this size of the intersection is always polynomial in this distance (quadratic in dimension 2). Clearly the distance between the two anchors, $u_O - tv$ and $(RT -t)v$, does not depend on t. Therefore we only have to bound the distance between u_0 and $RT\,v$.

Because the vertex v is limiting, and because the proportion of the input is well chosen (the same proportion as v), we know that $RT\,v = (n, m)$. It is then rather easy to see that the distance between the anchors is $\|(c.log(n), c.log(m))\|$.

The size of $K(t)$ is therefore bounded by some $O(\log(n) + \log(m))$ Thus the lemma is proved.

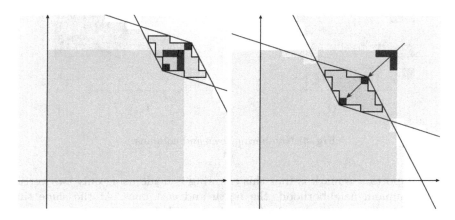

Fig. 3. The set U is in red, both $W(t)$ and $W'(t)$ are in green. The intersection is therefore in darker green. The left picture is taken at time $t = 0$ and the right one at $t = 2$. (Color figure online)

The number of equivalence classes is therefore bounded by $2^{O(RT(\log(n)+\log(m))} = 2^{O((n+m)(\log(n)+log(m))}$ which contradicts the fact that there is at least $2^{\frac{nm}{T}}$ classes.

4   Incomparability Between Neighborhoods

The previous section introduced a convenient class of languages that are not real time recognizable with some neighborhoods. In this section we will construct specific languages in this class to prove that the real time recognition classes of many neighborhoods cannot be compared. Before proving our theorem we shall recall another result from V. Terrier.

Theorem 16 (V. Terrier 1999). *The language L can be recognized in real time with von Neumann neighborhood.*

Proof. The automaton will have two tasks to perform in parallel. The first one is numbering lines and columns, starting from the north east corner. Therefore in all the rest of the section, the ith column (row) denotes the ith one starting from the right (top) of the picture. In particular the first column denotes the rightmost one and the first row denotes the topmost one. V. Terrier explains how the numbering can be done such that the ith bit of the number of the jth column (or row) can be known at time $i + j - 1$ (see Fig. 4).

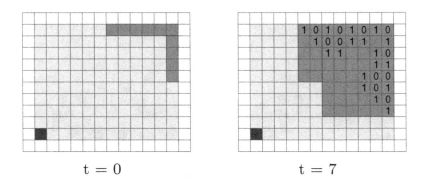

$$t = 0 \qquad\qquad t = 7$$

Fig. 4. Numbering rows and columns.

One important remark is that this counting is made using only two vertices of von Neumann neighborhood, the north and east ones. At the same time, the automaton starts copying the first column and the first row on every other column and row. This copying is made on another layer of the automaton, not destroying any information. At time t, the bit of least weight of the number of the tth column (row) is known, and the last column (row) is copied on the tth one. The automaton immediately starts a comparison between the code and the number of the column (row). This comparison ends at the end of the code (which has to be recognizable). If the two numbers correspond, a signal is sent, waiting to cross a row signal of same nature. When the two signals cross each other, if the cell hosting the crossing contains a 1, a specific signal is sent towards the origin in order the make the automaton accept the word. Otherwise no signal is sent and the origin stays in a reject state. Once more all this work is done using only the same two vertices of von Neumann neighborhood.

Theorem 17. *If $\mathcal{N}$ is a neighborhood with a limiting vertex $v = (a, b)$ and $\mathcal{N}'$ is another neighborhood with no vertex (c, d) having slope $\alpha = \frac{c}{d} = \frac{a}{b}$, then $\mathrm{CA}_{\mathcal{N}'}(\mathrm{RT}) \nsubseteq \mathrm{CA}_{\mathcal{N}}(\mathrm{RT})$.*

The idea here is to create a specific encoding, and therefore a language that cannot be recognized in real time with $\mathcal{N}$, which "helps" $\mathcal{N}'$.

Proof. We will focus on the neighborhood $\mathcal{N}'$, as the condition for being not recognizable with $\mathcal{N}$ are, with the previous theorem, rather easy to achieve.

Consider v_1 the first vertex of $\mathcal{N}'$ of slope greater than α, v_2 the last one of slope lower than α (see Fig. 5(a)). If v_1 is not in the north east quarter, we will consider n'_y instead, and n'_x instead of v_2 if it is outside the north east quarter similarly. These two vectors are a linear combination of v_1 and v_2 in that particular case. However, they might not be in $\mathcal{N}'$. If so we consider the smallest power of $\mathcal{N}'$ for which they are, which exists because $\mathcal{N}'$ is supposed to be complete. The construction will be presented for this neighborhood, as an automaton with $\mathcal{N}'$ can simulate its behavior in real time.

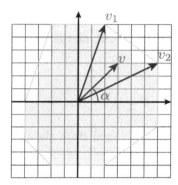 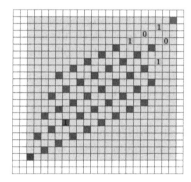

(a) The neighborhood $\mathcal{N}'$ in blue have no vertex of slope α. the two vertices v_1 and v_2 are the first ones after and before α.

(b) A word with a grid marked in dark green (we do not show the marking of the borders), and a code depicted in blue.

Fig. 5. (a) The neighborhood $\mathcal{N}'$ in blue have no vertex of slope α. The two vertices v_1 and v_2 are the first ones after and before α. (b) A word with a grid marked in dark green (we do not show the marking of the borders), and a code depicted in blue. (Color figure online)

The language we define is constructed on two layers: one is used to mark some specific grid, and the other should contain some geometric transformation of a word of L.

The first layer needs 6 symbols, one for cells outside the grid, one for cells in the grid and 4 to mark each border of the grid. To build an acceptable grid for a word w, pick a cell $c = k_1 v_1 + k_2 v_2$ near the north-east corner of the word (that is neither $c + v_1$ nor $c + v_2$ are in w), and mark every cell $i = i_1 v_1 + i_2 v_2 = c - j_1 v_1 - j_2 v_2$. Those cells are exactly the ones accessible from the origin by v_1 and v_2, and that can access c with the same vectors.

We can easily construct a bijection between this grid and a rectangle of $\mathbb{Z}^2$, associating v_1 to $(0, 1)$ and v_2 to $(1, 0)$. That is, we can associate to each word w with such a grid a smaller word w'.

The new language we consider is the set of words w with the described grid on one layer, and such that it is associated to a word of L (considering only the second layer of the associated word).

Moreover we restrict this language to pictures of size (n, m) with ($\frac{n}{m} = \alpha$. Without any restriction $\mathcal{N}'$ would not be able to recognize our language.

One example of a word in this language is depicted in Fig. 5(b).

First, we prove that this language is not recognizable in real time with $\mathcal{N}$. Because of the "maximality" of the cell c delimiting the grid, the codes are contained in a small area near the north-east corner. Moreover the image of the code contains every cell of the grid, which is a linear subset of w. Therefore by the Theorem 13, this language is not recognizable in real time with $\mathcal{N}$.

Now we have to check that it is recognizable with $\mathcal{N}'$. To do that we construct an automaton with $\mathcal{N}'$ the following way: the automaton will at the same time

- check that the input has the right proportion,
- check that the grid is correctly marked,
- check that the word on the grid is in L.

The first verification can be done by sending a signal from the north-east corner, at maximum speed, with the right slope α. As $\mathcal{N}'$ have no vertex of slope α, we will use a combination of v_1 and v_2 to approximate it. The signal then have an average slope of α but will differ from the correct path by a constant value from time to time. As α is rational, this difference is easy to correct near the origin.

The second point can be done locally in one step. If an error is detected by a cell on the grid, it is sent to the origin is less than the real time, using vertices v_1 and v_2. However it is possible that some cells located outside the main grid, are also marked as cells in the grid. As all the interactions are made accordingly to v_1 and v_2 nothing happening outside the main grid (the one started by the origin cell, and closed by marked borders) can interfere with the result of the computation.

To do the last step remark that, by associating v_1 to $(0, 1)$ and v_2 to $(1, 0)$ the automaton will, on the grid, simulate any automaton with von Neumann neighborhood working on the associated word. In particular we can simulate the one that recognizes L in real time. This real time is less than the real time with only v_1 and v_2 as vertices (it is the optimal time for the cell c to send its information to the origin). Because $\mathcal{N}'$ does not have any vertex between v_1 and v_2, its real time is no less than this one.

Corollary 18. *Any two neighborhoods with incomparable sets of limiting vertices induce incomparable real time classes.*

5   Conclusion

In this paper we investigated the link between real time computational power and neighborhood in two dimensions. One first observation made by Victor Poupet in [2] was that only the shape of the convex hull matters. This is why, in this paper, every result is stated using this notion only. The notion of limiting vertex appears to be particularly relevant when looking at the limits of real time computation. This vertex is the witness (or the cause) of a weakness of the neighborhood. V. Terrier already used similar notions in [9] in order to answer to the question of the equality between real time and linear time. She showed that those classes are different for many neighborhoods. However it remains open for some others, as von Neumann neighborhood for example.

In our work, we adapted those results to get differences between real time classes for different neighborhoods. Indeed, although the equality of the convex hulls implies the equality of the real time class, the difference does not always imply the difference of those classes. Some cases of equality had already been showed in [3] for example. Some trivial inclusions are also known. Our result of incomparability is the first of its kind.

Even though we now better understand the differences between neighborhoods, many questions remain open, especially concerning von Neumann neighborhood. Moreover some neighborhoods which do not fulfill the hypothesis of our theorem are still poorly understood, mainly the one with some vertices, but no limiting ones.

References

1. Choffrut, C., Culik II, K.: On real-time cellular automata and trellis automata. Acta Inf. **21**, 393–407 (1984). http://dx.doi.org/10.1007/BF00264617
2. Delacourt, M., Poupet, V.: Real time language recognition on 2D cellular automata: dealing with non-convex neighborhoods. In: Kučera, L., Kučera, A. (eds.) MFCS 2007. LNCS, vol. 4708, pp. 298–309. Springer, Heidelberg (2007). doi:10.1007/978-3-540-74456-6_28
3. Grandjean, A.: Constant acceleration theorem for extended von neumann neighbourhoods. In: Cook, M., Neary, T. (eds.) AUTOMATA 2016. LNCS, vol. 9664, pp. 149–158. Springer, Cham (2016). doi:10.1007/978-3-319-39300-1_12
4. Kutrib, M., Malcher, A.: Transductions computed by iterative arrays. Journees Automates Cellulaires **2010**, 156–167 (2010)
5. von Neumann, J.: Theory of Self-reproducing Automata. University of Illinois Press, Urbana (1966)
6. Poupet, V.: Cellular automata: real-time equivalence between one-dimensional neighborhoods. In: Proceedings of STACS 2005, pp. 133–144 (2005)
7. Smith III, A.R.: Simple computation-universal cellular spaces. J. ACM **18**(3), 339–353 (1971)
8. Terrier, V.: Two-dimensional cellular automata recognizer. Theor. Comput. Sci. **218**, 325–346 (1999)
9. Terrier, V.: Two-dimensional cellular automata and their neighborhoods. Theor. Comput. Sci. **312**, 203–222 (2004)

A Medvedev Characterization of Recognizable Tree Series

Luisa Herrmann[(✉)]

Faculty of Computer Science, Technische Universität Dresden,
Nöthnitzer Str. 46, 01062 Dresden, Germany
luisa.herrmann@tu-dresden.de

Abstract. We introduce representable tree series over commutative semirings, which extend representable sets [10] to the weighted setting. We prove that restricted representable tree series are exactly those tree series that can be recognized by weighted tree automata. Moreover, we investigate the relation between unrestricted representable tree series and weighted monadic second-order logic.

1 Introduction

Recognizable languages and recognizable tree languages are well-developed and robust language classes as they can be described by many equivalent devices, such as grammars, automata, logics and many more. One such device are the *representable sets* introduced by Medvedev [10]. Representable sets form a class of languages that contains particular simple languages, called *elementary sets*, and that is closed under particular operations, the *elementary operations*. As Medvedev showed, the class of representable sets coincides with the class of recognizable languages.

Later on representable sets were lifted to the tree case by Costich [4]. Moreover, he could reduce the four elementary sets and five elementary operations defined by Medvedev to three and four, respectively. Again, he achieved an equivalence result and, hence, obtained a *Medvedev characterization of recognizable tree languages*.

Also in the weighted setting, recognizable series [2] and recognizable tree series [1] form language classes that can be described by several formalisms. However, in the literature a quantitative version of Medvedev's result is still open. We close this gap here by introducing *representable tree series* that are built from *elementary tree series* and *elementary operations*. In comparison to Costich, we can again reduce the number of elementary tree series by one. However, to obtain a characterization for recognizable tree series, we have to restrict our representable tree series: We limit the application of one particular elementary operation to recognizable step functions. With the help of this restriction we obtain a *Medvedev characterization of recognizable tree series*.

Supported by DFG Graduiertenkolleg 1763 (QuantLA).

É. Charlier et al. (Eds.): DLT 2017, LNCS 10396, pp. 210–221, 2017.
DOI: 10.1007/978-3-319-62809-7_15

Moreover, we investigate the relation between unrestricted representable tree series and weighted monadic second-order logic [5]. Their relation is interesting because, to obtain a characterization of recognizable tree series, MSO-formulas also have to be restricted (by avoiding universal second-order quantification and by restricting universal first-order quantifications to recognizable step functions) [7]. Here we will prove that the class of representable tree series is a proper subclass of the class of tree series definable by weighted MSO-formulas.

2   Trees, Tree Series, and Tree Automata

Let $\mathbb{N}$ be the set of natural numbers including zero. We let $[n] = \{1, \ldots, n\}$ for $n \in \mathbb{N}$. Let A be a set. The set of all subsets of A is denoted by $\mathcal{P}(A)$. For sets $A_1, \ldots, A_n$ and $a = (a_1, \ldots, a_n) \in A_1 \times \ldots \times A_k$ we let $(a)_i = a_i$ for each $i \in [n]$.

A *ranked alphabet* is a tuple (Σ, rk) where Σ is a finite set and $\mathrm{rk} \colon \Sigma \to \mathbb{N}$. For every $n \in \mathbb{N}$, we denote the set of all symbols of Σ with rank n by Σ_n. By writing $\{\gamma_1, \ldots, \gamma_l\}_n$ we mean that $\gamma_1, \ldots, \gamma_l$ have rank n.

Let U be a set disjoint from Σ. The set of *trees over Σ and U*, denoted by $T_\Sigma(U)$, is the smallest set T such that (i) $\Sigma_0 \cup U \subseteq T$ and (ii) if $\sigma \in \Sigma_n$ with $n \geq 1$ and $\xi_1, \ldots, \xi_n \in T$, then $\sigma(\xi_1, \ldots, \xi_n) \in T$. We denote $T_\Sigma(\emptyset)$ by T_Σ and identify $\sigma()$ with σ. Each set $L \subseteq T_\Sigma$ is called a *tree language*.

Let $\xi \in T_\Sigma$. The set of *positions of ξ* is denoted by $\mathrm{pos}(\xi) \subseteq \mathbb{N}^*$ and for each $\rho \in \mathrm{pos}(\xi)$ we let $\xi(\rho)$ denote the symbol of ξ at position ρ. We abbreviate $|\mathrm{pos}(\xi)|$ by $|\xi|$ and $|\{\rho \in \mathrm{pos}(\xi) \mid \xi(\rho) = \gamma\}|$ by $|\xi|_\gamma$. We denote by $\sqsubseteq_\xi$ the lexicographic order on $\mathrm{pos}(\xi)$. Moreover, the set of *subtrees of ξ* is denoted by $\mathrm{sub}(\xi)$ and for each $\rho \in \mathrm{pos}(\xi)$ we let $\xi_{|\rho}$ denote the subtree of ξ at position ρ.

A *semiring* is an algebraic structure $(K, +, \cdot, 0, 1)$ such that $(K, +, 0)$ is a commutative monoid and $(K, \cdot, 1)$ is a monoid, $\cdot$ distributes over $+$, and $a \cdot 0 = 0 \cdot a = 0$ for every $a \in K$. We say that K is *commutative* if $k \cdot k' = k' \cdot k$ for every $k, k' \in K$. We often identify $(K, +, \cdot, 0, 1)$ with its carrier set K.

Note: *From here on let Σ and Δ be ranked alphabets and let K be a commutative semiring.*

A *tree series* (over Σ and K) is a mapping $r \colon T_\Sigma \to K$ and we denote the set of all such tree series by $K\langle\langle T_\Sigma \rangle\rangle$. For each $L \subseteq T_\Sigma$ we define the *characteristic tree series* $\mathbb{1}_L \colon T_\Sigma \to K$ by letting $\mathbb{1}_L(\xi) = 1$ if $\xi \in L$ and 0 otherwise. Now let $r, s \in K\langle\langle T_\Sigma \rangle\rangle$. We let $\mathrm{supp}(r) = \{\xi \in T_\Sigma \mid r(\xi) \neq 0\}$ and $\mathrm{im}(r) = \{r(\xi) \mid \xi \in T_\Sigma\}$. The sum $r + s$ and the Hadamard product $r \odot s$ are defined pointwise for each $\xi \in T_\Sigma$ as follows: $(r + s)(\xi) = r(\xi) + s(\xi)$ and $(r \odot s)(\xi) = r(\xi) \cdot s(\xi)$; $(a \cdot r)(\xi) = a \cdot r(\xi)$ for $a \in K$.

Let $X = \{x_1, x_2, \ldots\}$ be a set of variables and let $h \colon \Sigma \to T_\Delta(X)$ be a mapping such that $h(\Sigma_n) \subseteq T_\Delta(\{x_1, \ldots, x_n\})$. In the usual way, we can extend h to a tree homomorphism $h \colon T_\Sigma \to T_\Delta$. We say that h is *alphabetic* if for each $\sigma \in \Sigma_n$ we have $h(\sigma) = \delta(x_1, \ldots, x_n)$ for some $\delta \in \Delta_n$. In this case we extend h to a mapping $h \colon K\langle\langle T_\Sigma \rangle\rangle \to K\langle\langle T_\Delta \rangle\rangle$ by letting $(h(s))(\xi) = \sum_{\xi' \in h^{-1}(\xi)} s(\xi')$.

A *weighted tree automaton over Σ and K* $((\Sigma, K)$-*wta*$)$ is a tuple $\mathcal{A} = (Q, \delta, F)$ where Q is a finite set of *states*, $\delta = (\delta_\sigma \mid \sigma \in \Sigma)$ is a family of *transition mappings* $\delta_\sigma\colon Q^k \times Q \to K$ with $k \in \mathbb{N}$, $\sigma \in \Sigma_k$, and $F\colon Q \to K$ assigns *root weights*. We call $\mathcal{A}$ *(total) deterministic* (or a (Σ, K)-*(t)dwta*) if for each $\sigma \in \Sigma$, $q_1, \ldots, q_{\mathrm{rk}(\sigma)} \in Q$ there is at most (exactly) one $q \in Q$ such that $\delta_\sigma(q_1 \ldots q_{\mathrm{rk}(\sigma)}, q) \neq 0$. We say $\mathcal{A}$ has *Boolean transition weights* or *Boolean root weights* if $\mathrm{im}(\delta_\sigma) \subseteq \{0,1\}$ for each $\sigma \in \Sigma$ or $\mathrm{im}(F) \subseteq \{0,1\}$, resp.

Let $\xi \in \mathrm{T}_\Sigma$. A *run of* $\mathcal{A}$ *on* ξ is a mapping $\kappa\colon \mathrm{pos}(\xi) \to Q$ and we denote by $\mathrm{Run}_\mathcal{A}(\xi)$ the set of all runs of $\mathcal{A}$ on ξ. We define the mapping $\mathrm{wt}_\xi^\mathcal{A}\colon \mathrm{Run}_\mathcal{A}(\xi) \to K$ by letting for each $\kappa \in \mathrm{Run}_\mathcal{A}(\xi)$

$$\mathrm{wt}_\xi^\mathcal{A}(\kappa) = \prod\nolimits_{\rho \in \mathrm{pos}(\xi)} \delta_{\xi(\rho)}\Big(\kappa(\rho 1) \ldots \kappa(\rho\, \mathrm{rk}(\xi(\rho))), \kappa(\rho)\Big).$$

Then the *tree series recognized by* $\mathcal{A}$ is the mapping $[\![\mathcal{A}]\!]\colon \mathrm{T}_\Sigma \to K$ given for each $\xi \in \mathrm{T}_\Sigma$ by

$$[\![\mathcal{A}]\!](\xi) = \sum\nolimits_{\kappa \in \mathrm{Run}_\mathcal{A}(\xi)} F(\kappa(\varepsilon)) \cdot \mathrm{wt}_\xi^\mathcal{A}(\kappa).$$

We say that a tree series r is *(deterministically)* (Σ, K)-*recognizable* if there is a (Σ, K)-(d)wta $\mathcal{A}$ such that $[\![\mathcal{A}]\!] = r$.

Now let $\mathcal{A} = (Q, \delta, F)$ be a $(\Sigma, \mathbb{B})$-wta where $\mathbb{B}$ denotes the Boolean semiring. Then we identify δ with the family $\delta' = (\delta'_\sigma \mid \sigma \in \Sigma)$ where for each $k \in \mathbb{N}$, $\sigma \in \Sigma_k$ we let $\delta'_\sigma \subseteq Q^k \times Q$ and we identify F with the set $\mathrm{supp}(F)$. We call $\mathcal{A}$ a *tree automaton over* Σ (or a Σ-*ta*) and we define the *language recognized by* $\mathcal{A}$ as the set $L(\mathcal{A}) = \mathrm{supp}([\![\mathcal{A}]\!])$. Moreover, if $\mathcal{A}$ is total and deterministic (or short a Σ-*dta*), we also write $\delta_\sigma(q_1, \ldots, q_{\mathrm{rk}(\sigma)}) = q$.

3   Representable Tree Series

The aim of this section is to define a weighted version of the representable sets introduced by Medvedev, called representable tree series[1]. These tree series are built up from particular operations, called elementary operations, that are applied to particular tree series, called elementary tree series.

Definition 1. We call the following tree series over Σ and K *elementary*:

(i) for each $\sigma \in \Sigma$ and $a \in K$ the tree series $\mathrm{RT}_{\sigma,a} \in K\langle\!\langle \mathrm{T}_\Sigma \rangle\!\rangle$ defined for each $\xi \in \mathrm{T}_\Sigma$ by $\mathrm{RT}_{\sigma,a}(\xi) = a$ if $\xi(\varepsilon) = \sigma$ and 0 otherwise, and

(ii) for each $n \geq 1$, $\gamma_1, \ldots, \gamma_n \in \Sigma$, and $a \in K$ the tree series $\mathrm{NXT}_{\gamma_1 \ldots \gamma_n, a} \in K\langle\!\langle \mathrm{T}_\Sigma \rangle\!\rangle$ defined for each $\xi \in \mathrm{T}_\Sigma$ by $\mathrm{NXT}_{\gamma_1 \ldots \gamma_n, a}(\xi) = a$ if $\xi(\varepsilon) \in \Sigma_n$ and $\xi(i) = \gamma_i$ for each $i \in [n]$, and 0 otherwise.

Definition 2. We call the following operations on tree series *elementary*: (i) $+$, (ii) $\odot$, (iii) alphabetic tree homomorphisms, (iv) the *restriction mapping* $\mathrm{RST}\colon K\langle\!\langle \mathrm{T}_\Sigma \rangle\!\rangle \to K\langle\!\langle \mathrm{T}_\Sigma \rangle\!\rangle$ which is defined for each $s \in K\langle\!\langle \mathrm{T}_\Sigma \rangle\!\rangle$ and $\xi \in \mathrm{T}_\Sigma$ by

$$(\mathrm{RST}(s))(\xi) = \prod\nolimits_{\rho \in \mathrm{pos}(\xi)} s(\xi|_\rho).$$

---

[1] Note that here the notion of representable tree series is different from the one in [3].

Note that in the case $\Sigma = \Sigma_0 \cup \Sigma_1$ our definition of the restriction mapping coincides with $(\mathrm{RST}(s))(\xi) = \prod_{t \in \mathrm{sub}(\xi)} s(t)$ for each $s \in K\langle\langle \mathrm{T}_\Sigma \rangle\rangle$ and $\xi \in \mathrm{T}_\Sigma$.

The class of K-*representable tree series*, denoted by $\mathrm{REPR}(K)$, is the smallest class of tree series that contains for each ranked alphabet Σ the elementary tree series over Σ and K and that is closed under elementary operations. Moreover, for each ranked alphabet Σ, the class $\mathrm{REPR}(\Sigma, K)$ of (Σ, K)-*representable tree series* is the subclass of $\mathrm{REPR}(K)$ containing all tree series of type $\mathrm{T}_\Sigma \to K$.

A term made up of elementary tree series and elementary operations that results in a (Σ, K)-representable tree series is called a (Σ, K)-*representation*. Clearly, each (Σ, K)-representation e can be seen as a tree by considering elementary series as nullary symbols and elementary operations as unary respectively binary symbols. Then we denote by $\mathrm{ht}(e)$ the *height of the tree associated to* e, defined as usual.

Example 3. Let $\Sigma = \{\alpha\}_0 \cup \{\gamma, \delta\}_1$ and $K = (\mathbb{N}, +, \cdot, 0, 1)$. The tree series s mapping each $\xi \in \mathrm{T}_\Sigma$ to $2^{|\xi|_\gamma}$ can be expressed by the $(\Sigma, \mathbb{N})$-representation $e = \mathrm{RST}(\mathrm{RT}_{\gamma,2} + (\mathrm{RT}_{\delta,1} + \mathrm{RT}_{\alpha,1}))$. Its height is $\mathrm{ht}(e) = 3$.

The next property of the restriction function follows directly from the definition of the set of positions: For each $n \in \mathbb{N}$, $\sigma \in \Sigma_n$, $\xi_1, \ldots, \xi_n \in \mathrm{T}_\Sigma$, and $s \in K\langle\langle \mathrm{T}_\Sigma \rangle\rangle$ we have that $(\mathrm{RST}(s))(\sigma(\xi_1, \ldots, \xi_n)) = s(\sigma(\xi_1, \ldots, \xi_n)) \cdot (\mathrm{RST}(s))(\xi_1) \cdot \ldots \cdot (\mathrm{RST}(s))(\xi_n)$.

4   Medvedev Characterization

In this section we want to characterize recognizable tree series by means of (Σ, K)-representations. However, we cannot do this directly, as not all tree series that are representable are also recognizable, as illustrated by the following example.

Example 4. Let $\Sigma = \{\alpha\}_0 \cup \{\gamma\}_1$ and $K = (\mathbb{N}, +, \cdot, 0, 1)$. Moreover, let $r_{\exp} \colon \mathrm{T}_\Sigma \to K$ be the tree series mapping each tree $\gamma^n(\alpha) \in \mathrm{T}_\Sigma$ to $2^{(n+1)^2}$ for $n \in \mathbb{N}$. It is well-known that this tree series is not recognizable [7, p. 236]. Now consider the (Σ, K)-representation

$$e_{\exp} = \mathrm{RST}(\mathrm{RT}_{\gamma,2} + \mathrm{RT}_{\alpha,2}) \odot \mathrm{RST}(\mathrm{RST}(\mathrm{RT}_{\gamma,4} + \mathrm{RT}_{\alpha,1})).$$

Since for each $m \geq 1$ and $x \in \{2, 4\}$ it holds that $\mathrm{RT}_{\gamma,x}(\gamma^m(\alpha)) = x$ we obtain

$$\mathrm{RST}(\mathrm{RT}_{\gamma,2} + \mathrm{RT}_{\alpha,2})(\gamma^n(\alpha)) = \prod_{t \in \mathrm{sub}(\gamma^n(\alpha))} (\mathrm{RT}_{\gamma,2} + \mathrm{RT}_{\alpha,2})(t) = 2^{(n+1)}$$

and, with a similar argument, $\mathrm{RST}(\mathrm{RT}_{\gamma,4} + \mathrm{RT}_{\alpha,1})(\gamma^n(\alpha)) = 4^n$. Using the Gaussian sum it follows that

$$\mathrm{RST}(\mathrm{RST}(\mathrm{RT}_{\gamma,4} + \mathrm{RT}_{\alpha,1}))(\gamma^n(\alpha)) = \prod_{t \in \mathrm{sub}(\gamma^n(\alpha))} \mathrm{RST}(\mathrm{RT}_{\gamma,4} + \mathrm{RT}_{\alpha,1})(t)$$

$$= 1 \cdot 4^1 \cdot \ldots \cdot 4^n = 4^{\frac{n^2+n}{2}} = 2^{(n^2+n)}.$$

Then following the definition of $\odot$ we obtain that $e_{\exp}(\gamma^n(\alpha)) = 2^{(n+1)} \cdot 2^{(n^2+n)} = 2^{(n+1)^2} = r_{\exp}(\gamma^n(\alpha))$ and hence that $r_{\exp}$ is (Σ, K)-representable.

To obtain a characterization of recognizable tree series, we therefore have to restrict representable tree series. For this we use the concept of recognizable step functions that describe almost Boolean tree series. We say that a tree series $r \in K\langle\!\langle T_\Sigma \rangle\!\rangle$ is a *recognizable step function* if $r = \sum_{i=1}^{n} k_i \cdot \mathbb{1}_{L_i}$ for some $n \geq 1$, $k_i \in K$, and recognizable tree languages L_i, $i \in [n]$. It is rather folklore that recognizable step functions can be recognized by very restricted weighted tree automata as stated in the next lemma.

Lemma 5. *Let $r \in K\langle\!\langle T_\Sigma \rangle\!\rangle$ be a recognizable step function. Then there exists a (Σ, K)-tdwta $\mathcal{A}$ with Boolean transition weights such that $[\![\mathcal{A}]\!] = r$.*

Proof (sketch). We can assume that each step language L_i can be recognized by a Σ-dta $\mathcal{A}_i$. We construct a (Σ, K)-tdwta $\mathcal{A}$ that uses as states the Cartesian product of the states of $A_1, \ldots, A_n$. When assigning a run to a tree ξ, $\mathcal{A}$ simulates in the ith component of its states the run of $\mathcal{A}_i$ on ξ. The weight of a final state is the sum of all weights k_j where $\mathcal{A}_j$ results in a final state. $\square$

Now we will restrict our (Σ, K)-representations by limiting the usage of the restriction function. A tree series $r \in K\langle\!\langle T_\Sigma \rangle\!\rangle$ is *restricted (Σ, K)-representable* if it can be expressed by a (Σ, K)-representation where the restriction mapping RST is only applied to recognizable step functions.

In the rest of this section we will prove the following theorem which extends the Medvedev characterization for tree languages to the weighted setting.

Theorem 6. *Let K be a commutative semiring and $r \in K\langle\!\langle T_\Sigma \rangle\!\rangle$. Then r is restricted (Σ, K)-representable if and only if r is (Σ, K)-recognizable.*

Proof. This follows directly from Lemmas 11 and 12.

4.1   Restricted Representable Implies Recognizable

First we wish to prove that each restricted representable tree series is recognizable. For this, we show that the elementary tree series are recognizable and that elementary operations preserve recognizability. In fact, we even prove that the elementary tree series are recognizable step functions as we will need this property in the proof of the opposite direction.

Lemma 7. *Let $a \in K$ and $\sigma \in \Sigma$. Then $\mathrm{RT}_{\sigma,a}$ is a recognizable step function.*

Proof. It is clear that the language L containing all trees ξ with $\xi(\varepsilon) = \sigma$ is Σ-recognizable. Then $\mathrm{RT}_{\sigma,a} = a \cdot \mathbb{1}_L$ is a recognizable step function. $\square$

Lemma 8. *Let $a \in K$, $n \geq 1$, and $\gamma_1, \ldots, \gamma_n \in \Sigma$. Then $\mathrm{NXT}_{\gamma_1 \ldots \gamma_n, a}$ is a recognizable step function.*

Proof. We construct the tree automaton $\mathcal{A} = (Q, \delta, F)$ with $Q = \{q_0, q_f\} \cup \{q_{\gamma_i} \mid i \in [n]\}$ and $F = \{q_f\}$ as follows. For each $k \in \mathbb{N}$, $\sigma \in \Sigma^{(k)}$, and $q_1, \ldots, q_k \in Q$ we let $(q_1 \ldots q_k, q_{\gamma_i}) \in \delta_\sigma$ if $\sigma = \gamma_i$ for some $i \in [n]$ and $(q_1 \ldots q_k, q_0) \in \delta_\sigma$ otherwise. Moreover, if $k = n$, then we let $(q_{\gamma_1} \ldots q_{\gamma_n}, q_f) \in \delta_\sigma$. It is not hard to see that $L(\mathcal{A}) = \{\xi \in T_\Sigma \mid \xi(\varepsilon) \in \Sigma_n, \xi(i) = \gamma_i, i \in [n]\}$. Therefore, $\mathrm{NXT}_{\gamma_1 \ldots \gamma_n, a} = a \cdot \mathbb{1}_{L(\mathcal{A})}$, which is a recognizable step function. $\square$

It is well-known that sum, Hadamard product, and alphabetic tree homomorphisms preserve recognizability. Moreover, the recognizable step functions are closed under the first two operations.

Lemma 9 ([9, Theorem 3.8], [6, Lemma 5.9]). *Let $h\colon T_\Sigma \to T_\Delta$ be an alphabetic tree homomorphism and let r, $s \in K\langle\!\langle T_\Sigma \rangle\!\rangle$.*

(1) If r and s are (Σ, K)-recognizable, then $r + s$ and $r \odot s$ are (Σ, K)-recognizable, and $h(r)$ is (Δ, K)-recognizable.

(2) If r and s are recognizable step functions, then $r + s$ and $r \odot s$ are so.

It remains to prove that the restriction function preserves recognizability provided it is applied to a recognizable step function.

Lemma 10. *Let $r \in K\langle\!\langle T_\Sigma \rangle\!\rangle$ be a recognizable step function. Then $\mathrm{RST}(r)$ is (Σ, K)-recognizable.*

Proof. Let $\mathcal{A} = (Q, \delta, F)$ be a (Σ, K)-wta recognizing r. By Lemma 5 we can assume that $\mathcal{A}$ is total deterministic and has Boolean transition weights; in this case the weight of a tree only comes from final state weights. We construct a wta $\mathcal{B}$ that simulates $\mathcal{A}$ on each input, but assigns to each transition a final state weight of $\mathcal{A}$. Formally, we construct the wta $\mathcal{B}$ recognizing $\mathrm{RST}(r)$ as follows. We let $\mathcal{B} = (Q, \delta', F')$ where $F'(q) = 1$ for each $q \in Q$. Moreover, the family δ' of transitions is defined as follows. For each $\sigma \in \Sigma$, $q, q_1, \ldots, q_{\mathrm{rk}(\sigma)} \in Q$ we let

$$\delta'_\sigma(q_1 \ldots q_{\mathrm{rk}(\sigma)}, q) = F(q) \cdot \delta_\sigma(q_1 \ldots q_{\mathrm{rk}(\sigma)}, q).$$

Clearly, $\mathcal{B}$ is deterministic. Since $\mathcal{A}$ is total deterministic and has Boolean transition weights, for each $\xi \in T_\Sigma$ there exists exactly one $\kappa \in \mathrm{Run}_\mathcal{A}(\xi)$ with $\mathrm{wt}_\xi^\mathcal{A}(\kappa) = 1$, we denote this κ by κ_ξ. Moreover, since $\mathrm{supp}(\delta) \supseteq \mathrm{supp}(\delta')$, κ_ξ is the only run in $\mathrm{Run}_\mathcal{B}(\xi)$ which can have a non-zero weight.

Next we prove by structural induction that $(\mathrm{RST}(\llbracket\mathcal{A}\rrbracket))(\xi) = \llbracket\mathcal{B}\rrbracket(\xi)$ for each $\xi \in T_\Sigma$.

First, let $\xi = \alpha$ for some $\alpha \in \Sigma_0$. Note that $(\mathrm{RST}(\llbracket\mathcal{A}\rrbracket))(\alpha) = \llbracket\mathcal{A}\rrbracket(\alpha)$. Then

$$
\begin{aligned}
\mathrm{RST}(\llbracket\mathcal{A}\rrbracket))(\alpha) &= \textstyle\sum_{\kappa \in \mathrm{Run}_\mathcal{A}(\alpha)} F(\kappa(\varepsilon)) \cdot \delta_\alpha(\varepsilon, \kappa(\varepsilon)) \\
&= \textstyle\sum_{\kappa \in \mathrm{Run}_\mathcal{B}(\alpha)} F'(\kappa(\varepsilon)) \cdot \delta'_\alpha(\varepsilon, \kappa(\varepsilon)) \\
&= \llbracket\mathcal{B}\rrbracket(\alpha).
\end{aligned}
$$

Next let $\xi = \sigma(\xi_1, \ldots, \xi_n)$ for some $n \geq 1$, $\sigma \in \Sigma_n$, and $\xi_1, \ldots, \xi_n \in T_\Sigma$. Then

$$
\begin{aligned}
(\mathrm{RST}(\llbracket\mathcal{A}\rrbracket))(\xi) &= \llbracket\mathcal{A}\rrbracket(\xi) \cdot (\mathrm{RST}(\llbracket\mathcal{A}\rrbracket))(\xi_1) \cdot \ldots \cdot (\mathrm{RST}(\llbracket\mathcal{A}\rrbracket))(\xi_n) \\
&= \llbracket\mathcal{A}\rrbracket(\xi) \cdot \llbracket\mathcal{B}\rrbracket(\xi_1) \cdot \ldots \cdot \llbracket\mathcal{B}\rrbracket(\xi_n) \qquad\qquad\text{(IH)} \\
&= \big(\textstyle\sum_{\kappa \in \mathrm{Run}_\mathcal{A}(\xi)} F(\kappa(\varepsilon)) \cdot \mathrm{wt}_\xi^\mathcal{A}(\kappa)\big) \cdot \llbracket\mathcal{B}\rrbracket(\xi_1) \cdot \ldots \cdot \llbracket\mathcal{B}\rrbracket(\xi_n) \\
&= \big(F(\kappa_\xi(\varepsilon)) \cdot \mathrm{wt}_\xi^\mathcal{A}(\kappa_\xi)\big) \cdot \llbracket\mathcal{B}\rrbracket(\xi_1) \cdot \ldots \cdot \llbracket\mathcal{B}\rrbracket(\xi_n) \\
&= \big(F(\kappa_\xi(\varepsilon)) \cdot \delta_\sigma(\kappa_\xi(1) \ldots \kappa_\xi(n), \kappa_\xi(\varepsilon))\big) \cdot \llbracket\mathcal{B}\rrbracket(\xi_1) \cdot \ldots \cdot \llbracket\mathcal{B}\rrbracket(\xi_n) \\
&= \delta'_\sigma(\kappa_\xi(1) \ldots \kappa_\xi(n), \kappa_\xi(\varepsilon)) \cdot \llbracket\mathcal{B}\rrbracket(\xi_1) \cdot \ldots \cdot \llbracket\mathcal{B}\rrbracket(\xi_n) \\
&\overset{(*)}{=} \llbracket\mathcal{B}\rrbracket(\xi),
\end{aligned}
$$

where $(*)$ holds since $\mathcal{B}$ is deterministic and $F'(q) = 1$ for each $q \in Q$, and $[\![\mathcal{B}]\!](\xi_i) = \mathrm{wt}^{\mathcal{B}}_{\xi_i}(\kappa_{\xi_i})$ for each $i \in [n]$. This proves that $\mathrm{RST}([\![\mathcal{A}]\!]) = [\![\mathcal{B}]\!]$.   □

Using Lemmas 7–10 we can prove the next lemma by induction on the structure of restricted (Σ, K)-representations.

Lemma 11. *Let K be a commutative semiring and let $r \in K\langle\!\langle \mathrm{T}_\Sigma \rangle\!\rangle$. If r is restricted (Σ, K)-representable, then r is (Σ, K)-recognizable.*

4.2   Recognizable Implies Restricted Representable

Now we prove that each recognizable tree series is restricted representable by constructing a restricted (Σ, K)-representation. As the restriction function can only be applied to recognizable step functions, at this place we need the closure of the recognizable step functions under sum and Hadamard product that was shown beforehand.

Lemma 12. *Let K be a commutative semiring and let $r \in K\langle\!\langle \mathrm{T}_\Sigma \rangle\!\rangle$. If r is (Σ, K)-recognizable, then r is restricted (Σ, K)-representable.*

Proof. Let $\mathcal{A} = (Q, \delta, F)$ be a (Σ, K)-wta such that $[\![\mathcal{A}]\!] = r$. Then let Ω be a new ranked alphabet such that $\Omega_n = \{(\sigma, q) \mid \sigma \in \Sigma_n, q \in Q\}$ for each $n \in \mathbb{N}$. Moreover, we define the alphabetic tree homomorphism $h \colon \mathrm{T}_\Omega \to \mathrm{T}_\Sigma$ by letting $h((\sigma, q)) = \sigma(x_1, \ldots, x_n)$ for each $(\sigma, q) \in \Omega_n$, $n \in \mathbb{N}$. Now we construct the three tree series

$$s_1 = \sum_{\substack{(\sigma, q) \in \Omega: \\ q \in \mathrm{supp}(F)}} \mathrm{RT}_{(\sigma, q), F(q)}, \qquad s_2 = \sum_{(\sigma, q) \in \Omega_0} \mathrm{RT}_{(\sigma, q), \delta_\sigma(\varepsilon, q)}, \text{ and}$$

$$s_3 = \sum_{\substack{n \geq 1, (\sigma, q) \in \Omega_n, \\ (\sigma_i, q_i) \in \Omega, i \in [n]}} \left(\mathrm{NXT}_{(\sigma_1, q_1) \ldots (\sigma_n, q_n), 1} \odot \mathrm{RT}_{(\sigma, q), \delta_\sigma(q_1 \ldots q_n, q)} \right)$$

and we let $s = h(s_1 \odot \mathrm{RST}(s_2 + s_3))$. By Lemmas 7, 8, and 9(2) we have that $s_2 + s_3$ is a recognizable step function. Thus, s is restricted (Σ, K)-representable. Next we show that $[\![\mathcal{A}]\!] = s$.

Intuitively, a tree t in T_Ω can be seen as a tree $\xi \in \mathrm{T}_\Sigma$ extended by labeling each node additionally with the appropriate state from some run in $\mathrm{Run}_\mathcal{A}(\xi)$. Formally, for each tree $\xi \in \mathrm{T}_\Sigma$ we define a bijection $\mathrm{run}_\xi \colon \mathrm{Run}_\mathcal{A}(\xi) \to \mathrm{T}_\Omega$ such that for each $\kappa \in \mathrm{Run}_\mathcal{A}(\xi)$ and $\rho \in \mathrm{pos}(\xi)$ we have $\mathrm{run}_\xi(\kappa)(\rho) = (\xi(\rho), \kappa(\rho))$. Then we can prove by structural induction over ξ that for each $\kappa \in \mathrm{Run}_\mathcal{A}(\xi)$:

$$\mathrm{wt}^{\mathcal{A}}_\xi(\kappa) = (\mathrm{RST}(s_2 + s_3))(\mathrm{run}_\xi(\kappa)). \tag{$*$}$$

We proceed with

$$
\begin{aligned}
h(s_1 \odot \mathrm{RST}(s_2 + s_3))(\xi) &= \sum_{t \in h^{-1}(\xi)} s_1(t) \cdot \mathrm{RST}(s_2 + s_3)(t) \\
&= \sum_{\kappa \in \mathrm{Run}_\mathcal{A}(\xi)} s_1(\mathrm{run}_\xi(\kappa)) \cdot \mathrm{RST}(s_2 + s_3)(\mathrm{run}_\xi(\kappa)) \\
&\overset{(\dagger)}{=} \sum_{\kappa \in \mathrm{Run}_\mathcal{A}(\xi)} F(\kappa(\varepsilon)) \cdot \mathrm{RST}(s_2 + s_3)(\mathrm{run}_\xi(\kappa)) \\
&\overset{(*)}{=} \sum_{\kappa \in \mathrm{Run}_\mathcal{A}(\xi)} F(\kappa(\varepsilon)) \cdot \mathrm{wt}^{\mathcal{A}}_\xi(\kappa) = [\![\mathcal{A}]\!](\xi).
\end{aligned}
$$

where (†) holds since $s_1(\mathrm{run}_\xi(\kappa)) = \mathrm{RT}_{(\xi(\varepsilon),\kappa(\varepsilon)),F(\kappa(\varepsilon))}(\mathrm{run}_\xi(\kappa)) = F(\kappa(\varepsilon))$. This proves that $[\![\mathcal{A}]\!] = s$ and, therefore, each (Σ,K)-recognizable tree series is restricted (Σ,K)-representable.   □

5   Comparison with Unrestricted MSO

This section investigates the relation between unrestricted (Σ,K)-representations and weighted monadic second-order logic. We will prove that MSO-formulas are more expressive than representations.

Let $\mathcal{V}$ be a finite set of first-order variables (often denoted by x, y, or z) and second-order variables (as X, Y, or Z). The set $\mathrm{MSO}(\Sigma,K)$ of *weighted MSO-formulas over Σ and K* is given by the EBNF

$$\psi ::= \mathrm{label}_\sigma(x) \mid \mathrm{edge}_i(x,y) \mid x \in X \mid x \sqsubseteq y$$

$$\varphi ::= k \mid \psi \mid \neg\psi \mid \varphi \vee \varphi \mid \varphi \wedge \varphi \mid \exists x.\varphi \mid \exists X.\varphi \mid \forall x.\varphi \mid \forall X.\varphi$$

where $k \in K$, $\sigma \in \Sigma$, and $i \in [\mathrm{maxrk}(\Sigma)]$. As usual, we use the macro $\mathrm{edge}(x,y) = \mathrm{edge}_1(x,y) \vee \ldots \vee \mathrm{edge}_{\mathrm{maxrk}(\Sigma)}(x,y)$.

Let $\varphi \in \mathrm{MSO}(\Sigma,K)$. We denote the set of free variable of φ by $\mathrm{Free}(\varphi)$. Recall that for each set $\mathcal{V}$ of variables and tree $\xi \in T_\Sigma$, a $\mathcal{V}$-assignment for ξ is a function mapping each first-order variable in $\mathcal{V}$ to a position of ξ and each second-order variable in $\mathcal{V}$ to a subset of positions of ξ. We denote the set of all $\mathcal{V}$-assignments for ξ by $\Phi_{\mathcal{V},\xi}$. For each $\rho \in \Phi_{\mathcal{V},\xi}$, $i \in \mathrm{pos}(\xi)$, and $I \subseteq \mathrm{pos}(\xi)$, the assignment updates $\rho[x \mapsto i]$ and $\rho[X \mapsto I]$ are defined as usual. In addition, we define the update $\rho + i$ by letting $(\rho + i)(x) = i\rho(x)$ and $(\rho + i)(X) = \{iw \mid w \in \rho(X)\}$ for each $x,X \in \mathcal{V}$. Moreover, we recall the usual technique of encoding a pair (ξ,ρ) as a tree over the ranked alphabet $\Sigma_\mathcal{V} = \Sigma \times \mathcal{P}(\mathcal{V})$ preserving the ranks of Σ; we identify $\Sigma_\emptyset$ with Σ. Then a tree $\zeta \in T_{\Sigma_\mathcal{V}}$ is called *valid* if for each first-order variable $x \in \mathcal{V}$ there exists exactly one position $i \in \mathrm{pos}(\zeta)$ such that $x \in (\zeta(i))_2$; we denote the set of all valid trees in $T_{\Sigma_\mathcal{V}}$ by $T_{\Sigma_\mathcal{V}}^{\mathrm{v}}$. As usual, we will not distinguish between $T_{\Sigma_\mathcal{V}}^{\mathrm{v}}$ and $\{(\xi,\rho) \mid \xi \in T_\Sigma, \rho \in \Phi_{\mathcal{V},\xi}\}$.

Now let $\varphi \in \mathrm{MSO}(\Sigma,K)$ and let $\mathcal{V}$ be a finite set of variables containing $\mathrm{Free}(\varphi)$. The *semantics of φ with respect to $\mathcal{V}$* is the tree series $[\![\varphi]\!]_\mathcal{V} : T_{\Sigma_\mathcal{V}} \to K$ such that $\mathrm{supp}([\![\varphi]\!]_\mathcal{V}) \subseteq T_{\Sigma_\mathcal{V}}^{\mathrm{v}}$ and inductively defined as in Fig. 1. As usual, we abbreviate $[\![\varphi]\!]_{\mathrm{Free}(\varphi)}$ by $[\![\varphi]\!]$. A tree series $s : T_\Sigma \to K$ is called (Σ,K)-*definable* if there is a $\varphi \in \mathrm{MSO}(\Sigma,K)$ with $[\![\varphi]\!] = s$.

Let $K = \mathbb{B}$ be the Boolean semiring. Then $\mathrm{MSO}(\Sigma,\mathbb{B})$ reduces to the classical unweighted MSO formulas where we define for each $\varphi \in \mathrm{MSO}(\Sigma,\mathbb{B})$ the *set of models* $L_\mathcal{V}(\varphi) = \mathrm{supp}([\![\varphi]\!]_\mathcal{V})$. We abbreviate $\mathrm{MSO}(\Sigma,\mathbb{B})$ by $\mathrm{MSO}(\Sigma)$.

Now let $\psi \in \mathrm{MSO}(\Sigma)$. We call a formula $\varphi \in \mathrm{MSO}(\Sigma,K)$ an *unambiguous formula representing ψ* if $[\![\varphi]\!] = \mathbb{1}_{L(\psi)}$.

Proposition 13 ([8, Proposition 5.3]). *For each $\psi \in \mathrm{MSO}(\Sigma)$ we can effectively construct an unambiguous formula $\varphi \in \mathrm{MSO}(\Sigma,K)$ representing ψ, i.e., such that $[\![\varphi]\!] = \mathbb{1}_{L(\psi)}$.*

$[\![k]\!]_{\mathcal{V}}(\xi, \rho) = k$

$[\![\mathrm{label}_\sigma(x)]\!]_{\mathcal{V}}(\xi, \rho) = \begin{cases} 1 & \text{if } \xi_{|\rho(x)} = \sigma \\ 0 & \text{otherwise} \end{cases}$

$[\![\mathrm{edge}_i(x, y)]\!]_{\mathcal{V}}(\xi, \rho) = \begin{cases} 1 & \text{if } \rho(y) = \rho(x)i \\ 0 & \text{otherwise} \end{cases}$

$[\![x \in X]\!]_{\mathcal{V}}(\xi, \rho) = \begin{cases} 1 & \text{if } \rho(x) \in \rho(X) \\ 0 & \text{otherwise} \end{cases}$

$[\![x \sqsubseteq y]\!]_{\mathcal{V}}(\xi, \rho) = \begin{cases} 1 & \text{if } \rho(x) \sqsubseteq_\xi \rho(y) \\ 0 & \text{otherwise} \end{cases}$

$[\![\neg\varphi]\!]_{\mathcal{V}}(\xi, \rho) = \begin{cases} 1 & \text{if } [\![\varphi]\!]_{\mathcal{V}}(\xi, \rho) = 0 \\ 0 & \text{otherwise} \end{cases}$

$[\![\varphi \vee \psi]\!]_{\mathcal{V}}(\xi, \rho) = [\![\varphi]\!]_{\mathcal{V}}(\xi, \rho) + [\![\psi]\!]_{\mathcal{V}}(\xi, \rho)$

$[\![\varphi \wedge \psi]\!]_{\mathcal{V}}(\xi, \rho) = [\![\varphi]\!]_{\mathcal{V}}(\xi, \rho) \cdot [\![\psi]\!]_{\mathcal{V}}(\xi, \rho)$

$[\![\exists x.\varphi]\!]_{\mathcal{V}}(\xi, \rho) = \sum_{i \in \mathrm{pos}(\xi)} [\![\varphi]\!]_{\mathcal{V} \cup \{x\}}(\xi, \rho[x \mapsto i])$

$[\![\exists X.\varphi]\!]_{\mathcal{V}}(\xi, \rho) = \sum_{I \subseteq \mathrm{pos}(\xi)} [\![\varphi]\!]_{\mathcal{V} \cup \{X\}}(\xi, \rho[X \mapsto I])$

$[\![\forall x.\varphi]\!]_{\mathcal{V}}(\xi, \rho) = \prod_{i \in \mathrm{pos}(\xi)} [\![\varphi]\!]_{\mathcal{V} \cup \{x\}}(\xi, \rho[x \mapsto i])$

$[\![\forall X.\varphi]\!]_{\mathcal{V}}(\xi, \rho) = \prod_{I \subseteq \mathrm{pos}(\xi)} [\![\varphi]\!]_{\mathcal{V} \cup \{X\}}(\xi, \rho[X \mapsto I])$

Fig. 1. The semantics of a formula $\varphi \in \mathrm{MSO}(\Sigma, K)$ with respect to $\mathcal{V}$.

In [8] the construction for a syntactically unambiguous formula was given. We will not recall it here, but assume that the unambiguous formula $\varphi \in \mathrm{MSO}(\Sigma, K)$ representing ψ results from this construction and denote it by ψ^+.

Now we will prove that elementary operations preserve (Σ, K)-definability.

Lemma 14. *Let s_1, s_2 be (Σ, K)-definable tree series. Then $s_1 + s_2$ and $s_1 \odot s_2$ are (Σ, K)-definable.*

It is not hard to see that also definable tree series are closed under alphabetic tree homomorphisms, even if they are not recognizable.

Lemma 15. *Let s be a (Σ, K)-definable tree series and $h : T_\Sigma \to T_\Delta$ an alphabetic tree homomorphism. Then $h(s)$ is (Δ, K)-definable.*

Proof. Let $\varphi \in \mathrm{MSO}(\Sigma, K)$ such that $s = [\![\varphi]\!]$. We enumerate $\Sigma = \{\sigma_1, \ldots, \sigma_n\}$ for some $n \in \mathbb{N}$. We let $\mathcal{V} = \{X_\sigma \mid \sigma \in \Sigma\}$ be a set of second-order variables. Then let

$$\psi = \exists X_{\sigma_1} \ldots \exists X_{\sigma_n}.(\varphi' \wedge \psi^+_{\mathrm{part}} \wedge \psi^+_{\mathrm{check}})$$

with the following intuition. For each tree $\zeta \in T_\Delta$, $\exists X_{\sigma_1} \ldots \exists X_{\sigma_n}$ guesses an assignment of positions of ζ to symbols from Σ. Then ψ_{part} checks whether this assignment forms a partitioning, i.e., each position is assigned to exactly one symbol from Σ, and ψ_{check} ensures that the assignment encodes a preimage ξ of h. Additionally, φ' simulates φ on ξ. Formally,

- φ' is obtained from φ by replacing each occurrence of $\mathrm{label}_\sigma(x)$ by $(x \in X_\sigma)$,
- $\psi_{\mathrm{part}} = \forall x.\left(\bigvee_{i \in [n]} ((x \in X_{\sigma_i}) \wedge \bigwedge_{j \in [n] : j \neq i} \neg(x \in X_{\sigma_j})) \right)$, and
- $\psi_{\mathrm{check}} = \forall x.\left(\bigwedge_{i \in [n]} (\neg(x \in X_{\sigma_i}) \vee \mathrm{label}_{h(\sigma_i)}(x)) \right)$.

It is not hard to see that $[\![\psi]\!] = h([\![\varphi]\!])$. $\qquad\square$

Lemma 16. *Let s be a (Σ, K)-definable tree series. Then $\mathrm{RST}(s)$ is (Σ, K)-definable.*

Proof. Let $\varphi \in \text{MSO}(\Sigma, K)$ such that $s = [\![\varphi]\!]$. We define the following formulas in $\text{MSO}(\Sigma)$ modeling paths between positions:[2] We let $\text{closed}(X) = \forall x. \forall y. (\neg \text{edge}(x, y) \lor \neg(x \in X) \lor (y \in X))$ and

- $P(x, y) = \forall X. (\neg \text{closed}(X) \lor \neg(x \in X) \lor (y \in X))$, and
- $P(x, Y) = \forall y. (\neg(y \in Y) \lor P(x, y))$.

Intuitively, $P(x, y)$ holds if there is a path from x to y (and y is below x) and $P(x, Y)$ holds if there is for each $y \in Y$ such a path from x to y.

Now let z be a new variable not occurring in φ. Then we define the mapping $\pi_z \colon \text{MSO}(\Sigma, K) \to \text{MSO}(\Sigma, K)$ inductively on the structure of φ as follows:

$$\pi_z(\psi) = \psi \text{ for an atom } \psi, \qquad \pi_z(\exists x. \psi) = \exists x. P(z, x)^+ \land \pi_z(\psi),$$

$$\pi_z(\neg \psi) = \neg \psi \text{ for an atom } \psi, \qquad \pi_z(\forall x. \psi) = \forall x. (\neg P(z, x))^+ \lor (P(z, x)^+ \land \pi_z(\psi)),$$

$$\pi_z(\psi_1 \land \psi_2) = \pi_z(\psi_1) \land \pi_z(\psi_2), \qquad \pi_z(\exists X. \psi) = \exists X. P(z, X)^+ \land \pi_z(\psi),$$

$$\pi_z(\psi_1 \lor \psi_2) = \pi_z(\psi_1) \lor \pi_z(\psi_2), \qquad \pi_z(\forall X. \psi) = \forall X. (\neg P(z, X))^+ \lor (P(z, X)^+ \land \pi_z(\psi))$$

Intuitively, $\pi_z(\varphi)$ restricts the evaluation of φ on a tree ξ to the subtree of ξ at the position assigned to z. Then we construct the formula $\varphi' = \forall z. \pi_z(\varphi)$.

Next we show that $\text{RST}([\![\varphi]\!]) = [\![\varphi']\!]$. For this, we prove the following statement (†): For all $\xi \in T_\Sigma$, $\mathcal{V} \supseteq \text{Free}(\varphi)$, $i \in \text{pos}(\xi)$, and $\rho \in \Phi_{\mathcal{V}, \xi_{|i}}$: $[\![\varphi]\!]_{\mathcal{V}}(\xi_{|i}, \rho) = [\![\pi_z(\varphi)]\!]_{\mathcal{V} \cup \{z\}}(\xi, (\rho + i)[z \mapsto i])$. We prove (†) by structural induction on φ.

Let $\varphi = \text{label}_\sigma(x)$ and note that $[\![\varphi]\!]$ can only be 1 or 0. Let $\zeta = \xi_{|i}$. Then

$$\begin{aligned}
[\![\text{label}_\sigma(x)]\!]_{\mathcal{V}}(\zeta, \rho) = 1 \quad &\Leftrightarrow \quad \zeta_{|\rho(x)} = \sigma \\
&\Leftrightarrow \quad \xi_{|(\rho+i)(x)} = \sigma \\
&\Leftrightarrow \quad [\![\text{label}_\sigma(x)]\!]_{\mathcal{V}}(\xi, \rho + i) = 1 \\
&\Leftrightarrow \quad [\![\pi_z(\text{label}_\sigma(x))]\!]_{\mathcal{V}}(\xi, \rho + i) = 1 \\
&\Leftrightarrow \quad [\![\pi_z(\text{label}_\sigma(x))]\!]_{\mathcal{V} \cup \{z\}}(\xi, (\rho + i)[z \mapsto i]) = 1.
\end{aligned}$$

All other cases of φ being an atom or of the form $\neg \psi$ for some atom ψ can be proved analogously.

Now let $\varphi = \exists x. \psi$ and assume that the statement holds for ψ. Then

$$\begin{aligned}
[\![\exists x. \psi]\!]_{\mathcal{V}}(\zeta, \rho) &= \sum_{k \in \text{pos}(\zeta)} [\![\psi]\!]_{\mathcal{V} \cup \{x\}}(\zeta, \rho[x \mapsto k]) \\
&= \sum_{k \in \text{pos}(\zeta)} [\![\pi_z(\psi)]\!]_{\mathcal{V} \cup \{x, z\}}(\xi, (\rho[x \mapsto k] + i)[z \mapsto i]) \qquad \text{(IH)} \\
&= \sum_{k \in \text{pos}(\xi) \colon \exists k' \colon k = ik'} [\![\pi_z(\psi)]\!]_{\mathcal{V} \cup \{x, z\}}(\xi, (\rho + i)[x \mapsto k][z \mapsto i]) \\
&= \sum_{k \in \text{pos}(\xi) \colon \exists k' \colon k = ik'} [\![\pi_z(\psi)]\!]_{\mathcal{V} \cup \{x, z\}}(\xi, (\rho + i)[z \mapsto i][x \mapsto k]) \\
&= \sum_{k \in \text{pos}(\xi)} \big([\![P(z, x)^+]\!]_{\mathcal{V} \cup \{x, z\}}(\xi, (\rho + i)[z \mapsto i][x \mapsto k]) \\
&\qquad \cdot [\![\pi_z(\psi)]\!]_{\mathcal{V} \cup \{x, z\}}(\xi, (\rho + i)[z \mapsto i][x \mapsto k]) \big) \qquad (*)
\end{aligned}$$

---

[2] In $\text{MSO}(\Sigma)$, $\neg$ can appear anywhere.

$$= \textstyle\sum_{k\in\mathrm{pos}(\xi)} [\![P(z,x)^+ \wedge \pi_z(\psi)]\!]_{\mathcal{V}\cup\{x,z\}}(\xi,(\rho+i)[z\mapsto i][x\mapsto k])$$
$$= [\![\exists x.P(z,x)^+ \wedge \pi_z(\psi)]\!]_{\mathcal{V}\cup\{x,z\}}(\xi,(\rho+i)[z\mapsto i])$$
$$= [\![\pi_z(\exists x.\psi)]\!]_{\mathcal{V}\cup\{z\}}(\xi,(\rho+i)[z\mapsto i]) \hspace{2cm} \text{(Constr.)}$$

where (*) holds since $[\![P(z,x)^+]\!]_{\mathcal{V}\cup\{x,z\}}(\xi,(\rho+i)[z\mapsto i][x\mapsto k]) = 1$ if $k = ik'$ for some $k' \in \mathbb{N}^*$ and 0 otherwise. It is not hard to see that for all other cases of φ we can argue in a similar way. Therefore, these cases are omitted here. This finishes the proof of the statement (†).

Now let $\xi \in T_\Sigma$. Then

$$\mathrm{RST}([\![\varphi]\!])(\xi) = \textstyle\prod_{i\in\mathrm{pos}(\xi)} [\![\varphi]\!](\xi_{|i})$$
$$\overset{(†)}{=} \textstyle\prod_{i\in\mathrm{pos}(\xi)} [\![\pi_z(\varphi)]\!]_{\{z\}}(\xi,[z\mapsto i]) = [\![\forall z.\pi_z(\varphi)]\!](\xi)$$

Thus, $\mathrm{RST}([\![\varphi]\!])$ is (Σ, K)-definable. $\hspace{2cm}\square$

Using the subsequent lemmas we can prove the first main result of this section.

Theorem 17. *Let K be a commutative semiring and let $r \in K\langle\langle T_\Sigma \rangle\rangle$. If r is (Σ, K)-representable, then r is (Σ, K)-definable.*

Proof. Clearly, elementary tree series are (Σ, K)-definable. Using now Lemmas 14, 15, and 16, we can prove the theorem by induction on the structure of (Σ, K)-representations. $\hspace{2cm}\square$

However, we obtain that (Σ, K)-expression are weaker than MSO-formulas as it is shown in the next theorem.

Theorem 18. *Let K be a commutative semiring. There is a (Σ, K)-definable tree series r that is not (Σ, K)-representable.*

Proof. Consider the ranked alphabet $\Sigma = \{\alpha\}_0 \cup \{\gamma\}_1$, the semiring $(\mathbb{N}, +, \cdot, 0, 1)$ and the formula $\varphi = \forall X.2$ in $\mathrm{MSO}(\Sigma, \mathbb{N})$. Clearly, for each $\xi \in T_\Sigma$ we have $[\![\varphi]\!](\xi) = 2^{(2^{|\xi|})}$. On the other hand, we will show that every $(\Sigma, \mathbb{N})$-representable tree series is bounded exponentially.

For this, we define $\hat{e}(n) = \max(\{e(\xi) \mid \xi \in T_\Sigma, |\xi| = n\})$ for each $(\Sigma, \mathbb{N})$-representation e and each $n \in \mathbb{N}$. Now we prove by structural induction that for each $(\Sigma, \mathbb{N})$-expression e and $n \in \mathbb{N}$ we have $\hat{e}(n) \in \mathcal{O}(2^{(n^{\mathrm{ht}(e)})})$.

Let, for each $n \in \mathbb{N}$, $\xi_n = \mathrm{argmax}_{\zeta \in T_\Sigma: |\zeta|=n} e(\zeta)$. First, let $e = \mathrm{RT}_{\gamma,k}$ for some $\gamma \in \Sigma$, $k \in \mathbb{N}$. Clearly, $\hat{e}(n) \in \mathcal{O}(1) \subseteq \mathcal{O}(2^{(n^{\mathrm{ht}(e)})})$. This holds for all elementary series.

Now, let $e = \mathrm{RST}(e_1)$ and assume that the statement holds for e_1. Let $\mathrm{sub}(\xi_n) = \{t_1, \ldots, t_n\}$ with $\mathrm{ht}(t_i) < \mathrm{ht}(t_{i+1})$ for $i \in [n-1]$. Then

$$\hat{e}(n) = e(\xi_n) = e_1(t_1) \cdot \ldots \cdot e_1(t_n) \leq \hat{e}_1(1) \cdot \ldots \cdot \hat{e}_1(n)$$
$$\in \mathcal{O}(2^{(1^{\mathrm{ht}(e_1)})} \cdot \ldots \cdot 2^{(n^{\mathrm{ht}(e_1)})}) \subseteq \mathcal{O}(2^{n \cdot (n^{\mathrm{ht}(e_1)})}) = \mathcal{O}(2^{(n^{\mathrm{ht}(e)})}).$$

All other cases of e can be proved with similar arguments. This finishes the proof of the induction statement. So if $\llbracket\varphi\rrbracket$ were $(\Sigma,\mathbb{N})$-representable, then for each $\xi \in T_\Sigma$ we had $\llbracket\varphi\rrbracket(\xi) \in \mathcal{O}(2^{|\xi|^c})$ for some constant c, which clearly is a contradiction.
$\square$

As a last point we conjecture that also the fragment of $\mathrm{MSO}(\Sigma,K)$ that uses no second-order universal quantification is more expressive than (Σ,K)-representations. For this, consider the ranked alphabet $\Sigma = \{\alpha\}_0 \cup \{\gamma,\delta\}_1$, the semiring $(\mathbb{N},+,\cdot,0,1)$, and the tree series s given by the semantics of the formula $\varphi = \forall y.P_\gamma(y) \to 2$. It is easy to see that for each tree $\xi \in T_\Sigma$ we have $\llbracket\varphi\rrbracket(\xi) = 2^{|\xi|_\gamma}$ and, as shown in Example 3, this tree series is $(\Sigma,\mathbb{N})$-representable. Now consider the tree series $\llbracket\forall x.\varphi\rrbracket$ that maps each tree $\xi \in T_\Sigma$ to $2^{|\xi|\cdot|\xi|_\gamma}$. We believe that this tree series is not $(\Sigma,\mathbb{N})$-representable anymore with the following intuition: We can only obtain a tree series with this growth by nesting at least two restriction functions. But restriction functions do not only consider the number of γs in ξ but also the positions of their occurrences and, thus, may map trees with the same height and number of γs to different values. However, a proof of this conjecture is left as an open problem.

Acknowledgement. The author thanks Tobias Denkinger and Johannes Osterholzer for several useful discussions concerning the content of this work.

References

1. Berstel, J., Reutenauer, C.: Recognizable formal power series on trees. Theor. Comput. Sci. **18**(2), 115–148 (1982)
2. Berstel, J., Reutenauer, C.: Rational Series and Their Languages. EATCS Monographs on Theoretical Computer Science, vol. 12. Springer, Heidelberg (1988)
3. Bozapalidis, S.: Representable tree series. Fundam. Inf. **21**(4), 367–389 (1994)
4. Costich, O.L.: A Medvedev characterization of sets recognized by generalized finite automata. Math. Syst. Theor. **6**(1–2), 263–267 (1972)
5. Droste, M., Gastin, P.: Weighted automata and weighted logics. Theor. Comput. Sci. **380**(1), 69–86 (2007)
6. Droste, M., Götze, D., Märcker, S., Meinecke, I.: Weighted tree automata over valuation monoids and their characterization by weighted logics. In: Kuich, W., Rahonis, G. (eds.) Algebraic Foundations in Computer Science. LNCS, vol. 7020, pp. 30–55. Springer, Heidelberg (2011). doi:10.1007/978-3-642-24897-9_2
7. Droste, M., Vogler, H.: Weighted tree automata and weighted logics. Theor. Comput. Sci. **366**, 228–247 (2006)
8. Droste, M., Vogler, H.: Weighted logics for unranked tree automata. Theor. Comput. Syst. **48**(1), 23–47 (2011)
9. Fülöp, Z., Vogler, H.: Weighted tree automata and tree transducers. In: Droste, M., Kuich, W., Vogler, H. (eds.) Handbook of Weighted Automata, pp. 313–403. Springer, Heidelberg (2009)
10. Medvedev, Y.T.: On the class of events representable in a finite automaton. Automata Studies (1956). Also in Sequential machines - Selected papers (translated from Russian), pp. 215–227. Addison-Wesley (1964)

On the Descriptive Complexity of $\overline{\Sigma^*\overline{L}}$

Michal Hospodár$^{(\boxtimes)}$, Galina Jirásková, and Peter Mlynárčik

Mathematical Institute, Slovak Academy of Sciences,
Grešákova 6, 040 01 Košice, Slovakia
hosmich@gmail.com, jiraskov@saske.sk, mlynarcik1972@gmail.com

Abstract. We examine the descriptive complexity of the combined unary operation $\overline{\Sigma^*\overline{L}}$ and investigate the trade-offs between various models of finite automata. We consider complete and partial deterministic finite automata, nondeterministic finite automata with single or multiple initial states, alternating, and boolean finite automata. We assume that the argument and the result of this operation are accepted by automata belonging to one of these six models. We investigate all possible trade-offs and provide a tight upper bound for 32 of 36 of them. The most interesting result is the trade-off from nondeterministic to deterministic automata given by the Dedekind number $M(n-1)$. We also prove that the nondeterministic state complexity of $\overline{\Sigma^*\overline{L}}$ is 2^{n-1} which solves an open problem stated by Birget [1996, The state complexity of $\overline{\Sigma^*\overline{L}}$ and its connection with temporal logic, Inform. Process. Lett. 58, 185–188].

1 Introduction

Formal languages may be recognized by several kinds of formal systems. Different classes of formal systems can be compared either from the point of view of their computational power, or from the descriptive complexity point of view. As for computational power, for example, deterministic and nondeterministic finite automata recognize the same class of languages, while the class of languages recognized by deterministic pushdown automata is strictly included in the class of languages recognized by nondeterministic ones. However, from the descriptive complexity point of view, there is an exponential gap between the cost of description of regular languages by deterministic and nondeterministic finite automata [14, 16–18, 20].

Descriptive complexity, which measures the cost of description of languages by different formal systems, was deeply investigated in last three decades (cf. [1, 7, 15, 22]) mostly in the class of regular languages. Several kinds of finite automata were proposed and the trade-offs between the costs of description in different

M. Hospodár et al.—Research supported by grant VEGA 2/0084/15 and grant APVV-15-0091. This work was conducted as a part of PhD study of Michal Hospodár and Peter Mlynárčik at the Faculty of Mathematics, Physics and Informatics of the Comenius University.

© Springer International Publishing AG 2017
E. Charlier et al. (Eds.): DLT 2017, LNCS 10396, pp. 222–234, 2017.
DOI: 10.1007/978-3-319-62809-7_16

classes of automata were examined. Let us mention at least the exact trade-off $\binom{2n}{n+1}$ for the conversion of two-way nondeterministic automata to one-way nondeterministic automata [11], and the exact trade-off for the conversion of self-verifying automata to deterministic automata given by the function that counts the maximal number of maximal cliques in a graph with n vertices [10].

In 1996, Jean-Camille Birget [2] answered the following question of Jean-Éric Pin. Let L be a regular language over an alphabet Σ recognized by a nondeterministic finite automaton (NFA) or a deterministic finite automaton (DFA) with n states. How many states are sufficient and necessary in the worst case for an NFA (DFA) to recognize the language $\overline{\Sigma^*\overline{L}}$? The notation $\overline{L}$ stands for the complement of L. Birget provided the exact trade-off from DFAs to NFAs, and lower and upper bounds for the nondeterministic state complexity of $\overline{\Sigma^*\overline{L}}$.

The motivation of Pin's question came from the word model of Propositional Temporal Logic [5]. The set of all models of a formula φ over a fixed alphabet Σ is a formal language $L(\varphi)$ over Σ which has the non-trivial property of being regular and aperiodic. Some of the temporal operators used in this logic are ∘ ("next") and ◇ ("eventually", or "at some moment in the future"); there are also the usual boolean operations $-$, ∧, ∨. A natural dual to the "eventually" operator is the "forever" (or, "always in the future") operator □, defined to be $-\diamond-$ ("not eventually not"). Formulas and their models are related as follows: $L(\overline{\varphi}) = \overline{L(\varphi)}$, $L(\varphi \wedge \psi) = L(\varphi) \cap L(\psi)$, $L(\varphi \vee \psi) = L(\varphi) \cup L(\psi)$, $L(\circ\varphi) = \Sigma L(\varphi)$, $L(\diamond\varphi) = \Sigma^* L(\varphi)$. Thus $L(\square\varphi) = L(\overline{\diamond\overline{\varphi}}) = \overline{\Sigma^*\overline{L(\varphi)}}$. Hence in [2], Birget studied the state complexity of the "forever" operator.

Here we continue this research by investigating the complexity of the forever operator for different models of finite automata. We consider complete and partial deterministic finite automata, nondeterministic automata with a single or multiple initial states, and boolean automata with a single initial state, called *alternating* finite automata in [6], or with an initial function [4]. Similarly as Jean-Éric Pin, we ask the following question: If a language L is represented by an n-state automaton of some model, how many states are sufficient and necessary in the worst case for an automaton of some other model to accept $\overline{\Sigma^*\overline{L}}$?

We study all the possible 36 trade-offs, and except for four cases, we always get tight upper bounds. In particular, we are able to prove that the upper bound on the nondeterministic state complexity of $\overline{\Sigma^*\overline{L}}$ is 2^{n-1}. This improves Birget's upper bound $2^{n+1}+1$ and meets his lower bound for DFA-to-NFA trade-off. The most interesting result of this paper is the tight upper bound for the NFA-to-DFA trade-off given by the Dedekind number $M(n-1)$; recall that the Dedekind number $M(n)$ counts the number of antichains of subsets of an n-element set. To get lower bounds, we describe languages over a fixed alphabet, except for four cases where the alphabet grows exponentially with n. In most cases our worst-case examples are binary and unary, and these alphabets are always optimal.

2 Preliminaries

We assume that the reader is familiar with basic notions in automata theory. For details, we refer to [19, 21].

We use standard models of (complete) deterministic finite automata (DFAs), partial deterministic finite automata (PFAs), nondeterministic finite automata with a single initial state (NFAs), nondeterministic finite automata with multiple initial states (NNFAs), boolean finite automata (BFAs), and boolean finite automata with a single initial state (AFAs).

We call a state of an NNFA $A = (Q, \Sigma, \cdot, I, F)$ *sink state* if it has a loop on every input symbol. For a symbol a and states p and q, we say that (p, a, q) is a transition in the NNFA A if $q \in p \cdot a$, and for a string w, we write $p \xrightarrow{w} q$ if $q \in p \cdot w$. We also say that the state q has an *in-transition* on symbol a, and the state p has an *out-transition* on symbol a.

Let q be a state of a DFA A. To *omit* the state q means to remove it from the state set and to remove also all its in-transitions and out-transitions. To *replace* the state q with a sink state means to remove each its out-transition (q, a, p) and add a loop (q, a, q) for each a.

The reverse of a string is defined as $\varepsilon^R = \varepsilon$ and $(wa)^R = aw^R$ for each symbol a and string w. The reverse of a language L is the language $L^R = \{w^R \mid w \in L\}$. The reverse of an NNFA $A = (Q, \Sigma, \cdot, I, F)$ is an NNFA A^R obtained from A by reversing all the transitions and by swapping the roles of initial and final states. The NNFA A^R recognizes the reverse of $L(A)$.

Every NNFA $A = (Q, \Sigma, \cdot, I, F)$ can be converted to an equivalent DFA $\mathcal{D}(A) = (2^Q, \Sigma, \cdot, I, F')$ where $F' = \{S \in 2^Q \mid S \cap F \neq \emptyset\}$. We call the DFA $\mathcal{D}(A)$ the *subset automaton* of the NNFA A. We use the following proposition to prove reachability of states in a subset automaton in some cases.

Proposition 1. *In the subset automaton of the NFA shown in Fig. 1 (left), each subset containing 0 is reachable from $\{0\}$, and in the subset automaton of the NFA shown in Fig. 1 (right), each subset is reachable from $\{0, 1, \ldots, n-1\}$.* □

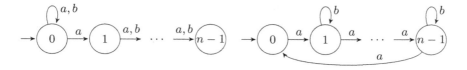

Fig. 1. The NFAs used in Proposition 1

To prove distinguishability, we use the following notions and observations. A state q of an NFA $A = (Q, \Sigma, \cdot, s, F)$ is called *uniquely distinguishable* (cf. [3]) if there is a string w which is accepted by A from and only from the state q. A transition (p, a, q) in the NFA A is called a *unique in-transition* if there is no state r of A such that $r \neq p$ and (r, a, q) is a transition in A. A state q is *uniquely reachable* from a state p, if there is a sequence of unique in-transitions (p_{i-1}, a_i, p_i) $(1 \leq i \leq k)$ such that $p_0 = p$ and $p_k = q$.

Proposition 2 (cf. [3]). *If there is a uniquely distinguishable state of an* NFA *A that is uniquely reachable from any other state of A, then the subset automaton $\mathcal{D}(A)$ does not have equivalent states.*   □

A *boolean finite automaton* (BFA, cf. [4]) is a quintuple $A = (Q, \Sigma, \delta, g_s, F)$, where Q is a finite non-empty set of states such that $Q = \{q_1, \ldots, q_n\}$, Σ is an input alphabet, δ is the transition function that maps $Q \times \Sigma$ into the set $\mathcal{B}_n$ of boolean functions with variables $\{q_1, \ldots, q_n\}$, $g_s \in \mathcal{B}_n$ is the initial boolean function, and $F \subseteq Q$ is the set of final states. The transition function δ is extended to the domain $\mathcal{B}_n \times \Sigma^*$ as follows: For all g in $\mathcal{B}_n$, a in Σ, and w in Σ^*, we have $\delta(g, \varepsilon) = g$; if $g = g(q_1, \ldots, q_n)$, then $\delta(g, a) = g(\delta(q_1, a), \ldots, \delta(q_n, a))$; $\delta(g, wa) = \delta(\delta(g, w), a)$. Next, let $f = (f_1, \ldots, f_n)$ be the boolean vector with $f_i = 1$ iff $q_i \in F$. The language accepted by the BFA A is the set of strings $L(A) = \{w \in \Sigma^* \mid \delta(g_s, w)(f) = 1\}$. A boolean finite automaton is called *alternating* (AFA, cf. [6]) if the initial function is a projection $g(q_1, \ldots, q_n) = q_i$.

We use the following observations for trade-offs between various automata throughout this paper. We provide the proof of case (e) since all the remaining cases are either well known, or follow from [4,13], [6, Theorem 4.1 and Corollary 4.2], and [9, Lemmas 1 and 2]. We use the claim in Lemma 3(a) quite often in the paper without referring to Lemma 3(a) again and again.

Lemma 3 (Properties of Finite Automata). *Let L be a regular language.*

(a) *The language L is accepted by an n-state BFA (AFA) if and only if L^R is accepted by a DFA of 2^n states (of which 2^{n-1} are final, respectively).*

(b) *Let L^R be a regular language accepted by a minimal n-state DFA. Then every BFA for L has at least $\lceil \log n \rceil$ states.*

(c) *If the minimal DFA for L^R has more than 2^{n-1} final states, then every AFA for L has at least $n + 1$ states.*

(d) *Let L be unary. Then L is accepted by an n-state BFA (AFA) if and only if L is accepted by a DFA of 2^n states (of which 2^{n-1} are final).*

(e) *If L is accepted by an n-state BFA (AFA), then $\overline{L}$ is accepted by an n-state BFA (AFA, respectively).*

(f) *If L is accepted by an n-state BFA, then L is accepted by an AFA of at most $n + 1$ states, and by an NNFA of at most 2^n states.*

(g) *If L is accepted by an n-state NNFA, then L is accepted by an NFA of at most $n + 1$ states and by a PFA of at most $2^n - 1$ states. If L is accepted by an n-state PFA, then L is accepted by a DFA of at most $n + 1$ states.*

Proof. (e) Let L be accepted by an n-state BFA (AFA). Then, by (a), the language L^R is accepted by a DFA of 2^n states (of which 2^{n-1} are final). Then the complement $\overline{L^R}$ is also accepted by a DFA of 2^n states (of which 2^{n-1} are final). Since $\overline{L^R} = \overline{L}^R$, the claim follows again by (a).   □

If u, v, and w are strings over Σ such that $w = uv$, then u is a *prefix* of w and v is a *suffix* of w. A language L is *prefix-closed* (*suffix-closed*) if $w \in L$ implies that every prefix (suffix) of w is in L.

In 1996, Birget [2] studied the state complexity of the "forever" operator $\overline{\Sigma^* \overline{L}}$ on DFAs and NFAs. Here we continue this research and to simplify the exposition, we use the following notation:

$$b(L) = \overline{\Sigma^* \overline{L}}. \tag{1}$$

3   Results

We start with an investigation of some properties of the "forever" operator.

Lemma 4 (Properties of $\overline{\Sigma^* \overline{L}}$). *Let L be a regular language and $b(L) = \overline{\Sigma^* \overline{L}}$.*

(a) $b(L) = \{w \in L \mid$ every suffix of w is in $L\}$.
(b) $b(L) = \emptyset$ if and only if $\varepsilon \notin L$.
(c) $b(L) = L$ if and only if L is suffix-closed.
(d) If L^R is accepted by a DFA A, then $b(L)^R$ is accepted by a DFA obtained from A by replacing each non-final state of A with a non-final sink state. $\square$

In what follows we consider six models of finite automata: DFAs, PFAs, NFAs, NNFAs, AFAs, and BFAs. We try to answer the following question. If a language L is represented by an n-state automaton of some model, how many states are sufficient and necessary in the worst case for an automaton of some other model to accept the language $b(L) = \overline{\Sigma^* \overline{L}}$? We first consider upper bounds. Although we have 36 possible trade-offs, it is enough to prove only some of them. The remaining trade-offs follow either from inclusions of some models of finite automata or from Lemma 3. For the (N)NFA-to-(P)DFA trade-offs, we need the Dedekind number $M(n)$ which counts the number of antichains of subsets of an n-element set. The number $M(n)$ lies in the order of magnitude $2^{2^{\Theta(n)}}$ [12]:

$$2^{n - \log n} \leq \binom{n}{\lfloor n/2 \rfloor} \leq \log_2 M(n) \leq \binom{n}{\lfloor n/2 \rfloor}\left(1 + O\left(\frac{\log n}{n}\right)\right) \leq 2^{n+1-(\log n)/2}.$$

It follows that $\log_2 M(n)$ lies in the order of magnitude $2^{n-\Theta(\log n)}$. Moreover, we assume that $\varepsilon \in L$ and $L \neq \Sigma^*$ in the statement of the next theorem because otherwise $b(L)$ is empty or equals Σ^* by Lemma 4(b) and (c).

Theorem 5 (Upper Bounds). *Let L be a regular language such that $\varepsilon \in L$ and $L \neq \Sigma^*$. Let L be accepted by a finite automaton A of n states.*

(1) If A is a DFA, then $b(L)$ is accepted by a DFA of at most 2^{n-1} states.
(2) If A is a PFA, then $b(L)$ is accepted by a PFA of at most 2^{n-1} states.
(3) If A is an NFA, then $b(L)$ is accepted by
   (a) an NFA of at most 2^{n-1} states;
   (b) a PFA of at most $M(n-1) - 1$ states.

(4) If A is an NNFA, then $b(L)$ is accepted by
  (a) an NNFA of at most $2^n - 2$ states.
  (b) a PFA of at most $M(n) - 1$ states.
(5) If A is an AFA, then $b(L)$ is accepted by
  (a) an AFA of at most n states;
  (b) an NNFA of at most 2^{n-1} states.
(6) If A is a BFA, then $b(L)$ is accepted by
  (a) a BFA of at most n states;
  (b) an NNFA of at most $2^n - 1$ states.

Proof (1). We first interchange final and non-final states in A to get the DFA $\overline{A}$ for $\overline{L}$. Then we add a loop on every input symbol in the initial state of $\overline{A}$ to get an NFA N for $\Sigma^*\overline{L}$. In $\mathcal{D}(N)$, only subsets containing the initial state are reachable. Finally, we again interchange the final and non-final states of $\mathcal{D}(N)$.

(2) Let $A = (Q, \Sigma, \cdot, s, F)$ be an n-state PFA for L. It is enough to show that the language $\Sigma^*\overline{L}$ is accepted by a DFA of at most $2^{n-1} + 1$ states, one of which is final sink state. To get an $(n+1)$-state DFA $\overline{A}$ for $\overline{L}$, we first add a new non-final sink state q_d to A. Then, for each transition which is undefined in A, we add the corresponding transition to q_d. Finally, we interchange final and non-final states of the resulting automaton. We construct an $(n+1)$-state NFA N for $\Sigma^*\overline{L}$ from DFA $\overline{A}$, by adding a loop on each input symbol in the initial state s. In the corresponding subset automaton, each reachable subset must contain s. Moreover, the state q_d is a final sink state. It follows that each string is accepted by N from q_d, and therefore each subset containing q_d, is equivalent to $\{q_d\}$. In total, we get at most $2^{n-1} + 1$ reachable and pairwise distinguishable states.

(3a) Let $A = (Q, \Sigma, \cdot, s, F)$ be an n-state NFA for L. We have $s \in F$ since $\varepsilon \in L$. We reverse A to get an n-state NNFA A^R for L^R with a unique final state s. In the subset automaton $\mathcal{D}(A^R)$, we omit all the non-final subsets, that is, all subsets not containing s, to get a 2^{n-1}-state PFA B with the initial state F. All states of B are final, and all of them contain s. We have two cases. If there is a final subset of $\mathcal{D}(A^R)$ which is not reachable in B, then we reverse B and add a new initial state to get an NFA for $b(L)$ of at most 2^{n-1} states. Otherwise, we modify PFA B as follows. We make all states of B non-final, except for $\{s\}$. Next, we add the ε-transition to $\{s\}$ from any other state in B. Denote the resulting ε-NFA by B'. We can show that $L(B') = L(B)$. This means that B' is a 2^{n-1}-state ε-NFA with one final state for $b(L)^R$. By reversing B' and removing ε-transitions, we get a 2^{n-1}-state NFA for $b(L)$.

(3b) It is enough to show that $\Sigma^*\overline{L}$ is accepted by a DFA of at most $M(n-1)$ states, one of which is a final sink state. Let $A = (Q, \Sigma, \cdot, s, F)$ be an n-state NFA for L, and B be the 2^n-state subset automaton of A. We interchange the final and non-final states in B, to get a 2^n-state DFA $\overline{B}$ for $\overline{L}$. To get a 2^n-state NFA N for $\Sigma^*\overline{L}$, we add a loop on each input symbol in the initial state of the DFA $\overline{B}$. Finally, let C be the subset automaton of N. Then C is a DFA for $\Sigma^*\overline{L}$. Formally, we have

$$B = \mathcal{D}(A) = (2^Q, \Sigma, \cdot, \{s\}, F_B) \text{ where } F_B = \{X \subseteq Q \mid X \cap F \neq \emptyset\});$$
$$\overline{B} = (2^Q, \Sigma, \cdot, \{s\}, F_{\overline{B}}) \text{ where } F_{\overline{B}} = 2^Q \setminus F_B = \{X \subseteq Q \mid X \subseteq Q \setminus F\});$$

$N = (2^Q, \Sigma, \circ, \{s\}, F_{\overline{B}})$ where for each X in 2^Q and each a in Σ,
$$\{s\} \circ a = \{\{s\}, \{s\} \cdot a\}, \text{ and}$$
$$X \circ a = \{X \cdot a\} \text{ if } X \neq \{s\};$$
$C = \mathcal{D}(N) = (2^{2^Q}, \Sigma, \circ, \{\{s\}\}, F_C)$ where $F_C = \{X \in 2^{2^Q} \mid X \cap F_{\overline{B}} \neq \emptyset\}$.
Thus, the states of C are sets of subsets of Q, and a state $\mathcal{S} = \{S_1, S_2, \ldots, S_k\}$ is final if there is an i such that $S_i \subseteq Q \setminus F$. Our aim is to show that C has at most $M(n-1)$ reachable and pairwise distinguishable states. We first show that each state of C is equivalent to an antichain in 2^Q. Let $S \subseteq T \subseteq Q$ and w be accepted by N from the state T. We can show that w is accepted by N also from S. Thus if in a state $\mathcal{S} = \{S_1, S_2, \ldots, S_k\}$ of C we have $S_i \subseteq S_j$ for some i and j, then $\mathcal{S}$ is equivalent to $\mathcal{S} \setminus \{S_j\}$. It follows that each state of C is equivalent to an antichain in 2^Q. Moreover, since N has a loop on each symbol in its initial state $\{s\}$, and C is the subset automaton of N, each reachable state of C must contain the set $\{s\}$, that is, each reachable antichain has a form $\{\{s\}, S_2, S_3, \ldots, S_k\}$, where $k \geq 1$, and $\{S_2, S_3, \ldots, S_k\}$ is an antichain in $2^{Q \setminus \{s\}}$. This gives the upper bound $M(n-1)$. Notice that the empty antichain corresponds to the initial state $\{\{s\}\}$. We also have to count the antichain $\{\emptyset\}$ which is unreachable final sink state, but it is equivalent to the reachable state $\{\{s\}, \emptyset\}$.

(4a) If all the states of a given NNFA are initial, then L is suffix-closed, and therefore $b(L) = L$ by Lemma 4(c). Otherwise, L^R is accepted by a PFA which has $2^n - 1$ states, and at least one of them is non-final. Omit all the non-final states to get a PFA for $b(L)^R$ (cf. Lemma 4(d)), and reverse the resulting PFA to get the desired NNFA for $b(L)$.

(4b) Similarly as in (3b), we prove that only states $\mathcal{S} = \{I, S_1, S_2, \ldots, S_k\}$ where $\{S_1, S_2, \ldots, S_k\}$ is an antichain in 2^Q are pairwise distinguishable.

(5a) If L is accepted by an n-state AFA, then L^R is accepted by a DFA of 2^n states of which 2^{n-1} are final. Replace each non-final state with a non-final sink state to get a DFA for $b(L)^R$ of 2^n states of which 2^{n-1} are final. Hence $b(L)$ is accepted by an n-state AFA.

(5b) In the DFA for $b(L)^R$ obtained as in case (5a), we omit the non-final sink states to get an equivalent PFA of 2^{n-1} states. By reversing this PFA, we get a 2^{n-1}-state NNFA for $b(L)$.

(6a) If A is an n-state BFA, then L^R is accepted by a DFA of 2^n states. Replace each non-final state with a non-final sink state to get a DFA for $b(L)^R$ of 2^n states given by Lemma 4(d). Hence $b(L)$ is accepted by an n-state BFA.

(6b) In the DFA for $b(L)^R$ obtained as in case (6a), we omit the non-final sink states to get an equivalent PFA of at most $2^n - 1$ states; recall that $L \neq \Sigma^*$. By reversing this PFA, we get the desired NNFA for $b(L)$.   □

Now we turn our attention to lower bounds. We again need to prove only some of them and all the remaining bounds follow from the inclusions of models or from Lemma 3. However, in some cases, we use witnesses over a smaller alphabet for the bound that follows from some other trade-off.

In 32 of 36 cases, our lower bounds meet the upper bounds given by Theorem 5. The remaining four cases are the trade-offs from NNFA to DFA, PFA, NFA, and NNFA. With the exception of four trade-offs, our witness languages are defined over a fixed alphabet of size one, two, three, or four. The binary case is always optimal in the sense that there is no unary language meeting the upper bound (and the unary alphabet is always optimal :-).

Theorem 6 (Lower Bounds). *There exists a regular language L accepted by an n-state finite automaton A such that A is*

(1) a ternary DFA and every BFA for $b(L)$ has at least n states;
(2) a ternary DFA and every NNFA for $b(L)$ has at least 2^{n-1} states;
(3) a binary DFA and every PFA for $b(L)$ has at least 2^{n-1} states;
(4) a quaternary PFA and every DFA for $b(L)$ has at least $2^{n-1} + 1$ states;
(5) an NFA and every DFA for $b(L)$ has at least $M(n-1)$ states;
(6) a binary NNFA and every AFA for $b(L)$ has at least $n + 1$ states;
(7) a unary AFA and
  (a) every BFA for $b(L)$ has at least n states;
  (b) every NNFA for $b(L)$ has at least 2^{n-1} states;
(8) a binary AFA and every NFA for $b(L)$ has at least $2^{n-1} + 1$ states;
(9) a binary AFA and every DFA for $b(L)$ has at least $2^{2^{n-1}}$ states;
(10) a unary BFA and
  (a) every AFA for $b(L)$ has at least $n + 1$ states;
  (b) every NNFA for $b(L)$ has at least $2^n - 1$ states;
(11) a binary BFA and every NFA for $b(L)$ has at least 2^n states;
(12) a binary BFA and every DFA for $b(L)$ has at least $2^{2^n - 1}$ states.

Proof. (1) Let L be the language accepted by the DFA A shown in Fig. 2. We reverse A to get an NFA A^R for L^R. Using Propositions 1 and 2, we can prove that in the minimal DFA for L^R we have 2^{n-1} final states and one non-final sink state, so the language L^R is prefix-closed. Therefore L is suffix-closed, so $b(L) = L$. Since the minimal DFA for L^R has $2^{n-1} + 1$ states, every BFA for L, so for $b(L)$, has at least n states.

(2) This case follows from the proof of [2, Theorem 2(a)].

(3) Let L be accepted by DFA $A = (\{0, \ldots, n-1\}, \{a, b\}, \cdot, 0, \{0, 1, \ldots, n-2\})$, where $i \cdot a = (i + 1) \bmod n$, $0 \cdot b = 0$, and $i \cdot b = (i + 1) \bmod n$ if $i \neq 0$.

We construct an n-state NFA N for $\Sigma^* \overline{L}$ by interchanging final and non-final states in A and by adding the transition $(0, a, 0)$. It is enough to prove that the subset automaton $\mathcal{D}(N)$ has at least 2^{n-1} reachable and pairwise distinguishable states. We prove reachability by using Proposition 1. To prove distinguishability, notice that the state $n - 1$ is uniquely distinguishable by ε in N and it is uniquely reachable from any other state through unique in-transitions on a. By Proposition 2, the subset automaton $\mathcal{D}(N)$ does not have equivalent states. Since $\mathcal{D}(N)$ has no non-final sink state, it is also a minimal PFA. Notice that the lower bound 2^{n-1} for a DFA accepting $b(L)$ follows from the proof. In [2, Proof of Theorem 2(b)], it is claimed that this bound is met by the binary language $a\{a, b\}^{n-2}$. However, the minimal DFA for this language has $n + 1$ states.

(4) Let L be the language accepted by the PFA A shown in Fig. 3. We construct an $(n + 1)$-state NFA N for $\Sigma^* \overline{L}$ as follows. First, we add a new non-final

sink state n and the transitions on a, b, c from $n - 1$ to n. Then we make state n final, and all the remaining states non-final.

Finally, we add transitions $(0, a, 0)$ and $(0, d, 0)$. By using Propositions 1 and 2 we can show that $\mathcal{D}(N)$ has $2^{n-1} + 1$ reachable and pairwise distinguishable states.

(5) Let L be accepted by the n-state NFA $A = (Q, \Sigma, \cdot, 0, F)$, where $Q = \{0, 1, \ldots, n - 1\}$, $\Sigma = \{a_X, b_X \mid X \subseteq Q\}$, $F = Q \setminus \{n - 1\}$, and the transition function is defined as follows:

$0 \cdot a_X = X$ and $i \cdot a_X = \{i\}$ if $i \neq 0$,

$$i \cdot b_X = \begin{cases} \{n - 1\}, & \text{if } i \in X; \\ \{0\}, & \text{if } i \notin X. \end{cases}$$

Then $B = \mathcal{D}(A) = (2^Q, \Sigma, \cdot, \{0\}, 2^Q \setminus \{\{n - 1\}, \emptyset\})$;

$\overline{B} = (2^Q, \Sigma, \cdot, \{0\}, \{\{n - 1\}, \emptyset\})$;

$N = (2^Q, \Sigma, \circ, \{0\}, \{\{n - 1\}, \emptyset\})$ where

$\{0\} \circ a = \{0\} \cup \{0\} \cdot a$,

$X \circ a = X \cdot a$ if $X \neq \{0\}$;

$C = \mathcal{D}(N) = (2^{2^Q}, \Sigma, \circ, \{\{0\}\}, \{X \in 2^{2^Q} \mid X \cap \{\{n - 1\}, \emptyset\} \neq \emptyset\})$. Our aim is to show that C has at least $M(n - 1)$ reachable and distinguishable states. Let $S_1, S_2, \ldots, S_k$ be subsets of Q such that $0 \notin S_i$ for every i. Then in C we have

$$\{\{0\}\} \xrightarrow{a_{S_1}} \{\{0\}, S_1\} \xrightarrow{a_{S_2}} \{\{0\}, S_1, S_2\} \xrightarrow{a_{S_3}} \ldots \xrightarrow{a_{S_k}} \{\{0\}, S_1, S_2, \ldots, S_k\}.$$

It follows that every state $\mathcal{S} = \{\{0\}, S_1, S_2, \ldots, S_k\}$ where $\{S_1, S_2, \ldots, S_k\}$ is an antichain of subsets of $\{1, 2, \ldots, n - 1\}$ is reachable. We can prove that two distinct antichains are distinguishable. It follows that C has at least $M(n - 1)$ reachable and distinguishable states.

(6) Let L be accepted by the NNFA A shown in Fig. 4. Since each state of A is initial, L is suffix-closed, so $b(L) = L$. We can show that the minimal DFA for L^R has more than 2^{n-1} final states. It follows that every AFA for L, so for $b(L)$, has at least $n + 1$ states.

(7) Let $L = \{a^i \mid 0 \leq i \leq 2^{n-1} - 1\}$. Then L is a unary language accepted by a 2^n-state DFA with 2^{n-1} final states. So L is accepted by an n-state AFA. Since L is suffix-closed, $b(L) = L$. (a) Since the minimal DFA for L has $2^{n-1} + 1$ states, every BFA for L has at least n states. (b) The longest string in L is of length $2^{n-1} - 1$, and therefore every NNFA for L has at least 2^{n-1} states.

(8) Let K be accepted by the 2^n-state DFA A shown in Fig. 5; notice that A has 2^{n-1} final states. Set $L = K^R$. Then L is accepted by an n-state AFA. By Lemma 4(d), if we omit all non-final states of A, we get a PFA C for $b(L)^R$ of 2^{n-1} states, all of them final. It is shown in [8, Theorem 2] that every NFA for $L(C)^R$ has at least $2^{n-1} + 1$ states. Since $L(C)^R = b(L)$, the claim follows.

(9) Let K be accepted by the 2^n-state DFA A shown in Fig. 6; notice that A has 2^{n-1} final states. Set $L = K^R$. Then the language L is accepted by an n-state AFA. By Lemma 4(d), if we omit all non-final states of A, we get a PFA C for $b(L)^R$ of 2^{n-1} states, all of them final. Next, we reverse the PFA C to get an NNFA $N = (\{0, 1, \ldots, 2^{n-1} - 1\}, \{a, b\}, \cdot^R, \{0, 1, \ldots, 2^{n-1} - 1\}, \{0\})$ for $b(L)$.

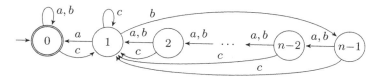

Fig. 2. The DFA for L such that every BFA for $\overline{\Sigma^*\overline{L}}$ has n states

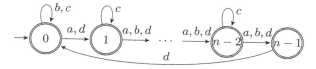

Fig. 3. The PFA for L such that every DFA for $\overline{\Sigma^*\overline{L}}$ has $2^{n-1} + 1$ states

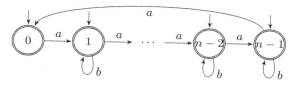

Fig. 4. The NNFA for L such that every AFA for $\overline{\Sigma^*\overline{L}}$ has $n + 1$ states

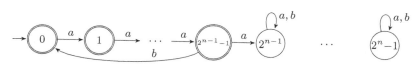

Fig. 5. The reverse of the witness for the AFA-to-NFA trade-off

By Proposition 1, the subset automaton $\mathcal{D}(N)$ has $2^{2^{n-1}}$ reachable states. To prove distinguishability, notice that the state 0 is uniquely distinguishable by ε, and it is uniquely reachable from any other state through unique in-transitions on symbol a. By Proposition 2, the subset automaton has no equivalent states.

(10) Let $L = \{a^i \mid 0 \le i \le 2^n - 2\}$. Then L is a unary language accepted by a minimal 2^n-state DFA A, so L is accepted by a n-state BFA. Since L is suffix-closed, $b(L) = L$. (a) Every AFA accepting L has at least $n+1$ states since the number of final states in A is greater than 2^{n-1}. (b) The longest string in L is of length $2^n - 2$, and therefore every NNFA for L has at least $2^n - 1$ states.

(11) Let K be accepted by the 2^n-state DFA A shown in Fig. 7. Set $L = K^R$. Then L is accepted by an n-state BFA. Now the proof goes exactly the same way as in the case (8) and it results in the lower bound 2^n.

(12) Let K be accepted by the 2^n-state DFA A shown in Fig. 8. Set $L = K^R$. Then L is accepted by an n-state BFA. Now the proof goes exactly the same way as in the case (9) and it results in the lower bound 2^{2^n-1}.   □

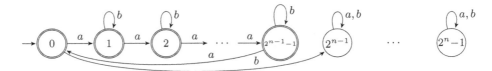

Fig. 6. The reverse of the witness for the AFA-to-DFA trade-off

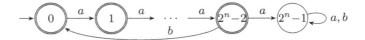

Fig. 7. The reverse of the witness for the BFA-to-NFA trade-off

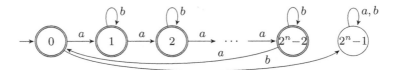

Fig. 8. The reverse of the witness for the BFA-to-DFA trade-off

Table 1. The complexity of $\overline{\Sigma^*\overline{L}}$ for various types of finite automata

$L\backslash b(L)$	DFA	$|\Sigma|$	PFA	$|\Sigma|$	NFA	$|\Sigma|$	NNFA	$|\Sigma|$	AFA	$|\Sigma|$	BFA	$|\Sigma|$
DFA	2^{n-1}	2	2^{n-1}	2	2^{n-1}[2]	3	2^{n-1}[2]	3	n	3	n	3
PFA	$2^{n-1}+1$	4	2^{n-1}	2	2^{n-1}	3	2^{n-1}	3	n	3	n	3
NFA	$M(n-1)$	2^{n+1}	$M(n-1)-1$	2^{n+1}	2^{n-1}	3	2^{n-1}	3	n	3	n	3
NNFA	$\geq M(n-1)$ $\leq M(n)$	2^{n+1}	$\geq M(n-1)-1$ $\leq M(n)-1$	2^{n+1}	$\geq 2^{n-1}$ $\leq 2^n-1$	3	$\geq 2^{n-1}$ $\leq 2^n-2$	3	$n+1$	2	n	2
AFA	$2^{2^{n-1}}$	2	$2^{2^{n-1}}-1$	2	$2^{n-1}+1$	2	2^{n-1}	1	n	1	n	1
BFA	2^{2^n-1}	2	$2^{2^n-1}-1$	2	2^n	2	2^n-1	1	$n+1$	1	n	1

4   Conclusions

We investigated the descriptive complexity of $\overline{\Sigma^*\overline{L}}$ over complete and partial deterministic, nondeterministic, alternating, and boolean finite automata. For each trade-off, except for those starting with NNFAs, we provided tight upper bounds for complexity of $\overline{\Sigma^*\overline{L}}$ depending on the complexity of L. The most interesting result is the tight upper bound on NFA-to-DFA trade-off given by the Dedekind number $M(n-1)$. However, we used a growing alphabet of size 2^{n+1} to get the lower bound in this case. Except for (N)NFA-to-(P)DFA trade-offs, all witnesses are described over an alphabet of fixed size. Moreover, binary and unary alphabets are optimal for their respective cases. Whenever we have a larger alphabet, we do not know whether or not it is optimal. The precise complexity for NNFA-to-(P)DFA and NNFA-to-(N)NFA trade-offs remains open as well (Table 1).

References

1. Birget, J.: Intersection and union of regular languages and state complexity. Inform. Process. Lett. **43**(4), 185–190 (1992)
2. Birget, J.: The state complexity of $\overline{\Sigma^* \overline{L}}$ and its connection with temporal logic. Inform. Process. Lett. **58**(4), 185–188 (1996)
3. Brzozowski, J., Jirásková, G., Liu, B., Rajasekaran, A., Szykuła, M.: On the state complexity of the shuffle of regular languages. In: Câmpeanu, C., Manea, F., Shallit, J. (eds.) DCFS 2016. LNCS, vol. 9777, pp. 73–86. Springer, Cham (2016). doi:10.1007/978-3-319-41114-9_6
4. Brzozowski, J.A., Leiss, E.L.: On equations for regular languages, finite automata, and sequential networks. Theoret. Comput. Sci. **10**, 19–35 (1980)
5. Cohen, J., Perrin, D., Pin, J.: On the expressive power of temporal logic. J. Comput. Syst. Sci. **46**(3), 271–294 (1993)
6. Fellah, A., Jürgensen, H., Yu, S.: Constructions for alternating finite automata. Int. J. Comput. Math. **35**(1–4), 117–132 (1990)
7. Gao, Y., Moreira, N., Reis, R., Yu, S.: A survey on operational state complexity. CoRR abs/1509.03254 (2015). http://arxiv.org/abs/1509.03254
8. Jirásková, G.: State complexity of some operations on binary regular languages. Theoret. Comput. Sci. **330**(2), 287–298 (2005)
9. Jirásková, G.: Descriptional complexity of operations on alternating and boolean automata. In: Hirsch, E.A., Karhumäki, J., Lepistö, A., Prilutskii, M. (eds.) CSR 2012. LNCS, vol. 7353, pp. 196–204. Springer, Heidelberg (2012). doi:10.1007/978-3-642-30642-6_19
10. Jirásková, G., Pighizzini, G.: Optimal simulation of self-verifying automata by deterministic automata. Inform. Comput. **209**(3), 528–535 (2011)
11. Kapoutsis, C.: Removing bidirectionality from nondeterministic finite automata. In: Jędrzejowicz, J., Szepietowski, A. (eds.) MFCS 2005. LNCS, vol. 3618, pp. 544–555. Springer, Heidelberg (2005). doi:10.1007/11549345_47
12. Kleitman, D., Markowsky, G.: On Dedekind's problem: the number of isotone boolean functions. II. Trans. Amer. Math. Soc. **213**, 373–390 (1975)
13. Leiss, E.L.: Succint representation of regular languages by boolean automata. Theoret. Comput. Sci. **13**, 323–330 (1981)
14. Lupanov, O.B.: A comparison of two types of finite automata. Problemy Kibernetiki **9**, 321–326 (1963). (in Russian) German translation: Über den Vergleich zweier Typen endlicher Quellen. Probleme der Kybernetik **6**, 328–335 (1966)
15. Maslov, A.N.: Estimates of the number of states of finite automata. Soviet Math. Doklady **11**, 1373–1375 (1970)
16. Meyer, A.R., Fischer, M.J.: Economy of description by automata, grammars, and formal systems. In: Proceedings of the 12th Annual Symposium on Switching and Automata Theory, pp. 188–191. IEEE Computer Society Press (1971)
17. Moore, F.R.: On the bounds for state-set size in the proofs of equivalence between deterministic, nondeterministic, and two-way finite automata. IEEE Trans. Comput. **C-20**, 1211–1219 (1971)
18. Rabin, M.O., Scott, D.: Finite automata and their decision problems. IBM J. Res. Develop. **3**, 114–125 (1959)
19. Sipser, M.: Introduction to the Theory of Computation. Cengage Learning (2012)
20. Yershov, Yu.L.: On a conjecture of V. A. Uspenskii. Algebra i Logika (Seminar) **1**, 45–48 (1962). (in Russian)

21. Yu, S.: Chapter 2: Regular languages. In: Rozenberg, G., Salomaa, A. (eds.) Handbook of Formal Languages, vol. I, pp. 41–110. Springer, Heidelberg (1997)
22. Yu, S., Zhuang, Q., Salomaa, K.: The state complexities of some basic operations on regular languages. Theoret. Comput. Sci. **125**(2), 315–328 (1994)

Variations of Checking Stack Automata: Obtaining Unexpected Decidability Properties

Oscar H. Ibarra[1] and Ian McQuillan[2(✉)]

[1] Department of Computer Science, University of California,
Santa Barbara, CA 93106, USA
ibarra@cs.ucsb.edu
[2] Department of Computer Science, University of Saskatchewan,
Saskatoon, SK S7N 5A9, Canada
mcquillan@cs.usask.ca

Abstract. We introduce a model of one-way language acceptors (a variant of a checking stack automaton) and show the following decidability properties:

1. The deterministic version has a decidable membership problem but has an undecidable emptiness problem.
2. The nondeterministic version has an undecidable membership problem and emptiness problem.

There are many models of accepting devices for which there is no difference with these problems between deterministic and nondeterministic versions, i.e., the membership problem for both versions are either decidable or undecidable, and the same holds for the emptiness problem. As far as we know, the model we introduce above is the first one-way model to exhibit properties (1) and (2). We define another family of one-way acceptors where the nondeterministic version has an undecidable emptiness problem, but the deterministic version has a decidable emptiness problem. We also know of no other model with this property in the literature. We also investigate decidability properties of other variations of checking stack automata (e.g., allowing multiple stacks, two-way input, etc.). Surprisingly, two-way deterministic machines with multiple checking stacks and multiple reversal-bounded counters are shown to have a decidable membership problem, a very general model with this property.

Keywords: Checking stack automata · Pushdown automata · Decidability · Reversal-bounded counters

1 Introduction

The deterministic and nondeterministic versions of most known models of language acceptors exhibit the same decidability properties for each of the membership and emptiness problems. For example, for one-way models, it is easy to

The research of O.H. Ibarra was supported, in part, by NSF Grant CCF-1117708.
The research of I. McQuillan was supported, in part, by Natural Sciences and Engineering Research Council of Canada Grant 2016-06172.

© Springer International Publishing AG 2017
E. Charlier et al. (Eds.): DLT 2017, LNCS 10396, pp. 235–246, 2017.
DOI: 10.1007/978-3-319-62809-7_17

show (by coding the "transition rules" on the input string) that the emptiness problem is decidable for the deterministic version if and only if it is decidable for the nondeterministic version. For a formal proof of this, it is possible to either use the model of Abstract Families of Acceptors (AFAs) from [3], or a type of abstract store types used in this paper. For the membership problem, as far as we know, no one-way model has been shown to exhibit different decidability properties for deterministic and nondeterministic versions.

A one-way checking stack automaton [4] is similar to a pushdown automaton that cannot erase its stack, but can enter and read the stack in two-way read-only mode, but once this mode is entered, the stack cannot change. Here, we introduce a new model of one-way language acceptors that exhibits the decidability properties above. It is defined by augmenting a checking stack automaton with reversal-bounded[1] counters, and the deterministic and nondeterministic versions are denoted by DCSACM and NCSACM, respectively. The models with two-way input (with end-markers) are called 2DCSACM and 2NCSACM. These are generalized further to models with k checking stacks: k-stack 2DCSACM and k-stack 2NCSACM.

We show the following results concerning membership and emptiness:

1. The membership and emptiness problems for NCSACMs are undecidable, even when there are only two reversal-bounded counters.
2. The emptiness problem for DCSACM is decidable when there is only one reversal-counter but undecidable when there are two reversal-bounded counters.
3. The membership problem for k-stack 2DCSACMs is decidable for any k.

We define another family of one-way acceptors where the deterministic version has a decidable emptiness problem, but the nondeterministic version has an undecidable emptiness problem. Further, we introduce a new family with decidable emptiness, containment, and equivalence problems, which is one of the most powerful families to have these properties (one-way deterministic machines with one reversal-bounded counter and a checking stack that can only read from the stack at the end of the input). We also investigate the decidability properties of other variations of checking stack automata (e.g., allowing multiple stacks, two-way input, etc.).

Many proofs are omitted due to space constraints, but are found in online in [9].

2   Preliminaries

This paper requires basic knowledge of automata and formal languages [7].

We use a variety of machine models here, mostly built on top of the checking stack. It is possible to define each machine model directly. An alternate approach

---

[1] A counter is reversal-bounded if there is a bound on the number of changes between increasing and decreasing.

is to define "store types" first, which describes just the behavior of the store, including instructions that can change the store, and the manner in which the store can be read. This can capture standard types of stores studied in the literature, such as a pushdown, or a counter. Defined generally enough, it can also define a checking stack, or an l-reversal-bounded counter (one that makes at most l alternations between increasing and decreasing). Then, machines using one or more store types can be defined, in a standard fashion. A $(\Omega_1, \ldots, \Omega_k)$ machine is a machine with k stores, where Ω_i describes each store. This is the approach taken here, in a similar fashion to the one taken in [2] or [3] to define these same types of automata. This generality will also help in illustrating what is required to obtain certain decidability properties; see e.g. Lemma 1 and Proposition 2 which are proven generally for arbitrary store types.

Due to space constraints, we will omit such definitions and refer to [9] (similar definitions also appear in [8]). Here, all machine models will only be described informally. A checking stack is similar to a pushdown, although initially, it is only possible to push to the checking stack (without popping). Then, at some point, it is possible to read from within the checking stack (without erasing it), in a two-way read-only fashion. However, once this reading has started, then the stack can no longer be changed. Thus, there are two phases, a "writing phase", where it can push or stay (no pop), and a "reading phase", where it enters the stack in read-only mode. But once it starts reading, it cannot change the stack again.

The different *machine modes* are combinations of either one-way or two-way, deterministic or nondeterministic, and r-head for some $r \geq 1$, For example, one-way, 1-head, deterministic, is a machine mode. Given a sequence of store types $\Omega_1, \ldots, \Omega_k$ and a machine mode, one can study the set of all $(\Omega_1, \ldots, \Omega_k)$ machines with this mode. The set of all such machines with a mode is said to be *complete*. Any strict subset is said to be *incomplete*. Given a set of (complete or incomplete) machines $\mathcal{M}$ of this type, the family of languages accepted by these machines is denoted $\mathcal{L}(\mathcal{M})$. For example, the set of all one-way deterministic pushdown automata is complete. But consider the set of all one-way deterministic pushdown automata that can only decrease the size of the stack when scanning the right end-marker. Then, the instructions available to such machines depend on the location of the input (whether it has reached the end of the input or not). Therefore, this is an incomplete set of automata. Later in the paper, we will consider variations of checking stack automata such as one called *no-read*, which means that they do not read from the inside of the checking stack before hitting the right input end-marker. This is similarly an incomplete set of automata since the instructions allowed differs depending on the input position.

The class of one-way deterministic (resp. nondeterministic) checking stack automata is denoted by DCSA (resp., NCSA) [4]. The class of deterministic (resp. nondeterministic), finite automata is denoted by DFA (resp., NFA) [7].

For $k, l \geq 1$, the class of one-way deterministic (resp. nondeterministic) l-reversal-bounded k-counter machines is denoted by DCM(k, l) (resp. NCM(k, l)). If only one integer is used, e.g. NCM(k), this class contains all l-reversal-bounded

k counter machines, for some l, and if the integers is omitted, e.g., NCM and DCM, they contain all l-reversal-bounded k counter machines, for some k, l. Note that a counter that makes l reversals can be simulated by $\lceil \frac{l+1}{2} \rceil$ 1-reversal-bounded counters [10]. Closure and decidable properties of various machines augmented with reversal-bounded counters have been studied in the literature (see, e.g., [10]). For example, it is known that the membership and emptiness problems are decidable for NCM [10].

Also, here we will study the following new classes of machines that have not been studied in the literature: one-way deterministic (resp. nondeterministic) machines defined by stores consisting of one checking stack and k l-reversal-bounded counters, denoted by DCSACM(k, l) (resp. NCSACM(k, l)), those with k-reversal-bounded counters, denoted by DCSACM(k) (resp. NCSACM(k)), and those with some number of reversal-bounded counters, denoted by DCSACM (resp. NCSACM).

All models above also have two-way versions of the machines defined, denoted by preceding them with 2, e.g. 2DCSA, 2NCSA, 2NCM(1), 2DFA, 2NFA, etc.

We will also define models with k checking stacks for some k, which we will precede with the phrase "k-stack", e.g. k-stack 2DCSA, k-stack 2NCSA, k-stack 2DCSACM, k-stack 2NCSACM, etc. When $k = 1$, then this corresponds with omitting the phrase "k-stack".

3   A Checking Stack with Reversal-Bounded Counters

Before studying a new type of store and machine model, we determine several properties that are equivalent for any complete set of machines. This helps to demonstrate what is required to potentially have a machine model where the deterministic version has a decidable membership problem with an undecidable emptiness problem, while both problems are undecidable for the nondeterministic version.

First, we examine a machine's behavior on one word.

Lemma 1. *Let M be a one- or two-way, r-head, for some $r \geq 1$, $(\Omega_1, \ldots, \Omega_k)$-machine, and let $w \in \Sigma^*$. We can effectively construct another $(\Omega_1, \ldots, \Omega_k)$-machine M_w that is one-way and 1-head which accepts λ if and only if M accepts w. Furthermore, M_w is deterministic if M is deterministic.*

Proof. The input w is encoded in the state of M_w, and M_w on input λ, simulates the computation of M and accepts λ if and only if M accepts w. This uses a subset of the sequence of transitions used by M. Since M_w is only reading λ, two-way input is not needed in M_w, and the r-heads are simulated completely in the finite control.   □

Then, for all machines with the same store types, the following decidability problems are equivalent:

Proposition 2. *Consider store types $(\Omega_1, \ldots, \Omega_k)$. The following problems are equivalently decidable, for the stated complete sets of automata:*

1. *the emptiness problem for one-way deterministic $(\Omega_1, \ldots, \Omega_k)$-machines,*
2. *the emptiness problem for one-way nondeterministic $(\Omega_1, \ldots, \Omega_k)$-machines,*
3. *membership problem for one-way nondeterministic $(\Omega_1, \ldots, \Omega_k)$-machines,*
4. *acceptance of λ, for one-way nondeterministic $(\Omega_1, \ldots, \Omega_k)$-machines,*
5. *the membership problem for two-way r-head (for $r \geq 1$) nondeterministic $(\Omega_1, \ldots, \Omega_k)$-machines.*

It is important to note that this proposition is not necessarily true for incomplete sets of automata, as the machines constructed in the proof need to be present in the set. We will see some natural restrictions later where this is not the case, such as sets of machines where there is a restriction on what instructions can be performed on the store based on the position on the input. And indeed, to prove the equivalence of (1) and (2) above, the deterministic machine created reads a letter for every transition of the nondeterministic machine applied. So, consider a set of automata that is only allowed to apply a strict subset of store instructions before the end-marker. Let M be a nondeterministic machine of this type, and say that M applies some instruction on the end-marker that is not available to the machine before the end-marker. But the deterministic machine M' created from M in Proposition 2 reads an input letter when every instruction is applied, even those applied on the end-marker of M. But since M' is reading an input letter during this operation, it would violate the instructions allowed by M' before the end-marker.

The above proposition indicates that for complete sets of one-way machines, membership for nondeterminism, emptiness for nondeterminism, and emptiness for determinism are equivalent. Thus, the only one that can potentially differ is membership for deterministic machines. Yet we know of no existing model where it differs from the other three properties. We examine one next.

Next, we will study NCSACMs and DCSACMs, which are NCSAs and DCSAs (nondeterministic and deterministic checking stack automata) respectively, augmented by reversal-bounded counters. First, two examples will be shown, demonstrating a language that can be accepted by a DCSACM.

Example 3. Consider the language $L = \{(a^n \#)^n \mid n \geq 1\}$. A DCSACM M with one 1-reversal-bounded counter can accept L as follows: M when given an input w (we may assume that the input is of the form $w = a^{n_1} \# \cdots a^{n_k} \#$ for some $k \geq 1$ and $n_i \geq 1$ for $1 \leq i \leq k$, since the finite control can check this), copies the first segment a^{n_1} to the stack while also storing number n_1 in the counter. Then M goes up and down the stack comparing n_1 to the rest of the input to check that $n_1 = \cdots = n_k$ while decrementing the counter by 1 for each segment it processes. Clearly, $L(M) = L$ and M makes only 1 reversal on the counter. We will show in Proposition 5 that L cannot be accepted by an NCSA.

Example 4. Let $L = \{a^i b^j c^k \mid i, j \geq 1, k = i \cdot j\}$. We can construct a DCSACM(1) M to accept L as follows. M reads a^i and stores a^i in the stack. Then it reads b^j and increments the counter by j. Finally, M reads c^k while moving up and down the stack containing a^i and decrementing the counter by 1 every time the stack has moved i cells, to verify that k is divisible by i and $k/i = j$. Then M

accepts L, and M needs only one 1-reversal counter. We will see in Proposition 29 that L cannot be accepted by a 2DCM(1).

The following shows that, in general, NCSACMs and DCSACMs are computationally more powerful than NCSAs and DCSAs, respectively.

Proposition 5. *There are languages in $\mathcal{L}(DCSACM(1,1)) - \mathcal{L}(NCSA)$. Hence, $\mathcal{L}(DCSA) \subsetneq \mathcal{L}(DCSACM(1,1))$, and $\mathcal{L}(NCSA) \subsetneq \mathcal{L}(NCSACM(1,1))$.*

Proof. Consider the language $L = \{(a^n\#)^n \mid n \geq 1\}$ from Example 3. L cannot be accepted by an NCSA; otherwise, $L' = \{a^{n^2} \mid n \geq 1\}$ can also be accepted by an NCSA (since NCSA languages are closed under homomorphism), but it was shown in [4] that L' cannot be accepted by any NCSA. However, Example 3 showed that L can be accepted by a DCSACM.   □

We now proceed to show that the membership problem for DCSACMs is decidable. In view of Lemma 1, our problem reduces to deciding, given a DCSACM M, whether it accepts λ. So we only need to worry about the operation of the checking stack and the counters. For acceptance of λ, the next lemma provides a normal form.

Lemma 6. *Let M be a DCSACM. We can effectively construct a DCSACM M' such that:*

- *all counters of M' are 1-reversal-bounded and each must return to zero before accepting,*
- *M' always writes on the stack at each step during the writing phase,*
- *the stack head returns to the left end of the stack before accepting,*

whereby M' accepts λ if and only if M accepts λ.

In view of Lemma 6, we may assume that a DCSACM writes a symbol at the end of the stack at each step during the writing phase. This is important for deciding the following problem.

Lemma 7. *Let M be a DCSACM satisfying the assumptions of Lemma 6. We can effectively decide whether or not M on λ input has an infinite writing phase (i.e., will keep on writing).*

From this, decidability of acceptance of λ is straightforward.

Lemma 8. *It is decidable, given a DCSACM M satisfying the assumptions of Lemma 6, whether or not M accepts λ.*

From Lemmas 1, 6, and 8:

Proposition 9. *For $r \geq 1$, the membership problem for r-head 2DCSACM is decidable.*

We now give some undecidability results. The proofs will use the following result in [10]:

Proposition 10 [10]. *It is undecidable, given a 2DCM(2) M over a letter-bounded language, whether $L(M)$ is empty.*

Proposition 11. *The membership problem for NCSACM(2) is undecidable.*

Proof. Let M be a 2DCM(2) machine over a letter-bounded language. Construct from M an NCSACM M' which, on λ - input (i.e. the input is fixed), guesses an input w to M and writes it on its stack. Then M' simulates the computation of M by using the stack and two reversal-bounded counters and accepts if and only if M accepts. Clearly, M' accepts λ if and only if $L(M)$ is not empty which is undecidable by Proposition 10. □

By Propositions 2 and 11, the following is true:

Corollary 12. *The emptiness problem for DCSACM(2) is undecidable.*

The next restriction serves to contrast this undecidability result. Consider an NCSACM where during the reading phase, the stack head crosses the boundary of any two adjacent cells on the stack at most d times for some given $d \geq 1$. Call this machine a d-crossing NCSACM. Then we have:

Proposition 13. *It is decidable, given a d-crossing NCSACM M, whether or not $L(M) = \emptyset$.*

Proof. Define a d-crossing NTMCM to be an nondeterministic Turing machine with a one-way read-only input tape and a d-crossing read/write worktape (i.e., the worktape head crosses the boundary between any two adjacent worktape cells at most d times) augmented with reversal-bounded counters. Note that a d-crossing NCSACM can be simulated by a d-crossing NTMCM. It was shown in [6] that it is decidable, given a d-crossing NTMCM M, whether $L(M) = \emptyset$. The proposition follows. □

Although we have been unable to resolve the open problem as to whether the emptiness is decidable for both NCSACM and DCSACM with one reversal-bounded counter, as with membership for the nondeterministic version, we show they are all equivalent to an open problem in the literature.

Proposition 14. *The following are equivalent:*

1. *the emptiness problem is decidable for 2NCM(1),*
2. *the emptiness problem is decidable for NCSACM(1),*
3. *the emptiness problem is decidable for DCSACM(1),*
4. *the membership problem is decidable for r-head 2NCSACM(1),*
5. *it is decidable if λ is accepted by a NCSACM(1).*

It is indeed a longstanding open problem as to whether the emptiness problem for 2NCM(1) is decidable [10].

Now consider the following three restricted models, with k counters: For $k \geq 1$, a DCSACM(k) (NCSACM(k)) machine is said to be:

- *no-read/no-counter* if it does not read the checking stack nor use any counter before hitting the right input end-marker,
- *no-read/no-decrease* if it does not read the checking stack nor decrease any counter before hitting the right input end-marker,
- *no-read* if it does not read the checking stack before hitting the right input end-marker.

We will consider the families of DCSACM(k) (NCSACM(k)) machines satisfying each of these three conditions.

Proposition 15. *For any $k \geq 1$, every 2DCM(k) machine can be effectively converted to an equivalent no-read/no-decrease DCSACM(k) machine, and vice-versa.*

From this, the following is immediate, since emptiness for 2DCM(1) is known to be decidable [11].

Corollary 16. *The emptiness problem for no-read/no-decrease DCSACM(1) is decidable.*

In the first part of the proof of Proposition 15, the DCSACM(k) machine created from a 2DCM(k) machine was also no-read/no-counter. Therefore, the following is immediate:

Corollary 17. *For $k \geq 1$, the family of languages accepted by the following three sets of machines coincide:*

- *all no-read/no-decrease DCSACM(k) machines,*
- *all no-read/no-counter DCSACM(k) machines,*
- *2DCM(k).*

One particularly interesting corollary of this result is the following:

Corollary 18. *1. The family of languages accepted by no-read/no-decrease (respectively no-read/no-counter) DCSACM(1) is effectively closed under union, intersection, and complementation.*
2. Containment and equivalence are decidable for languages accepted by no-read/no-decrease DCSACM(1) machines.

This follows since this family is equal to 2DCM(1), and these results hold for 2DCM(1) [11]. Something particularly noteworthy about closure of languages accepted by no-read/no-decrease 2DCSACM(1) under intersection, is that, the proof does not follow the usual approach for one-way machines. Indeed, it would be usual to simulate two machines in parallel, each requiring its own counter (and checking stack). But here, only one counter is needed to establish intersection, by using a result on two-way machines. Later, we will show that Corollary 18, part 2 still holds for no-read DCSACM(1)s.

Also, since emptiness is undecidable for 2DCM(2), even over letter-bounded languages [10], the following is true:

Corollary 19. *The emptiness problem for languages accepted by no-read/no-counter* DCSACM(2) *is undecidable, even over letter-bounded languages.*

Turning now to the nondeterministic versions, from the first part of Proposition 15, it is immediate that for any $k \geq 1$, every 2NCM(k) can be effectively converted to an equivalent no-read/no-decrease NCSACM(k). But, the converse is not true combining together the following two facts:

Proposition 20. *1. For every $k \geq 1$, the emptiness problem for languages accepted by* 2NCM(k) *over a unary alphabet is decidable.*
2. The emptiness problem for languages accepted by no-read/no-counter (or no-read/no-decrease) NCSACM(2) *over a unary alphabet is undecidable.*

Proof. The first part was shown in [11]. For the second part, it is known that the emptiness problem for 2DCM(2) M (even over a letter-bounded language) is undecidable by Proposition 10. We construct a no-read/no-counter NCSACM(2) M' which, on a unary input, nondeterministically writes some string w on the stack. Then M' simulates M using w. The result follows since $L(M') = \emptyset$ if and only if $L(M) = \emptyset$. □

In contrast to part 2 of this proposition:

Proposition 21. *For any $k \geq 1$, the emptiness problem for languages accepted by no-read/no-decrease* DCSACM(k) *machines over a unary alphabet, is decidable.*

Proof. If M is a no-read/no-decrease DCSACM(k) over a unary alphabet, we can effectively construct an equivalent 2DCM(k) M (over a unary language) from Proposition 15. The result follows since the emptiness problem for 2NCM(k) over unary languages is decidable [11]. □

Combining these two results yields the following somewhat strange contrast:

Corollary 22. *Over a unary input alphabet and for all $k \geq 2$, the emptiness problem for no-read/no-counter* NCSACM(k)*s is undecidable, but decidable for no-read/no-counter* DCSACM(k)*s.*

As far as we know, this demonstrates the first known example of a family of one-way acceptors where the nondeterministic version has an undecidable emptiness problem, but the deterministic version has a decidable emptiness problem. This presents an interesting contrast to Proposition 2, where it was shown that for complete sets of automata for any store types, the emptiness problem of the deterministic version is decidable if and only if it is decidable for the nondeterministic version. However, the set of unary no-read/no-counter NCSACM(k) machines can be seen to not be a **complete** set of automata, as a complete set of machines contains every possible machine involving a store type. This includes those machines that read input letters while performing read instructions on the checking stack. And indeed, to prove the equivalence of (1) and (2) in Proposition 2, the deterministic machine created reads a letter for every transition applied, which can produce machines that are not of the restriction no-read/no-counter.

When there is only one counter, decidability of the emptiness problem for no-read/no-decrease NCSACM(1), and for no-read/no-counter NCSACM(1) can be shown to be equivalent to all problems listed in Proposition 14. This is because (2) of Proposition 14 implies each immediately, and each implies (1) of Proposition 14, as a 2NCM(1) machine M can be converted to a no-read/no-decrease, or no-read/no-counter NCSACM(1) machine where the input is copied to the stack, and then the 2NCM(1) machine simulated.

Therefore, it is open as to whether the emptiness problem for no-read/no-decrease (or no-read/no-counter) NCSACM(1) is decidable, as this is equivalent to the emptiness problem for 2NCM(1). One might again suspect that decidability of emptiness for no-read/no-decrease DCSACM(1) implies decidability of emptiness for no-read/no-decrease NCSACM(1) by Proposition 2. However, it is again important to note that Proposition 2 only applies to complete sets of machines, including those machines that read input letters while performing read instructions on the checking stack, again violating the 'no-read/no-decrease' condition.

Even though it is open as to whether the emptiness problem is decidable for no-read/no-decrease NCSACM(1)s, we have the following result, which contrasts Corollary 18, part 2:

Proposition 23. *The universe problem is undecidable for no-read/no-counter NCSACM(1)s. (Thus, containment and equivalence are undecidable.)*

Proof. It is known that the universe problem for a one-way nondeterministic 1-reversal-bounded one-counter automaton M is undecidable [1]. Clearly, we can construct a no-read/no-counter NCSACM(1) M' to simulate M.   □

In the definition of a no-read/no-decrease DCSACM, we imposed the condition that the counters can only decrement when the input head reaches the end-marker. Consider the weaker condition no-read, i.e., the only requirement is that the machine can only enter the stack when the input head reaches the end-marker, but there is no constraint on the reversal-bounded counters. It is an interesting open question about whether no-read DCSACM(k) languages are also equivalent to a 2DCM(k) (we conjecture that they are equivalent). However, the following stronger version of Corollary 16 can be proven.

Proposition 24. *The emptiness problem is decidable for no-read DCSACM(1)s.*

We can further strengthen Proposition 24 somewhat. Define a restricted no-read NCSACM(1) to be a no-read NCSACM(1) which is only nondeterministic during the writing phase. Then the proof of Proposition 24 applies to the following, as the sequence of transition symbols used in the proof can be simulated deterministically:

Proposition 25. *The emptiness problem is decidable for languages accepted by restricted no-read NCSACM(1)s.*

While we are unable to show that the intersection of two no-read DCSACM(1) languages is a no-read DCSACM(1) language, we can prove:

Proposition 26. *It is decidable, given two no-read DCSACM(1)s M_1 and M_2, whether $L(M_1) \cap L(M_2) = \emptyset$.*

One can show that no-read DCSACM(1) languages are effectively closed under complementation. Thus, from Proposition 26:

Corollary 27. *The containment and equivalence problems are decidable for no-read DCSACM(1)s.*

No-read DCSACM(1) is indeed quite a large family for which emptiness, equality, and containment are decidable. The proof of Proposition 26 also applies to the following:

Proposition 28. *It is decidable, given two restricted no-read NCSACM(1)s M_1 and M_2, whether $L(M_1) \cap L(M_2) = \emptyset$.*

Finally, consider the general model DCSACM(1) (i.e., unrestricted). While it is open whether no-read DCSACM(1) is equivalent to 2DCM(1), we can prove:

Proposition 29. $\mathcal{L}(2DCM(1)) \subsetneq \mathcal{L}(DCSACM(1))$.

Proof. It is obvious that any 2DCM(1) can be simulated by a DCSACM(1) (in fact by a no-read/no-counter DCSACM(1)). Now let $L = \{a^i b^j c^k \mid i, j \geq 1, k = i \cdot j\}$. We can construct a DCSACM(1) M to accept L by Example 4. However, it was shown in [5] that L cannot be accepted by a 2DCM(1) by a proof that shows that if L can be accepted by a 2DCM(1), then one can use the decidability of the emptiness problem for 2DCM(1)s to show that Hilbert's Tenth Problem is decidable. □

4 Multiple Checking-Stacks with Reversal-Bounded Counters

In this section, we will study deterministic and nondeterministic k-checking-stack machines. These are defined by using multiple checking stack stores. Implied from this definition is that each stack has a "writing phase" followed by a "reading phase", but these phases are independent for each stack.

A k-stack DCSA (NCSA respectively) is the deterministic (nondeterministic) version of this type of machine. The two-way versions (with input end-markers) are called k-stack 2DCSA and k-stack 2NCSA, respectively. These k-stack models can also be augmented with reversal-bounded counters and are called k-stack DCSACM, k-stack NCSACM, k-stack 2DCSACM, and k-stack 2NCSACM.

Consider a k-stack DCSACM M. By Lemma 1, for the membership problem, we need only investigate whether λ is accepted. Also, as in Lemma 6, we may assume that each stack pushes a symbol at each move during its writing phase, and that all counters are 1-reversal-bounded.

We say that M has an infinite writing phase (on λ input) if no stack enters a reading phase. Thus, all stacks will keep on writing a symbol at each step. If M has a finite writing phase, then directly before a first such stack enters its reading phase, all the stacks would have written strings of the same length.

Lemma 30. *Let $k \geq 1$ and M be a $(k + 1)$-stack DCSACM M satisfying the assumption of Lemma 6.*

1. *We can determine if M has an infinite writing phase. If so, M does not accept λ.*
2. *If M has a finite writing phase, we can construct a k-stack DCSACM M'' such that M'' accepts λ if and only if M accepts λ.*

Notice that M'' has fewer stacks than M. Then, from Proposition 9 (the result for a single stack) and using Lemma 30 recursively:

Proposition 31. *The membership problem for k-stack DCSACMs is decidable.*

Then, by Lemma 1:

Corollary 32. *The membership problem for r-head k-stack 2DCSACM is decidable.*

References

1. Baker, B.S., Book, R.V.: Reversal-bounded multipushdown machines. J. Comput. Syst. Sci. **8**(3), 315–332 (1974)
2. Engelfriet, J.: The power of two-way deterministic checking stack automata. Inf. Comput. **80**(2), 114–120 (1989)
3. Ginsburg, S.: Algebraic and Automata-Theoretic Properties of Formal Languages. North-Holland Publishing Company, Amsterdam (1975)
4. Greibach, S.: Checking automata and one-way stack languages. J. Comput. Syst. Sci. **3**(2), 196–217 (1969)
5. Gurari, E.M., Ibarra, O.H.: Two-way counter machines and diophantine equations. J. ACM **29**(3), 863–873 (1982)
6. Harju, T., Ibarra, O., Karhumäki, J., Salomaa, A.: Some decision problems concerning semilinearity and commutation. J. Comput. Syst. Sci. **65**(2), 278–294 (2002)
7. Hopcroft, J.E., Ullman, J.D.: Introduction to Automata Theory, Languages, and Computation. Addison-Wesley, Reading (1979)
8. Ibarra, O., McQuillan, I.: On store languages of language acceptors (2017, submitted). A preprint appears in https://arxiv.org/abs/1702.07388
9. Ibarra, O., McQuillan, I.: Variations of checking stack automata: Obtaining unexpected decidability properties (2017). Full version of current paper with all definitions and proofs. https://arxiv.org/abs/1705.09732
10. Ibarra, O.H.: Reversal-bounded multicounter machines and their decision problems. J. ACM **25**(1), 116–133 (1978)
11. Ibarra, O.H., Jiang, T., Tran, N., Wang, H.: New decidability results concerning two-way counter machines. SIAM J. Comput. **23**(1), 123–137 (1995)

The Generalized Rank of Trace Languages

Michal Kunc[(⊠)] and Jan Meitner

Department of Mathematics and Statistics, Faculty of Science,
Masaryk University, Kotlářská 2, 611 37 Brno, Czech Republic
{kunc,xmeitner}@math.muni.cz

Abstract. The notion of rank of a language with respect to an independence alphabet is generalized from concatenations of two words to an arbitrary fixed number of words. It is proved that in the case of free commutative monoids, as well as in the more general case of direct products of free monoids, sequences of ranks of regular languages are exactly non-decreasing sequences that are eventually constant. On the other hand, by uncovering a relationship between rank sequences of regular languages and rational series over the min-plus semiring, it is shown that already for free products of free commutative monoids, rank sequences need not be eventually periodic.

Keywords: Trace language · Rank · Regular language · Rational series · Tropical semiring

1 Introduction

It is a classical observation that the commutative closure of a regular set of words is not necessarily regular; for instance, the closure of $(ab)^*$ consists of all words where both letters a and b have the same number of occurrences. In the more general setting of trace alphabets, where some pairs of letters commute and some do not, the most useful technique for proving regularity of the closure of a given regular set of words under the trace equivalence is based on the notion of rank, introduced by Hashiguchi [4]. The idea of this notion is to cut an arbitrary word from the closure into a prefix and a suffix, and then ask into how many segments the prefix and the suffix have to be split, so that some word from the original set can be obtained by shuffling independent segments. In the above mentioned case of the language $(ab)^*$, when the word $a^n b^n$ is cut right in the middle, each copy of the letter a has to be moved separately into a different position between copies of b on the right to obtain a word from $(ab)^*$; this shows that the rank of $(ab)^*$ is unbounded. On the other hand, if a regular language $L \subseteq \Sigma^*$ has finite rank n, one can consider for each word all possible decompositions to n segments, and in this way obtain an equivalence relation on Σ^* of finite index such that equivalent words reach the same state in the minimal automaton of the closure

This work was supported by grant 15-02862S of the Czech Science Foundation.

E. Charlier et al. (Eds.): DLT 2017, LNCS 10396, pp. 247–259, 2017.
DOI: 10.1007/978-3-319-62809-7_18

of L; in other words, this equivalence relation is finer than the syntactic right congruence of the closure. In particular, this implies that the closure is regular.

The crucial property of the rank is that its finiteness is preserved by several basic language-theoretic operations, most importantly concatenation of arbitrary languages and Kleene star of languages consisting solely of connected words (the behaviour of the rank with respect to operations was studied in detail by Klunder et al. [5]). This property makes rank the best tool for proving characterizations of recognizable sets in trace monoids by certain rational expressions (see [2]). The appropriate generalization of rank was used by Droste and Kuske [3] to obtain such a characterization of recognizable sets also for a larger class of monoids, so-called divisibility monoids. The notion of rank turned out to be useful also in another more general setting of semi-commutations. In particular, it was used by Ochmański and Wacrenier [7] to solve the problem of characterizing pairs of semi-commutation relations such that for every regular language closed under the first relation, its closure under the second relation is again regular.

In spite of a significant interest in this notion, answers to many basic questions about the rank are not known. For instance, while it is known due to Sakarovitch [8] that trace alphabets for which one can algorithmically decide whether the closure of a given regular language is regular are precisely those that define free products of free commutative monoids, decidability of finiteness of the rank of a regular language is an open problem.

Once it is known that a language has a finite rank, one can ask whether there also exists a bound on the required number of segments when a word from the closure is cut into three or more pieces instead of just two. Such decompositions are often studied in the theory of regular languages; in particular, the definition of a syntactic monoid is based on decomposing words into triples, and if the number of segments were bounded also in this case, it would be possible to construct an equivalence relation of finite index finer than the syntactic congruence of the closure, similarly to the construction of an equivalence relation finer than the syntactic right congruence described above. However, it turns out that there exist regular languages with finite rank, where decompositions into triples require an unbounded number of segments. This leads us to consider the generalized notion of rank for an arbitrary fixed number of cuts. In this way, for each independence alphabet, every language is assigned a sequence of non-negative integers (possibly including infinity), expressing how the required number of segments increases with respect to the number of allowed cuts.

This paper studies basic properties of these rank sequences. The generalized rank is defined in Sect. 2. In Sect. 3 it is shown that both in the case of free commutative monoids and in the case of direct products of free monoids (that is, trace alphabets with a transitive dependence relation), rank sequences of regular languages are exactly all non-decreasing and eventually constant sequences. However, it turns out that, in general, rank sequences need not be even eventually periodic; more precisely, in Sect. 4, it is proved that every sequence obtained from some rational series over the min-plus semiring by taking the nth element equal to the maximum of all coefficients of words of length at most n, is the rank sequence of some regular language with respect to a transitive independence relation (that is, for a free product of free commutative monoids).

2   Definition

The paper deals with the monoid Σ^* consisting of all words over a finite alphabet Σ. Any subset L of Σ^* is called a *word language*. The number of occurrences of a letter $a \in \Sigma$ in a word $w \in \Sigma^*$ is referred to as $|w|_a$. For $w \in \Sigma^*$ and $\Gamma \subseteq \Sigma$, the notation $\pi_\Gamma(w)$ stands for the word obtained from w by deleting all letters not belonging to Γ. The empty word is denoted by ε.

Trace monoids are determined by prescribing which pairs of letters commute. This is expressed by an irreflexive and symmetric binary relation I on Σ, called an *independence relation*. Its complement $D = (\Sigma \times \Sigma) \setminus I$ is the corresponding *dependence relation*. The pair (Σ, I) is called an *independence alphabet*. Let $\sim_I$ be the congruence of Σ^* generated by the set $\{ (ab, ba) \mid (a, b) \in I \}$. The *trace monoid* $M(\Sigma, I)$ is then defined as the quotient monoid $\Sigma^* / \sim_I$. Thus, elements of $M(\Sigma, I)$, called *traces*, are precisely classes of words that can be obtained from each other by successively commuting some neighbouring independent letters.

Trace languages are arbitrary subsets of $M(\Sigma, I)$. Trace languages can be also viewed as sets of words closed under $\sim_I$. In order to describe a trace language T, it is sufficient to provide an arbitrary word language L such that T consists precisely of those traces which have some representative in L; viewed as a set of words, the trace language T is then equal to the closure of L under $\sim_I$, that is $\{ w \in \Sigma^* \mid \exists u \in L : w \sim_I u \}$, which will be denoted by $\overline{L}$.

The definition of the rank of a word language L with respto an independence alphabet is based on the following characterization of equivalence of a concatenation of several words to a given word.

Lemma 1. *Let (Σ, I) be an independence alphabet, m a positive integer, and $w, u_0, \ldots, u_m \in \Sigma^*$ words. The equivalence $u_0 \ldots u_m \sim_I w$ holds if and only if there exist a non-negative integer n and words $u_{0,0}, \ldots, u_{0,n}, \ldots, u_{m,0}, \ldots, u_{m,n} \in \Sigma^*$ such that*

1. *$u_k \sim_I u_{k,0} \ldots u_{k,n}$ for all $k \in \{0, \ldots, m\}$,*
2. *$u_{0,0} \ldots u_{m,0} \ldots u_{0,n} \ldots u_{m,n} = w$,*
3. *for all $k, \ell \in \{0, \ldots, m\}$ satisfying $k > \ell$ and $i, j \in \{0, \ldots, n\}$ satisfying $i < j$, all letters of $u_{k,i}$ are independent of all letters of $u_{\ell,j}$.*

The rank of L expresses into how many segments words $u_0, \ldots, u_m$ in Lemma 1 have to be cut, so that the segments can be shuffled to get some word $w \in L$.

Definition 2. *Let (Σ, I) be an independence alphabet and m a non-negative integer. Assume that $L \subseteq \Sigma^*$ and $u_0, \ldots, u_m \in \Sigma^*$ are such that $u_0 \ldots u_m \in \overline{L}$. Define $\mathrm{Rank}(L; u_0, \ldots, u_m)$ as the smallest non-negative integer n such that there exist words $w \in L$ and $u_{0,0}, \ldots, u_{m,n} \in \Sigma^*$ satisfying conditions 1–3 of Lemma 1.*

The m-rank of L is defined as the supremum of all values Rank
$(L; u_0, \ldots, u_m)$ *for admissible words* $u_0, \ldots, u_m \in \Sigma^*$*, that is,* $\text{Rank}_m(L)$ *is equal to*

$$
\begin{cases}
\max\{\, \text{Rank}(L; u_0, \ldots, u_m) \mid u_0 \ldots u_m \in \overline{L} \,\} & \text{if the maximum exists,} \\
0 & \text{if } L = \emptyset, \\
\omega & \text{otherwise.}
\end{cases}
$$

The original rank introduced by Hashiguchi [4] coincides with the 1-rank of the above definition. For $m = 0$, one obtains $\text{Rank}_0(L) = 0$ for every language L, and consequently, we will be interested only in $\text{Rank}_m(L)$ for positive integers m; however, 0-rank will be employed in formulas for calculating the rank of more complex languages. Note that the value $\text{Rank}(L; u_0, \ldots, u_m)$ depends only on non-empty words in the sequence $u_0, \ldots, u_m$, that is, it does not change if some empty words are added or omitted. This in particular implies that for every language L, the rank sequence $(\text{Rank}_m(L))_{m=1}^{\infty}$ is non-decreasing. Another consequence is that the rank sequence of a finite language L is eventually constant.

Example 3. Consider the word language $L = a^* b^* a^*$ over the alphabet $\Sigma = \{a, b\}$, with $(a, b) \in I$. Its closure $\overline{L}$ is the whole set Σ^*. The value of $\text{Rank}_1(L)$ is 0, because for arbitrary words $u_0, u_1 \in \Sigma^*$, the words $u_{0,0} = a^{|u_0|_a} b^{|u_0|_b}$ and $u_{1,0} = b^{|u_1|_b} a^{|u_1|_a}$ satisfy the conditions of Lemma 1 for $n = 0$ and $w = a^{|u_0|_a} b^{|u_0 u_1|_b} a^{|u_1|_a}$.

Now take three words $u_0 = u_2 = b$ and $u_1 = a$, satisfying $u_0 u_1 u_2 \in \overline{L}$. These words are the only representatives of their traces. Therefore, bab is the only word that can be obtained from $u_0 u_1 u_2$ by reordering letters within the factors u_0, u_1 and u_2. However, this word does not belong to L, which shows that $\text{Rank}_2(L) > 0$. In order to show that $\text{Rank}_m(L) = 1$ for all $m \geq 2$, it is sufficient to provide appropriate decompositions of arbitrary words $u_0, \ldots, u_m \in \Sigma^*$, for instance $u_{k,0} = a^{|u_k|_a}$ and $u_{k,1} = b^{|u_k|_b}$ for $k \in \{0, \ldots, m\}$.

3   Direct Products of Free Monoids

This section is devoted to the description of rank sequences of arbitrary word languages over trace alphabets whose dependence relation is transitive; the corresponding trace monoids are direct products of free monoids over the connected components of the graph (Σ, D). It turns out that all these rank sequences are eventually constant, that is, there exists some positive integer m_0 such that for all $m \geq m_0$, the values of $\text{Rank}_m(L)$ are equal. Since rank sequences are always non-decreasing, this is actually equivalent to saying that one of the following two possibilities arises: either there exists some m for which $\text{Rank}_m(L) = \omega$, or the sequence is bounded. The following definition describes the property of a language that decides whether the former or the latter possibility is true.

Definition 4. *Let* (Σ, I) *be an independence alphabet, with D transitive. The alternation complexity of a word language $L \subseteq \Sigma^*$ with respect to I is defined* *as the least non-negative integer n such that for each word $w \in \overline{L}$ there exist words $v_0, \ldots, v_n \in \Sigma^*$ such that $v_0 \ldots v_n \in L$, $v_0 \ldots v_n \sim_I w$ and for each $i \in \{0, \ldots, n\}$, all letters occurring in v_i are pairwise dependent. If such an integer n does not exist, then we say that the alternation complexity is infinite.*

Proposition 5. *Let (Σ, I) be an independence alphabet, with D transitive, and let $L \subseteq \Sigma^*$ be a word language. If the alternation complexity of L is an integer n, then $\mathrm{Rank}_m(L) \leq n$ for every m. If the alternation complexity of L is infinite, then $\mathrm{Rank}_{d-1}(L) = \omega$, where d is the number of connected components of (Σ, D).*

Proof. First assume that the alternation complexity of L is an integer n. Consider arbitrary words $u_0, \ldots, u_m \in \Sigma^*$ such that $u_0 \ldots u_m \in \overline{L}$. The definition of alternation complexity provides us with words $v_0, \ldots, v_n \in \Sigma^*$ such that $v_0 \ldots v_n \in L$, $u_0 \ldots u_m \sim_I v_0 \ldots v_n$ and for each $i \in \{0, \ldots, n\}$, all letters in v_i are pairwise dependent. Each of the required words $u_{j,i}$ can then be chosen so that it consists precisely of occurrences of letters common to u_j and v_i.

Now assume that $\mathrm{Rank}_{d-1}(L)$ is equal to a positive integer n, and let us prove that the alternation complexity of L is finite. Let $\Sigma_0 \cup \ldots \cup \Sigma_{d-1}$ be the decomposition of Σ to components of the graph (Σ, D), that is, $\Sigma_k \times \Sigma_k \subseteq D$ and $\Sigma_k \times \Sigma_\ell \subseteq I$ for $k \neq \ell$. Take any $v \in \overline{L}$ and consider words $u_k = \pi_{\Sigma_k}(v)$ for $k = 0, \ldots, d-1$. The equality $\mathrm{Rank}_{d-1}(L) = n$, applied to the sequence $u_0, \ldots, u_{d-1}$, provides us with certain words $u_{0,0}, \ldots, u_{d-1,n} \in \Sigma^*$ such that the word $w = u_{0,0} \ldots u_{d-1,0} \ldots u_{0,n} \ldots u_{d-1,n}$ belongs to L. Since all letters in different words u_k are pairwise independent, the word v represents the same trace as the word $u_0 \ldots u_{d-1}$, which is $\sim_I$-equivalent to w by conditions 1 and 3 of Lemma 1. Because all letters of each word $u_{k,i}$ are pairwise dependent, this shows that the alternation complexity of L is at most $d \cdot (n+1) - 1$. □

In the rest of this section, it will be shown that every non-decreasing and eventually constant sequence of non-negative integers and infinity is in fact a rank sequence of some regular language over an alphabet with all letters independent. Such languages will be constructed by combining several languages whose rank sequences begin with finitely many zeros, then jump to a prescribed value n, either a positive integer or infinity, and continue as constant sequences forever. With this aim, let m be a positive integer and consider the independence alphabet (Σ, I) with $\Sigma = \{a_0, a_1, \ldots, a_m\}$ and all letters independent.

First, we deal with the simpler case of $n = \omega$. In this case, the language with the required properties is $L_m = \bigcup_{\ell=0}^m (L_{m,\ell})^m$, where

$$L_{m,\ell} = \prod \{ a_i^* (a_\ell a_i)^* \mid i = 0, \ldots, m, \ i \neq \ell \}.$$

The basic idea is that for any given words $u_0, \ldots, u_{m-1} \in \Sigma^*$, there exists $\ell \in \{0, \ldots, m\}$ such that in each word u_j, there is a letter different from a_ℓ with at least the same number of occurrences. This allows one to produce a word from $L_{m,\ell}$ by commuting letters in u_j. On the other hand, such an index ℓ does not exist in the case of words $u_j = a_j^n$, for $j = 0, \ldots, m$.

Lemma 6. *For* $k < m$, $\mathrm{Rank}_k(L_m) = 0$, *and for* $k \geq m$, $\mathrm{Rank}_k(L_m) = \omega$. *Moreover,* $\mathrm{Rank}(L_m; a_0^n, \ldots, a_m^n) = n - 1$ *holds for every positive integer* n.

A regular language, whose rank sequence begins with finitely many zeros and continues to infinity with values equal to a given positive integer n, is constructed by adding some words to the language L_m in order to decrease the infinite values of ranks. These additional words are prescribed by the language

$$K_{m,n} = \{\, w \in \Sigma^* \mid |w|_{a_0} < n \text{ or } |w|_{a_1} < n \,\}$$
$$\cup \{\, w \in \Sigma^* \mid \pi_{a_0,a_1}(w) \in \{a_0, a_1\}^* a_0 a_1^+ (a_0 a_1)^{n-1} \,\}$$
$$\cup \{\, w \in \Sigma^* \mid \pi_{a_0,a_1}(w) \in \{a_0, a_1\}^* a_1 a_0^+ (a_1 a_0)^{n-1} \,\}.$$

Lemma 7. *The languages* L_m *and* $K_{m,n}$ *satisfy* $\overline{L_m \cup K_{m,n}} = \Sigma^*$, $\mathrm{Rank}_k(L_m \cup K_{m,n}) = 0$ *for* $k < m$ *and* $\mathrm{Rank}_k(L_m \cup K_{m,n}) = n - 1$ *for* $k \geq m$. *Moreover,* $\mathrm{Rank}(L_m \cup K_{m,n}; a_0^h, \ldots, a_m^h) = n - 1$ *holds for every* $h \geq n$.

Theorem 8. *For every infinite sequence whose each member is a non-negative integer or* ω, *the following statements are equivalent.*

1. *The sequence is the rank sequence of some language with respect to a trace alphabet with a transitive dependence relation.*
2. *The sequence is the rank sequence of a regular language with respect to a trace alphabet with all distinct letters pairwise independent.*
3. *The sequence is non-decreasing and eventually constant.*

Proof. The first statement implies the third one by Proposition 5. In order to prove that the third statement implies the second one, observe that every non-decreasing and eventually constant sequence can be obtained by taking the maximum of finitely many sequences, each of which consists of finitely many zeros, followed by an infinite constant sequence. According to Lemmata 6 and 7, every such sequence is the rank sequence of a regular language with respect to a trace alphabet with all letters independent. A regular language for the original sequence can then be obtained as the union of these languages, considered over disjoint alphabets. The proof of the theorem is complete, since the first statement is a trivial consequence of the second one. □

4   Free Products of Free Commutative Monoids

This section deals with rank sequences of regular languages with respect to trace alphabets defining free products of free commutative monoids, that is, such that the relation $I \cup \mathrm{id}_\Sigma$ is transitive. The goal of this section is to show that for these trace alphabets there is a close relationship between rank sequences of some regular languages and rational series over the min-plus semiring. This relationship is used to prove that there exist regular languages such that their rank sequences with respect to these trace alphabets are not eventually periodic, as their growth rate is sublinear.

4.1   Rational Series Over the Min-plus Semiring

Let us begin by introducing notation for rational series; for a more detailed introduction to rational series the reader is referred to the book by Berstel and Reutenauer [1]. Recall that the min-plus semiring $\mathbb{T}$ consists of all non-negative integers together with the additional element ∞. In our setting, it is more natural to work with the extended min-plus semiring $\mathbb{T}_\omega$, as introduced by Leung [6], which additionally contains the element ω. In this semiring, the minimum operation is defined according to the ordering $0 < 1 < \ldots < \omega < \infty$, and the addition operation is defined in the usual way, with the additional rules $\omega + n = \max\{\omega, n\}$ for all $n \in \mathbb{T}_\omega$.

A $(\min, +)$-automaton $\mathcal{S} = (\Gamma, Q, E, \iota, \tau)$ over an alphabet Γ consists of a finite set of states Q, a weighted transition relation $E : Q \times \Gamma \times Q \to \mathbb{T}$ and mappings $\iota, \tau : Q \to \mathbb{T}$ determining initial and terminal weights of states, respectively. Here, labelling a transition with the element ∞ is used to express that the transition actually does not exist, while the element ω will serve to express that certain sets of finite weights are unbounded. Since we are not interested in the constant coefficient of the series, we can assume, without loss of generality, that $\iota, \tau : Q \to \{0, \infty\}$, and we will view both ι and τ as sets of states, consisting precisely of states mapped to 0. The automaton $\mathcal{S}$ defines a rational series, which is a mapping $\mathcal{S} : \Gamma^+ \to \mathbb{T}$. For a positive integer n and letters $\gamma_1, \ldots, \gamma_n \in \Gamma$, the value $\mathcal{S}(\gamma_1 \ldots \gamma_n)$ is defined as the minimum of the sums

$$E(q_0, \gamma_1, q_1) + E(q_1, \gamma_2, q_2) + \ldots + E(q_{n-1}, \gamma_n, q_n),$$

over all choices of states $q_0, \ldots, q_n \in Q$ with $q_0 \in \iota$ and $q_n \in \tau$, if such a choice exists, and as ∞ otherwise.

In what follows we will be particularly interested in the growth of the coefficients of the series $\mathcal{S}$. This will be expressed by the notation $\mathcal{S}_{\max}(m)$, for a positive integer m, which stands for the maximum of the values $\mathcal{S}(w)$, for non-empty words w of length at most m and such that $\mathcal{S}(w) \neq \infty$; if $\mathcal{S}(w) = \infty$ for all w of length at most m, then $\mathcal{S}_{\max}(m)$ is defined as 0.

For our purposes, it will be useful to view the $(\min, +)$-automaton $\mathcal{S}$ as a homomorphism $\varphi_{\mathcal{S}}$ from Γ^+ to the monoid $\mathcal{M}_Q$ of $Q \times Q$ matrices over the extended min-plus semiring $\mathbb{T}_\omega$ (with matrix multiplication as operation), defined by the rule $\varphi_{\mathcal{S}}(\gamma)(p, q) = E(p, \gamma, q)$ for $\gamma \in \Gamma$ and $p, q \in Q$. For every matrix $M \in \mathcal{M}_Q$, let the notation $\min_{\iota, \tau}(M)$ stand for the value $\min\{ M(p, q) \mid p \in \iota, q \in \tau \}$. The formula for calculating the coefficients of the series $\mathcal{S}$ can now be written as $\mathcal{S}(w) = \min_{\iota, \tau}(\varphi_{\mathcal{S}}(w))$. Consequently, the value $\mathcal{S}_{\max}(m)$ is equal to the maximum of all values $\min_{\iota, \tau}(M)$ smaller than ∞, over all matrices M that can be obtained as a product of at most m matrices from the set $\varphi_{\mathcal{S}}(\Gamma)$.

In the following construction, we need a generalization of the monoid of $(\min, +)$-matrices $\mathcal{M}_Q$ that allows to provide every matrix with a certain price that has to be paid each time this matrix is used. This generalization $\mathcal{N}_Q$ is obtained as a direct product of $\mathcal{M}_Q$ and the additive monoid of non-negative integers $(\mathbb{N}_0, +)$. The projection homomorphisms from $\mathcal{N}_Q$ to $\mathcal{M}_Q$ and $\mathbb{N}_0$ will be denoted by μ and π, respectively. A $(\min, +)$-automaton $\mathcal{S}$ can be viewed as a homomorphism $\psi_S : \Gamma^+ \to \mathcal{N}_Q$ defined by $\psi_S(\gamma) = (\varphi_S(\gamma), 1)$ for $\gamma \in \Gamma$. Words having an accepting path in $\mathcal{S}$ are precisely those whose image under this homomorphism belongs to the subset $\mathcal{N}_Q^{(\iota,\tau)\text{-acc}}$ of $\mathcal{N}_Q$, which consists of all elements (M, m) such that $M(p, q) \neq \infty$ for some $p \in \iota$ and $q \in \tau$. Using ψ_S, the value $\mathcal{S}_{\max}(m)$ can be calculated as the maximum of values $\min_{\iota,\tau}(\mu(\psi_S(w)))$, for all $w \in \Gamma^+$ such that $\psi_S(w) \in \mathcal{N}_Q^{(\iota,\tau)\text{-acc}}$ and $\pi(\psi_S(w)) \leq m$. Equivalently, the formula for calculating $\mathcal{S}_{\max}(m)$ can be written in terms of the subsemigroup $\langle \psi_S(\Gamma) \rangle$ of $\mathcal{N}_Q$ generated by images of letters under ψ_S:

$$\mathcal{S}_{\max}(m) = \max\{ \min_{\iota,\tau}(\mu(N)) \mid N \in \langle \psi_S(\Gamma) \rangle \cap \mathcal{N}_Q^{(\iota,\tau)\text{-acc}}, \ \pi(N) \leq m \}. \quad (1)$$

4.2   Representing Rank Sequences by Rational Series

Assume that (Σ, I) is an independence alphabet, with $I \cup \mathrm{id}_\Sigma$ transitive. This means that Σ is a disjoint union of subalphabets Σ_i for $i = 1, \ldots, c$ such that $\Sigma_i \times \Sigma_i \subseteq I \cup \mathrm{id}_\Sigma$ and $\Sigma_i \times \Sigma_j \subseteq D$ for $i \neq j$. Every non-empty word u over Σ can be uniquely factorized as $u = u_1 \ldots u_\ell$, with u_k a non-empty word for $k \in \{1, \ldots, \ell\}$, where each word u_k is a maximal factor of u consisting of letters from the same subalphabet Σ_{i_k}, and $i_k \neq i_{k+1}$. Then all words from the trace of u are obtained by arbitrarily changing the order of letters within each factor u_k.

Assume that a language $L \subseteq \Sigma^*$ is given by a generalized non-deterministic automaton $\mathcal{A} = (\Sigma, Q, F, \iota, \tau)$, with each edge labelled by an ε-free regular language over one of the subalphabets Σ_i, and such that no two consecutive edges are labelled by languages over the same subalphabet. In other words, the transition relation is $F : Q \times Q \to \mathcal{P}(\Sigma^+)$, with each $F(p, q)$ a regular subset of some Σ_i^+, and it satisfies the condition that if $\emptyset \neq F(p, q) \subseteq \Sigma_i^+$ and $\emptyset \neq F(q, r) \subseteq \Sigma_j^+$, then $i \neq j$. Note that such an automaton exists for every regular language L.

Example 9. Let the independence alphabet be $(\{a, b, c\}, I)$, with only a and b independent. The language $L = (abc)^*$ is defined by the generalized automaton

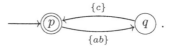

Our goal is to reformulate the calculation of ranks for the language L and $(m + 1)$-tuples of words over Σ in terms of ranks for the languages $F(p, q)$ and sequences of words over the corresponding subalphabets Σ_i. This is achieved by introducing a one-to-one correspondence that relates every sequence of words over Σ with the sequence of its maximal subsequences consisting of letters over the same subalphabet Σ_i, as demonstrated by the following example.

Example 9 (continued). For the sequence $abcb, a, c$ over $\{a, b, c\}$, the associated sequence of sequences over $\{a, b\}$ and $\{c\}$ is $(ab), (c), (b, a), (\varepsilon, c)$. Note that one can recover the original sequence from the new one by concatenating all words between commas inside particular sequences: $ab \cdot c \cdot b, a \cdot \varepsilon, c$.

Ranks of sequences of words over Σ_i are encoded into elements of $\mathcal{N}_Q$: for $i \in \{1, \ldots, c\}$ and $w_0, \ldots, w_m \in \Sigma_i^*$, with $m \geq 0$ and $w_m \neq \varepsilon$, let $N(w_0, \ldots, w_m)$ denote the element (M, m) of $\mathcal{N}_Q$, where the matrix M is defined by the rule

$$M(p, q) = \begin{cases} \mathrm{Rank}(F(p, q); w_0, \ldots, w_m) & \text{if } w_0 \ldots w_m \in \overline{F(p, q)}, \\ \infty & \text{otherwise.} \end{cases}$$

Let $G_{\mathcal{A}}$ be the subset of $\mathcal{N}_Q$ consisting of all such elements $N(w_0, \ldots, w_m)$.

Example 9 (continued). Let us describe all elements of $G_{\mathcal{A}}$ for $L = (abc)^*$. If the word $w_0 \ldots w_m$ belongs neither to $\overline{\{ab\}} = \{ab, ba\}$, nor to $\overline{\{c\}} = \{c\}$, then $N(w_0, \ldots, w_m) = \left(\left(\begin{smallmatrix} \infty & \infty \\ \infty & \infty \end{smallmatrix} \right), m \right)$. If $w_0 \ldots w_m = ab$, then $N(w_0, \ldots, w_m) = \left(\left(\begin{smallmatrix} \infty & 0 \\ \infty & \infty \end{smallmatrix} \right), m \right)$. The same result is obtained if $w_0 \ldots w_m = ba$ and both letters belong to the same word w_i. If $w_0 \ldots w_m = ba$ and letters a and b belong to different words w_i, then $N(w_0, \ldots, w_m) = \left(\left(\begin{smallmatrix} \infty & 1 \\ \infty & \infty \end{smallmatrix} \right), m \right)$ (note that m is at least 1 in this case). Finally, if $w_0 \ldots w_m = c$, then $N(w_0, \ldots, w_m) = \left(\left(\begin{smallmatrix} \infty & \infty \\ 0 & \infty \end{smallmatrix} \right), m \right)$.

Proposition 10. *For every positive integer m,*

$$\mathrm{Rank}_m(L) = \sup\{ \min_{\iota, \tau}(\mu(N)) \mid N \in \langle G_{\mathcal{A}} \rangle \cap \mathcal{N}_Q^{(\iota, \tau)\text{-acc}}, \ \pi(N) \leq m \}. \quad (2)$$

The main drawback of formula (2) is that the set $G_{\mathcal{A}}$ is infinite. If the rank sequence of every language $F(p, q)$ is bounded, then $G_{\mathcal{A}}$ can be replaced with its finite subset consisting of elements minimal with respect to a suitable ordering of $\mathcal{N}_Q$. Moreover, this set can be further modified, so that the second component of each element is equal to 1: elements with second component 0 can be joined with other elements, while elements with second component greater than 1 have to be split, which can be achieved by extending the set of states Q.

Example 9 (continued). The minimal elements of $G_{\mathcal{A}}$ are

$$\left(\left(\begin{smallmatrix} \infty & \infty \\ \infty & \infty \end{smallmatrix} \right), 0 \right), \ \left(\left(\begin{smallmatrix} \infty & 0 \\ \infty & \infty \end{smallmatrix} \right), 0 \right), \ \left(\left(\begin{smallmatrix} \infty & 1 \\ \infty & \infty \end{smallmatrix} \right), 1 \right) \text{ and } \left(\left(\begin{smallmatrix} \infty & \infty \\ 0 & \infty \end{smallmatrix} \right), 0 \right).$$

After multiplying the only element having non-zero second coordinate by other elements, and then removing them, this list changes to

$$\left(\left(\begin{smallmatrix} \infty & 1 \\ \infty & \infty \end{smallmatrix} \right), 1 \right), \left(\left(\begin{smallmatrix} \infty & \infty \\ \infty & \infty \end{smallmatrix} \right), 1 \right), \left(\left(\begin{smallmatrix} \infty & \infty \\ \infty & 1 \end{smallmatrix} \right), 1 \right), \left(\left(\begin{smallmatrix} 1 & \infty \\ \infty & \infty \end{smallmatrix} \right), 1 \right). \quad (3)$$

The only elements of the subsemigroup generated by these elements that belong to $\mathcal{N}_Q^{(\iota, \tau)\text{-acc}}$ are $\left(\left(\begin{smallmatrix} m & \infty \\ \infty & \infty \end{smallmatrix} \right), m \right)$ for positive integers m. Formula (2) now gives $\mathrm{Rank}_m(L) = m$. Note that these values of ranks come from products alternating elements $N(b, a)$ and $N(c)$, which expresses the necessity of interchanging the order of a and b in each factor ba split between consecutive words u_k and u_{k+1}.

Once the modified set $G_{\mathcal{A}}$ in formula (2) consists of finitely many elements whose second component is 1, this formula becomes essentially the same as formula (1), giving the following result.

Theorem 11. *Let (Σ, I) be an independence alphabet with $I \cup \mathrm{id}_\Sigma$ transitive. If $L \subseteq \Sigma^*$ is a regular language given by a generalized non-deterministic automaton $\mathcal{A}$ such that*

- *each edge of $\mathcal{A}$ is labelled by an ε-free regular language over one of the components of the graph (Σ, I) with all ranks finite, and*
- *every two consecutive edges are labelled by languages over different components of (Σ, I),*

then there exists a $(\min, +)$-automaton $\mathcal{S}$ such that $\mathrm{Rank}_m(L) = \mathcal{S}_{\max}(m)$ for all positive integers m.

Example 9 (continued). For $L = (abc)^*$, all elements in (3) have the second component equal to 1, which allows one to directly construct the $(\min, +)$-automaton $\mathcal{S}$ over an alphabet $\Gamma = \{\gamma, \delta, \varsigma, \eta\}$, whose letters correspond to matrices in (3):

4.3   Encoding Rational Series into Ranks

The aim of the rest of the section is to construct, for any given $(\min, +)$-automaton $\mathcal{S} = (\Gamma, Q, E, \iota, \tau)$, a generalized automaton $\mathcal{A}$ accepting a language L such that $\mathrm{Rank}_m(L) = \mathcal{S}_{\max}(m)$ for every positive integer m. This automaton will be obtained by choosing one subalphabet consisting of independent letters for every letter of the alphabet Γ and replacing each edge labelled by this letter with an appropriate regular language whose all ranks are equal to the weight of this edge. However, when the construction of Subsect. 4.2 is applied to the resulting generalized automaton in order to reconstruct the original $(\min, +)$-automaton, it inevitably introduces new letters, whose corresponding values in $\mathcal{N}_Q$ have the second component equal to zero. This is why the automaton $\mathcal{S}$ has to be adjusted, so that these undesired letters do not influence the resulting sequence, which is calculated by formula (2) instead of formula (1). Each of these new letters can be viewed as a shadow of the original letter, in the sense that it performs the same actions on the automaton, but it is neither counted towards the length of the word, nor has any influence on the weight of the computation.

Let $\mathcal{S} = (\Gamma, Q, E, \iota, \tau)$ be a $(\min, +)$-automaton. For every letter $\gamma \in \Gamma$, introduce a copy $\tilde{\gamma}$ of this letter, called its *shadow*, and let Γ_{sh} denote the extended alphabet $\Gamma \cup \{\tilde{\gamma} \mid \gamma \in \Gamma\}$. Extend the mapping E to the alphabet Γ_{sh} by setting $E(p, \tilde{\gamma}, q) = 0$ if $E(p, \gamma, q) \neq \infty$, and $E(p, \tilde{\gamma}, q) = \infty$ if $E(p, \gamma, q) = \infty$. The sequence defined by the automaton $\mathcal{S}$ is thus extended to all words over Γ_{sh}. The corresponding sequence $\mathcal{S}_{\max}^{\mathrm{sh}}$ representing the growth of the coefficients of the series is then defined by setting $\mathcal{S}_{\max}^{\mathrm{sh}}(m)$ equal to the supremum of the coefficients $\mathcal{S}(w)$, for words $w \in (\Gamma_{\mathrm{sh}})^+$ such that $\mathcal{S}(w) \neq \infty$ and the number of occurrences of the original letters from Γ in w is at most m.

Proposition 12. *For every* $(\min, +)$*-automaton* $\mathcal{S} = (\Gamma, Q, E, \iota, \tau)$ *there exists a* $(\min, +)$*-automaton* $\mathcal{T} = (\Gamma, Q', E', \iota, \tau)$ *such that* $\mathcal{S}_{\max}(m) = \mathcal{T}_{\max}(m) = \mathcal{T}^{sh}_{\max}(m)$ *for all positive integers* m.

The automaton $\mathcal{T}$ is constructed by adding new paths to the automaton $\mathcal{S}$, called *jumps*. Each of these jumps is a copy of an existing path of length at most $2 \cdot 2^{|Q|^2} - 1$ in $\mathcal{S}$, as far as letters labelling the edges are concerned; however, on these new paths all weights of the individual edges of the original path are concentrated to one edge, leaving all the other edges with weight zero. The idea of this construction is to allow one to use jumps whenever there are shadows available within a word, so that the original letters can be used on transitions with weight zero, while shadows are matched to the high-valued edges of jumps, which are consequently not counted towards the weight of the computation. It can be proved that in this way, the weight of all words with shadows becomes smaller than the weight of some shorter words without shadows.

Theorem 13. *For every* $(\min, +)$*-automaton* $\mathcal{S}$, *there exists an independence alphabet* (Σ, I), *with* $I \cup \mathrm{id}_\Sigma$ *transitive, and a word language* $L \subseteq \Sigma^*$ *such that* $\mathcal{S}_{\max}(m) = \mathrm{Rank}_m(L)$ *for all positive integers* m.

Proof. According to Proposition 12, we can assume that the $(\min, +)$-automaton $\mathcal{S} = (\Gamma, Q, E, \iota, \tau)$ is such that $\mathcal{S}_{\max}(m) = \mathcal{S}^{sh}_{\max}(m)$ for all positive integers m. Elements of $\mathcal{N}_Q$ describing the behaviour of shadows are determined by the mapping $\theta_\mathcal{S} : \Gamma \to \mathcal{N}_Q$ defined by the rule $\theta_\mathcal{S}(\gamma) = (M_\gamma, 0)$, where $M_\gamma(p, q)$ is obtained from $E(p, \gamma, q)$ by replacing all values other than ∞ with 0. In terms of elements of $\mathcal{N}_Q$, the above assumption means that $\mathcal{S}_{\max}(m)$ is equal to

$$\max\{\min_{\iota, \tau}(\mu(N)) \mid N \in \langle \psi_\mathcal{S}(\Gamma) \cup \theta_\mathcal{S}(\Gamma) \rangle \cap \mathcal{N}_Q^{(\iota, \tau)\text{-acc}}, \ \pi(N) \leq m \}. \quad (4)$$

Let the independence alphabet (Σ, I) consist of a two-letter component $\Sigma_\gamma = \{a_{0,\gamma}, a_{1,\gamma}\}$ for each letter $\gamma \in \Gamma$ and one additional singleton component $\{b\}$. The language L will be described by a generalized non-deterministic automaton $\mathcal{A}$ constructed basically by labelling each edge of $\mathcal{S}$ with a suitable ε-free regular language, instead of a letter and a weight. However, in order to ensure that consecutive edges of $\mathcal{A}$ are labelled with languages over different components, each state of $\mathcal{S}$ has to be split into two states connected by an edge labelled with the language $\{b\}$. Accordingly, let $Q' = \{ q' \mid q \in Q \}$ be a disjoint copy of Q and let $\mathcal{A} = (\Sigma, Q \cup Q', F, \iota, \tau)$, with $F(q, q') = \{b\}$ for all $q \in Q$. The other edges of $\mathcal{A}$ are labelled by languages studied in Lemma 7: let $L_{\gamma, n}$ be a copy of the language $(L_1 \cup K_{1,n}) \setminus \{\varepsilon\}$, with letters a_0 and a_1 replaced by $a_{0,\gamma}$ and $a_{1,\gamma}$, respectively. For every edge of $\mathcal{S}$ from p to q labelled with γ, let $F(p', q) = L_{\gamma, E(p, \gamma, q) + 1}$.

The ranks of the language defined by the automaton $\mathcal{A}$ will be calculated using Proposition 10. In order to do this, it is not necessary to describe the whole set $G_\mathcal{A}$, since any subset of $G_\mathcal{A}$ containing all of its minimal elements can be used instead. As all ranks of the language $\{b\}$ are equal to zero, the only minimal element of $G_\mathcal{A}$ derived from words over the subalphabet $\{b\}$ is

$N(b) = (M_b, 0)$, where $M_b(q, q') = 0$ for all $q \in Q$, and all the other entries of M_b are ∞.

Now consider some $\gamma \in \Gamma$. Note that since $\overline{L_{\gamma,n}} = (\Sigma_\gamma)^+$ for all n by Lemma 7, all matrices of elements of G_A derived from words over Σ_γ have exactly the same entries equal to ∞. Words over Σ_γ determine only one potential minimal element whose second coordinate is 0, namely $N(a_{0,\gamma}) = (M_{\gamma,0}, 0)$, where $M_{\gamma,0}(p', q) = 0$ for $p, q \in Q$ such that $E(p, \gamma, q) \neq \infty$, and all the other entries of $M_{\gamma,0}$ are ∞.

It remains to describe minimal elements of G_A contributed by words over Σ_γ that have a non-zero second coordinate. According to Lemma 7, for all positive integers k, the value of $\mathrm{Rank}_k(L_{\gamma,E(p,\gamma,q)+1})$ is $E(p, \gamma, q)$. Moreover, Lemma 7 states that these values are reached by the same pair of words $(a_{0,\gamma}^h, a_{1,\gamma}^h)$, where h is an arbitrary integer greater than all weights $E(p, \gamma, q)$ of edges of S labelled with γ. Therefore, the only potential minimal element is $N(a_{0,\gamma}^h, a_{1,\gamma}^h) = (M_{\gamma,1}, 1)$, where $M_{\gamma,1}(p', q) = E(p, \gamma, q)$ for $p, q \in Q$ such that $E(p, \gamma, q) \neq \infty$, and all the other entries of $M_{\gamma,1}$ are equal to ∞.

The formula for calculating ranks in Proposition 10 takes into account only elements of the subsemigroup $\langle G_A \rangle$ that belong to the set $\mathcal{N}_Q^{(\iota,\tau)\text{-acc}}$. The only way to produce such elements is to alternately multiply $N(b)$ with elements $(M_{\gamma,0}, 0)$ or $(M_{\gamma,1}, 1)$, beginning with $N(b)$ and ending with one of the other generators. In other words, such elements are obtained as products of elements $(M_b \cdot M_{\gamma,0}, 0)$ and $(M_b \cdot M_{\gamma,1}, 1)$. However, these elements are equal to $\theta_S(\gamma)$ and $\psi_S(\gamma)$, respectively, apart from having matrices on their first coordinates extended to the dimensions $(Q \cup Q') \times (Q \cup Q')$ with entries equal to ∞. This shows that the formula of Proposition 10 gives the same result as formula (4), which proves the equality $\mathcal{S}_{\max}(m) = \mathrm{Rank}_m(L)$. ∎

As it was proved by Simon [9] that for every positive integer n there exists a $(\min, +)$-automaton S such that the sequence $(\mathcal{S}_{\max}(m))_{m=1}^\infty$ grows asymptotically as $\sqrt[n]{m}$, Theorem 13 implies that there exists a word language whose rank sequence with respect to an independence alphabet with transitive independence relation has such an asymptotic growth rate; in particular, it is not eventually periodic.

Acknowledgments. We are grateful to Jacques Sakarovitch and Sylvain Lombardy for pointing us to the result of Simon [9].

References

1. Berstel, J., Reutenauer, C.: Noncommutative Rational Series with Applications. Cambridge University Press, Cambridge (2011)
2. Diekert, V., Métivier, Y.: Partial commutation and traces. In: Rozenberg, G., Salomaa, A. (eds.) Handbook of Formal Languages. Beyond Words, vol. 3, pp. 457–533. Springer, Berlin (1997)
3. Droste, M., Kuske, D.: Recognizable languages in divisibility monoids. Math. Struct. Comput. Sci. **11**(6), 743–770 (2001)

4. Hashiguchi, K.: Recognizable closures and submonoids of free partially commutative monoids. Theoret. Comput. Sci. **86**(2), 233–241 (1991)
5. Klunder, B., Ochmański, E., Stawikowska, K.: On star-connected flat languages. Fund. Inform. **67**(1–3), 93–105 (2005)
6. Leung, H.: An algebraic method for solving decision problems in finite automata theory. Ph.D. thesis, Pennsylvania State University (1987)
7. Ochmański, E., Wacrenier, P.-A.: On regular compatibility of semi-commutations. In: Lingas, A., Karlsson, R., Carlsson, S. (eds.) ICALP 1993. LNCS, vol. 700, pp. 445–456. Springer, Heidelberg (1993). doi:10.1007/3-540-56939-1_93
8. Sakarovitch, J.: The "last" decision problem for rational trace languages. In: Simon, I. (ed.) LATIN 1992. LNCS, vol. 583, pp. 460–473. Springer, Berlin (1992)
9. Simon, I.: The nondeterministic complexity of a finite automaton. In: Lothaire, M. (ed.) Mots, Mélanges offerts à M.-P. Schützenberger, pp. 384–400. Hermès, Paris (1990)

Two-Variable First Order Logic with Counting Quantifiers: Complexity Results

Kamal Lodaya[1] and A.V. Sreejith[2(✉)]

[1] The Institute of Mathematical Sciences, CIT Campus, Chennai 600113, India
[2] Chennai Mathematical Institute, Siruseri, Kelambakkam 603103, India
sreejithav@cmi.ac.in

Abstract. Etessami et al. [5] showed that satisfiability of two-variable first order logic $FO^2[<]$ on word models is NEXPTIME-complete. We extend this upper bound to the slightly stronger logic $FO^2[<, succ, \equiv]$, which allows checking whether a word position is congruent to r modulo q, for some divisor q and remainder r. If we allow the more powerful modulo counting quantifiers of Straubing, Thérien et al. [22] (we call this two-variable fragment $FOMOD^2[<, succ]$), satisfiability becomes EXPSPACE-complete. A more general counting quantifier, $FOUNC^2[<, succ]$, makes the logic undecidable.

1 Introduction

It is well known that first order logic cannot express counting properties like modulo counting. Two-variable logic cannot express threshold counting. One option is to add counting quantifiers, but there are many ways of doing this, see for example Paris and Wilkie [14], Cai et al. [3], Straubing et al. [22], Schweikardt [16]. In this work, we look at extensions of first order logic with a few counting quantifiers.

In the most general setting FOUNC, we have counting terms and allow their comparison. This logic can define addition [9] and is equivalent to the well studied *majority logic* [7]. Hence, by an old result of Robinson [15], in the presence of the unary predicates (for $|\Sigma| \geq 2$), satisfiability is undecidable.

In the more restricted FOMOD, counting can only be done modulo a number. From Büchi's theorem [2], the satisfiability problem can be decided by building an automaton and checking for its emptiness. This is, like FO, nonelementary on the quantifier depth.

So it is of interest to study the counting quantifiers in a weaker framework, such as the two-variable sublogic, studied in Grädel et al. [6], Pacholski et al. [13], Straubing and Thérien [21]. Etessami et al. [5], showed that satisfiability of $FO^2[<]$ over finite words is NEXPTIME-complete. This was further extended to words over arbitrary linear orderings in [11]. On the other hand satisfiability of $FO^2[<]$ over constant alphabet is NP-complete as shown by Weis and Immerman [26]. With only two variables, having successor as an atomic formula increases

---

Affiliated to Homi Bhabha National Institute, Anushaktinagar, Mumbai *400094*.

© Springer International Publishing AG 2017
E. Charlier et al. (Eds.): DLT 2017, LNCS 10396, pp. 260–271, 2017.
DOI: 10.1007/978-3-319-62809-7_19

expressiveness and the logic $\mathrm{FOMOD}^2[<, succ]$ is a well studied class. Straubing et al. [20] give an algebraic characterisation of this logic. In another paper Straubing and Thérien [21] show that any formula in this logic is equivalent to a formula with all modulo quantifiers inside existential quantifiers. Moreover they show that $\mathrm{FOMOD}^2[<, succ] = \Sigma_2\mathrm{MOD}[<, succ] \cap \Pi_2\mathrm{MOD}[<, succ]$. Interestingly the language $(ab)^*$ can be expressed in the logic $\mathrm{FOMOD}^2[<]$ and $\mathrm{FO}[<]$, but not in $\mathrm{FO}^2[<]$ [21]. Tesson and Thérien [23] review the connections between $\mathrm{FOMOD}^2[<, succ]$, algebra and circuit complexity (Table 1).

Table 1. Complexity of satisfiability of various fragments of two variable logics. Those not cited are the results of this paper.

Logic	$|\Sigma| \geq 2$
$\mathrm{FO}^2[<]$	NEXPTIME-complete [5] (over fixed alphabet, NP-complete [26])
$\mathrm{FO}^2[<, succ]$	NEXPTIME-complete [5]
$\mathrm{FO}^2[<, \equiv]$	NEXPTIME-complete
$\mathrm{FO}^2[<, succ, \equiv]$	NEXPTIME-complete
$\mathrm{FOMOD}^2[<]$	EXPSPACE-complete
$\mathrm{FOMOD}^2[<, succ, \equiv]$	EXPSPACE-complete
$\mathrm{FOUNC}^2[<]$	Undecidable

Our contribution. What is known about the satisfiability problems for these logics over word models? The complexities in the table are tight. In this paper we show that the Etessami, Vardi, Wilke upper bound for $\mathrm{FO}^2[<, succ]$ [5] extends to the slightly stronger $\mathrm{FO}^2[<, succ, \equiv]$, and also that their lower bound holds for $\mathrm{FO}^2[<, \equiv]$ for a constant alphabet size; recall that $\mathrm{FO}^2[<]$ is NP-complete for a constant alphabet [26]. Secondly, we show that $\mathrm{FOMOD}^2[<, succ]$ is EXPSPACE-complete. Our upper bound results assume that the integers in modulo quantifiers and modulo predicates are in binary, whereas our lower bound results assume they are in unary. Thus the complexity does not depend on the representation of integers. Our third contribution is to show that the two variable fragment of counting logic $\mathrm{FOUNC}^2[<]$ is undecidable. The PhD thesis [19] contains many of these results. It also shows that *two-variable* Presburger arithmetic (where $y = x + 1$ and $y = x + x$ are definable) in the presence of unary predicates is undecidable.

Structure of the paper. In the next section, we formally introduce the counting quantifiers and the various logics we look into. In Sect. 3, we give the upper bound results, namely the EXPSPACE upper bound for $\mathrm{FOMOD}^2[<, succ]$ and the NEXPTIME algorithm for $\mathrm{FO}^2[<, succ, \equiv]$. Section 4 gives corresponding lower bounds and the undecidability of $\mathrm{FOUNC}^2[<]$.

2   Preliminaries

We denote by $\mathbb{N}$ the set of all natural numbers $\{0, 1, 2, \dots\}$. An alphabet Σ is a finite set of symbols. Each letter a of Σ is also the name of a unary predicate which holds at positions which have that letter. We will use a left end-marker for a word, which is outside Σ. The set of nonempty subsets of Σ is denoted by $\mathcal{P}(\Sigma)$. The set of all finite words over Σ is denoted by Σ^*. The length of a word w is denoted by $|w|$. For a word $w \in \Sigma^*$ the notation $w(i)$ denotes the ith letter in w, i.e. $w = w(0)w(1)w(2)\dots w(|w|)$, here $w(0)$ is the left end-marker, $w(1)\dots w(|w|)$ are the letter positions. Let $\mathcal{V} = \{x_1, x_2, \dots\}$ be a set of variables. A *word model* over $(\Sigma, \mathcal{V})$ is a pair (u, s), where $u \in \Sigma^*$ and $s : \mathcal{V} \to \{0, \dots, |u|\}$.

First order logic (FO[<]) over a finite alphabet Σ is a logic which can be inductively built using the following operations.

$$a(x), \ a \in \Sigma \mid x < y \mid x = y \mid \alpha_1 \vee \alpha_2 \mid \neg\alpha \mid \exists x \ \alpha$$

We will also consider other *regular* relations like (a) *succ* : where $succ(x, y)$ says that $y = x + 1$ and (b) $x \equiv r \mod q$, where $q > 1$.

We use the superscript 2 to denote the sublogics which (perhaps repeatedly) use only two variables. For example, the two-variable fragment of FO[<] is denoted by FO^2. Over finite words, FO^2 can talk about occurrences of letters and also about the order in which they appear [17]. The satisfiability problem for a formula α checks if there is a model (in our case, a word model) for it. Complexity of satisfiability problems for two-variable logics are the focus of this study.

Counting quantifiers. We now introduce the syntax for the counting capabilities which extend FO. In the most general setting $\mathrm{FOUNC}[<]$, we have counting terms and allow their comparison. The additional syntax is

$$\#x(\alpha) \sim x_j \mid \#x(\alpha) \sim \max \mid \#x(\alpha) \sim n, \ n \in \mathbb{N}$$

Here α is an inductively defined $\mathrm{FOUNC}[<]$ formula, max denotes the last position of a word, $x, x_j \in \mathcal{V}$ and $\sim$ is in $\{<, =, >\}$. The interpretation of the counting term $\#x(\alpha)$ is $|\{i \mid w, s[x \mapsto i] \models \alpha, 1 \le i \le |w|\}|$. Quantification is over letter positions and not over the end-marker. The positions $0, 1, \dots, |w|$ interpret the counts $0, 1, \dots, |w|$ respectively. Counts greater than $|w|$ do not have an interpretation on this word. Hence in the semantics below we do not require a different sort for numbers, a formal difficulty pointed out to us by Anand Pillay.

$$(w, s) \models \#x(\alpha) \sim x_j \Leftrightarrow |\{i \mid (w, s[x \mapsto i]) \models \alpha, 1 \le i \le |w|\}| \sim s(x_j)$$
$$(w, s) \models \#x(\alpha) \sim \max \Leftrightarrow |\{i \mid (w, s[x \mapsto i]) \models \alpha, 1 \le i \le |w|\}| \sim |w|$$
$$(w, s) \models \#x(\alpha) \sim n \Leftrightarrow |\{i \mid (w, s[x \mapsto i]) \models \alpha, 1 \le i \le |w|\}| \sim n, \ n \in \mathbb{N}$$

Modulo counting quantifiers. In the more restricted $\mathrm{FOMOD}[<]$, counting terms cannot be compared with variables, but only compared modulo a number. The extended syntax is:

$$\#x(\alpha) \equiv r \mod q$$

The semantics is given as follows for $r, q \in \mathbb{N}$, $q > 1$.

$$(w, s) \models \#x(\alpha) \equiv r \mod q \Leftrightarrow |\{i \mid (w, s[x \mapsto i]) \models \alpha, 1 \leq i \leq |w|\}| \equiv r \mod q$$

For example, if a count in a word is 0, it maps to the position 0, and this is interpreted as even because 0, an even number, maps to this position. For later convenience, we define the abbreviated quantifier **Odd** $y(\alpha)$ to stand for $\#y(\alpha) \equiv 1 \mod 2$, and similarly **Even** $y(\alpha)$. Counting the parity of a letter a requires an FOMOD formula, such as **Even** $y(a(y))$.

We will use MOD[$<$] when first order quantifiers are not allowed. Also, we use FOMOD(q)[$<$] (respectively MOD(q)[$<$]) and FOMOD(D)[$<$] (resp. MOD(D)[$<$]) when the divisors in the modulo counting is restricted to be q or from the set of divisors D.

Modulo counting positions. The logic FO[$<, \equiv$] extends FO[$<$] with the following unary relations.

$$x_i \equiv r \mod q, \ r, q \in \mathbb{N}, \ q > 1$$

which is true iff $s(x_i)$ when divided by q leaves a remainder of $r \mod q$. This allows comparison of positions.

Examples. Let us look at some example languages definable in the logics. Even length words over alphabet $\{a, b\}$ can be expressed in $\text{FO}^2[<, \equiv]$ by $max \equiv 0 \mod 2$, and in $\text{MOD}^2(2)[<]$ by $\#x(a(x) \vee b(x)) \equiv 0 \mod 2$. By refining this sentence with the one below, the language $(ab + ba)^*$ can also be described in $\text{FO}^2[<, \equiv]$.

$$\forall x \forall y \Big(\big(succ(x, y) \wedge x \equiv 1 \mod 2 \big) \Rightarrow \big((a(x) \Rightarrow b(y)) \wedge (b(x) \Rightarrow a(y)) \big) \Big)$$

The regular language which allows at most k more a's than b's in every prefix can be defined in first order logic [24]. The simple $\text{FOUNC}[<, +k]$ sentence below can be written using the successor relation and one additional variable. We do not know whether the language can be defined in $\text{FOUNC}^2[<, succ]$. A similar example was brought to our notice by Diego Figueira.

$$\forall x \exists y \Big(\big(\#y(y \leq x \wedge a(y)) = y \big) \wedge \big(y \leq \#y(y \leq x \wedge b(y)) + k \big) \Big)$$

The $\text{FOUNC}^2[<]$ sentence below defines the nonregular context-free language $\{a^n b^n \mid n \geq 1\}$.

$$\exists x \exists y \Big(\big(y = \#y(y \leq x \wedge a(y)) \wedge y = \#y(y \leq x \wedge (a(y) \vee b(y))) \big) \wedge$$
$$\big(y = \#y(y > x \wedge b(y)) \wedge y = \#y(y > x \wedge (a(y) \vee b(y))) \big) \Big)$$

Sizes. The size of a formula is defined inductively as usual. We use binary-size and unary-size to mean that the length of natural number constants is counted as written in binary and unary respectively. For example the number ten (10 in the decimal notation we use) has binary-size 4 and unary-size 10.

3   Upper Bounds via Linear Temporal Logic

Temporal logic. The temporal logic UTL over the set of propositions A is the logic with the set of formulas closed under Boolean operations, and including a when a is a letter in A, and $\mathsf{F}\varphi, \mathsf{P}\varphi, \mathsf{X}\varphi, \mathsf{Y}\varphi$ when φ is a formula. To state the semantics fix a word $u \in \mathcal{P}(A)^*$. A position $i \leq |u|$ satisfies the formula a if i is labeled with the letter a, the formula $\mathsf{X}\varphi$ (*resp.* $\mathsf{Y}\varphi$) if position $i + 1$ (*resp.* $i - 1$) satisfies formula φ, and the formula $\mathsf{F}\varphi$ (*resp.* $\mathsf{P}\varphi$) if there is a position $i \leq j \leq |u|$ (*resp.* $i \geq j$) that satisfies the formula φ. The semantics for Boolean connectives are defined in the usual way. The language of the formula φ is the set of all $u \in \mathcal{P}(A)^*$ that satisfy φ.

Modulo counting temporal logic. The logic UTLMOD extends UTL with the following modulo counting operators.

$$MOD_{r,q}^P \varphi \mid MOD_{r,q}^F \varphi$$

For a word $u \in \mathcal{P}(A)^*$, formula $MOD_{r,q}^P \varphi$ (*resp.* $MOD_{r,q}^F \varphi$) is satisfied at a position $i \leq |u|$ if the number of positions $j \leq i$ (*resp.* $j \geq i$) which satisfy φ is $r \mod q$.

The logic UTLMOD gives a linear time translation into $\text{FOMOD}^2[<, succ]$. There is an *exponential time* translation in the reverse direction.

Lemma 1. [5,18] *For an* $\text{FOMOD}^2[<, \text{succ}]$ *formula* α *of quantifier depth* d *and binary-size* n *there exists an* UTLMOD *formula* α' *of operator depth* $2d$ *and binary-size at most* $O(2^n)$, *such that* α *and* α' *accept the same set of word models. Moreover this translation can be done in exponential time.*

From our earlier work, we know that UTLMOD satisfiability is in PSPACE.

Theorem 2. [10] *Satisfiability of* UTLMOD *is* PSPACE*-complete.*

Combining Lemma 1 with the above Theorem we get:

Theorem 3. $\text{FOMOD}^2[<, succ]$ *satisfiability is in* EXPSPACE.

Length counting temporal logic. For $\text{FO}^2[<, succ, \equiv]$ we can do better. For this, we introduce the logic UTLLEN that extends UTL with a restricted modulo counting operator.

$$MOD_{r,q}^P \ true \mid MOD_{r,q}^F \ true$$

Note that in this logic, we can measure the distance of a position (modulo some number) from the start or end of a word.

We next show that a satisfiable UTLLEN formula φ has a model which is exponential in the modality depth and number of propositions (the size of the formula is irrelevant) and polynomial in $lcm(\varphi)$. Here, $lcm(\varphi)$ stands for the least common multiple of the integers q which occur as divisors in a modulo predicate in φ. Following Etessami et al. [5] we define an UTLLEN formula φ to be of depth (k, k') if its $\{\mathsf{F}, \mathsf{P}\}$ depth is k and its $\{\mathsf{X}, \mathsf{Y}\}$ depth is k'.

For a word w and a position i, the (k, k', m)-type of i in w is the set of all UTLLEN(m) formulas of depth (k, k') that hold in w at i. The following lemma says that the question of satisfiability can be reduced to counting the number of (k, k', m)-types possible in a word. Let $T(k, k', m)$ be the maximum number of distinct (k, k', m)-types possible in a word. It can be counted inductively.

Lemma 4. *For all* k, k', m, $T(k + 1, k', m) \leq (2T(k, k', m) + 1)T(0, k', m)$.

Proof (following [5], *Lemmas 2,5).* Let w be a word. Then the $(k+1, k', m)$-type at position i in w is uniquely given by the $(0, k', m)$-type at i, the (k, k', m)-types that occur to its right and the (k, k', m)-types that occur to its left.   □

A "snipping" lemma gives a small model property for formulas in UTLLEN.

Lemma 5. *Let* $m \in \mathbb{N}$ *and let* φ *be an* UTLLEN(m) *formula of depth* (k, k'). *Then if* φ *is satisfiable, it is satisfiable in a model of size* $T(k, k', m) + 1$.

Proof (following [5], *Lemma 4).* Let $w = u_0 u_1 \ldots u_n$ be a model for φ and let $n > T(k, k', m) + 1$. Then there exists $i < j \leq n$ such that the (k, k', m)-type at positions i and j are the same. The word $\hat{w} = u_0 u_1 \ldots u_i u_{j+1} \ldots u_n$ obtained by removing the intervening portion continues to be a model for φ.   □

Lemma 6 now shows that a satisfiable UTLLEN formula φ has a model which is exponential in the operator depth and number of propositions (the size of the formula is irrelevant). The divisors contribute a multiplicative factor.

Lemma 6. *Let* φ *be an* UTLLEN *formula of operator depth* d *and number of propositions* p. *If* φ *is satisfiable, it has a satisfying model of size* $O(lcm(\varphi)^{2d^2} 2^{2pd^2})$.

Proof. Let $m = lcm(\varphi)$ and let w be a word model over p propositions and let the depth of φ be (k, k'). The lemma follows from Lemma 5 if we can show that the number of (k, k', m)-types is bounded by $m^{2d^2} 2^{2pd^2}$. The $(0, k', m)$-type at a position depends on the current position and k' positions to its left and k' positions to its right. Each position satisfies some subset of propositions and $(i \mod m)$ for some $i < m$. Thus $T(0, k', m)$ is bounded by $(m2^p)^{2k'+1}$. Hence $T(k, k', m) \leq (2T(k-1, k', m) + 1)(m2^p)^{2k'+1} = O((m2^p)^{2kk'})$.

The above small model property for UTLLEN gives us that:

Theorem 7. FO^2$[<, \mathrm{succ}, \equiv]$ *satisfiability is in* NEXPTIME.

Proof. Lemma 1 shows that for every FO$^2[<, succ, \equiv]$ formula α, there exists an UTLLEN formula α' such that α and α' have the same satisfying models. Moreover, if the quantifier depth of α is d, then the operator depth of α' is $2d$. Lemma 6 shows that every satisfying UTLLEN formula α' of operator depth $2d$ has a satisfying model of size $s = O(lcm(\alpha)^{4d^2} 2^{4pd^2})$. A NEXPTIME machine can guess this model and verify it in time $s^2 \times |\alpha|$.

4   Lower Bounds via Tiling Problems

The lower bound results in this section are shown by reducing from Tiling problems. We define the required Tiling problems now.

Tiling problems. A tiling system is a tuple $\mathfrak{S} = (\mathfrak{T}, \mathfrak{R}, \mathfrak{D})$, where $\mathfrak{T}$ is a finite set of tiles, $\mathfrak{R} \subseteq \mathfrak{T} \times \mathfrak{T}$ and $\mathfrak{D} \subseteq \mathfrak{T} \times \mathfrak{T}$ are, respectively, the right (horizontal) and down (vertical) adjacency relations. A tiling problem is the tuple $(\mathfrak{S}, n, top_1, ..., top_n, bot)$, where $n \in \mathbb{N}$ and $top_1, ..., top_n, bot \in \mathfrak{T}$. A tiling of an $m \times k$ grid $G \subseteq \mathbb{N}^2$ is a mapping $\tau : G \to \mathfrak{T}$ respecting the right and down relations, that is, whenever $(i, j+1)$ or $(i+1, j)$ is in G, we have $\mathfrak{R}(\tau(i, j+1), \tau(i, j))$ or $\mathfrak{D}(\tau(i+1, j), \tau(i, j))$, as the case may be.

We give below two versions of the tiling problem $(\mathfrak{S}, n, top_1, ..., top_n, bot)$ corresponding to EXPSPACE and NEXPTIME Turing machines respectively [4].

Rectangle tiling problem. Do there exist an m and a tiling of an $m \times 2^n$ grid such that the first n tiles in the top row are $top_1, ..., top_n$ in order and there exists a tile bot in the bottom row?

Proposition 8. [4]   *There exists a tiling system $\mathfrak{S} = (\mathfrak{T}, \mathfrak{R}, \mathfrak{D})$, such that its Rectangle tiling problem $(\mathfrak{S}, n, top_1, ..., top_n, bot)$ is* EXPSPACE*-complete.*

Square tiling problem. Does there exist a tiling of a $2^n \times 2^n$ grid, such that the first n tiles in the top row are $top_1, ..., top_n$ in order and there exists a tile bot in the bottom row?

Proposition 9. [4]   *There exists a tiling system $\mathfrak{S} = (\mathfrak{T}, \mathfrak{R}, \mathfrak{D})$, such that its Square tiling problem $(\mathfrak{S}, n, top_1, ..., top_n, bot)$ is* NEXPTIME*-complete.*

4.1   Modulo Counting is Expspace-Hard

We show that satisfiability of $\text{FOMOD}^2[<]$ is EXPSPACE-hard by reducing from the EXPSPACE-complete Rectangle tiling problem. The following lemma shows that $x \equiv y \mod 2^n$ is definable by a $\text{MOD}^2[<]$ formula, which allows us to assert the down relation of the tiling system.

Lemma 10. *There is a polynomial time algorithm which given an $n \in \mathbb{N}$ outputs $\text{Cong}_1^n(x, y)$ in $MOD^2[<]$, where $\text{Cong}_1^n(x, y)$ is of binary-size $O(n)$ and quantifier depth 2 such that $\text{Cong}_1^n(x, y)$ is true if and only if $x \equiv y \mod 2^n$.*

There is also a formula $\text{Cong}_2^n(x, y) \in MOD^2(2)[<]$ of unary-size $O(n^2)$ and quantifier depth n^2, such that $\text{Cong}_2^n(x, y)$ is true iff $x \equiv y \mod 2^n$.

Proof. We first give Cong_1^n which is of quantifier depth 2. For all $i \leq n$, we give formulas $lsb_i(x)$ such that $lsb_i(x)$ is true if and only if the ith least significant bit of x is 1. $lsb_1(x)$ is true if x is odd and is given by the formula **Odd** $y(y \leq x)$. For all $i \geq 2$:

$$lsb_i(x) := \textbf{Odd}\ y\Big(y < x \wedge \big(\#x(x \leq y) \equiv (2^{i-1} - 1) \mod 2^{i-1}\big)\Big)$$

The claim is now proved by induction on the number of positions y which satisfy the conditions

$$y < x \text{ and } y \equiv \left(2^{i-1} - 1\right) \mod 2^{i-1}. \tag{1}$$

When this number is 0 there is no y which satisfies property (1). This means the ith lsb is 0. For the induction step, assume the claim to be true for a number k. Consider the first position z where the count is $k + 1$. Then we know that $z - 1 \equiv \left(2^{i-1} - 1\right) \mod 2^{i-1}$. This implies that if we add 1 to $z - 1$ the ith bit toggles. Since the claim is true when the count is $k + 1$, we have by induction that $lsb_1(x)$ is true if x is odd. Now $\text{Cong}_1^n(x, y) := \bigwedge_{i=1}^n \left(lsb_i(x) \Leftrightarrow lsb_i(y)\right)$. The binary-size of $\text{Cong}_1^n(x, y)$ is $O(n)$.

Note that the unary-size of Cong_2^n is exponential in n, because we need to encode numbers 2^i. This can be reduced as follows. We can replace the subformula $x \equiv \left(2^{i-1} - 1\right) \mod 2^{i-1}$ in lsb_i by $\text{Cong}_2^{i-1}(x, 2^{i-1} - 1)$. An inductive replacement will give us a formula $\text{Cong}_2^n \in \text{MOD}^2(2)[<]$ of size $O(n^2)$ and quantifier depth n^2.   □

In the above lemma, Cong_1^n has quantifier depth 2 and in binary notation, and Cong_2^n is in unary notation and has quantifier depth polynomial in n. If we want both unary notation and constant quantifier depth, we need to introduce modulo counting over primes and use Chinese remaindering.

Lemma 11. *For every $n > 1$, there is a number $q > 2^n$ and a formula $\text{Cong}_3^n(x, y)$ in $MOD^2[<]$ of unary-size $O(n^4)$ and quantifier depth 1 such that $\text{Cong}_3^n(x, y)$ is true if and only if $x \equiv y \mod q$.*

Proof. Let $p_1, ..., p_n$ be the first n primes and $q = \prod_{i=1}^n p_i$ be their product. Clearly $q \geq 2^n$. For all numbers $x, y \in \mathbb{N}$, Chinese remaindering says that the vector $(x \mod p_1, \ldots, x \mod p_n) = (y \mod p_1, \ldots, y \mod p_n)$ if and only if $x \equiv y \mod q$. The following formula asserts this

$$\text{Cong}_3^n(x, y) := \bigwedge_{j \leq n} \bigvee_{r_j < p_j} \left((x \equiv r_j \mod p_j) \wedge (y \equiv r_j \mod p_j) \right)$$

Note that $(x \equiv r_j \mod p_j)$ can be asserted by the following $\text{MOD}^2[<]$ formula, $\left(\#y(y \leq x) \equiv r_j \mod p_j\right)$. By the prime number theorem, asymptotically there are n primes within the first $n \log n$ numbers and hence one can generate the first n primes in time polynomial in n. Therefore, the unary-size of Cong_3^n is $\sum_{j \leq n} \sum_{i \leq p_j} i \leq \sum_{k \leq q} k \leq n^4$.   □

We will now go to our EXPSPACE-hardness result. Assume $|\Sigma| \geq 2$. Below we show the hardness for three classes of logics, each depending on the formula Cong_i^n, where $i \in \{1, 2, 3\}$ we choose from the above Lemmas.

Theorem 12. *The satisfiability problem for the following logics over a constant alphabet is EXPSPACE-hard.*

1. $\text{FOMOD}^2(2)[<]$ formulas (using unary notation).
2. $\text{FOMOD}^2(D)[<]$ formulas (using binary notation) of quantifier depth 3, where $D = \{2^i \mid i \in \mathbb{N}\}$.
3. $\text{FOMOD}^2[<]$ formulas (using unary notation) of quantifier depth 3.

Proof. The proof of the three claims differ only on the use of Cong^n formula and therefore we follow the same proof for all the three claims. We show EXPSPACE hardness by reducing from the EXPSPACE-complete Rectangle tiling problem $\mathfrak{I} = (\mathfrak{S}, n, top_1, \ldots, top_n, bot)$ where $\mathfrak{S} = (\mathfrak{T}, \mathfrak{R}, \mathfrak{D})$ and $\mathfrak{T} = \{T_1, \ldots, T_t\}$ given by Proposition 8.

We give a polynomial time algorithm which when given the tiling problem $\mathfrak{I}$ outputs the formula $\psi_{\mathfrak{I}}$ such that there is a tiling for $\mathfrak{I}$ if and only if $\psi_{\mathfrak{I}}$ is satisfiable. The alphabet for $\psi_{\mathfrak{I}}$ is $\Sigma = \mathfrak{T} \times \mathcal{P}(\mathfrak{T}^{Dn}) \times \mathcal{P}(\mathfrak{T}^{Rt})$, where $\mathfrak{T}^{Dn} = \{T_1^{Dn}, \ldots, T_t^{Dn}\}$ and $\mathfrak{T}^{Rt} = \{T_1^{Rt}, \ldots, T_t^{Rt}\}$ are two copies of $\mathfrak{T}$. Note that we are overriding the symbol $\mathfrak{T}$ to mean both tiles and part of the alphabet. It will be clear from the context of the proof what we refer to.

We associate a word model $w_\tau \in \Sigma^*$ with a tiling τ such that τ is a tiling for $\mathfrak{I}$ iff $w_\tau \models \psi_{\mathfrak{I}}$. In fact every position in w_τ contains atmost 2 letters from $\mathfrak{T}^{Dn}$ and atmost 2 letters from $\mathfrak{T}^{Rt}$. We denote by $w_\tau(i, j)$ the letter at the $(i-1)2^n + j^{th}$ position in w_τ. We will ensure that w_τ will satisfy the property $\tau(i, j) = T_l \Leftrightarrow w_\tau(i, j) \in T_l \times \mathcal{P}(\mathfrak{T}^{Dn}) \times \mathcal{P}(\mathfrak{T}^{Rt})$.

The formula ψ_I is written as a conjunction of the formulas $\psi_{init}, \psi_{final}, \psi_{next}$ and $\psi_{constraints}$ describing the initial configuration, the final configuration, the next move, and the tiling constraints respectively. The formula for the initial configuration, ψ_{init} is the conjunction of $\alpha_1, \ldots, \alpha_n$, where α_i says that the ith cell in the first row contains the tile top_i. This is encoded by saying that the first location x which satisfies $x \equiv i \pmod{2^n}$ is the ith cell in the first row.

$$\alpha_i := \forall x \left(\left(\text{Cong}^n(x, i) \wedge \forall y < x \, \neg\text{Cong}^n(y, i) \right) \implies top_i(x) \right)$$

Cong^n denotes one of Cong_j^n, where $j \in \{1, 2, 3\}$ (comes from either Lemma 10 or Lemma 11). Similarly, ψ_{final} is given by saying $\exists y \, bot(y)$. We also need to ensure that there is exactly one tile $T_k \in \mathfrak{T}$ in a cell. This is asserted by a sentence $\psi_{constraints}$ in $\text{FO}^2[<]$. The hardest part of the reduction is to ensure that the relations down $\mathfrak{D}$ and right $\mathfrak{R}$ are respected in the word model. This is given by the sentence ψ_{next} which is a conjunction of the formulae ψ_{down} and ψ_{right}.

We will now explain how the down constraints are respected. Let us assume $T_k \in \tau(i, j)$ and the down constraint $\mathfrak{D}(T_k, T_l)$ is true. We need to now assert that $w_\tau(i+1, j)$ contains T_l. The idea is to count modulo 2, the number of occurrences of T_l^{Dn} in all cells above i and in the same column. That is, we count the size of the set $\{k \mid T_l^{Dn} \in \tau(k, j), k < i\}$. If this count is even then we force T_l^{Dn} to be true at $w_\tau(i, j)$, otherwise we force T_l^{Dn} to be false. This ensures that the count $\{k \mid T_l^{Dn} \in \tau(k, j), k \leq i\}$ is odd. For other tiles we ensure that the count is even. We preserve this invariant at every cell. Hence the tile at $(i+1, j)$ can be determined by looking at the counts for every tile T_l^{Dn}

and setting that tile whose count is odd. The following formula says that in the column strictly above x, there is an even number of occurrences of T_l^{Dn}.

$$\phi_l(x) := \mathbf{Even}\ y\Big(T_l^{Dn}(y) \wedge y < x \wedge \mathrm{Cong}^n(x,y)\Big)$$

Now if we want to transfer the information that the cell right below x has to contain letter T_l, we set the count of T_l^{Dn} on this column above and including position x to be odd. This can be asserted by the formula, $\phi_l(x) \Leftrightarrow T_l(x)$. The following formula $\psi_1(x)$ transfers this information by taking into consideration the down constraints.

$$\psi_1(x) := \bigwedge_{k=1}^{t}\left(T_k(x) \implies \left(\bigvee_{(T_k,T_l)\in\mathfrak{D}}\Big(\phi_l(x) \wedge \bigwedge_{j\neq l}\neg\phi_j(x)\Big)\right)\right)$$

Now we need to set the tiles at $w_\tau(i+1,j)$ by looking at the count of T_l^{Dn} strictly above and in the same column as x. The following formula $\psi_2(x)$ says that if you see an odd number of occurrences of the letter $T_l^{Dn} \in \mathfrak{T}^{Dn}$ in the column strictly above x, then we set letter $T_l \in \mathfrak{T}$ to be true at x.

$$\psi_2(x) := \bigwedge_{l=1}^{t}\Big(\mathbf{Odd}\ y\big(T_l^{Dn}(y) \wedge y < x \wedge \mathrm{Cong}^n(x,y)\big)\Big) \implies T_l(x)$$

The formula ψ_{down} is a conjunction of the formulas ψ_1 and ψ_2. A similar formula ψ_{right} using the letters $\mathfrak{T}^{Rt}$ can assert that the right relations $\mathfrak{R}$ are ensured. $\square$

4.2 Modulo Predicates are Harder Than Linear Order

We now show that $\mathrm{FO}^2[<,\equiv]$ is NEXPTIME-hard even for a constant alphabet, as opposed to $\mathrm{FO}^2[<]$ being NP-complete [26].

Theorem 13. $\mathrm{FO}^2[<,\equiv]$ *satisfiability is* NEXPTIME-*hard (constant alphabet size).*

Proof. We reduce from the NEXPTIME-complete Square tiling problem given by Proposition 9. We introduce $2n$ distinct primes, $p_1,...,p_n$ (for encoding row index), and $q_1,...,q_n$ (for encoding column index). These primes can encode any cell (i,j). One now writes a formula $\alpha_{down}(x)$ which asserts there exists a y such that the row index of y is one more than that of x and the column index of x and y are the same. The formula can also specify that y should satisfy the down constraints. Similarly one can write a formula to force the right constraints. It is easy to write the initial and final conditions. $\square$

4.3 General Counting is Undecidable

We show that the satisfiability problem of $\mathrm{FOUNC}[<,succ]$ with just two variables is undecidable. Grädel et al. [6] had showed this over graphs. Here we show that it is undecidable even over words.

Theorem 14. *Satisfiability of* $FOUNC^2[<, succ]$ *over words is undecidable.*

Proof. Recall that a 2-counter automaton $M = (Q, \rightarrow, q_0, q_n)$ is a finite automaton over the alphabet $A = \{inc_k, dec_k, zero_k \mid k = 1, 2\}$ with both counters initially set to zero. Reachability over 2-counter automata is undecidable [12]. A valid run $q_0 \xrightarrow{a_0} q_1 \xrightarrow{a_1} \ldots \xrightarrow{a_n} q_n$ on the word $a_0 a_1 \ldots a_n$ satisfies the property that for all i from 0 to n, there is an a_i-labeled transition from q_{i-1} to q_i and a_i is enabled at q_{i-1}. Here are the enabling conditions for $k = 1, 2$ which enforce the semantics of counters: inc_k is always enabled, $zero_k$ is enabled if the count of inc_k labels for $j < i$ equals the count of dec_k labels for $j < i$, and dec_k is enabled if the count of inc_k labels for $j < i$ exceeds that of dec_k labels for $j < i$. A zero test on counter 1 at position x is written as follows $\exists y(y = \#y(y < x \wedge inc_1(y)) \wedge y = \#y(y < x \wedge dec_1(y)))$. Given these conditions, it is easy to see that reachability from q_0 to q_n can be expressed by an $FOUNC^2[<, succ]$ formula over the monadic predicates $Q \cup A$.   □

5   Outlook

In an earlier paper [10], we studied the effect of adding modulo counting to linear temporal logic LTL. In this paper we carried out the same effort for two-variable first order logic FO^2, also in the presence of counting quantifiers and unary predicates.

Over words, the logic $FOMOD$ [1] is a strict subset of monadic second order logic; the latter is pleasant to use and has a well-developed theory [2, 25]. The main advantage of the modulo counting logics is that they directly represent numbers using standard binary notation and the two variable fragment provide an elementary decision procedure. This is also the case if we add threshold counting quantifiers [8]. We can also add both kinds of quantifiers at the cost of an extra exponent, but we do not know whether this is necessary. We also leave open the complexity of decidability in modulo counting logic $MOD[<]$.

In this paper we show that once we add unary predicates (in other words, a small alphabet of letters), even over two variables, general counting quantifiers bring undecidability. In the absence of unary predicates, Presburger logic is well known to be decidable, also in the presence of counting quantifiers [16].

Acknowledgments. We would like to thank four DLT referees and the DLT program committee for their suggestions to improve this paper.

References

1. Barrington, D.A.M., Immerman, N., Straubing, H.: On uniformity within NC^1. J. Comp. Syst. Sci. **41**(3), 274–306 (1990)
2. Büchi, J.R.: Weak second-order arithmetic and finite automata. Z. Math. Logik Grundl. Math. **6**, 66–92 (1960)
3. Cai, J., Fürer, M., Immerman, N.: An optimal lower bound on the number of variables for graph identification. Combinatorica **12**(4), 389–410 (1992)

4. van Emde Boas, P.: The convenience of tilings. In: Sorbi, A. (ed.) Complexity, Logic and Recursion Theory, pp. 331–363. CRC Press (1997)
5. Etessami, K., Vardi, M.Y., Wilke, T.: First-order logic with two variables and unary temporal logic. Inf. Comput. **179**(2), 279–295 (2002)
6. Grädel, E., Otto, M., Rosen, E.: Undecidability results on two-variable logics. Arch. Math. Log. **38**, 213–354 (1999)
7. Krebs, A.: Typed semigroups, majority logic, and threshold circuits. Ph.D. thesis, Universität Tübingen (2008)
8. Krebs, A., Lodaya, K., Pandya, P., Straubing, H.: Two-variable logic with a between relation. In: Proceeding of 31st LICS (2016)
9. Lautemann, C., McKenzie, P., Schwentick, T., Vollmer, H.: The descriptive complexity approach to LOGCFL. J. Comp. Syst. Sci. **62**(4), 629–652 (2001)
10. Lodaya, K., Sreejith, A.V.: LTL can be more succinct. In: Bouajjani, A., Chin, W.-N. (eds.) ATVA 2010. LNCS, vol. 6252, pp. 245–258. Springer, Heidelberg (2010). doi:10.1007/978-3-642-15643-4_19
11. Manuel, A., Sreejith, A.: Two-variable logic over countable linear orderings. In: 41st MFCS, Kraków, pp. 66:1–66:13 (2016)
12. Minsky, M.L.: Computation: Finite and Infinite Machines. Prentice-Hall, New Jersy (1967)
13. Pacholski, L., Szwast, W., Tendera, L.: Complexity of two-variable logic with counting. In: 12th LICS, Warsaw, pp. 318–327. IEEE (1997)
14. Paris, J., Wilkie, A.: Counting Δ_0 sets. Fundam. Math. **127**, 67–76 (1986)
15. Robinson, R.M.: Restricted set-theoretical definitions in arithmetic. Proc. Am. Math. Soc. **9**, 238–242 (1958)
16. Schweikardt, N.: Arithmetic, first-order logic, and counting quantifiers. ACM Trans. Comput. Log. **6**(3), 634–671 (2005)
17. Schwentick, T., Thérien, D., Vollmer, H.: Partially-ordered two-way automata: A new characterization of DA. In: Kuich, W., Rozenberg, G., Salomaa, A. (eds.) DLT 2001. LNCS, vol. 2295, pp. 239–250. Springer, Heidelberg (2002). doi:10.1007/3-540-46011-X_20
18. Sreejith, A.V.: Expressive completeness for LTL with modulo counting and group quantifiers. Electr. Notes Theor. Comput. Sci. **278**, 201–214 (2011)
19. Sreejith, A.V.: Regular quantifiers in logics. Ph.D. thesis, HBNI (2013)
20. Straubing, H., Tesson, P., Thérien, D.: Weakly iterated block products and applications to logic and complexity. Int. J. Alg. Comput. **20**(2), 319–341 (2010)
21. Straubing, H., Thérien, D.: Regular languages defined by generalized first-order formulas with a bounded number of bound variables. Theor. Comput. Syst. **36**(1), 29–69 (2003)
22. Straubing, H., Thérien, D., Thomas, W.: Regular languages defined with generalized quantifiers. Inf. Comput. **118**(3), 389–301 (1995)
23. Tesson, P., Thérien, D.: Logic meets algebra: the case of regular languages. Log. Meth. Comp. Sci. **3**(1), 1–26 (2007)
24. Thomas, W.: Classifying regular events in symbolic logic. J. Comput. Syst. Sci. **25**(3), 360–376 (1982)
25. Thomas, W.: Languages, automata and logic. In: Rozenberg, G., Salomaa, A. (eds.) Handbook of Formal Language Theory, vol. III, pp. 389–455. Springer, Heidelberg (1997)
26. Weis, P., Immerman, N.: Structure theorem and strict alternation hierarchy for FO^2 on words. In: Duparc, J., Henzinger, T.A. (eds.) CSL 2007. LNCS, vol. 4646, pp. 343–357. Springer, Heidelberg (2007). doi:10.1007/978-3-540-74915-8_27

Deleting Deterministic Restarting Automata with Two Windows

František Mráz[1] and Friedrich Otto[2(⊠)]

[1] Department of Computer Science, Faculty of Mathematics and Physics,
Charles University, Malostranské nám. 25, 118 00 Praha 1, Czech Republic
frantisek.mraz@mff.cuni.cz
[2] Fachbereich Elektrotechnik/Informatik, Universität Kassel, 34109 Kassel, Germany
otto@theory.informatik.uni-kassel.de

Abstract. We study deterministic restarting automata with two windows. In each cycle of a computation, these det-2-RR-automata can perform up to two delete operations, one with each of their two windows. We study the class of languages accepted by these automata, comparing it to other well-known language classes and exploring closure properties.

Keywords: Restarting automaton · Language class · Closure property

1 Introduction

The *restarting automaton* introduced in [9] and many of its later variants are motivated by techniques and problems from linguistics. The original model of the restarting automaton was presented in order to model the so-called 'analysis by reduction,' which is a technique used in linguistics to analyze sentences of natural languages that have free word order. This technique consists in a stepwise simplification of an extended sentence such that the (in)correctness of the sentence is not affected (see, e.g., [9,11,20]). Accordingly, a restarting automaton M consists of a flexible tape with end markers, a read/write window of a fixed size $k \geq 1$, and a finite-state control. It works in *cycles*, where each cycle begins with the window at the left end of the tape and M being in its initial state. During a cycle M scans the current tape contents from left to right and executes a single length-reducing rewrite step. The cycle ends with a restart that takes the window back to the left end of the tape and resets M to its initial state. A computation is completed by a *tail computation* that is similar to a cycle but that ends with accepting or rejecting the input. In its rewrite steps M may introduce non-input symbols, so-called *auxiliary symbols*. This type of restarting automaton is called an RRWW-automaton. By placing certain restrictions on the definition, we obtain various subclasses of restarting automata (see, e.g., [16]).

In order to investigate the complexity of the *word order* of languages, the *freely rewriting restarting automaton* (FRR-automaton) was introduced in [17]

F. Mráz was supported by the Czech Science Foundation under the project 15-04960S.

É. Charlier et al. (Eds.): DLT 2017, LNCS 10396, pp. 272–283, 2017.
DOI: 10.1007/978-3-319-62809-7_20

(see also [15]), which is defined like an RRWW-automaton, but which is allowed to perform an arbitrary positive number of rewrite steps in a cycle. Then in [18], a restricted type of FRR-automata is shown to be closely related to *parallel communicating grammar systems with regular control*. Further, in [13] a model of the restarting automaton is studied which has a single window that can move in both directions along the tape, and that can perform an arbitrary number of delete operations in a cycle. In [19, 20] this model is even extended to produce structured output in the form of a (dependency) tree for the given input.

A serious problem in analysis of (some) free word order languages is caused by *long-distance scrambling*, which induces a dependency of items across a large distance (see, e.g., [21]). Certainly, the use of two independent heads or windows can help in this matter. And indeed, in [7] an ad-hoc variant of the restarting automaton is defined that has two windows: one window can only delete a single symbol in a rewrite step, and the other window may rewrite a non-empty word into a word of length at most one. It is shown in [7] that a certain restricted type of these restarting automata corresponds to *sorted dependency insertion grammars with regular selectors* that work in top-down derivation style. However, restarting automata with two windows have not yet been studied in a systematic way.

Here we present the first such study, concentrating on a rather restricted type, called *det-2-RR-automaton*. Such an automaton is deterministic, and the rewrite operations that it can perform are only deletions. In addition, we require that within a cycle, each of the two windows executes at most one such operation.

Obviously, the det-2-RR-automaton is an extension of the deterministic two-head finite-state acceptor. However, it is much more powerful, as it accepts, for example, the language $L_{pal} = \{ ww^R \mid w \in \{a, b\}^* \}$ of palindromes of even length, which is not accepted by any multi-head finite-state acceptor [8], and it accepts the copy language $L_{copy} = \{ ww \mid w \in \{a, b\}^* \}$, which is not even growing context-sensitive [2].

We show that with respect to the size of the windows, the det-2-RR-automata yield a strictly increasing infinite hierarchy of language classes that are all incomparable to the (deterministic) context-free languages, the Church-Rosser languages [14], and the growing context-sensitive languages [6] with respect to inclusion. On the other hand, the union of all these classes strictly includes the deterministic context-free languages. Further, all these language classes are closed under the operation of complementation, but they are not closed under intersection (with regular sets), union, non-erasing morphisms, or inverse morphisms. Finally, while the membership problem is decidable in quadratic time for each language that is accepted by a det-2-RR-automaton, emptiness, finiteness, universality, inclusion, and equivalence are all undecidable for det-2-RR-automata. In fact, these problems are not even recursively enumerable.

The paper is structured as follows. In Sect. 2, we present the definition of the det-2-RR-automaton and some examples illustrating its expressive power. In Sect. 3, we show that the simple language of a det-2-RR-automaton is accepted by a nondeterministic 2-head writing finite automaton [22]. Then, in Sect. 4, we establish the announced hierarchy and in Sect. 5, we study closure and non-closure properties for the language classes accepted by det-2-RR-automata. Finally, we address decision problems for det-2-RR-automata in short.

2   Definitions

For a finite alphabet Σ, Σ^* is the set of all words over Σ, and $\Sigma^+ = \Sigma^* \setminus \{\lambda\}$, where λ denotes the empty word. The length of a word w is written as $|w|$.

A det-2-RR-*automaton* consists of a finite-state control, a flexible tape with endmarkers, and two windows of a fixed finite size. Formally, it is defined by a 7-tuple $M = (Q, \Sigma, \mathateg, \$, \delta, q_0, k)$, where Q is a finite set of states, Σ is a finite input alphabet, $\mathateg$ and $\$$ are special letters that mark the left and right border of the tape (these letters are not elements of Σ), $q_0 \in Q$ is the initial state, $k \geq 1$ is the size of the windows, and $\delta : Q \times \mathcal{PC}^k \times \mathcal{PC}^k \to \mathcal{OP}$ is the transition function. Here $\mathcal{PC}^k$ denotes the set of all possible words that can occur in a window of M, and $\mathcal{OP}$ is the set of possible operations, which are defined as follows:

- An operation of the form $\delta(q, u_1, u_2) = (q', \mathsf{MVR}_i)$, $i \in \{1, 2\}$, moves window i one position to the right and changes the state to q'. However, such an operation is only applicable if $u_i \neq \$$.
- An operation of the form $\delta(q_1, u_1, u_2) = (q', \mathsf{DEL}_i(v))$, $i \in \{1, 2\}$, replaces the factor u_i contained in window i by the word v, where v is a proper scattered subword of u_i, moves this window by $|u_i| - |v|$ steps to the right (which means that window i is refilled from the right), and changes the state to q'. Here some restrictions apply in that the delimiters $\mathateg$ and $\$$ must not be deleted and that the window cannot be moved across the right sentinel $\$$. In addition, if one or more of the letters deleted are currently also contained in the other window, then this window is also refilled from the right.
- An operation of the form $\delta(q, u_1, u_2) = \mathsf{Restart}$ moves both windows to the left end of the tape such that the first letter they contain is the left sentinel $\mathateg$, and resets the state to the initial state q_0.
- An operation of the form $\delta(q, u_1, u_2) = \mathsf{Accept}$ causes M to halt and accept.

If $\delta(q, u_1, u_2)$ is undefined, then M halts without accepting. It is required that before the first restart operation and between any two successive restart operations, each window can execute at most one delete operation, that at least one delete operation is indeed executed, and that after the last restart operation in a computation, none, one, or two delete operations (of different windows) can be performed. For each $k \geq 1$, we use det-2-RR(k) to denote the class of deterministic 2-RR-automata with windows of size k.

The notions of *cycle* and *tail* computations carry over from RR-automata (see, e.g., [16]). However, as a det-2-RR-automaton has two windows we use the additional symbols H_1, H_2, and $H_{1,2}$ to denote the positions of these windows, where H_1 (H_2) is placed immediately to the left of the first letter that is currently inside window one (two), and $H_{1,2}$ is used if both windows are at the same position. Then the language accepted by M is the set

$L(M) = \{w \in \Sigma^* \mid (q_0, H_{1,2}cw\$) \vdash_M^* \text{Accept}\}$, where $\vdash_M$ denotes the single-step computation relation that M induces on its set of configurations, and $\vdash_M^*$ denotes its reflexive and transitive closure. By $\mathcal{L}(\text{det-2-RR}(k))$ we denote the class of languages that are accepted by det-2-RR-automata with windows of size k, and $\mathcal{L}(\text{det-2-RR}) := \bigcup_{k \geq 1} \mathcal{L}(\text{det-2-RR}(k))$.

Example 1. A det-2-RR(1)-automaton M for the marked copy language $L'_{\text{copy}} = \{wcw \mid w \in \{a,b\}^*\}$ is given through the following transition function, where $x, x' \in \{a, b\}$, $y \in \{a, b, c\}$, and $z \in \{a, b, \$\}$:

$$
\begin{aligned}
\delta(q_0, c, c) &= (q_1, \text{MVR}_1), & \delta(q_1, y, c) &= (q_2, \text{MVR}_2), \\
\delta(q_2, c, c) &= (q_3, \text{MVR}_1), & \delta(q_3, \$, c) &= \text{Accept}, \\
\delta(q_2, x, x') &= (q_2, \text{MVR}_2), & \delta(q_2, x, c) &= (q_4, \text{MVR}_2), \\
\delta(q_4, x, x) &= (q_5, \text{DEL}_1(\lambda)), & \delta(q_5, y, x) &= (q_6, \text{DEL}_2(\lambda)), \\
\delta(q_6, y, z) &= \text{Restart.}
\end{aligned}
$$

Example 2. Let L_∞ denote the following language on $\Sigma = \{a, b, \#, \&\}$:

$$L_\infty = \{w_1 \# w_2 \# \cdots \# w_n \& w_n \# \cdots \# w_2 \# w_1 \mid n \geq 1, w_1, w_2, \ldots, w_n \in \{a, b\}^*\}$$

It is known that this language is not accepted by any multi-head PDA [3,4]. However, based on Example 1 it can be shown that L_∞ is accepted by a deterministic 2-RR(1)-automaton. □

It is easily seen that the marked palindromic language $L'_{\text{pal}} = \{wcw^R \mid w \in \{a, b\}^*\}$ is accepted by some det-2-RR(1)-automaton. However, as shown in [22] this language is not accepted by any two-head writing NFA (see Sect. 3), and, as pointed out in [8], it is not accepted by any multi-head NFA. In fact, also the language $L_{\text{pal}} = \{ww^R \mid w \in \{a, b\}^*\}$ of palindromes of even length without a middle marker is accepted by a det-2-RR(2)-automaton that, in each cycle, deletes the first and the last letter, if they coincide.

Example 3. A det-2-RR(2)-automaton M_{copy} for the copy language $L_{\text{copy}} = \{ww \mid w \in \{a, b\}^*\}$ is given through the following transition function, where $d, e, f, g \in \{a, b\}$:

$$
\begin{aligned}
&(1)\ \delta(q_0, cd, cd) = (q_1, \text{MVR}_2), & &(9)\ \ \delta(q_2, de, fg) = (q_0, \text{MVR}_1), \\
&(2)\ \delta(q_0, de, fg) = (q_1, \text{MVR}_2), & &(10)\ \delta(q_2, de, f\$) = (q_3, \text{MVR}_1), \\
&(3)\ \delta(q_1, cd, de) = (q_2, \text{MVR}_2), & &(11)\ \delta(q_3, de, d\$) = (q_4, \text{DEL}_1(e)), \\
&(4)\ \delta(q_1, de, fg) = (q_2, \text{MVR}_2), & &(12)\ \delta(q_4, de, f\$) = (q_5, \text{DEL}_2(\$)), \\
&(5)\ \delta(q_2, cd, ef) = (q_0, \text{MVR}_1), & &(13)\ \ \delta(q_5, de, \$) = \quad \text{Restart,} \\
&(6)\ \delta(q_2, cd, e\$) = (q_3, \text{MVR}_1), & &(14)\ \delta(q_0, c\$, c\$) = \quad \text{Accept,} \\
&(7)\ \delta(q_5, \$, \$) \quad = \text{Restart,} & &(15)\ \ \delta(q_5, d\$, \$) = \quad \text{Restart.} \\
&(8)\ \delta(q_4, d\$, d\$) = (q_5, \text{DEL}_2(\$)),
\end{aligned}
$$

Given an input of the form $w = a_1 a_2 \cdots a_n b_1 b_2 \cdots b_n$ of length $2n$, M_{copy} cycles through the states q_0, q_1, and q_2, in each round moving window 2 two letters to the right and window 1 only one letter to the right. Hence, when window 2 reaches the suffix $b_n\$$, then window 1 contains the factor $a_n b_1$. Now M_{copy} enters state q_3 and checks whether $a_n = b_n$ holds. In the affirmative, the letters a_n and b_n are deleted and M_{copy} restarts; otherwise, M_{copy} gets stuck and therewith rejects. It follows that indeed $L(M_{\text{copy}}) = L_{\text{copy}}$. □

Observe that the language L_{copy} is not even growing context-sensitive [2], and so it is not accepted by any weakly monotone RRWW-automaton [12].

3   Simple Languages of Det-2-RR-Automata

In [22] the *one-way multi-head writing finite automaton* is introduced. A one-way n-head writing finite automaton (wNFA(n)) is essentially defined like an n-head NFA that, in addition to moving its heads from left to right, can execute rewrite steps that replace the current symbol under a head by another symbol.

It is known that the language class $\mathcal{L}(\text{wNFA}(2))$ is closed under the operations of union, product, Kleene star, intersection with regular languages, λ-free morphisms, and inverse morphisms, but that the language $L'_{\text{pal}} = \{wcw^R \mid w \in \{a,b\}^*\}$ is not accepted by any wNFA(2) [22]. Here we are interested in wNFA(2) because of their ability to simulate tail computations of det-2-RR-automata.

Definition 4. *Let $M = (Q, \Sigma, ¢, \$, \delta, q_0, k)$ be a det-2-RR-automaton. The simple language $L_{\text{sim}}(M)$ of M is the sublanguage of $L(M)$ that consists of all those words $w \in \Sigma^*$ that M accepts through tail computations, that is,*

$$L_{\text{sim}}(M) = \{w \in \Sigma^* \mid (q_0, H_{1,2}¢w\$) \vdash_M^* \text{Accept without using a restart}\}.$$

Lemma 5. *For each det-2-RR(1)-automaton M, $L_{\text{sim}}(M) \in \mathcal{L}(\text{wDFA}(2))$.*

Proof. Let $M = (Q, \Sigma, ¢, \$, \delta, q_0, 1)$ be a det-2-RR-automaton with window size 1. A wDFA(2) A can simulate the tail computations of M by using its two heads to simulate the two windows of size 1 of M. In addition, a delete operation $a \mapsto \lambda$ of M is simulated by the rewrite operation $a \to \#_a$, where $\{\#_a \mid a \in \Sigma\}$ are new auxiliary symbols. When a head of A reaches one of these auxiliary symbols, then it just moves on to the next symbol without changing the state. It now follows easily that $L(A) = L_{\text{sim}}(M)$. □

However, for det-2-RR-automata M with window size $k \geq 2$, the situation is more complicated. First of all, a wDFA(2) A has only two heads that each see a single symbol only. Hence, each head must make k move-right steps until it has seen the current content of the corresponding window of M. Next, A may have to simulate a delete operation of M, which deletes between 1 and k letters from the k letters inside one of the windows. For simulating these deletions, A would have to move back to the left, which it cannot do. Thus, instead of first reading all the k letters inside the corresponding window of M, A guesses the last $k-1$ letters when it sees only the first letter. This means that the simulating 2-head writing finite automaton will be nondeterministic. After guessing the next $k-1$ letters (for each of its heads), A simulates the next operation of M, realizing deletions of the form $a \mapsto \lambda$ by rewrites of the form $a \mapsto \#_a$. Here, the information on the deleted letters is important for the following reason. A window of M may delete some letters that are also contained in the other window. Now for the det-2-RR(k)-automaton M that does not cause any problems, since M has seen both window contents before executing these deletions, but for the simulating wNFA(2) A, this would cause a problem as it must still verify the correctness of the letters guessed. However, by simulating a delete operation $a \mapsto \lambda$ by the rewrite operation $a \mapsto \#_a$, this information is saved, allowing A to perform the required verification. Accordingly, we get the following weaker result.

Lemma 6. *For each det-2-RR(k)-automaton M, $L_{\text{sim}}(M) \in \mathcal{L}(\text{wNFA}(2))$.*

4   The Hierarchy

Next we will show that the det-2-RR-automata yield a strictly increasing infinite hierarchy of language classes based on the window size.

Let $\varphi : \{a, b, c\}^* \to \{a', b', c\}^*$ be the morphism that is defined through $a \mapsto a'$, $b \mapsto b'$, and $c \mapsto c$, and let $L_{\text{pal},\varphi} = \{wc(\varphi(w))^R \mid w \in \{a, b\}^*\}$ be a variant of the language of marked palindromes. It is easily seen that $L_{\text{pal},\varphi}$ is accepted by a det-2-RR(1)-automaton, but not by any 2-head writing NFA.

Now, for an integer $k \geq 2$, let $\psi_k : \{a, b, c, a', b'\}^* \to \{a_1, a_2, b_1, b_2, c\}^*$ be the morphism that is defined through $a \mapsto a_1 a_2^{k-1}$, $b \mapsto b_1 b_2^{k-1}$, $c \mapsto c$, $a' \mapsto a_2^{k-1} a_1$, and $b' \mapsto b_2^{k-1} b_1$. Then

$$L_{\text{pal},\varphi}^{(k)} := \psi_k(L_{\text{pal},\varphi}) = \{\psi_k(w)c(\psi_k(w))^R \mid w \in \{a, b\}^*\}.$$

Proposition 7. *For all $k \geq 2$, $L_{\text{pal},\varphi}^{(k)} \in \mathcal{L}(\text{det-2-RR}(k)) \setminus \mathcal{L}(\text{det-2-RR}(k-1))$.*

Proof. The language $L_{\text{pal},\varphi}^{(k)}$ is obtained from $L_{\text{pal},\varphi} = \{wc(\varphi(w))^R \mid w \in \{a, b\}^*\}$ by replacing each letter a or b by the block $a_1 a_2^{k-1}$ or $b_1 b_2^{k-1}$ and by replacing each letter a' or b' by the block $a_2^{k-1} a_1$ or $b_2^{k-1} b_1$. Thus, from the det-2-RR(1)-automaton for $L_{\text{pal},\varphi}$ we easily obtain a det-2-RR(k)-automaton for $L_{\text{pal},\varphi}^{(k)}$.

Assume now that M is a det-2-RR-automaton with windows of size $k-1$ such that $L(M) = L_{\text{pal},\varphi}^{(k)}$, and let $x = \psi_k(u)c(\psi_k(u))^R \in L_{\text{pal},\varphi}^{(k)}$. Then M has an accepting computation on input x. Assume that this computation begins with a cycle that rewrites x into the word y. In the corresponding rewrite steps, M deletes between 1 and $2k - 2$ letters, that is, $1 \leq |x| - |y| \leq 2k - 2$. From the form of the words in the language $L_{\text{pal},\varphi}^{(k)}$ it then follows that $y \notin L_{\text{pal},\varphi}^{(k)}$, which contradicts the correctness preserving property for M (see, e.g., [16]).

Hence, the accepting computation of M on input x must be a tail computation, which implies that $L_{\text{pal},\varphi}^{(k)}$ is accepted by a wNFA(2). However,

$$\psi_k^{-1}(L_{\text{pal},\varphi}^{(k)}) = \{x \in \{a, b, c, a', b'\}^* \mid \psi_k(x) \in L_{\text{pal},\varphi}^{(k)}\}$$
$$= \{x \in \{a, b, c, a', b'\}^* \mid \exists w \in \{a, b\}^* : \psi_k(x) = \psi_k(w)c(\psi_k(w))^R\}$$
$$= \{wc(\varphi(w))^R \mid w \in \{a, b\}^*\} = L_{\text{pal},\varphi},$$

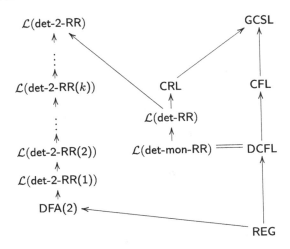

Fig. 1. Relating the det-2-RR-automata to the Chomsky hierarchy.

and as $\mathcal{L}(\mathrm{wNFA}(2))$ is closed under inverse morphisms (see above), it follows that the language $L_{\mathrm{pal},\varphi}$ is accepted by a wNFA(2), a contradiction. Thus, we see that $L_{\mathrm{pal},\varphi}^{(k)}$ is not accepted by any det-2-RR($k-1$)-automaton.   □

This result yields the following infinite hierarchy.

Corollary 8. $\mathcal{L}(\text{det-2-RR}(k)) \subsetneq \mathcal{L}(\text{det-2-RR}(k+1))$ *for all* $k \geq 1$.

As for all $k \geq 1$, the language $L_{\mathrm{pal},\varphi}^{(k)}$ is deterministic context-free, we obtain the following incomparability results from Example 1 and Proposition 7. Here (D)CFL denotes the (deterministic) context-free languages, CRL the Church-Rosser languages [14], and GCSL the growing context-sensitive languages [6].

Corollary 9. *For* $k \geq 1$, *the language class* $\mathcal{L}(\text{det-2-RR}(k))$ *is incomparable to the classes DCFL, CFL, CRL, and GCSL with respect to inclusion.*

Concerning the deterministic 2-RR-automata with windows of arbitrary size, however, we have the following result.

Corollary 10. $\mathrm{DCFL} = \mathcal{L}(\text{det-mon-RR}) \subsetneq \mathcal{L}(\text{det-RR}) \subsetneq \mathcal{L}(\text{det-2-RR})$.

The equality $\mathrm{DCFL} = \mathcal{L}(\text{det-mon-RR})$ was shown in [11], and in [10] an example of a non-context-free language is given that is accepted by a det-RR-automaton. The properness of the last inclusion follows from Example 1. The diagram in Fig. 1 summarizes all the inclusion relations derived above. Below we will encounter a Church-Rosser language that is not accepted by any det-2-RR-automaton, but it remains open whether CFL is contained in $\mathcal{L}(\text{det-2-RR})$.

5   Closure Properties

We begin with a simple positive result that is easily proved using standard techniques.

Proposition 11. *The language classes* $\mathcal{L}(\text{det-2-RR}(k))$, $k \geq 1$, *and* $\mathcal{L}(\text{det-2-RR})$ *are closed under complementation.*

The language $L_{\text{pal},\varphi}^{(k)}$ is obtained as the image of the language $L_{\text{pal},\varphi}$ under the morphism ψ_k. From Proposition 7 we know that $\psi_2(L_{\text{pal},\varphi}) \notin \mathcal{L}(\text{det-2-RR}(1))$, while $L_{\text{pal},\varphi} \in \mathcal{L}(\text{det-2-RR}(1))$. Let $\psi : \{a_1, a_2, b_1, b_2, c\}^* \rightarrow \{a_1, a_2, b_1, b_2, c\}^*$ be the morphism that is defined through $a_1 \mapsto a_1$, $a_2 \mapsto a_2^2$, $b_1 \mapsto b_1$, $b_2 \mapsto b_2^2$, and $c \mapsto c$. Then $\psi \circ \psi_k = \psi_{2k-1}$, and hence, we see from Proposition 7 that, for all $k \geq 2$, $L_{\text{pal},\varphi}^{(k)} \in \mathcal{L}(\text{det-2-RR}(k))$, but that $\psi(L_{\text{pal},\varphi}^{(k)}) = L_{\text{pal},\varphi}^{(2k-1)} \notin \mathcal{L}(\text{det-2-RR}(k))$. Hence, we have the following non-closure property.

Proposition 12. *For each* $k \geq 1$, *the language class* $\mathcal{L}(\text{det-2-RR}(k))$ *is not closed under* λ-*free morphisms.*

Let $\psi_* : \{a, b, a', b', c\}^* \rightarrow 2^{\{a_1, a_2, b_1, b_2, c\}^*}$ be the regular substitution that is defined through $a \mapsto a_1 \cdot a_2^*$, $b \mapsto b_1 \cdot b_2^*$, $a' \mapsto a_2^* \cdot a_1$, $b' \mapsto b_2^* \cdot b_1$, and $c \mapsto c$. By $L_{\text{pal},\varphi}^{(*)}$ we denote the language $L_{\text{pal},\varphi}^{(*)} = \psi_*(L_{\text{pal},\varphi})$.

Lemma 13. $L_{\text{pal},\varphi}^{(*)} \in \mathcal{L}(\text{det-2-RR}(1))$.

Proof. A det-2-RR(1)-automaton M_* for the language $L_{\text{pal},\varphi}^{(*)}$ works in essentially the same way as the det-2-RR(1)-automaton M for the language $L'_{\text{pal}} = \{wcw^R \mid w \in \{a_1, b_1\}^*\}$ (see the remark after Example 2). However, each time M_* encounters an occurrence of a letter a_2 or b_2, it simply deletes that letter and restarts. Observe that in this situation, M_* can use its second window to first check that the tape content is from the regular set $\psi_*(\{a, b\}^*) \cdot c \cdot \psi_*(\{a', b'\}^*)$. It is now easily seen that M_* accepts the language $L_{\text{pal},\varphi}^{(*)}$.   □

For $k \geq 1$, let R_k be the language that is given through the regular expression $R_k = \big((a_1 \cdot a_2^k) \cup (b_1 \cdot b_2^k)\big)^* \cdot c \cdot \big((a_2^k \cdot a_1) \cup (b_2^k \cdot b_1)\big)^*$. Then $L_{\text{pal},\varphi}^{(k+1)} = L_{\text{pal},\varphi}^{(*)} \cap R_k$. Hence, we see from Proposition 7 that $L_{\text{pal},\varphi}^{(*)} \cap R_k$ is not accepted by any det-2-RR(k)-automaton, which yields the following non-closure property.

Proposition 14. *For each* $k \geq 1$, *the language class* $\mathcal{L}(\text{det-2-RR}(k))$ *is not closed under intersection with regular languages.*

As each regular language is accepted by a det-2-RR(1)-automaton, this result has the following consequences in combination with Proposition 11.

Corollary 15. *For each* $k \geq 1$, *the language class* $\mathcal{L}(\text{det-2-RR}(k))$ *is neither closed under union nor under intersection.*

Finally, we consider inverse morphisms. Let $\Gamma_8 = \{a, b, c, d, x_1, x_2, x_3, x_4\}$, $\Gamma_9 = \Gamma_8 \cup \{x_0\}$, let D_8^* be the Dyck language over $\Gamma_8 \cup \overline{\Gamma}_8$ (see, e.g., [1]), let D_9^* be the Dyck language over $\Gamma_9 \cup \overline{\Gamma}_9$, and let $L_D = D_9^* \cap (x_0 \cdot D_8^* \cdot \bar{x}_0)$. The following is easily seen.

Lemma 16. $L_D \in \mathcal{L}(\text{det-2-RR}(1))$.

Let $\Sigma = \{a_0, a_1, a_2, a_3, a_4, b_0, b_1, b_2, b_3, b_4, c, d_1, d_2, e_1, e_2\}$, let p, q be two distinct primes, and let $\psi_{p,q} : \Sigma^* \to (\Gamma_9 \cup \overline{\Gamma}_9)^*$ be the morphism that is given through the following table:

$$a_0 \mapsto x_0 a x_1, \ a_2 \mapsto \bar{x}_2 a x_1, \ a_3 \mapsto \bar{x}_3 \bar{a} x_4, \ b_4 \mapsto \bar{x}_4 \bar{b} x_3, \ d_2 \mapsto \bar{x}_2 d^p x_2,$$
$$a_1 \mapsto \bar{x}_1 a x_2, \ b_2 \mapsto \bar{x}_2 b x_1, \ b_3 \mapsto \bar{x}_3 \bar{b} x_4, \ c \ \mapsto \bar{x}_1 c \bar{c} x_3, \ e_1 \mapsto \bar{x}_3 \bar{d}^q x_3,$$
$$b_1 \mapsto \bar{x}_1 b x_2, \ b_0 \mapsto \bar{x}_3 \bar{a} \bar{x}_0, \ a_4 \mapsto \bar{x}_4 \bar{a} x_3, \ d_1 \mapsto \bar{x}_1 d^p x_1, \ e_2 \mapsto \bar{x}_4 \bar{d}^q x_4.$$

Now let $L_{p,q} = \psi_{p,q}^{-1}(L_D) = \{w \in \Sigma^* \mid \psi_{p,q}(w) \in L_D\}$. Further, let $\pi : \{a_0, a_1, a_2, a_3, a_4, b_0, b_1, b_2, b_3, b_4, c\}^* \to \{a, b, c, d\}^*$ be the morphism defined through $a_0 \mapsto d$, $b_0 \mapsto d$, $c \mapsto c$, and $a_i \mapsto a$, $b_i \mapsto b$ for $i = 1, 2, 3, 4$. Then $w \in \Sigma^*$ belongs to the language $L_{p,q}$ if and only if

$$w = a_0 d_1^{i_0} \alpha_1^{(1)} d_2^{i_1} \alpha_2^{(2)} d_1^{i_2} \cdots d_2^{i_{2k-1}} \alpha_2^{(2k)} d_1^{i_{2k}} c e_1^{j_{2k}} \gamma_3^{(2k)} e_2^{j_{2k-1}} \gamma_4^{(2k-1)} \cdots \gamma_4^{(1)} e_1^{j_0} b_0$$

for some $k \geq 0$, $i_0, i_1, \ldots, i_{2k}, j_0, j_1, \ldots, j_{2k} \geq 0$ satisfying $i_r \cdot p = j_r \cdot q$ for all $0 \leq r \leq 2k$, and $\alpha_1^{(1)}, \alpha_1^{(3)}, \ldots, \alpha_1^{(2k-1)} \in \{a_1, b_1\}$, $\alpha_2^{(2)}, \alpha_2^{(4)}, \ldots, \alpha_2^{(2k)} \in \{a_2, b_2\}$, $\gamma_3^{(2)}, \gamma_3^{(4)}, \ldots, \gamma_3^{(2k)} \in \{a_3, b_3\}$, and $\gamma_4^{(1)}, \gamma_4^{(3)}, \ldots, \gamma_4^{(2k-1)} \in \{a_4, b_4\}$ such that $\pi(\alpha_1^{(1)} \alpha_2^{(2)} \cdots \alpha_2^{(2k)}) = \pi(\gamma_4^{(1)} \gamma_3^{(2)} \cdots \gamma_3^{(2k)})$.

Lemma 17. $L_{p,q} \notin \mathcal{L}(\text{wNFA}(2))$.

Proof. Let $R = a_0 \cdot (\{a_1, b_1\} \cdot \{a_2, b_2\})^* \cdot c \cdot (\{a_3, b_3\} \cdot \{a_4, b_4\})^* \cdot b_0$. Then

$$L_{p,q} \cap R = \{a_0 \alpha_1^{(1)} \alpha_2^{(2)} \cdots \alpha_2^{(2k)} c \gamma_3^{(2k)} \gamma_4^{(2k-1)} \cdots \gamma_4^{(1)} b_0 \mid$$
$$\alpha_1^{(1)}, \alpha_1^{(3)}, \ldots, \alpha_1^{(2k-1)} \in \{a_1, b_1\}, \alpha_2^{(2)}, \alpha_2^{(4)}, \ldots, \alpha_2^{(2k)} \in \{a_2, b_2\},$$
$$\gamma_3^{(2)}, \gamma_3^{(4)}, \ldots, \gamma_3^{(2k)} \in \{a_3, b_3\}, \text{ and } \gamma_4^{(1)}, \gamma_4^{(3)}, \ldots, \gamma_4^{(2k-1)} \in \{a_4, b_4\}$$
$$\text{such that } \pi(\alpha_1^{(1)} \alpha_2^{(2)} \cdots \alpha_2^{(2k)}) = \pi(\gamma_4^{(1)} \gamma_3^{(2)} \cdots \gamma_3^{(2k)})\},$$

and $\pi(L_{p,q} \cap R) = \{ducu^R d \mid u \in (\{a, b\}^2)^*\}$.

Now assume that $L_{p,q}$ is accepted by a wNFA(2). As $\mathcal{L}(\text{wNFA}(2))$ is closed under intersection with regular languages and λ-free morphisms [22], it follows that also the language $\pi(L_{p,q} \cap R)$ is accepted by a wNFA(2). However, from the proof of the fact that the language L'_{pal} of marked palindromes is not accepted by any wNFA(2) [22], it follows that $\pi(L_{p,q} \cap R) = \{ducu^R d \mid u \in (\{a, b\}^2)^*\}$ is not accepted by a wNFA(2), either. This contradiction shows that $L_{p,q}$ is not accepted by any wNFA(2). $\square$

Based on this result we can now derive the following.

Proposition 18. $L_{p,q} \notin \mathcal{L}(\text{det-2-RR}(k))$ *for any* $k < \max\{p, q\}$.

Proof. Assume that M is a det-2-RR(k)-automaton such that $L(M) = L_{p,q}$. By Lemma 17, $L_{p,q}$ is not accepted by any wNFA(2), which implies by Lemma 6 that $L_{p,q}$ is not the simple language $L_{\text{sim}}(M)$. Now consider a word $w \in L_{p,q}$ of the form

$$w = a_0 d_1^q \alpha_1^{(1)} d_2^q \alpha_2^{(2)} \cdots \alpha_2^{(2r)} d_1^q c e_1^p \gamma_3^{(2r)} e_2^p \gamma_4^{(2r-1)} \cdots \gamma_4^{(1)} e_1^p b_0,$$

where $r \geq 1$ is sufficiently large such that w is not accepted in a tail computation. Then on input w, M executes a cycle that rewrites the word w into a word z, where $z \in L_{p,q}$ and $|w| - 1 \geq |z| \geq |w| - 2k$. In this cycle M cannot delete two corresponding factors $\alpha_\mu^{(i)}$ and $\gamma_{5-\mu}^{(i)}$, where $\mu \in \{1, 2\}$ and $1 \leq i \leq 2r$, as in that case z would contain the factor $d_\mu^q d_{3-\mu}^q$. Hence, M must delete two corresponding factors d_ν^q and e_ν^p, which implies that $k \geq \max\{p, q\}$. □

From the proof above it follows easily that $L_{p,q}$ is accepted by a det-2-RR-automaton with windows of size $\max\{p, q\}$. Lemma 16 and Proposition 18 together yield the following non-closure property, as $L_{p,q} = \psi^{-1}(L_D)$.

Proposition 19. *For each* $k \geq 1$, *the language class* $\mathcal{L}(\text{det-2-RR}(k))$ *is not closed under inverse morphisms.*

We suppose that the language classes $\mathcal{L}(\text{det-2-RR}(k))$ are not closed under product and Kleene star, either. For $k \geq 2$, we expect that the language $L_{\text{pal}}^2 = \{uu^R vv^R \mid u, v \in \{a, b\}^*\}$ is not accepted by any det-2-RR(k)-automaton, although L_{pal} is. This would prove non-closure under product for all $k \geq 2$.

Next we extend the above non-closure properties to the class $\mathcal{L}(\text{det-2-RR})$. In [10] a det-R-automaton M with a window of size five is presented that accepts a language $L(M) \subseteq \{a, b\}^*$ such that $L(M) \cap (ab)^* = \{(ab)^{2^n} \mid n \geq 0\}$, where M deletes exactly one symbol within each cycle. Hence, we can transform M into a det-2-RR(1)-automaton M_1 such that $L(M_1) = L(M)$. From this observation the following non-closure property can be derived.

Proposition 20. *The language class* $\mathcal{L}(\text{det-2-RR})$ *is not closed under intersection with regular sets.*

From Propositions 11 and 20 we obtain the following non-closure property.

Corollary 21. *The language class* $\mathcal{L}(\text{det-2-RR})$ *is not closed under union.*

Finally, let $\psi : a^* \to \{a, b\}^*$ be the morphism defined by $a \mapsto ab$. Then $\psi^{-1}(L(M)) = \psi^{-1}(L(M) \cap (ab)^*) = \psi^{-1}(\{(ab)^{2^n} \mid n \geq 0\}) = \{a^{2^n} \mid n \geq 0\}$, which is a Church-Rosser language [14] that can be shown to be not accepted by any det-2-RR-automaton by using the same kind of reasoning as in the proof of Proposition 20. This then yields the following.

Corollary 22. *The language class* $\mathcal{L}(\text{det-2-RR})$ *is not closed under inverse morphisms.*

It remains to show that $\mathcal{L}(\text{det-2-RR})$ is not closed under λ-free morphisms, product, and Kleene star, either.

6 Conclusion

Finally, we shortly look at the algorithmic properties of det-2-RR(k)-automata.

Proposition 23. $\mathcal{L}(\text{det-2-RR}) \subseteq \text{DTIME}(n^2)$.

Proof. Let M be a det-2-RR-automaton. Given a word w of length n as input, the computation of M consists of at most n cycles and a tail. Each cycle consists of $O(n)$ many steps, as in each step, one of the two heads is moved at least one position to the right. This also holds for the tail computation. Hence, M can be simulated by a deterministic multi-tape Turing machine in time $O(n^2)$.   □

From Proposition 23 we see that the non-emptiness problem for det-2-RR-automata is recursively enumerable. In fact, this is the best we can do. Recall that the language VALID(T) of *valid computations* of a single-tape Turing machine T is accepted by a 2-head DFA [8]. In fact, one can effectively construct a 2-head DFA A for this language from T. Then $L(A)$ is empty (finite) iff $L(T)$ is empty (finite). As emptiness and finiteness are not even recursively enumerable for Turing machines, this yields the following result.

Proposition 24. *For all $k \geq 1$, emptiness, finiteness, regularity, context-freeness, universality, inclusion, and equivalence are not recursively enumerable for det-2-RR(k)-automata.*

To sum up, we have presented a deterministic version of the restarting automaton that has two windows which can delete symbols from the tape. We have seen that based on the window size, we obtain an infinite strictly increasing hierarchy of language classes the union of which includes the deterministic context-free languages. Unfortunately, these classes are not closed under most of the operations considered. In fact, they are actually *anti-AFLs*, if they are not closed under product and Kleene star, either. Anti-AFLs are sometimes referred to as 'unfortunate families of languages,' but there is linguistical evidence that such language families might be of crucial importance, since the family of natural languages is an anti-AFL, too [5].

References

1. Berstel, J.: Transductions and Context-Free Languages. Teubner Studienbücher, Teubner-Verlag, Stuttgart (1979)
2. Buntrock, G., Otto, F.: Growing context-sensitive languages and Church-Rosser languages. Inf. Comput. **141**, 1–16 (1998)
3. Chrobak, M.: Hierarchies of one-way multihead automata languages. Theor. Comput. Sci. **48**, 153–181 (1986)
4. Chrobak, M., Li, M.: $k + 1$ heads are better than k for PDAs. J. Comput. Syst. Sci. **37**, 144–155 (1988)
5. Culy, C.: Formal properties of natural language and linguistic theories. Linguist. Philos. **19**, 599–617 (1996)

6. Dahlhaus, E., Warmuth, M.: Membership for growing context-sensitive grammars is polynomial. J. Comput. Syst. Sci. **33**, 456–472 (1986)

7. Gramatovici, R., Martín-Vide, C.: Sorted dependency insertion grammars. Theor. Comput. Sci. **354**, 142–152 (2006)

8. Holzer, M., Kutrib, M., Malcher, A.: Complexity of multi-head finite automata: origins and directions. Theor. Comput. Sci. **412**, 83–96 (2011)

9. Jančar, P., Mráz, F., Plátek, M., Vogel, J.: Restarting automata. In: Reichel, H. (ed.) FCT 1995. LNCS, vol. 965, pp. 283–292. Springer, Heidelberg (1995). doi:10. 1007/3-540-60249-6_60

10. Jančar, P., Mráz, F., Plátek, M., Vogel, J.: On restarting automata with rewriting. In: Păun, G., Salomaa, A. (eds.) New Trends in Formal Languages. LNCS, vol. 1218, pp. 119–136. Springer, Heidelberg (1997). doi:10.1007/3-540-62844-4_8

11. Jančar, P., Mráz, F., Plátek, M., Vogel, J.: On monotonic automata with a restart operation. J. Autom. Lang. Comb. **4**, 287–311 (1999)

12. Jurdziński, T., Loryś, K., Niemann, G., Otto, F.: Some results on RWW- and RRWW-automata and their relation to the class of growing context-sensitive languages. J. Autom. Lang. Comb. **9**, 407–437 (2004)

13. Lopatková, M., Plátek, M., Sgall, P.: Towards a formal model for functional generative description - analysis by reduction and restarting automata. The Prague Bull. Math. Linguist. **87**, 7–26 (2007)

14. McNaughton, R., Narendran, P., Otto, F.: Church-Rosser Thue systems and formal languages. J. ACM **35**, 324–344 (1988)

15. Mráz, F., Otto, F., Plátek, M.: Free word-order and restarting automata. Fundam. Inf. **133**, 399–419 (2014)

16. Otto, F.: Restarting automata. In: Ésik, Z., Martín-Vide, C., Mitrana, V. (eds.) Recent Advances in Formal Languages and Applications. Studies in Computational Intelligence, vol. 25, pp. 269–303. Springer, Heidelberg (2006). doi:10.1007/ 978-3-540-33461-3_11

17. Otto, F., Plátek, M.: A two-dimensional taxonomy of proper languages of lexicalized FRR-automata. In: Martín-Vide, C., Otto, F., Fernau, H. (eds.) LATA 2008. LNCS, vol. 5196, pp. 409–420. Springer, Heidelberg (2008). doi:10.1007/ 978-3-540-88282-4_37

18. Pardubská, D., Plátek, M., Otto, F.: Parallel communicating grammar systems with regular control and skeleton preserving FRR automata. Theor. Comput. Sci. **412**, 458–477 (2011)

19. Plátek, M., Mráz, F., Lopatková, M.: (In)dependencies in functional generative description by restarting automata. In: Bordihn, H., Freund, R., Hinze, T., Holzer, M., Kutrib, M., Otto, F. (eds.) Proceedings of NCMA 2010. books@acg.at, vol. 263, pp. 155–170. Österreichische Computer Gesellschaft, Vienna (2010)

20. Plátek, M., Mráz, F., Lopatková, M.: Restarting automata with structured output and functional generative description. In: Dediu, A.-H., Fernau, H., Martín-Vide, C. (eds.) LATA 2010. LNCS, vol. 6031, pp. 500–511. Springer, Heidelberg (2010). doi:10.1007/978-3-642-13089-2_42

21. Rambow, O., Joshi, A.: A processing model for free word order languages. In: Clifton, C., Frazier, L., Rayner, K. (eds.) Perspectives on Sentence Processing, pp. 267–302. Lawrence Erlbaum Associates, Hillsdale (1994)

22. Sudborough, H.: One-way multihead writing finite automata. Inf. Control **30**, 1–20 (1976)

Relative Prefix Distance Between Languages

Timothy Ng$^{(\boxtimes)}$, David Rappaport, and Kai Salomaa

School of Computing, Queen's University, Kingston, ON K7L 3N6, Canada
{ng,daver,ksalomaa}@cs.queensu.ca

Abstract. The prefix distance between two words x and y is defined as the number of symbols in x and y that do not belong to their longest common prefix. The relative prefix distance from a language L_1 to a language L_2, if finite, is the smallest integer k such that for every word in L_1, there is a word in L_2 with prefix distance at most k. We study the prefix distance between regular, visibly pushdown, deterministic context-free, and context-free languages. We show how to compute the distance between regular languages and determine whether the distance is bounded. For deterministic context-free languages and visibly pushdown languages, we show that the relative prefix distance to and from regular languages is decidable.

1 Introduction

Distances on words are typically defined to compare the similarity between two words. The prefix distance between two words x and y is defined as the number of symbols in x and y that do not belong to their longest common prefix. In some sense, the prefix distance measures the distance of objects arranged in a hierarchical structure [19]. The edit distance, which counts the minimum number of insertions, deletions, and substitutions required to transform one word into another, is more commonly used for string comparisons. However, the prefix distance is often simpler to compute than the edit distance and may suffice for certain applications, such as defect measurement [12] and intrusion detection [5]. Beyond string comparisons, the prefix distance also has interesting topological properties and is used to characterize the subsequentiality of functions [2].

These distance measures can be extended to sets of words, or languages. The standard topological definition for a distance over words when extended to languages takes the minimum of the distances between a word in each language and has been well studied [10,11,13,14,17]. Choffrut and Pighizzini [8] consider an alternate definition for distance between languages, called the relative or Hausdorff distance. The relative distance from one language to another is defined as the supremum over all words in the first language of the distance to the second language and is non-symmetric. A symmetric distance can be attained by simply taking the minimum of the relative distance in each direction.

Choffrut and Pighizzini study the distance between languages from the point of view of relations on words. They study various distances on subclasses of deterministic rational relations, showing that questions about distances are decidable

© Springer International Publishing AG 2017
É. Charlier et al. (Eds.): DLT 2017, LNCS 10396, pp. 284–295, 2017.
DOI: 10.1007/978-3-319-62809-7_21

for recognizable relations and undecidable for deterministic rational relations with respect to the prefix, suffix, and subword distances.

The relative distance has applications in verification, as a generalization of the language inclusion problem. For instance, Benedikt et al. [3,4] consider the one-sided relative edit distance to measure the cost of repairing regular specifications. They give an algorithm for deciding when the distance between regular languages is bounded and give complexity results for computing the edit distance between regular languages. Chatterjee et al. [7] consider the same problems for a context-free language and a language belonging to a subclass of the context-free languages.

In this paper, we study the relative prefix distance between various classes of languages. We show that the relative prefix distance between DFAs (deterministic finite automata) can be computed in polynomial time while computing the relative prefix and suffix distance between NFAs (nondeterministic finite automata) is PSPACE-complete. We also consider the computational problem of computing the relative prefix distance between more general classes of languages. For a fixed value k, deciding whether the relative prefix distance from a context-free language to a DFA (respectively, an NFA) language is at most k can be done in polynomial time (respectively, is EXPTIME-complete). On the other hand, computing the relative prefix distance from a regular language to a context-free language is undecidable. We show that the prefix distance neighbourhood of a DCFL (deterministic context-free language) is deterministic and this yields an algorithm to compute the relative prefix distance from a regular language to a DCFL. Finally, we show that computing the relative prefix distance from one visibly pushdown language to another is EXPTIME-complete while computing the relative prefix distance from a DCFL to a visibly pushdown language, or vice versa, is undecidable.

2   Preliminaries

Here we briefly recall some definitions and notation used in the paper. For all unexplained notions on finite automata and regular languages the reader may consult the textbook by Shallit [18] or the survey by Yu [20]. A survey of distances is given by Deza and Deza [9].

In the following, Σ is always a finite alphabet, the set of words of Σ is denoted Σ^*, and ε denotes the empty word. The reversal of a word $w \in \Sigma^*$ is denoted by x^R. The length of a word w is denoted by $|w|$. The cardinality of a finite set S is denoted $|S|$ and the power set of S is 2^S. A word $w \in \Sigma^*$ is a *subword* or *factor* of x if and only if there exist words $u, v \in \Sigma^*$ such that $x = uwv$. If $u = \varepsilon$, then w is a *prefix* of x. If $v = \varepsilon$, then w is a *suffix* of x.

A *nondeterministic finite automaton* (NFA) is a tuple $A = (Q, \Sigma, \delta, q_0, F)$ where Q is a finite set of states, Σ is an alphabet, δ is a transition function $\delta : Q \times \Sigma \to 2^Q$, $q_0 \in Q$ is a set of initial states, and $F \subseteq Q$ is a set of final states. A word $w \in \Sigma^*$ is *accepted* by A if for some $q_0 \in Q_0$, $\delta(q_0, w) \cap F \neq \emptyset$ and the language recognized by A consists of all words accepted by A. An NFA is a *deterministic finite automaton* (DFA) if for all $q \in Q$ and $a \in \Sigma$, $\delta(q, a)$ either consists of one state or is undefined.

A *pushdown automaton* (PDA) is a tuple $P = (Q, \Sigma, \Gamma, \delta, q_0, F)$ where Q is a finite set of states, Σ is an alphabet, Γ is a stack alphabet, δ is a transition function $\delta : Q \times \Sigma \cup \{\varepsilon\} \times \Gamma \rightarrow 2^{Q \times \Gamma^*}$, $q_0 \in Q$ is an initial state, and $F \subseteq Q$ is a set of final states. For a transition $(q', \beta) \in \delta(q, a, \alpha)$, the PDA pops the symbol α from the top of the stack and pushes the symbols β onto the stack.

A configuration of a PDA is a triple (q, w, π) where $q \in Q$ is the current state, $w \in \Sigma^*$ is the remaining input, and $\pi \in \Gamma^*$ is the contents of the stack. For a transition $(q', \beta) \in \delta(q, a, \alpha)$, we write $(q, aw, \alpha\pi) \vdash (q', w, \beta\pi)$. We denote by $\vdash^*$ a sequence of transitions between the two configurations. Then a word w is accepted by a PDA if $(q_0, w, \varepsilon) \vdash^* (q_f, \varepsilon, \pi)$ for $q_f \in F$ and $\pi \in \Gamma^*$.

A PDA is a *deterministic pushdown automaton* (DPDA) if the transition function satisfies $|\delta(q, a, \alpha)| \leq 1$ for all $q \in Q$, $a \in \Sigma \cup \{\varepsilon\}$, and $\alpha \in \Gamma$, and that for all $q \in Q$ and $\alpha \in \Gamma$, if $\delta(q, \varepsilon, \alpha) \neq \emptyset$, then $\delta(q, a, \alpha) = \emptyset$ for all $a \in \Sigma$. It is well known that the class of deterministic context-free languages is a proper subclass of the context-free languages.

A *visibly pushdown automaton* is a tuple $V = (Q, \Sigma, \Gamma, \delta, q_0, F)$ as in a pushdown automaton, with the additional constraint that the alphabet Σ and transition function δ are partitioned into three sets

- call actions Σ_c with the transition function $\delta_c : Q \times \Sigma_c \rightarrow 2^{Q \times \Sigma}$,
- return actions Σ_r with the transition function $\delta_r : Q \times \Sigma_r \times \Gamma \rightarrow 2^Q$,
- internal actions Σ_i with the transition function $\delta_i : Q \times \Sigma_i \rightarrow 2^Q$.

The stack operations of a VPA are determined entirely by the input symbols. Specifically, upon reading a call action, the VPA must push to the stack and upon reading a return action, the VPA must pop from the stack. It is well known that the class of languages accepted by VPAs, the *visibly pushdown languages*, is a proper subclass of the deterministic context-free languages [1].

A *finite state transducer*, or transducer, is a tuple $T = (Q, \Sigma, \Delta, \delta, q_0, F)$, where Q is a finite set of states, Σ and Δ are finite alphabets, $\delta \subseteq Q \times \Sigma^* \times \Delta^* \times Q$ is a finite set of transitions, $q_0 \in Q$ is an initial state, and $F \subseteq Q$ is a set of accepting states. An accepting computation of T is a sequence of elements of δ

$$(q_0, x_1, y_1, q_1)(q_1, x_2, y_2, q_2) \cdots (q_{n-1}, x_n, y_n, q_n)$$

where $q_n \in F$. We say the transducer maps the input string $x = x_1 \cdots x_n$ to the output string $y = y_1 \cdots y_n$, which we denote by $x \rightarrow_T y$. The set $\{(x, y) \mid x \rightarrow_T y\}$ is the relation realized by T. We define the transduction realized by T by

$$T(x) = \{y \in \Delta^* \mid x \rightarrow_T y\}.$$

2.1   Distances

A function $d : \Sigma^* \times \Sigma^* \rightarrow \mathbb{N} \cup \{0\}$ is a *distance* if it satisfies for all $x, y \in \Sigma^*$

1. $d(x, y) = 0$ if and only if $x = y$,

2. $d(x, y) = d(y, x)$,
3. $d(x, z) \le d(x, y) + d(y, z)$ for $z \in \Sigma^*$.

A distance between words can be extended to a distance between a word $w \in \Sigma^*$ and a language $L \subseteq \Sigma^*$ by

$$d(w, L) = \min\{d(w, w') \mid w' \in L\}.$$

We define the relative distance [8] from a language L_1 to language L_2 to be

$$d(L_1 | L_2) = \sup\{d(w_1, L_2) \mid w_1 \in L_1\}.$$

In other words, $d(L_1 | L_2)$ is the value of the maximum distance from any word in L_1 to the language L_2. Note that under this definition, $d(L_1 | L_2)$ is not symmetric and can be unbounded.

The *prefix distance* of x and y counts the number of symbols which do not belong to the longest common prefix of x and y. It is defined by

$$d_p(x, y) = |x| + |y| - 2 \cdot \max_{z \in \Sigma^*}\{|z| \mid x, y \in z\Sigma^*\}.$$

The suffix distance and subword distance can be similarly defined by considering the number of symbols of x and y which do not belong to the longest common suffix (subword, respectively) of x and y.

The *neighbourhood* of a language L of radius k with respect to a distance d is the set of all words $v \in \Sigma^*$ such that $d(u, v) \le k$ for some $u \in L$ [6]. More formally,

$$E(L, d, k) = \{w \in \Sigma^* \mid d(w, L) \le k\}.$$

It has been shown that the neighbourhoods of a regular language with respect to the prefix, suffix, and subword distances are regular [16].

3   Relative Prefix Distance Between Regular Languages

Choffrut and Pighizzini [8] showed that the main questions about the almost-reflexivity of a recognizable relation is decidable. Here, we show that these questions are computable in polynomial time if the languages are given as DFAs and that they are PSPACE-complete for NFAs.

Since the relative distance can be unbounded, we would like to characterize when, for two given languages, the distance is finite. In the following result, we show that the distance is either bounded by a function of the state complexity of the languages, or it is unbounded. First, we establish a simple lemma.

Lemma 1. *Let A_1 and A_2 be two NFAs recognizing L_1 and L_2 with n_1 and n_2 states respectively. Suppose $u \in L_1, v \in L_2$ and let p be the longest word satisfying $u = pu', v = pv'$. Then there exists a word $pw \in L_2$ such that $|w| < n_2 - 1$.*

Proof. This follows directly from the Pumping Lemma [18].   □

Theorem 2. *Let L_1, L_2 be regular languages recognized by NFAs A_1 and A_2 with n_1 and n_2 states, respectively. Suppose $d_p(L_1|L_2)$ is bounded. Then,*

$$d_p(L_1|L_2) \le n_1 + n_2 - 2$$

Proof. Let $u \in L_1$ and $v \in L_2$ such that

$$k = d_p(u, v) = d_p(u, L_2) = d_p(L_1, L_2) < \infty$$

and suppose $k > n_1 + n_2 - 2$. We write $u = pu'$ and $v = pv'$, with p be being the longest common prefix of u and v. Since $d_p(u, v) = \min_{w \in L_2}\{d_p(u, w)\}$, by Lemma 1, we have $|v'| \le n_2 - 1$, which implies that $|u'| > n_1 - 1$.

By the Pumping Lemma [18], we can write $u' = xyz$ where $|yz| < n_1 - 1$, $|y| > 0$ and $pxy^i z \in L_1$ for any $i \in \mathbb{N}$. Consider a word $u_2 = pxy^2 z \in L_1$ and let $v_2 \in L_2$ be a word such that $d_p(u_2, v_2) = d_p(u_2, L_2)$. That is, v_2 is the word in L_2 which is closest to u_2. By our assumption, we have $d_p(u_2, v_2) \le k$. Now let q be the longest word such that $u_2 = qu'_2$ and $v_2 = qv'_2$.

First, suppose that $|q| \le |pxy|$. Let $\ell = |pxy| - |q|$. Then,

$$k \le d_p(u, v_2) = \ell + |z| + |v'_2| < \ell + |y| + |z| + |v'_2| = d_p(u_2, v_2) = k$$

which is a contradiction. Next, suppose that $|q| > |pxy|$. Recall that $u = pxyz$ and let q' be such that $v_2 = pxyq'v'_2$. By Lemma 1, let w be such that $|w| < n_2 - 1$ and $pxyw \in L_2$. Then we have

$$k \le d_p(u, pxyw) = |z| + |w| \le n_1 - 1 + n_2 - 1 < k$$

which again is a contradiction. Therefore, $k \le n_1 + n_2 - 2$.   □

Example 3. We will show that this bound is reachable. Let $\Sigma = \{a, b\}$ and let $L_1 = \Sigma^* a \Sigma^{n_1 - 2}$ and $L_2 = \Sigma^* b \Sigma^{n_2 - 2}$. Note that L_1 can be recognized by an NFA with n_1 states and L_2 can be recognized by an NFA with n_2 states. We observe that for any word in $w \in L_1$, we have $d_p(w, L_2) \le n_1 + n_2 - 2$.

If the distance is bounded, then it is possible construct a neighbourhood of finite radius with respect to the given distance.

Lemma 4. *Let L_1 and L_2 be languages. Then $d_p(L_1|L_2) \le k$ if and only if $L_1 \subseteq E(L_2, d_p, k)$.*

Theorem 5. *Let A_1 and A_2 be DFAs. Then it is decidable in polynomial time whether $d_p(L(A_1)|L(A_2))$ is bounded.*

Proof. By Theorem 2, we know that $d_p(L(A_1)|L(A_2)) \le n_1 + n_2 - 2$ if it is bounded. Otherwise, it is unbounded. Therefore, it is enough to check

$$L(A_1) \subseteq E(L(A_2), d_p, n_1 + n_2 - 2).$$

It is known that the size of a DFA for $E(L(A_2), d_p, n_1 + n_2 - 2)$ is at most $\frac{n_2(n_2-1)}{2} + n_1 + n_2 - 1$ states, which is polynomial in n_2 and can be constructed in polynomial time [16]. Then since the inclusion problem for DFAs is decidable in polynomial time, checking the above inclusion can also be done in polynomial time.   □

Theorem 6. *Let A_1 and A_2 be DFAs. Then $d_p(L(A_1)|L(A_2))$ is computable in polynomial time.*

We will now show that the same questions are PSPACE-complete when we are given nondeterministic finite automata. First, we will make use of the following observation.

Lemma 7. *Consider languages L_1 and L_2 over an alphabet Σ. Let $\#$ be a symbol not in Σ and $k \in \mathbb{N}$. Then*

$$d_p(L_1 \#^k | L_2) \leq k \text{ iff } L_1 \subseteq L_2.$$

Theorem 8. *Let $k \in \mathbb{N}$ be fixed. For given NFAs A_1 and A_2, deciding whether or not $d_p(L(A_1)|L(A_2)) \leq k$ is PSPACE-complete.*

Proof. First, we note that given an n-state NFA A, we can construct an NFA A' for $E(L(A), d_p, k)$ with at most $n + k$ states [16]. To see that the problem is in PSPACE, we note that the problem is equivalent to deciding

$$L(A_1) \subseteq E(L(A_2), d_p, k).$$

To see that the problem is PSPACE-hard, we reduce from NFA universality. Suppose we are given an NFA A. Then by Lemma 7, $L(A) = \Sigma^*$ if and only if $d_p(\Sigma^* \#^k | L(A)) \leq k$. $\square$

Corollary 9. *Let A_1 and A_2 be NFAs. Then the problem of deciding whether $d_p(L(A_1)|L(A_2))$ is bounded is PSPACE-complete.*

We can derive some results for the suffix distance as well, by using its symmetry with the prefix distance. One might assume that this means the complexity of questions regarding the relative suffix distance follow straightforwardly from our results on the relative prefix distance. However, computing the neighbourhood with respect to the suffix distance is much more difficult than for the prefix distance. First, as a corollary of the bound from Theorem 2, we get:

Corollary 10. *Let L_1, L_2 be regular languages recognized by NFAs A_1 and A_2 with n_1 and n_2 states, respectively. Then either $d_s(L_1|L_2) \leq n_1 + n_2 - 2$ or $d_s(L_1|L_2)$ is unbounded.*

Proposition 11. *Let A_1 and A_2 be DFAs with n_1 and n_2 states, respectively. Then deciding whether $d_s(L(A_1)|L(A_2))$ is bounded is in PSPACE.*

Proof. We can construct an NFA for the language $E(L(A_2), d_s, n_1 + n_2 - 2)$ that has at most $n_2 + (n_1 + n_2 - 2)$ states. Thus, we can decide the inclusion $L_1 \subseteq E(L_2, d_s, n_1 + n_2 - 2)$ in PSPACE. $\square$

We note that the current best known DFA construction for $E(L(A_2), d_s, n_1 + n_2 - 2)$ has at most $n_1 + 2^{n_2}$ states, and is therefore not known to be polynomial in n_2 [15].

Corollary 12. *Let A_1 and A_2 be NFAs. Then the problem of deciding whether $d_s(L(A_1)|L(A_2))$ is bounded is PSPACE-complete.*

4   Relative Prefix Distance and Context-Free Languages

Here, we consider the relative distance on non-regular languages. The distance from one context-free language to another is undecidable by Choffrut and Pighizzini [8]. Thus, we consider the distance between context-free languages and regular languages. First, we define the following useful finite-state transducer.

Let $P_k = (Q_k, \Sigma, \Sigma, \delta_k, I, F_k)$ be a finite state transducer, shown in Fig. 1, with $Q_k = \{0, \ldots, k\}$, $I = \{0\}$, $F_k = Q_k$, and transitions

- $(0, a, a, 0)$ for all $a \in \Sigma$,
- $(i, a, \varepsilon, i+1)$ for all $a \in \Sigma$ and $0 \le i \le k - 1$,
- $(i, \varepsilon, a, i+1)$ for all $a \in \Sigma$ and $0 \le i \le k - 1$,
- $(i, a, b, i+2)$ for all $a, b \in \Sigma$ with $a \ne b$ and $0 \le i \le k - 2$.

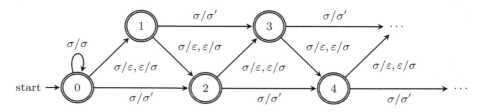

Fig. 1. The transducer P_k with $\sigma, \sigma' \in \Sigma$

Lemma 13. *Let $w \in \Sigma^*$. Then $P_k(w) = E(w, d_p, k)$.*

It is not difficult to see that for a language L, we have $P_k(L) = E(L, d_p, k)$. A similar transducer S_k with respect to the suffix distance can be defined by replacing the transition $(0, a, a, 0)$ for all $a \in \Sigma$ with (k, a, a, k) for all $a \in \Sigma^*$. We make use of the fact that context-free languages are closed under rational transductions to get the following result.

Proposition 14. *Let L be a context-free language. Then for every $k \ge 0$, the neighbourhood $E(L, d_p, k)$ is context-free.*

Proposition 15. *Let $k \in \mathbb{N}$ be fixed. Given a regular language L_1 and a context-free language L_2, determining whether or not $d_p(L_1|L_2) \le k$ is undecidable.*

On the other hand, computing the relative distance from a context-free language to a regular language becomes decidable, and when the regular language is given by a DFA, this problem can even be decided in polynomial time. First, we will state the following useful fact from Chatterjee et al. [7].

Proposition 16 ([7]). *Given a PDA P and an NFA A, the inclusion $L(P) \subseteq L(A)$ can be decided in EXPTIME. Given a deterministic PDA P and an NFA A, it is EXPTIME-hard to decide whether or not $L(P) \subseteq L(A)$.*

Proposition 17. *Let $k \in \mathbb{N}$ be fixed.*

1. *Given an NFA A and a PDA P, deciding whether or not $d_p(L(P)|L(A)) \leq k$ is EXPTIME-complete.*
2. *Given a DFA B and a PDA P, it can be decided in polynomial time whether or not $d_p(L(P)|L(B)) \leq k$.*

It is clear that by symmetry, we would attain the same results for the suffix distance between a context-free language and a regular language. However, if we consider the distance between a DCFL and a regular language, we get different results for the prefix and suffix distance. First, we show that the neighbourhood of a DCFL with respect to the suffix distance need not be a DCFL.

Lemma 18. *There exist a deterministic context-free language L and integer k for which $E(L, d_s, k)$ is not a deterministic context-free language.*

This lemma, together with Proposition 15, leads to some fairly straightforward results.

Proposition 19. *Let $k \in \mathbb{N}$ be fixed. Given a DPDA P and an NFA A,*

1. *deciding whether or not $d_s(L(A)|L(P)) \leq k$ is undecidable.*
2. *deciding whether or not $d_s(L(P)|L(A)) \leq k$ is EXPTIME-complete.*

Differing from the case of the suffix distance, we show that neighbourhoods of DCFLs with respect to the prefix distance are also DCFLs.

Theorem 20. *Let L be a deterministic context-free language. Then for every $k \geq 0$, the neighbourhood $E(L, d_p, k)$ is a deterministic context-free language.*

Proof. Given a DPDA A recognizing L, we will construct a DPDA A' that recognizes the neighbourhood $E(L, d_p, k)$. We need to determine whether the input word w has prefix distance at most k from some word in L. We can simulate a computation of w on A and based on the current state and the top k symbols of the pushdown stack, we can determine the length of a path to or from a closest final state of A. If such a path of length less than k exists, then w has a prefix distance of less than k from some word in L. This requires that we know what the top k symbols of the pushdown stack are, so we simulate the top of the stack via the finite state memory and store the rest of the stack on the pushdown stack as normal.

Let L be recognized by a DPDA $A = (Q, \Sigma, \Gamma, \delta, q_0, F)$. Then for each state $q \in Q$ and string of stack symbols $\pi \in \Gamma^{\leq k}$, we define the following function $\varphi_{A,k} : Q \times \Gamma^{\leq k} \to \mathbb{N}$,

$$\varphi_A(q, \pi) = \min_{w \in \Sigma^*} \left(\{ |w| \mid (q, w, \pi) \vdash^* (q', \varepsilon, \pi') \} \cup \{ k + 1 \} \right)$$

for some $q' \in F$ and $\pi' \in \Gamma^*$. The function $\varphi_{A,k}(q, \pi)$ gives the length of a shortest word w such that for any word x that reaches the state q with π on the top of the stack, we have $xw \in L$ if $|w| \leq k$. Based on this function, we can construct a DPDA $A' = (Q', \Sigma, \Gamma, \delta', q_0', F')$ that recognizes the language $E(L, d_p, k)$.

Let $Q' = Q \times \Gamma^{\leq k} \times \{0, \ldots, k+1\} \cup \{p_1, \ldots, p_k\}$. We set the initial state to be $q_0' = (q_0, \varepsilon, \varphi_A(q_0, \varepsilon))$. The set of final states is defined

$$F' = Q \times \Gamma^{\leq k} \times \{0, \ldots, k\} \cup \{p_1, \ldots, p_k\}.$$

We describe the transition function of A' by first describing the operation of the stack. We keep track of the top k symbols of the stack "in memory" in the states and store the rest of the stack as normal. Consider a state q, a symbol $a \in \Sigma$, and let $(q', \beta) = \delta(q, a, \alpha)$, where α is the top of the pushdown stack of A and $\beta = \beta_1 \cdots \beta_{|\beta|}$ is the symbols to be pushed onto the pushdown stack of A. Let $\pi = \gamma_1 \cdots \gamma_{|\pi|}$ be the top of the pushdown stack with $|\pi| \leq k$ to be kept in memory.

Consider the state (q, π, i) in A' with $\gamma_1 = \alpha$. First, we consider $\beta = \varepsilon$ to demonstrate a pop action. First, we consider when $|\pi| \leq k$. This occurs when the size of the in-memory portion of the stack has at most k elements and therefore, the size of the entire stack has at most k elements. In this case, the stack of A' is empty and the stack operations are performed only on the in-memory portion of the stack. Thus, for a transition $\delta(q, a, \alpha)$ in A, we have the transition $\delta'((q, \pi, i), a, \varepsilon) = ((q', \gamma_2 \cdots \gamma_{|\pi|}, j), \varepsilon)$.

If $|\pi| > k$ and the stack contains $m > k$ symbols, then the top k symbols of the stack of A are stored as π and the rest of the $m - k$ stack symbols are on the pushdown stack of A'. In this case, A' will pop the topmost symbol α' on its stack to append to the end of π and remove the first symbol γ_1 of π to simulate a pop from the top of the in-memory portion of the stack. Formally, the transition is $\delta'((q, \pi, i), a, \alpha') = (q', \gamma_2 \cdots \gamma_k \alpha', j), \varepsilon)$, where α' is the top of the pushdown stack of A'.

Now, for $\beta \neq \varepsilon$, we demonstrate the push action. Let $\pi' = \beta \cdot \gamma_2 \cdots \gamma_{|\pi|}$. First we consider when $|\pi'| \leq k$. In this case, γ_1 is popped as above and now we need to push the symbols onto the stack. Since the size of the stack is less than k, we store the entire contents in memory and we have $\delta'((q, \pi, i), a, \varepsilon) = ((q', \pi', j), \varepsilon)$. If $|\pi'| > k$, then we keep the first k symbols in memory and push the rest onto the stack. Let $\pi' = \eta_1 \cdots \eta_{|\pi'|}$. Then we have the transition $\delta'((q, \pi, i), a, \alpha') = ((q', \eta_1 \cdots \eta_k, j), \eta_{k+1} \cdots \eta_{|\pi'|} \alpha')$, where α' is the top of the pushdown stack of A'.

To see that this is all deterministic, we recall that each transition of A is uniquely determined by the current state q, input symbol a, and the top of the stack α. Each of the above transitions of A' is still uniquely determined by the same items, noting that α is γ_1, the first symbol of π and that the stack additional stack manipulations are determined by π, which is part of the state, and α', the top of the pushdown stack of A'.

Now, we consider the step counter in the third component of a state of A'. The counter either increments by one for each input symbol that is read, or takes on the value $\varphi_A(q, \pi)$ if it is smaller than the number of steps. Formally, for a transition $\delta'((q, \pi, i), a, \alpha) = ((q', \pi', j), \beta)$ of A', we define j by

$$j = \min(i + 1, \varphi_A(q', \pi')).$$

Finally, we consider transitions that were undefined in A. We define a chain of k new states p_i, $1 \leq i \leq k$. For each p_i and $1 \leq i \leq k - 1$, reading any symbol will transition to state p_{i+1} and there are no outgoing transitions from p_k. If on a state (q, π, j) with the top symbol of the pushdown stack α we ever read an input symbol a such that the transition $\delta(q, a, \alpha)$ is undefined, then based on the step counter component j, the machine A' enters the chain of k states at p_{j+1}.

Now we show that a word $w \in \Sigma^*$ reaches a state (q, π, i) if and only if there exists some word $x \in L$ such that $d_p(w, x) = i$ if $i \leq k$. First suppose that w reaches (q, π, i). Then this means w reaches the state q on the original machine. One of three cases is possible.

1. If $i \leq k$ and $\varphi_A(q, \pi) = i$, then there exists some suffix x' of length i such that $wx' \in L(A)$. This gives us $d_p(w, x) = d_p(w, wx') = |x'| = i$.
2. If $i \leq k$ and $\varphi_A(q, \pi) < i$, then on some prefix p of w, the first case applied. That is, for $w = pw'$, there exists a word $x = px'$ and $d_p(p, x) = |x'| = i - |w'|$. This implies that $d_p(w, x) = |x'| + |w'| = i$.
3. If $i > k$, then $i = k + 1$ and neither of the two cases above apply. Then there is no word x such that $d_p(w, x) \leq k$.

From the above, we observe that if w didn't reach a state in the original DPDA A, then, on some prefix p with $w = pw'$, reading p takes A' to the state (q, π, i). Then reading w' takes the machine to a state $p_{|w'|+i}$ if $i + |w'| \leq k$.

Since all states except for states of the form $(q, \pi, k + 1)$ are accepting states, we have $L(A') = E(L(A), d_p, k)$ and A' has $O(nk|\Gamma|^k)$ states. $\square$

Recall from Proposition 15 that the relative prefix distance from a regular language to a context-free language is undecidable. We get contrasting results for DCFLs using the construction from Theorem 20 and the fact that DCFLs are closed under complement.

Proposition 21. *Let $k \in \mathbb{N}$ be fixed.*

1. *Given an NFA A and a DPDA P, it can be decided in polynomial time whether or not $d_p(L(A)|L(P)) \leq k$.*
2. *Given a DFA B and a DPDA P, it can be decided in polynomial time whether or not $d_p(L(B)|L(P)) \leq k$.*

Now, we consider the class of visibly pushdown languages. The class of VPLs is known to be a proper subclass of DCFLs. First, we show that the relative prefix distance between a DCFL and VPL is undecidable.

Proposition 22. *Let $k \in \mathbb{N}$ be fixed. Given a visibly pushdown automaton A and a deterministic pushdown automaton P,*

1. *determining whether or not $d_p(L(A)|L(P)) \leq k$ is undecidable.*
2. *determining whether or not $d_p(L(P)|L(A)) \leq k$ is undecidable.*

Unlike DCFLs, the VPLs are closed under typical language operations and the standard questions involving VPLs are decidable. This will allow us to consider the problem of deciding whether the prefix distance from a VPL to another VPL is within k. First, we will show that the prefix neighbourhood of a VPL is also a VPL.

Theorem 23. *Let L be a visibly pushdown language. Then $E(L, d_p, k)$ is a visibly pushdown language for all $k \geq 0$.*

Proof. Let A be a visibly pushdown automaton that recognizes L. We will modify the construction of the prefix neighbourhood DPDA defined in the proof of Theorem 20 to construct a VPA A' that recognizes $E(L, d_p, k)$. To preserve the visibly pushdown property, we must push and pop from the stack as dictated by call and return symbols. This is only an issue when the stack of A has size less than k. In the DCFL construction, we simply ignored the stack, but we cannot do this for a VPA.

To solve this, we add a new symbol $\triangle$ to the stack alphabet. Let $q \in Q$, $a \in \Sigma_c$, and let $(q', \beta) = \delta_c(q, a)$. Consider the state (q, π, i) with $\pi = \gamma_1 \cdots \gamma_{|\pi|}$ and $|\pi| < k$. On this transition, the VPA A must push β onto the stack. In the VPA A', we add β to the top of the stack in memory and push a dummy symbol $\triangle$ onto the stack. Then our transition in A' is $\delta_c'((q, \pi, i), a) = ((q', \beta\pi, j), \triangle)$.

Now let $a \in \Sigma_r$ be a return action and consider the transition $\delta_r(q, a, \alpha) = q'$ of A. The VPA A must pop α from the stack, but if $|\pi| < k$, the stack of A' contains only $\triangle$s. Then the corresponding pop action on A' is to pop a $\triangle$ off of the stack, use γ_1 to determine the transition, and remove γ_1 from the in-memory portion of the stack. Then our corresponding transition in A' is $\delta_r'((q, \gamma_2 \cdots \gamma_{|\pi|}, i), a, \triangle) = q'$. Since a $\triangle$ is only pushed onto the stack whenever a call action is read and popped whenever a return action is read, we are guaranteed to have exactly as many $\triangle$s as there are symbols in the in-memory portion of the stack.

Once the in-memory portion of the stack reaches k symbols, the VPA behaves exactly like the DPDA that was constructed in Theorem 20 until the stack size becomes less than k again. Furthermore, since the construction preserves the determinism of the DCFL, if the VPA A is deterministic, then the VPA A' that is constructed via this process will also be deterministic.   □

Proposition 24. *Let $k \in \mathbb{N}$ be fixed. For given VPAs A_1 and A_2, deciding $d_p(L(A_1)|L(A_2)) \leq k$ is EXPTIME-complete.*

References

1. Alur, R., Madhusudan, P.: Adding nesting structure to words. J. ACM 56(3) (2009)
2. Béal, M.P., Carton, O., Prieur, C., Sakarovitch, J.: Squaring transducers: an efficient procedure for deciding functionality and sequentiality. Theor. Comput. Sci. **292**(1), 45–63 (2003)

3. Benedikt, M., Puppis, G., Riveros, C.: Bounded repairability of word languages. J. Comput. Syst. Sci. **79**(8), 1302–1321 (2013)
4. Benedikt, M., Puppis, G., Riveros, C.: The per-character cost of repairing word languages. Theor. Comput. Sci. **539**, 38–67 (2014)
5. Bruschi, D., Pighizzini, G.: String distances and intrusion detection: bridging the gap between formal languages and computer security. RAIRO Inform. Théor. et Appl. **40**, 303–313 (2006)
6. Calude, C.S., Salomaa, K., Yu, S.: Additive distances and quasi-distances between words. J. Univ. Comput. Sci. **8**(2), 141–152 (2002)
7. Chatterjee, K., Henzinger, T.A., Ibsen-Jensen, R., Otop, J.: Edit distance for pushdown automata. In: Halldórsson, M.M., Iwama, K., Kobayashi, N., Speckmann, B. (eds.) ICALP 2015. LNCS, vol. 9135, pp. 121–133. Springer, Heidelberg (2015). doi:10.1007/978-3-662-47666-6_10
8. Choffrut, C., Pighizzini, G.: Distances between languages and reflexivity of relations. Theor. Comput. Sci. **286**(1), 117–138 (2002)
9. Deza, M.M., Deza, E.: Encyclopedia of Distances. Springer, Heidelberg (2009)
10. Han, Y.-S., Ko, S.-K.: Edit-distance between visibly pushdown languages. In: Steffen, B., Baier, C., Brand, M., Eder, J., Hinchey, M., Margaria, T. (eds.) SOFSEM 2017. LNCS, vol. 10139, pp. 387–401. Springer, Cham (2017). doi:10.1007/978-3-319-51963-0_30
11. Han, Y.S., Ko, S.K., Salomaa, K.: The edit-distance between a regular language and a context-free language. Int. J. Found. Comput. Sci. **24**(07), 1067–1082 (2013)
12. Kutrib, M., Meckel, K., Wendlandt, M.: Parameterized prefix distance between regular languages. In: Geffert, V., Preneel, B., Rovan, B., Štuller, J., Tjoa, A.M. (eds.) SOFSEM 2014. LNCS, vol. 8327, pp. 419–430. Springer, Cham (2014). doi:10.1007/978-3-319-04298-5_37
13. Mohri, M.: Edit-distance of weighted automata: general definitions and algorithms. Int. J. Found. Comput. Sci. **14**(6), 957–982 (2003)
14. Ng, T.: Prefix distance between regular languages. In: Han, Y.-S., Salomaa, K. (eds.) CIAA 2016. LNCS, vol. 9705, pp. 224–235. Springer, Cham (2016). doi:10.1007/978-3-319-40946-7_19
15. Ng, T., Rappaport, D., Salomaa, K.: Descriptional complexity of error detection. In: Adamatzky, A. (ed.) Emergent Computation: A Festschrift for Selim G. Akl. ECC, vol. 24, pp. 101–119. Springer, Cham (2017). doi:10.1007/978-3-319-46376-6_6
16. Ng, T., Rappaport, D., Salomaa, K.: State complexity of prefix distance. Theor. Comput. Sci. **679**, 107–117 (2017)
17. Pighizzini, G.: How hard is computing the edit distance? Inf. Comput. **165**(1), 1–13 (2001)
18. Shallit, J.: A Second Course in Formal Languages and Automata Theory. Cambridge University Press, Cambridge (2009)
19. Skala, M.: Counting distance permutations. J. Discrete Algorithms **7**(1), 49–61 (2009)
20. Yu, S.: Regular languages. In: Rozenberg, G., Salomaa, A. (eds.) Handbook of Formal Languages, pp. 41–110. Springer, Heidelberg (1997). doi:10.1007/978-3-642-59136-5_2

On the Tree of Binary Cube-Free Words

Elena A. Petrova and Arseny M. Shur$^{(\boxtimes)}$

Institute of Natural Sciences and Mathematics, Ural Federal University,
Lenina str. 51, Ekaterinburg, Russia
{elena.petrova,arseny.shur}@urfu.ru

Abstract. We present two related results on the prefix tree of all binary cube-free words. First, we show that non-branching paths in this tree are short: such a path from a node of nth level has length $O(\log n)$. Second, we prove that the lower density of the set of branching points along any infinite path is at least 23/78. Our results are based on a technical theorem describing the mutual location of "almost cubes" in a cube-free word.

Keywords: Cube-free word · Power-free word · Prefix tree

1 Introduction

Power-free words and languages are studied in lots of papers starting with the seminal work by Thue [14], but the number of challenging open problems is still quite big. One group of problems concerns the internal structure of power-free languages. Such a language can be viewed as a poset with respect to prefix, suffix, or factor order; by "internal structure" we mean the structure of these posets. In the case of prefix or suffix order, the diagram of the poset is a tree; since power-free languages are closed under reversal, these two trees are isomorphic. Each node of a prefix tree generates a subtree and is a common prefix of its descendants. In this paper, we study the prefix tree of the language of binary cube-free words.

The structure of prefix trees of k-power-free languages is discussed in a series of papers. For all these languages, the subtree generated by any word has at least one leaf [1]. Further, it is always decidable whether a given word generates finite or infinite subtree, and every infinite subtree branches infinitely often [2,3]. All other results concern particular languages. Among these languages, the binary overlap-free language has the simplest structure due to its slow (polynomial) growth and immense connection to the Thue-Morse word. The finiteness problem for subtrees of the tree of binary overlap-free words was solved in [9] in a constructive way (the general solution [3] is non-constructive), together with the

E.A. Petrova—Supported by the grant 16-31-00212 of the Russian Foundation of Basic Research.

A.M. Shur—Supported by the grant 16-01-00795 of the Russian Foundation of Basic Research.

É. Charlier et al. (Eds.): DLT 2017, LNCS 10396, pp. 296–307, 2017.
DOI: 10.1007/978-3-319-62809-7_22

proof of existence of arbitrarily large finite subtrees. Furthermore, it is decidable (even in linear time!) whether the subtrees generated by two given words are isomorphic [12].

For the tree of ternary square-free words it is known that (a) finite subtrees of arbitrary depth can be built [7], (b) the share of branching points in an infinite path is at least $2/9$ [8], (c) if a node of depth n has a single descendant of depth $n + m$, then $m = O(\log n)$ [8]. If we take the subtree consisting of all infinite branches of the original tree, then the analog of (c) with the bound $m = O(n^{2/3})$ is known [10,11]. However, for the binary counterpart of this tree, namely, for the tree of binary cube-free words, only the analog of (a) is known [6]. The aim of this paper is to prove that the analogs of (b) (with the constant $23/78$) and of (c) hold for the prefix tree of binary cube-free words. Note that the property (b) shows that both mentioned trees are reasonably "uniform" in terms of branching. The exponential growth rate for the ternary square-free (binary cube-free) language is ≈ 1.30176 (resp., ≈ 1.45758) [13], and this is the average number of children of a node in the corresponding tree. The lower bound of $1 + 2/9$ (resp., $1 + 23/78$), stemming from (b), is not very far from this average.

The paper follows the same general line as [8], but the study of cube-free words requires different technique. In particular, the interaction between the "almost cubic" factors in cube-free words (see Theorem 1) is much more complicated than the interaction of "almost squares" in square-free words.

2   Preliminaries and Formulation of Main Results

We study words over the binary alphabet $\{a, b\}$. The empty word is denoted by λ. Finite [infinite] words over an alphabet Σ are treated as functions $w :$ $\{1, \ldots, n\} \to \Sigma$ [resp., $w : \mathbb{N} \to \Sigma$]. We write $[i \ldots j]$ for the range $i, i{+}1, \ldots, j$ of positive integers and $w[i \ldots j]$ for the factor of the word w occupying this range; $w[i \ldots i] = w[i]$ is just the ith letter of w. A word w has *period* $p < |w|$ if $w[1 \ldots |w|{-}p] = w[p{+}1 \ldots |w|]$. Two basic properties of periodic words (see, eg., [5]) will be widely used.

Lemma 1 (Lyndon, Schutzenberger). *If $uv = vw \neq \lambda$, then there are words $x \neq \lambda$, y and an integer n such that $u = xy, w = yx$ and $v = (xy)^n x$.*

Lemma 2 (Fine, Wilf). *If a word u has periods p and q and $|u| \geq p{+}q{-}\gcd$ (p, q) then u has period $\gcd(p, q)$.*

Standard notions of factor, prefix, and suffix are used. A *cube* is a nonempty word of the form uuu. A word is *cube-free* if it has no cubes as factors; a cube is *minimal* if it contains no other cubes as factors. A *p-cube* is a minimal cube of period p.

The set of binary cube-free words is infinite and can be represented by the *prefix tree* T, in which each cube-free word is a node, λ is the root, and any two nodes w and wx are connected by a directed edge labeled by the letter x. By a subtree in T we mean a tree consisting of a node and all its descendants in T. *Branching point* is a node with two children.

In a cube-free word w, we call a letter $w[i]$ *fixed by a p-cube* if $w[i-3p+1 \ldots i-1]y$ is a p-cube, where y is the letter distinct from $w[i]$. Note that $w[1 \ldots i]$ in this case is the only child of the word $w[1 \ldots i-1]$ in the tree T. A *fixed context* of w is any word v such that in the word wv each letter of v is fixed by some cube.

Theorem 3. *A fixed context of a binary cube-free word w has length $O(\log |w|)$.*

Theorem 4. *Let w be an infinite binary cube-free word and let $b(n)$ be the number of branching points in the path from the root to the vertex $w[1 \ldots n]$ in the tree T. Then $\liminf_{n \to \infty} \frac{b(n)}{n} \geq \frac{23}{78}$.*

3 Letters Fixed by Short Cubes

We call cubes with periods ≤ 8 *short*. Our first goal is to give an upper bound on the number of letters in a cube-free word that are fixed by short cubes. Namely, we will prove the following result.

Lemma 5. *Suppose that $t, l \geq 1$ are integers and w is a cube-free word such that either $|w| \geq t+l$ or w is infinite. Then the range $[t+1 \ldots t+l]$ contains at most $\frac{7l+24}{13}$ positions in which the letters of w are fixed by short cubes.*

Proof. The proof by hand would require a huge case analysis, so we used a computer search instead. Consider the language L of all binary words containing no cubes of period ≤ 8. It is regular, because is given by a finite set of forbidden factors; studying such "regular approximations" is a standard approach to power-free languages (see, e.g., [13, Sect. 3]). A recognizing partial deterministic automaton $\mathcal{A}$ for L can be built by a variation of the classical Aho-Corasick algorithm. The automaton accepts by any state and rejects if it cannot read the word. The situation when $\mathcal{A}$ reads $u[1 \ldots i-1]$ and cannot read $u[i]$ means exactly that $u[1 \ldots i]$ ends with a cube of period ≤ 8. Consider an accepting walk in $\mathcal{A}$ for a cube-free word w. The letter $w[i]$ is fixed by a short cube iff the reading of $w[1 \ldots i-1]$ ends in a vertex of outdegree 1 (we call these vertices *fixed*; the absent edge corresponds to the cube).

Let d be the maximum share of fixed vertices in a simple cycle in $\mathcal{A}$, c be the maximum difference between the numbers of fixed and non-fixed vertices in a simple path in $\mathcal{A}$. If $d \geq 1/2$, $l \geq c$, then the range $[t+1 \ldots t+l]$ contains at most $c + d(l - c)$ positions in which the letters of w are fixed by short cubes. Indeed, consider the walk corresponding to $w[t+1 \ldots t+l]$ when $\mathcal{A}$ reads w. This walk can be viewed as a simple path of length $k \geq 0$ with some cycles of total length $l - k$ attached to it; each cycle, in turn, is built from simple cycles. There are at most $\frac{k+c}{2}$ fixed vertices in the path (the origin of the path corresponds to $w[t]$ and thus does not count), and at most $d(l - k)$ such vertices in the cycles. For $d \geq 1/2$ we have $\frac{k+c}{2} + d(l - k) \leq c + d(l - c)$, as desired. The exhaustive search of simple cycles of $\mathcal{A}$ gives $d = 7/13$, $c = 4$. The lemma now follows.

Remark 6. To find cycles, we used Johnson's algorithm [4]. In our case it works in $O(|\mathcal{A}|C)$ time, where C is the number of cycles. But this number depends exponentially on the automaton's size and the automaton's size depends exponentially on the length of the longest forbidden word. So while we processed $\mathcal{A}$ in less than a minute, the same task for cubes of period ≤ 9 seems not feasible for a personal computer.

4   Long Cubes

The following theorem describe the restrictions on the cubes of similar length fixing closely located letters.

Theorem 7. *Suppose that $t, l \geq 1$, $p, q \geq 2$ are integers, w is a word of length $t+l$ such that $w[1\ldots t+l-1]$ is cube-free, the letter $w[t]$ is fixed by a p-cube, and w ends with a q-cube. Then q is outside the red zone in Fig. 1.*

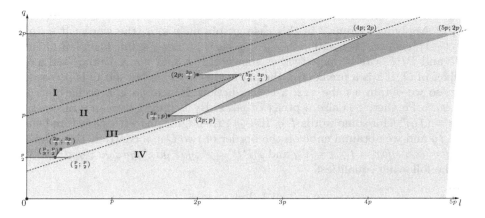

Fig. 1. The restrictions on fixing letters in a cube-free word. If $w[t]$ is fixed by a p-cube and $w[t+l]$ is fixed by a q-cube, then q (as a function of l with parameter p) must be outside the red polygon, including red border lines and red points. (Color figure online)

Remark 8. The restrictions of Theorem 7 are sharp in the sense that for any green point in Fig. 1 with rational coordinates one can find an instance word w (with p and q big enough); these examples can be derived from the corresponding constructions in the proof.

Proof. Let $P = w[t - 3p + 1\ldots t - 2p]$ and $P' = w[t - p + 1\ldots t]$. Then $w[t - 3p + 1\ldots t] = PPP'$. W.l.o.g. let P end with a, then P' ends with b. We denote the cube with period q at the end of w by $v = QQQ$.

There are four cases depending on the position of w where the word v begins (see Fig. 2):

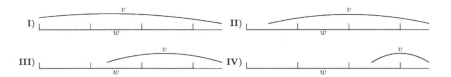

Fig. 2. Mutual location of factors w with period p and v with period q.

(I) $3q \geq 3p + l$;
(II) $2p + l \leq 3q < 3p + l$;
(III) $p + l \leq 3q < 2p + l$;
(IV) $3q \leq p + l$.

The corresponding areas are indicated in Fig. 1, separated by the dotted lines $q = l/3 + p$, $q = l/3 + 2p/3$, $q = l/3 + p/3$.

[I.] The word $w[t - 3p + 1 \ldots t-1]$ has periods p and q (or $q \geq 3p$, which agrees with Fig. 1). By Lemma 2 we obtain $3p - 1 < p + q - 1$ and then $q > 2p$, in agreement with Fig. 1.

[II.] Applying Lemma 2 in this case we obtain $p < q$ (the word $w[t - 2p + 1 \ldots t - 1]$ has periods p and q: $2p - 1 < p + q - 1$). Let $q < 2p$. Then left Q ends in P' and PPP' can be factorized as shown in Fig. 3, $x, z \neq \lambda$. Both y and z are prefixes of Q. If z is a prefix of y, then Q begins with zz; but left Q is preceeded by z, so we obtain a cube zzz, a contradiction. Hence, ya is a prefix of z; let $z = yaz'$. Further, z is also a prefix of $yaxz$. By Lemma 1 we can write $yax = fg, z = (fg)^n f$ for some words f, g. If $n \geq 1$ then Q begins with $fgfgf$ and ends with fg and we obtain a cube on the border of two Q's, a contradiction. Hence, $z = f = yaz', yax = fg, x = z'g$ and $PPP' = z'gyaz'ya\,z'gyaz'ya\,z'gyaz'yb$. We get the following equalities:

$$p = 2|z'| + 2|y| + |g| + 2,$$
$$q = 3|z'| + 3|y| + 2|g| + 3,$$
$$l = 5|z'| + 4|y| + 4|g| + 4.$$

Using these estimations of p, q, l we want to find

(a) $\alpha, \beta \geq 0$ such that $q \leq \alpha p + \beta l$ for any $|z'|, |y|, |g|$;
(b) $\bar{\alpha}, \bar{\beta} \geq 0$ such that $q \geq \bar{\alpha} p + \bar{\beta} l$ for any $|z'|, |y|, |g|$.

Fig. 3. The case $2p + l \leq 3q < 3p + l$, $q < 2p$

For $\alpha, \beta, \bar{\alpha}, \bar{\beta}$ we have vector inequalities with the coordinates $|z'|, |y| + 1, |g|$:

$$\begin{pmatrix} 3 \\ 3 \\ 2 \end{pmatrix} \leq \alpha \begin{pmatrix} 2 \\ 2 \\ 1 \end{pmatrix} + \beta \begin{pmatrix} 5 \\ 4 \\ 4 \end{pmatrix} ; \qquad \begin{pmatrix} 3 \\ 3 \\ 2 \end{pmatrix} \geq \bar{\alpha} \begin{pmatrix} 2 \\ 2 \\ 1 \end{pmatrix} + \bar{\beta} \begin{pmatrix} 5 \\ 4 \\ 4 \end{pmatrix}$$

Then all pairs (α, β) and $(\bar{\alpha}, \bar{\beta})$ are located in the first quadrant as in Fig. 4a.

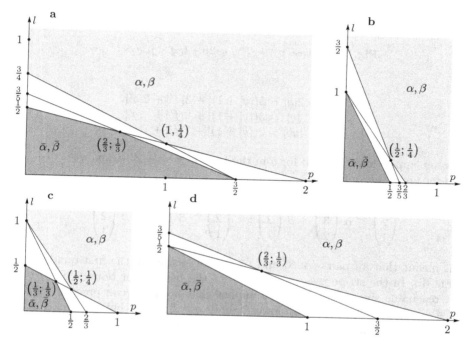

Fig. 4. Solutions for inequalities $q \leq \alpha p + \beta l$, $q \geq \bar{\alpha} p + \bar{\beta} l$.

Now we relate the bounds $q \leq \alpha p + \beta l, q \geq \bar{\alpha} p + \bar{\beta} l$ to the inequalities $2p + l \leq 3q < 3p + l$ defining the area II. The strongest bounds are on the boundaries; comparing them we have the best upper bound at the vertex $(1, 0.25)$: $q \leq p + \frac{l}{4}$. The best lower bound is at the vertex $(\frac{3}{2}, 0)$: $q \geq \frac{3}{2}p$. Finally the permitted values of q are inside the triangle with the sides $q = \frac{3p}{2}$ (not reachable as $|g| \neq 0$), $q = p + \frac{l}{4}$ (reachable when $|z'| = 0$), and $q = \frac{2p}{3} + \frac{l}{3}$ (the area boundary); the vertices are $(2p; \frac{3}{2}p), (\frac{5}{2}p; \frac{3}{2}p), (4p, 2p)$, as in Fig. 1.

III. In this case $q \neq p, q \neq \frac{p}{2}$ since $P \neq P'$. If $q < \frac{p}{2}$ there are no restrictions.

III.1: $q > l$, and then $q < p$ (also $q > \frac{p}{2}$ from above). Let QQQ start at ith position of the middle P. Then $P[i \ldots p]P[1 \ldots p - 1]$ is p- and q-periodic, so its length is less than $p + q - 1$ be Lemma 2. Then $|Q| > |P[i \ldots p]|$ and left Q ends in P'; we denote parts of w as in Fig. 5. Since $Q = xyaz = tx$, Lemma 1 implies

$x = (fg)^n f, t = fg, yaz = gf$ for some f, g. Since $tQ = ttx$ is cube-free, $n = 0$. If f is a prefix of g, then $Q^2 = fgffgf$ contains f^3. Similarly, if g is a prefix of f, then $tQ^2 = fgfgffgf$ contains $(fg)^3$. So f and g are not prefixes of each other. Last $Q = fgf$ begins with yb and in first Q, gf begins with ya, therefore, y is the longest common prefix of f and g: $f = ybf', g = yag'$. Estimate p, q and l:

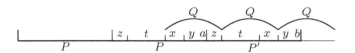

$$\text{\textbf{Fig. 5.} The case } p + l \leq 3q < 2p + l, \, q < p, q > l.$$

$$p = 3|f| + 2|g| = 5(|y| + 1)| + 3|f'| + 2|g'|,$$
$$q = 2|f| + |g| = 3(|y| + 1)| + 2|f'| + |g'|,$$
$$l = \quad q - |yb| = 2(|y| + 1)| + 2|f'| + |g'|.$$

We want upper [lower] bound for q in the form $q \leq \alpha p + \beta l$ [resp., $q \geq \bar{\alpha} p + \bar{\beta} l$], valid for any values of $|y|, |f'|, |g'|$. Then

$$\begin{pmatrix} 3 \\ 2 \\ 1 \end{pmatrix} \leq \alpha \begin{pmatrix} 5 \\ 3 \\ 2 \end{pmatrix} + \beta \begin{pmatrix} 2 \\ 2 \\ 1 \end{pmatrix}; \quad \begin{pmatrix} 3 \\ 2 \\ 1 \end{pmatrix} \geq \bar{\alpha} \begin{pmatrix} 5 \\ 3 \\ 2 \end{pmatrix} + \bar{\beta} \begin{pmatrix} 2 \\ 2 \\ 1 \end{pmatrix}$$

This means that all pairs (α, β) and $(\bar{\alpha}, \bar{\beta})$ are located in the first quadrant as in Fig. 4b. In the stripe $p + l < 3q < 2p + l$ the best upper bounds are $q \leq \frac{3l}{2}$ (not reachable since $|f'| = |g'| = 0$ implies that no letter can precede PPP'), $q \leq \frac{p}{2} + \frac{l}{4}$ (reachable if $|g'| = 0$); lower bounds $q \geq \frac{p}{2}, q \geq l$ (both not reachable). This gives us a quadrangle with vertices $(\frac{p}{3}; \frac{p}{2}); (\frac{2p}{5}; \frac{3p}{5}); (\frac{2p}{3}; \frac{2p}{3}); (\frac{p}{2}; \frac{p}{2})$.

III.2: $q \leq l, q < p$. As in III.1, the left Q ends in P' by Lemma 2. The parts of w are denoted in Fig. 6. We use Lemma 1 for tx and its prefix x: $t = fg, x = (fg)^n f$ for some f, g. Since ttx as a factor of PP, one has $n = 0$. Further, $Q = faz = fgfby; g = ag', z = g'fby$. Then $PP' = g'fbyfagfa \quad g'fbyfagfb$, $Q = fag'fby$. Estimate p, q and l:

$$p = 3(|f| + 1)| + 2|g'| + |y|,$$
$$q = 2(|f| + 1)| + |g'| + |y|,$$
$$l = 2(|f| + 1)| + |g'| + 2|y|.$$

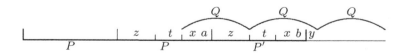

$$\text{\textbf{Fig. 6.} The case } p + l \leq 3q < 2p + l, \, q < p, q \leq l.$$

Using these equations, we find pairs (α, β), $(\bar{\alpha}, \bar{\beta})$ from vector inequalities

$$\begin{pmatrix} 2 \\ 1 \\ 1 \end{pmatrix} \le \alpha \begin{pmatrix} 3 \\ 2 \\ 1 \end{pmatrix} + \beta \begin{pmatrix} 2 \\ 1 \\ 2 \end{pmatrix} ; \quad \begin{pmatrix} 2 \\ 1 \\ 1 \end{pmatrix} \ge \bar{\alpha} \begin{pmatrix} 3 \\ 2 \\ 1 \end{pmatrix} + \bar{\beta} \begin{pmatrix} 2 \\ 1 \\ 2 \end{pmatrix}$$

The solutions are given in Fig. 4c. Similar to III.1, the best bounds leave for q a triangle with sides $q = l; q = \frac{l}{4} + \frac{p}{2}; q = \frac{l}{3} + \frac{p}{3}$ (all reachable) and vertices: $(\frac{p}{2}; \frac{p}{2}); (\frac{2p}{3}; \frac{2p}{3}); (2p; p)$. Together with the quadrangle from case III.1, we get the quadrangle with the vertices $(\frac{p}{3}; \frac{p}{2}); (\frac{p}{2}; \frac{p}{2}); (2p; p); (\frac{2p}{5}; \frac{3p}{5})$ from Fig. 1.

III.3: $q \le l, q > p$. Similar to the previous cases, we denote the parts of w (Fig. 7), write $t = fg, x = fgf$ by Lemma 1 and then $Q = fgfazt = fby, g = bg'$, $PP' = zfbg'fa \quad zfbg'fb$, $Q = fbg'fazt$. We find p, q, l and get vector inequalities:

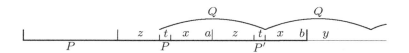

Fig. 7. The case $p + l \le 3q < 2p + l$, $q > p, q \le l$.

$$p = 2(|f| + 1) + \ |g'| + \ |z|,$$
$$q = 3(|f| + 1) + 2|g'| + \ |z|,$$
$$l = 5(|f| + 1) + 4|g'| + 2|z|;$$

$$\begin{pmatrix} 3 \\ 2 \\ 1 \end{pmatrix} \le \alpha \begin{pmatrix} 2 \\ 1 \\ 1 \end{pmatrix} + \beta \begin{pmatrix} 5 \\ 4 \\ 2 \end{pmatrix} ; \quad \begin{pmatrix} 3 \\ 2 \\ 1 \end{pmatrix} \ge \bar{\alpha} \begin{pmatrix} 2 \\ 1 \\ 1 \end{pmatrix} + \bar{\beta} \begin{pmatrix} 5 \\ 4 \\ 2 \end{pmatrix} .$$

The solutions are given in Fig. 4d. The best upper and lower bounds give a quadrangle with sides $q = p; q = \frac{3l}{5} q = \frac{l}{2}$ (all non-reachable); $q = \frac{l}{3} + \frac{2p}{3}$ (reachable with $|g'| = 0$), and vertices $(\frac{5p}{3}; p); (2p; p); (4p; 2p); (\frac{5p}{2}; \frac{3p}{2})$ and sides. This quadrangle is in Fig. 1 next to the triangle from Case II.

Thus we identified all "green" (and thus "red") parts of the areas I, II, III, getting the full picture from Fig. 1. (In area IV, the intersection of w and v is too small to impose a restriction.) Theorem 7 is proved.   □

Our aim is to find, using Theorem 7, the upper bound on the share of positions in which the letters of a cube-free word w are fixed by cubes with "close" periods.

Lemma 9. *Suppose that $t, l \ge 1$, $p \ge 2$ are integers, and w is a cube-free word such that either $|w| \ge t + l$ or w is infinite. Then the range $[t + 1 \ldots t + l]$ contains less than $\frac{3}{2} + \frac{3l}{4p}$ positions in which the letters of w are fixed by cubes with periods in the range $[p \ldots 2p - 1]$.*

Remark 10. The bound in Lemma 9 is not optimal; in fact, an asymptotically optimal ratio is $\frac{2l}{3p}$, not $\frac{3l}{4p}$. This ratio can be attained by using the periods p and $\lfloor \frac{3p}{2} \rfloor + 1$ interchangeably. But the proof we know for this better bound is tedious. Since this bound does not affect Theorem 3 and very slightly improve the constant in Theorem 4, we do not give the proof here.

Proof. Our aim is to place points into the horizontal stripe bounded by the lines $q = p$ and $q = 2p - 1$ (cf. Fig. 1; 0 on the l-axis corresponds to $t+1$) to achieve the maximum density of their l-coordinates (as the limit of the number of such coordinates in $[0 \ldots n]$ as $n \to \infty$). The points are ordered by their l-coordinate. The location of points is subject to the restriction that each point is outside the red polygons of all previous points; the set of points satisfying this restriction is called *valid*. Since we are interested in the upper bound, we allow non-integer coordinates of points. First note that for any point $(\bar{l}, \bar{q})$ the next point in a valid set is strictly to the right of the line $q = \frac{3}{2}(l - \bar{l})$ (the shortest red line segment in Fig. 1 belongs to this line). So for this next point we have $l - \bar{l} > \frac{2}{3}q \geq \frac{2}{3}p$.

Next we release the restrictions, replacing the red polygon in Fig. 1 with a smaller quadrangle with the borders $q = \frac{3}{8}l + \frac{1}{2}p$, $q = \frac{3}{4}l$ (these two lines intersect at the point $(\frac{4}{3}p, p)$), $q = 2p$ and $l = 0$; see Fig. 8, the borders are excluded. Clearly, since the restricted area became smaller, we can place points with either the same or bigger density; in particular, all valid sets will remain valid. The intersection of such a quadrangle built for the point $(\bar{l}, \bar{q})$ with the stripe $p \leq$

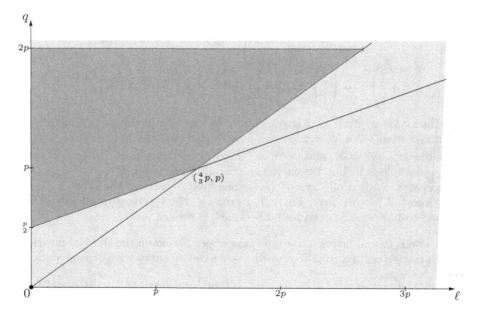

Fig. 8. The released restrictions on fixing letters in a cube-free word. Borders are excluded from restrictions.

$q \leq 2p - 1$ is called below the $(\bar{l}, \bar{q})$-*polygon*. We call two slanted segments of the boundary of the polygon the *lower* and *upper* segments, respectively.

Claim. Let $(l_1, q_1), (l_2, q_2)$ be two points in a valid set, $l_2 > l_1$. Then both slanted segments of the (l_2, q_2)-polygon are on the right of the corresponding segments of the (l_1, q_1)-polygon.

Proof. The upper segments are given by the equations $q = \frac{3}{4}(l - l_1)$ and $q = \frac{3}{4}(l - l_2)$, respectively. Since $l_2 > l_1$, the latter segment is of the right of the former one. For the lower segments note that the point (l', p) on the line $q = \frac{3}{8}(l - l_1) + \frac{1}{2}q_1$ satisfies $l' \leq l_2$, while the point (l'', p) on the line $q = \frac{3}{8}(l - l_2) + \frac{1}{2}q_2$ satisfies $l'' > l_2$. □

Let $S = \{(l_0, q_0), \ldots, (l_s, q_s)\}$ be a set of points from our stripe, $l_0 < \cdots < l_s$. We call S semi-valid if each (l_i, q) is outside all (l_j, q_j)-polygons with $j < i$. Claim implies a nice property: S is semi-valid iff each (l_i, q) is outside the (l_{i-1}, q_{i-1})-polygon.

We will modify a semi-valid schedule S to estimate $l_s - l_0$. First we replace (l_s, q_s) with (l_s, p). This "move down" obviously keeps the point outside the previous polygon. Then consider three successive points $(l', q'), (l'', q''), (\hat{l}, p)$ in a semi-valid set.

Note that $\hat{l} - l'' \geq \frac{8}{3}(p - \frac{q''}{2})$. Further, $l'' - l' \geq \frac{8}{3}(q'' - \frac{q'}{2})$ if $l'' \leq l'$ and $l'' - l' \geq \frac{4}{3}q''$ otherwise. In both cases we can replace (l'', q'') with the point $(\tilde{l}, p)$, where $\tilde{l} = l' + \frac{8}{3}(p - \frac{q'}{2})$, getting a semi-valid set again. Indeed, in both cases $(\tilde{l}, p)$ is on the border of the (l', q')-polygon (so $\tilde{l} \leq l''$), and the difference $\hat{l} - \tilde{l}$ is at least $\frac{4}{3}p$.

Now we proceed as follows: we replace all points in S, right to left, except for (l_0, s_0), by the points with the q-coordinate p. The difference between the l-coordinates of successive points is at least $\frac{4}{3}p$. Thus, $l_s - l_0 \geq \frac{4}{3}(s-1)p + (l^* - l_0)$, where l^* is the coordinate of the point replacing l_1. In addition we know that in a valid set the l-coordinates of successive points differ by at least $\frac{2}{3}p$. Gathering these results, we obtain

$$\frac{4}{3}p(s - 1) + \frac{2}{3}p + 1 \leq l,$$

implying $s + 1 < \frac{l - 2p/3}{4p/3}$. The result now follows. □

Proof (of Theorem 3). Let v be a fixed context of w, $|w| = n$, $|v| = l$. Thus, the cube-free word wv ends with l fixed letters. By Lemma 5, at most $\frac{7l + 24}{13}$ of these positions are fixed by short cubes (with periods ≤ 8). We partition the range $[9 \ldots \lfloor \frac{n+l}{3} \rfloor]$ of all possible longer periods into ranges of the form $[p \ldots 2p - 1]$ starting from the left (the last range can be incomplete). The number of ranges is the minimal number k such that $2^k \cdot 9 - 1 \geq \lfloor \frac{n+l}{3} \rfloor$, i.e., $k = \lceil \log \frac{1}{9} \lfloor \frac{n+l+3}{3} \rfloor \rceil$. By Lemma 9, the periods from the ith smallest range fix less than $\frac{3}{2} + \frac{3l}{9 \cdot 2^{i+1}}$ positions. Since the total number of fixed positions is l, we have

$$l \leq \frac{7l}{13} + \frac{24}{13} + \frac{3}{2}k + \frac{3l}{36}\left(\frac{1 - 1/2^k}{1 - 1/2}\right) = \frac{55l}{78} + \frac{24}{13} + \frac{3}{2}k - \frac{l}{6 \cdot 2^k}. \tag{1}$$

Observing that $k \leq \lceil \log \frac{n+l+3}{27} \rceil$, $2^k < \frac{n+l+3}{27}$, we get from (1)

$$\frac{23l}{78} < \frac{3}{2} \left\lceil \log \frac{n+l+3}{27} \right\rceil + \frac{24}{13} - \frac{9l}{2(n+l+3)},$$

and thus

$$l < \frac{117}{23} \left\lceil \log \frac{n+l+3}{27} \right\rceil + \frac{144}{23} - \frac{351l}{23(n+l+3)}. \tag{2}$$

If $l \geq n - 2$, then (2) implies $l < \frac{117}{23} \lceil \log \frac{2l+5}{27} \rceil + \frac{144}{23} - \frac{351l}{46l+115}$, but this inequality clearly has no positive integer solutions. Therefore $l \leq n - 3$ for any n, and we have, from (2),

$$l < \frac{117}{23} \left\lceil \log \frac{2n}{27} \right\rceil + \frac{144}{23} < \frac{117}{23} \log \frac{n}{27/4} + \frac{144}{23} < \frac{117}{23} \log n - 7$$

whence the result. $\qquad \square$

Proof (of Theorem 4). Assume to the contrary that $\liminf_{n \to \infty} \frac{b(n)}{n} < \frac{23}{78}$. Then there exist $\alpha < \frac{23}{78}$, $\{l_i\}_1^{\infty}$ such that $b(l_i) \leq \alpha l_i$. Consider the $(\lambda, w[1 \ldots l_i])$-path. The letters in it are partitioned in three sets: fixed by short cubes, fixed by long cubes, and not fixed. Lemmas 5 and 9 and our assumption give upper bounds for the numbers of letters in each part. Similar to (1), (2), we get the following formula (recall that now we have $n = 0$):

$$l_i < \frac{7l_i + 24}{13} + \frac{3}{2} \left\lceil \log \frac{l_i + 3}{27} \right\rceil + \frac{l_i}{6} - \frac{9l_i}{2(l_i + 3)} + \alpha l_i.$$

Then $\left(\frac{23}{78} - \alpha\right) l_i < \frac{3}{2} \log l_i - O(1)$. This inequality should hold for any l_i, but it has only finitely many integer solutions. This contradiction proves the theorem. $\qquad \square$

Remark 11. The increase of the upper bound on the periods of short cubes leads to better bounds in Theorem 4. For example, if we replace the maximum period 8, used in Lemma 5, with 7 we get $\frac{57}{208} \approx 0.274$ instead of $\frac{23}{78} \approx 0.295$.

5   Conclusion

This paper evens out the knowledge on the prefix trees of binary cube-free and ternary square-free words. The trees appear to have similar structure, which is not surprising since these languages have lot in common (in contrast with the language and the prefix tree of binary overlap-free words). The natural next steps in the study of these two trees are the answers to the following questions related to Theorem 3:

(1) is it true that every finite subtree in these trees has size $O(\log n)$, where n is the depth of its root?
(2) is it true that the same logarithmic upper bound works for the distance between branching nodes in the subtree consisting of all infinite branches of the original prefix tree?

References

1. Bean, D.A., Ehrenfeucht, A., McNulty, G.: Avoidable patterns in strings of symbols. Pac. J. Math. **85**, 261–294 (1979)
2. Currie, J.D.: On the structure and extendibility of k-power free words. Eur. J. Comb. **16**, 111–124 (1995)
3. Currie, J.D., Shelton, R.O.: The set of k-power free words over σ is empty or perfect. Eur. J. Comb. **24**, 573–580 (2003)
4. Johnson, D.B.: Finding all the elementary circuits in a directed graph. SIAM J. Comput. **4**(1), 77–84 (1975)
5. Lothaire, M.: Combinatorics on Words, Encyclopedia of Mathematics and Its Applications, vol. 17. Addison-Wesley, Reading (1983)
6. Petrova, E.A., Shur, A.M.: Constructing premaximal binary cube-free words of any level. Int. J. Found. Comput. Sci. **23**(8), 1595–1609 (2012)
7. Petrova, E.A., Shur, A.M.: Constructing premaximal ternary square-free words of any level. In: Rovan, B., Sassone, V., Widmayer, P. (eds.) MFCS 2012. LNCS, vol. 7464, pp. 752–763. Springer, Heidelberg (2012). doi:10.1007/978-3-642-32589-2_65
8. Petrova, E.A., Shur, A.M.: On the tree of ternary square-free words. In: Manea, F., Nowotka, D. (eds.) WORDS 2015. LNCS, vol. 9304, pp. 223–236. Springer, Cham (2015). doi:10.1007/978-3-319-23660-5_19
9. Restivo, A., Salemi, S.: Some decision results on non-repetitive words. In: Apostolico, A., Galil, Z. (eds.) Combinatorial Algorithms on Words. NATO ASI Series, vol. F12, pp. 289–295. Springer, Heidelberg (1985). doi:10.1007/978-3-642-82456-2_20
10. Shelton, R.: Aperiodic words on three symbols. II. J. Reine Angew. Math. **327**, 1–11 (1981)
11. Shelton, R.O., Soni, R.P.: Aperiodic words on three symbols. III. J. Reine Angew. Math. **330**, 44–52 (1982)
12. Shur, A.M.: Deciding context equivalence of binary overlap-free words in linear time. Semigroup Forum **84**, 447–471 (2012)
13. Shur, A.M.: Growth properties of power-free languages. Comput. Sci. Rev. **6**, 187–208 (2012)
14. Thue, A.: Über unendliche Zeichenreihen. Norske vid. Selsk. Skr. Mat. Nat. Kl. **7**, 1–22 (1906)

Limited Automata and Unary Languages

Giovanni Pighizzini$^{(\boxtimes)}$ and Luca Prigioniero

Dipartimento di Informatica, Università degli Studi di Milano, Milan, Italy
{pighizzini,prigioniero}@di.unimi.it

Abstract. Limited automata are one-tape Turing machines that are allowed to rewrite the content of any tape cell only in the first d visits, for a fixed constant d. When $d = 1$ these models characterize regular languages. An exponential gap between the size of limited automata accepting unary languages and the size of equivalent finite automata is proved. Since a similar gap was already known from unary context-free grammars to finite automata, also the conversion of such grammars into limited automata is investigated. It is proved that from each unary context-free grammar it is possible to obtain an equivalent 1-limited automaton whose description has a size which is polynomial in the size of the grammar. Furthermore, despite the exponential gap between the sizes of limited automata and of equivalent unary finite automata, there are unary regular languages for which d-limited automata cannot be significantly smaller than equivalent finite automata, for any arbitrarily large d.

1 Introduction

The investigation of computational models and of their computational power is a classical topic in computer science. For instance, characterizations of the classes in the Chomsky hierarchy by different types of computational devices are well-known. In particular, the class of context-free languages is characterized in terms of pushdown automata. A less known characterization of this class has been obtained in 1967 by Hibbard, in terms of Turing machines with rewriting restrictions, called *limited automata* [2]. For each integer $d \geq 0$, a d-limited automaton is a one-tape Turing machine which can rewrite the content of each tape cell *only in the first d visits*. For each $d \geq 2$, the class of languages accepted by these devices coincides with the class of context-free languages, while for $d = 1$ only regular languages are accepted [2,14].

More recently, limited automata have been investigated from the descriptional complexity point of view, by studying the relationships between the sizes of their descriptions and those of other equivalent formal systems. In [11], it has been proved that each 1-limited automaton M with n states can be simulated by a one-way deterministic automaton with a number of states double exponential in a polynomial in n. The upper bound reduces to a single exponential when M is deterministic. Furthermore, these bounds are optimal, namely, they cannot be reduced in the worst case. In [12], it has been shown how to transform each

© Springer International Publishing AG 2017
É. Charlier et al. (Eds.): DLT 2017, LNCS 10396, pp. 308–319, 2017.
DOI: 10.1007/978-3-319-62809-7_23

given 2-limited automaton M into an equivalent pushdown automaton, having a description of exponential size with respect to the description of M. Even for this simulation, the cost cannot be reduced in the worst case. On the other hand, the converse simulation is polynomial in size. In [4], it is proved that the cost of the simulation of d-limited automata by pushdown automata remains exponential even when $d > 2$. A subclass of 2-limited automata, which still characterizes context-free languages but whose members are polynomially related in size with pushdown automata, has been investigated in [10].

In all above mentioned results, lower bounds have been obtained by providing witness languages defined over a binary alphabet. In the unary case, namely in the case of languages defined over a one-letter alphabet, it is an open question if these bounds remain valid. It is suitable to point out that in the unary case the classes of regular and context-free languages collapse [1] and, hence, d-limited automata are equivalent to finite automata for each $d > 0$. The existence of unary 1-limited automata which require a quadratic number of states to be simulated by two-way nondeterministic finite automata has been proved in [11], while in [12] it has been shown that the set of unary strings of length multiple of 2^n, can be recognized by a 2-limited automaton of size $O(n)$, for any fixed $n > 0$. On the other hand, each (even two-way nondeterministic) finite automaton requires a number of states exponential in n to accept the same language. The investigation of the size of unary limited automata has been deepened in [5], where the authors stated several bounds for the costs of the simulations of different variants of unary limited automata by different variants of finite automata. Among these results, they proved the existence of languages accepted by $4n$-states 1-limited automata which require $n \cdot e^{\sqrt{n \ln n}}$ states to be accepted by two-way nondeterministic finite automata.

In this paper we improve these results, by obtaining an exponential gap between unary 1-limited automata and finite automata. We show that for each $n > 1$ the singleton language $\{a^{2^n}\}$ can be recognized by a deterministic 1-limited automaton having $2n + 1$ many states and a description of size $O(n)$. Since the same language requires $2^n + 1$ states to be accepted by a one-way nondeterministic automaton, it turns out that the state gap between deterministic 1-limited automata and one-way nondeterministic automata in the unary case is the same as in the binary case. We will also observe that the gap does not reduce if we want to convert unary deterministic 1-limited automata into two-way nondeterministic automata. However, when converting finite automata into limited automata, a size reduction corresponding to such a gap is not always achievable, even if we convert a unary finite automaton into a nondeterministic d-limited automaton for any arbitrarily large d.

In the second part of the paper, we consider unary context-free grammars. The cost of the conversion of these grammars into finite automata has been investigated in [9] by proving exponential gaps. Here, we study the conversion of unary context-free grammars into limited automata. With the help of a result presented in [8], we prove that each unary context-free grammar G can be converted into an equivalent 1-limited automaton whose description has a size which is polynomial in the size of G.

2   Preliminaries

In this section we recall some basic definitions useful in the paper. Given a set S, $\#S$ denotes its cardinality and 2^S the family of all its subsets. Given an alphabet Σ and a string $w \in \Sigma^*$, let us denote by $|w|$ the length of w and by ε the empty string.

We assume the reader familiar with notions from formal languages and automata theory, in particular with the fundamental variants of finite automata (1-DFAS, 1-NFAS, 2-DFAS, 2-NFAS, for short, where $1/2$ mean *one-way/two-way* and D/N mean *deterministic/nondeterministic*, respectively) and with *context-free grammars* (CFGS, for short). For further details see, e.g., [3].

Given an integer $d \geq 0$, a *d-limited automaton* (*d*-LA, for short) is a tuple $A = (Q, \Sigma, \Gamma, \delta, q_I, F)$, where Q is a finite set of states, Σ is a finite input alphabet, Γ is a finite *working alphabet* such that $\Sigma \cup \{\triangleright, \triangleleft\} \subseteq \Gamma$, $\triangleright$, $\triangleleft \notin \Sigma$ are two special symbols, called the *left* and the *right end-markers*, $\delta : Q \times \Gamma \rightarrow 2^{Q \times (\Gamma \setminus \{\triangleright, \triangleleft\}) \times \{-1, +1\}}$ is the transition function. At the beginning of the computation, the input is stored onto the tape surrounded by the two end-markers, the left end-marker being at the position zero. Hence, on input w, the right end-marker is on the cell in position $|w| + 1$. The head of the automaton is on cell 1 and the state of the finite control is the *initial state* q_I. In one move, according to δ and to the current state, A reads a symbol from the tape, changes its state, replaces the symbol just read from the tape by a new symbol, and moves its head to one position forward or backward. In particular, $(q, X, m) \in \delta(p, a)$ means that when the automaton in the state p is scanning a cell containing the symbol a, it can enter the state q, rewrite the cell content by X, and move the head to *left*, if $m = -1$, or to *right*, if $m = +1$. Furthermore, the head cannot violate the end-markers, except at the end of computation, to accept the input, as explained below. However, replacing symbols is allowed to modify the content of each cell only during the first d visits, with the exception of the cells containing the end-markers, which are never modified. For technical details see [12].

An automaton A is said to be *limited* if it is *d*-limited for some $d \geq 0$. A accepts an input w if and only if there is a computation path which starts from the initial state q_I with the input tape containing w surrounded by the two end-markers and the head on the first input cell, and which ends in a *final state* $q \in F$ after violating the right end-marker. The language accepted by A is denoted by $L(A)$. A is said to be *deterministic* (*d*-DLA, for short) whenever $\#\delta(q, \sigma) \leq 1$, for any $q \in Q$ and $\sigma \in \Gamma$.

In this paper we are interested to compare the size of the description of devices and formal systems. As customary, to measure the size of a finite automaton we consider the cardinality of the state set. For a context-free grammar we count the total number of symbols which are used to write down its productions.

For d-limited automata, the size depends on the number q of states and on the cardinality m of the working alphabet. In fact, given these two parameters, the possible number of transitions is bounded by $2q^2m^2$. Hence, if q and m are polynomial with respect to given parameters, also the size of the d-LA is polynomial.

The costs of the simulations of 1-LA by finite automata have been investigated in [11], proving the following result:

Theorem 1. *Let M be an n-state 1-LA. Then M can be simulated by a 1NFA with $n \cdot 2^{n^2}$ states and by a 1DFA with $2^{n \cdot 2^{n^2}}$ states. Furthermore, if M is deterministic then an equivalent 1DFA with no more than $n \cdot (n+1)^n$ states can be obtained.*

Using witness languages over a binary alphabet, it has been shown that the exponential gaps and the double exponential gap in Theorem 1 cannot be reduced [11].

3   On the Size of Unary Limited Automata

In this section we compare the sizes of unary limited automata with the sizes of equivalent finite automata. Our main result is that unary 1-LAS can be exponentially more succinct than finite automata even while comparing unary *deterministic* 1-LAS with *two-way nondeterministic* automata. However, there are unary regular languages that do not have any d-limited automaton which is significantly more succinct than finite automata, even for arbitrarily large d.

Let us start by showing that, for each $n > 1$, the language $L_n = \{a^{2^n}\}$, which requires $2^n + 1$ states to be accepted by a 1-NFA, can be accepted by a 1-DLA of size $O(n)$. Let us proceed by steps. In order to illustrate the construction, first it is useful to discuss how L_n can be accepted by a linear bounded automaton M_n (i.e., a Turing machine that can use as storage the tape which initially contains the input, by rewriting its cells an unbounded number of times).

M_n works in the following way:

i. Starting from the first input symbol, M_n scans the input tape from left to right by counting the as until the right end-marker is reached. Each odd-counted a is overwritten by X. A counter modulo 2 is enough to implement this step.

ii. The previous step is repeated n times, after moving backward the head until reaching the left end-marker. If in one of the first $n-1$ iterations M_n discovers that the number of as at the end of the iterations was odd, then M_n rejects. After the last iteration, M_n accepts if only one a is left on the tape.

It is possible to modify M_n, without any increasing of the number of the states, introducing a different kind of writing at step i.: at the first iteration, the machine uses the symbol 0 instead of X for rewriting, at the second one it uses the symbol 1, and so on. During the last check, the symbol n is written on the last cell of the input tape if it contains an a. If the original input is completely overwritten, then the automaton accepts. For example, in the case $n = 4$, at the end of computation, the content of the input tape will be 0102010301020104. If the original input is completely overwritten, then the automaton accepts. For example, in the case $n = 4$, at the end of computation, the content of the input tape will be 0102010301020104.

Considering the last extension of M_n, we are now going to introduce a 1-LA N_n, accepting the language L_n, based on the guessing of the final tape content of M_n.

In the first phase, N_n scans the input tape, replacing each a with a nondeterministically chosen symbol in $\{0, \ldots, n\}$. This requires one state. Next, the machine, after moving backward the head to the left end-marker, makes a scan from left to right for each $i = 0, \ldots, n-1$, where it checks if the symbol i occurs in all odd positions, where positions are counted ignoring the cells containing numbers less than i. This control phase needs three states for each value of i: one for moving backward the head and two for counting modulo 2 the positions containing symbols greater or equal to i. Finally, the automaton checks if only the last input cell contains a n (two states), in such case the input is accepted. The total number of states of N_n is $3(n+1)$, that, even in this case, is linear in the parameter n. This gives us a 1-LA of size $O(n)$ accepting L_n.

We are now going to prove that we can do better. In fact, we will show that switching to the deterministic case for the limited automata model, the size of the resulting device does not increase. Actually, we will slightly reduce the number of states, while using the same working alphabet.

Let us observe that the final tape content of M_n and of N_n, if the input is accepted, corresponds to the first 2^n elements of the *binary carry sequence* [13] defined as follows:

- The first two elements of the sequence are 0 and 1.
- The next elements of the sequence are recursively obtained concatenating the just constructed sequence to a copy of itself and by replacing the last element by its successor.

For example, from 01, concatenating itself and adding 1 to the last element of the obtained sequence, we get 0102, from which, iterating the last procedure it is possible to obtain 01020103, and so on.

Remark 2. Each symbol $0 \le i < n$ of the binary carry sequence occurs 2^{n-i-1} times, starting in position 2^i and at distance 2^{i+1}, i.e., it occurs in positions $2^i(2j-1)$, for $j = 1, \ldots, 2^{n-i-1}$. Instead, the symbol $i = n$ occurs in position 2^n only.

Remark 2 is a direct consequence of the recursive definition of the sequence. Consider, as an example, the sequence $x = 01020103$: reading x from left to the right, the symbol 0 appears for the first time in position 2^0 and then in positions $3, 5, 7$; 1 in positions $2, 6$; 2 in position 4; and, finally, 3 in position 8.

We will show that this sequence can be generated by a 1-DLA and written on its tape, without counting large indices as 2^i.

To this aim, we introduce the function BIS, that associates with a given sequence of integers $s = \sigma_1\sigma_2\cdots\sigma_j$, its *backward increasing sequence*, namely the longest strictly increasing sequence obtained taking the elements of s starting from the end, and using the greedy method. Formally, $BIS(\sigma_1\sigma_2\cdots\sigma_j) = (i_1, i_2, \ldots, i_r)$, $j, r > 0$, if and only if $i_1 = \sigma_{h_1}, i_2 = \sigma_{h_2}, \ldots, i_n = \sigma_{h_r}$ where $h_1 = j$, $h_t = \max\{h' < h_{t-1} \mid \sigma_{h'} > \sigma_{h_{t-1}}\}$ for $t = 2, \ldots, r$, and $\sigma_{h'} < \sigma_{h_r}$ for $0 < h' < h_r$.

For example, for $j = 11$, $BIS(01020103010) = (0, 1, 3)$. Notice that in the binary representation of j, namely 1011, the bits set to 1 occur, respectively, in position 0, 1, and 3. This fact is true for each j, as proved in the following lemma, i.e., the value of BIS, applied to the first j elements of the binary carry sequence, are the positions of bits equal to 1 in the binary representation of j, from the less significative bit.

Lemma 3. *Let $\sigma_1\sigma_2\cdots\sigma_j$ be the first j elements of the binary carry sequence, $j > 0$. If $BIS(\sigma_1\sigma_2\cdots\sigma_j) = (i_1, i_2, \ldots, i_r)$ then $j = \sum_{t=1}^{r} 2^{i_t}$.*

Proof. Considering the recursive definition of the binary carry sequence, we can proceed by induction on j, where the basis $j = 1, 2$ is trivial from the definition.

- If j is a power of 2, namely $j = 2^k$, for some $k \geq 0$, then, considering Remark 2 with $n = k$, k is the maximum number in the sequence and it occurs in position j only. So, $BIS(\sigma_1\sigma_2\cdots\sigma_j) = (k)$.
- If j is not a power of 2, namely $2^k < j < 2^{k+1}$, $j = 2^k + j'$ for some $k > 0$, $0 < j' < 2^k$, then, by construction, k is the maximum number which occurs in the sequence, and the elements $\sigma_{2^k+1}\cdots\sigma_j$ are equal to the first j' elements $\sigma_1\sigma_2\cdots\sigma_{j'}$. So, $BIS(\sigma_1\sigma_2\cdots\sigma_j)$ is obtained by appending k at the end of $BIS(\sigma_1\sigma_2\cdots\sigma_{j'})$.
  Let $BIS(\sigma_1\sigma_2\cdots\sigma_{j'}) = (i_1, \ldots, i_{r'})$, $r' = r - 1$. Then $BIS(\sigma_1\sigma_2\cdots\sigma_j) = (i_1, \ldots, i_{r'}, i_r)$, $i_r = k$. Furthermore, by induction hypothesis, $j' = \sum_{t=1}^{r'} 2^{i_t}$. Then $j = 2^k + j' = 2^k + \sum_{t=1}^{r'} 2^{i_t} = \sum_{t=1}^{r} 2^{i_t}$. $\square$

We are going to define a 1-DLA $A_n = (Q, \Sigma, \Gamma, \delta, q_1, F)$ accepting the language L_n. The automaton A_n works by replacing the content of cell i, in the first visit, by the symbol σ_i of the binary carry sequence.

Remark 4. Given the first $i-1$ elements of the binary carry sequence $\sigma_1\sigma_2\cdots\sigma_{i-1}$, it is possible to deterministically determine the next element σ_i only looking at the previous elements of the sequence. In particular, σ_i has to be determined so that $BIS(\sigma_1\sigma_2\cdots\sigma_i)$ is the positions of bits equal to 1 in the binary representation of i. This can be obtained computing from the binary representation of $i - 1$ the representation of i. Hence, σ_i is the smallest integer greater than or equal to 0 not occurring in $BIS(\sigma_1\sigma_2\cdots\sigma_{i-1})$.

The automaton A_n implements the procedure summarized in Algorithm 1 — note that, for ease of presentation, the algorithm assumes that the machine starts the computation with the head on the left end-marker — and it is defined as follows: $Q = \{q_I, q_F, q_1, \ldots, q_n, p_1, \ldots, p_{n-1}\}$, $\Sigma = \{a\}$, $\Gamma = \{0, \ldots, n\}$, q_I is the initial state and q_F is the unique final one. The transitions in δ are defined as follows (the remaining transitions are undefined):

  i. $\delta(q_I, a) = (p_1, 0, -1)$
  ii. $\delta(p_i, \sigma) = (p_i, \sigma, -1)$, for $i = 2, \ldots, n - 1$ and $\sigma < i - 1$
  iii. $\delta(p_i, i) = (p_{i+1}, i, -1)$, for $i = 1, \ldots, n - 2$
  iv. $\delta(p_i, \sigma) = (q_i, \sigma, +1)$, for $i = 1, \ldots, n - 1$ and $(\sigma > i$ or $\sigma = \triangleright)$
  v. $\delta(p_{n-1}, n - 1) = (q_n, n - 1, +1)$

Algorithm 1. Recognition of the language L_n

---

1 start with the head on the left end-marker
2 **while** *symbol under the head* $\neq n$ **do**
3 | move the head to the right
4 | write 0
5 | $j \leftarrow 0$
6 | **repeat**
7 | | **while** *symbol under the head* $\leq j$ *and* $\neq \triangleright$ **do**
8 | | | move the head to the left
9 | | $j \leftarrow j + 1$
10 | **until** *symbol under the head* $\neq j$
11 | **repeat**
12 | | move the head to the right
13 | **until** *symbol under the head* $= a$
14 | write j
15 move the head to the right
16 **if** *symbol under the head* $= \triangleleft$ **then** ACCEPT
17 **else** REJECT

---

vi. $\delta(q_i, \sigma) = (q_i, \sigma, +1)$, for $i = 1, \ldots, n$ and $\sigma < i$
vii. $\delta(q_i, a) = (q_I, i, +1)$, for $i = 1, \ldots, n - 1$
viii. $\delta(q_n, a) = (q_F, n, +1)$
ix. $\delta(q_F, \triangleleft) = (q_F, \triangleleft, +1)$

We finally observe that A_n has $2n + 1$ states, which is linear in the parameter n.

The machine starts in the initial state q_I. Since each symbol $\sigma \neq 0$ is preceded by 0 (a 0 occurs in each odd position), the automaton moves the head to the right and writes a 0 before of each symbol in $\Gamma \setminus \{0\}$ (transition i. – lines 3 and 4). Everytime the head is in a odd position p, the automaton has to look backward for the minimum integer j such that j is not in $BIS(\sigma_1, \ldots, \sigma_p)$. This is done with transitions from ii. to v. – lines from 6 to 10. After that, A_n moves its head to the right until the first a is reached (transitions vi. – lines from 11 to 13) and writes the symbol j (transitions vii. – line 14). This is repeated until the symbol n is written on the input tape. At this point is sufficient to verify if the next symbol on the input tape is the right end-marker: in this case, the automaton accepts (transitions viii. and ix. – lines from 15 to 17).

Hence we conclude that the language L_n is accepted by a 1-DLA with $O(n)$ many states, while it is an easy observation that each 1NFA accepting it requires $2^n + 1$ states. We can even obtain a stronger result by proving that between unary 1-DLAS and 2NFAS, there is the same gap. This gives the main result of this section:

Theorem 5. *For each integer $n > 1$ there exists a unary language K_n such that K_n is accepted by a deterministic 1-LA with $O(n)$ states and a working alphabet of size $O(n)$ while each 2NFA accepting it requires 2^n states.*

Proof (outline). With some minor changes, the above presented automaton A_n can accept $K_n = \{a^{2^n}\}^*$. From Theorem 9 in [7], each 2NFA requires at least 2^n many states to accept the same language. $\square$

We conclude this section, by proving that the exponential gap between unary limited automata and finite automata is not always achievable.

Theorem 6. *There exist constants c, n_0 such that for all integers $n \geq n_0$ there exists a unary 1DFA accepting a finite language L with at most n states, such that for any d-LA accepting L with $d > 0$, q states, and a working alphabet of m symbols, it holds that $qm \geq cn^{1/2}$.*

Proof. There are $2^{O(q^2 m^2)}$ different limited automata such that the cardinalities of the set of states and of the working alphabet are bounded by q and m, respectively. On the other hand, the number of different subsets of $\{a^0, a^1, \ldots, a^{n-1}\}$ is 2^n. Hence $kq^2 m^2 \geq n$ for a constant $k > 0$ and each sufficiently large n, which implies $qm \geq cn^{1/2}$, where $c = 1/k^{1/2}$. Note that each subset of $\{a^0, a^1, \ldots, a^{n-1}\}$ is accepted by a (possibly incomplete) 1DFA with at most n states. $\square$

Notice that the result in Theorem 6 does not depend on d, i.e., the lower bound holds even taking an arbitrarily large d. In the case $d = 1$, the argument in the proof can be refined to show that $qm^{1/2} \geq cn^{1/2}$.

4   Unary Grammars Versus Limited Automata

In Sect. 3 we proved an exponential gap between unary 1-LAS and finite automata. A similar gap was obtained between unary CFGs and finite automata [9]. Hence, it is natural to study the size relationships between unary CFGs and 1-LAS. Here, we prove that each context-free grammar specifying a unary language can be converted into an equivalent 1-LA which has a set of states and a working alphabet whose sizes are polynomial with respect to the description of the grammar.

Let us start by presenting some notions and preliminary results useful to reach our goal.

Definition 7. *A bracket alphabet Ω_b is a finite set containing an even number of symbols, say $2k$, with $k > 0$, where the first k symbols are interpreted as left brackets of k different types, while the remaining symbols are interpreted as the corresponding right brackets. The Dyck language D_{Ω_b} over Ω_b is the set of all sequences of balanced brackets from Ω_b.*

An extended bracket alphabet Ω is a nonempty finite set which is the union of two, possibly empty, sets Ω_b and Ω_n, where Ω_b, if not empty, is a bracket *alphabet, and Ω_n is a set of* neutral *symbols. The extended Dyck language $\widehat{D}_\Omega$ over Ω is the set of all the strings that can be obtained by arbitrarily inserting symbols from Ω_n in strings of D_{Ω_b}. Given an integer $d > 0$, the extended Dyck language with nesting depth bounded by d over Ω, denoted as $\widehat{D}_\Omega^{(d)}$ is the subset of $\widehat{D}_\Omega$ consisting of all strings where the nesting depth of brackets is at most d.*

Example 8. Let $\Omega_b = \{\,(\,,\,[\,,\,)\,,\,]\,\}, \Omega_n = \{\,|\,\}$, and $\Omega = \Omega_b \cup \Omega_n$. Then $(\,[\,[\,]\,]\,)\,[\,] \in D_{\Omega_b} \subset \widehat{D}_\Omega$, $|\,(\,|\,[\,[\,]\,]\,|\,)\,[\,|\,] \in \widehat{D}_\Omega^{(3)} \backslash \widehat{D}_\Omega^{(2)}$.

It is well-known that Dyck languages, and so extended Dyck languages, are context-free and nonregular. However, the subset obtained by bounding the nesting depth by each fixed constant is regular. We are interested in the recognition of such languages by "small" two-way automata:

Lemma 9. *Given an extended bracket alphabet with k types of brackets and an integer $d > 0$, the language $\widehat{D}_\Omega^{(d)}$ can be recognized by a 2DFA with $O(k \cdot d)$ many states.*

Proof. We can define a 1-DFA M which verifies the membership of its input w to $\widehat{D}_\Omega^{(d)}$ by making use of a counter c. In a first scan M checks whether or not the brackets are correctly nested, *regardless* their types. This is done as follows. Starting with 0 in c, M scans the input from left to right, incrementing the counter for each left bracket and decrementing it for each right bracket. If during this process the counter exceeds d or becomes negative then M rejects. M also rejects if at the end of this scan the value which is stored in the counter is positive.

In the remaining part of the computation, M verifies that the corresponding left and right brackets are of the same type. To this aim, starting from the left end-marker, M moves its head to the right, to locate a left bracket. Then, it moves to the right to locate the corresponding right bracket in order to check if they are of the same type. To this aim, M uses the counter c, which initially contains 0 and increments or decrements it for each left or right bracket to the right of the one under consideration. When a cell containing a right bracket is reached with 0 in c, M checks if it is matching with the left bracket. If this is not the case, then M stops and rejects. Otherwise, M should move back its head to the matched left bracket in order to continue the inspection. This can be done by moving the head to the left and incrementing or decrementing the counter for each right or left bracket, respectively, up to reach a cell containing a left bracket when 0 is in c.

This process is stopped when the right end-marker is reached and all pairs of brackets have been inspected. Notice that neutral symbols are completely ignored.

In its finite control, M keeps the counter c, that can assume $d + 1$ different values, and remembers the type of the left bracket, to verify the matching with the corresponding right bracket and then to move back the head to the left bracket. This gives $O(k \cdot d)$ many states. $\square$

The following nonerasing variant of the Chomsky-Schützenberger representation theorem for context-free languages, proved by Okhotin [8], is crucial to obtain our main result:

Theorem 10. *A language $L \subseteq \Sigma^*$ is context-free if and only if there exist an extended bracket alphabet Ω_L, a regular language $R_L \subseteq \Omega_L^*$ and a letter-to-letter homomorphism $h : \Omega_L \to \Sigma$ such that $L = h(\widehat{D}_{\Omega_L} \cap R_L)$.*

In [10], it was observed that the language R_L of Theorem 10 is local[1] and the size of the alphabet Ω is polynomial with respect to the size of a context-free grammar G generating L. This was used to prove that each context-free grammar G can be transformed into an equivalent *strongly limited automaton* (a special kind of 2-LA) whose description has polynomial size with respect to the description of G. In the following, when L is specified by a context-free grammar G, i.e., $L = L(G)$, we will write Ω_G and R_G instead of Ω_L and R_L, respectively.

Our goal, here, is to build 1-LAS of polynomial size from unary context-free grammars. To this aim, using the fact that factors in unary strings commute, by adapting the argument used to obtain Theorem 10, we prove the following result:

Theorem 11. *Let $L \subseteq \{a\}^*$ be a unary regular language and $G = (V, \{a\}, P, S)$ be a context-free grammar of size s generating it. Then, there exist an extended bracket alphabet Ω_G and a regular language $\widehat{R}_G \subseteq \Omega_G^*$ such that $L = h(\widehat{D}_{\Omega_G}^{(\#V)} \cap \widehat{R}_G)$, where:*

- $\widehat{D}_{\Omega_G}^{(\#V)}$ *is the extended Dyck language over Ω_G with nesting depth bounded by $\#V$,*
- h *is the letter-to-letter homomorphism from Ω_G to $\{a\}$.*

Furthermore, the size of Ω_G is polynomial in the size s of the grammar G and the language $\widehat{R}_G$ is recognized by a 2NFA with a number of states polynomial in s.

Proof (outline). Given a context-free grammar $G = (V, \{a\}, P, S)$ specifying a unary language L, we first obtain the representation in Theorem 10. According to Theorem 5.2 in [10], the size of the alphabet Ω_G is polynomial with respect to the size of the description of G. Each pair of brackets in Ω_G represents the root of a derivation tree of G, which starts from a certain variable of G and produces a terminal string.

If a sequence $w \in \Omega_G^*$ contains a pair of brackets corresponding to a variable A which is nested, at some level, in another pair corresponding to the same variable, then w can be replaced by a sequence w', of the same length, which is obtained by replacing the factor of w delimited by the outer pair of brackets corresponding

---

[1] A language L is *local* if there exist sets $\mathcal{A} \subseteq \Sigma \times \Sigma$, $\mathcal{I} \subseteq \Sigma$, and $\mathcal{F} \subseteq \Sigma$ such that $w \in L$ if and only if all factors of length 2 in w belong to $\mathcal{A}$ and the first and the last symbols of w belong to $\mathcal{I}$ and $\mathcal{F}$, respectively [6].

to A, by the factor delimited by the inner pair, and by moving the removed part at the end of w. For instance, $w = (s(A(B)B(C(A(B)B)A)C)A)s$ can be replaced by $w' = (s(A(B)B)A)s(A(B)B(C)C)A$, where, for the sake of simplicity, subscripts represent variables corresponding to brackets. In this way, each time the nesting depth is greater than $\#V$, it can be reduced by repeatedly moving some part to the end. So, from each string in $\widehat{D}_{\Omega_G}$, we can obtain an "equivalent" string of the same length in $\widehat{D}_{\Omega_G}^{(\#V)}$.

The regular language R_G should be modified accordingly. While in the representation in Theorem 10, the first and the last symbol of a string $w \in \widehat{D}_{\Omega_G} \cap R_G$ represent a matching pair corresponding to the variable S, after the above transformation, valid strings should correspond to sequences of blocks of brackets where the first block represents a derivation tree of a terminal string from S, while each of the subsequent blocks represents a *gap tree* from a variable A, namely a tree corresponding to a derivation of the form $A \overset{+}{\Rightarrow} a^i A a^j$, with $i + j > 0$, where A already appeared in some of the previous blocks. This condition, together with the conditions on R_G, can be verified by a 2NFA with a polynomial number of states. □

Notice that if we omit the state bound for the 2NFA accepting $\widehat{R}_G$, the statement of Theorem 11 becomes trivial: just take $L = R_G$ and $\widehat{\Omega}_G = \{a\}$ where a is a neutral symbol.

Using Theorem 11, we now prove the main result of this section:

Theorem 12. *Each context-free grammar of size s generating a unary language can be converted into an equivalent 1-LA having a size which is polynomial in s.*

Proof. Let $G = (V, \{a\}, P, S)$ be the given grammar, $L \subseteq \{a\}^*$ be the unary language generated by it, Ω_G be the extended bracket alphabet and $\widehat{R}_G$ be the regular language obtained from G according to Theorem 11.

We define a 1-LA M which works in the following steps:

1. M makes a complete scan of the input tape from left to right, by rewriting each input cell by a nondeterministically chosen symbol from Ω_G. Let $w \in \Omega_G^*$ be the string written on the tape at the end of this phase.
2. M checks whether or not $w \in \widehat{D}_{\Omega_G}^{(\#V)}$.
3. M checks whether or not $w \in \widehat{R}_G$.
4. M accepts if and only if the outcomes of steps 2 and 3 are both positive.

According to Lemma 9, step 2 can be done by simulating a 2DFA with $O(\#\Omega_G \cdot \#V)$ many states, hence a number polynomial in s. Furthermore, by Theorem 11, also step 3 can be performed by simulating a 2NFA with a number of states polynomial in s. Hence M has a size which is polynomial in s. □

We point out that from Theorem 12 and the exponential gap from unary CFGs to 1NFAs proved in [9], we can derive an exponential gap from unary *nondeterministic* 1-LAs to 1NFAs. In Sect. 3 we proved that the gap remains exponential if we restrict to unary *deterministic* 1-LAs and consider equivalent 2NFAs.

5 Conclusion

In [11], using languages defined over a binary alphabet, exponential size gaps have been proved for the conversion of 1-LAS into 2-NFAS and of 1-DLAS into 1DFAS. As a consequence of our results, these exponential gaps hold even if we restrict to unary languages. On the other hand, the size gap between 1-LAS and 1-DFAS is double exponential. Even in this case, the proof in [11] relies on witness languages defined over a binary alphabet. We leave as an open question to investigate whether or not a double exponential gap is possible between 1-LAS and 1-DFAS even in the unary case.

Another question we leave open is whether or not 1-LAS and CFGS are polynomially related in the unary case. While in Sect. 4 we proved that from each unary CFG we can build a 1-LA of polynomial size, at the moment we do not know the converse relationship.

References

1. Ginsburg, S., Rice, H.G.: Two families of languages related to ALGOL. J. ACM **9**(3), 350–371 (1962). http://doi.acm.org/10.1145/321127.321132
2. Hibbard, T.N.: A generalization of context-free determinism. Inf. Control **11**(1/2), 196–238 (1967)
3. Hopcroft, J.E., Ullman, J.D.: Introduction to Automata Theory, Languages and Computation. Addison-Wesley, Reading (1979)
4. Kutrib, M., Pighizzini, G., Wendlandt, M.: Descriptional complexity of limited automata. Inf. Comput. (to appear)
5. Kutrib, M., Wendlandt, M.: On simulation cost of unary limited automata. In: Shallit, J., Okhotin, A. (eds.) DCFS 2015. LNCS, vol. 9118, pp. 153–164. Springer, Cham (2015). doi:10.1007/978-3-319-19225-3_13
6. McNaughton, R., Papert, S.A.: Counter-Free Automata. M.I.T. Research Monograph, vol. 65. The MIT Press, Cambridge (1971)
7. Mereghetti, C., Pighizzini, G.: Two-way automata simulations and unary languages. J. Autom. Lang. Comb. **5**(3), 287–300 (2000)
8. Okhotin, A.: Non-erasing variants of the Chomsky–Schützenberger theorem. In: Yen, H.-C., Ibarra, O.H. (eds.) DLT 2012. LNCS, vol. 7410, pp. 121–129. Springer, Heidelberg (2012). doi:10.1007/978-3-642-31653-1_12
9. Pighizzini, G., Shallit, J., Wang, M.: Unary context-free grammars and pushdown automata, descriptional complexity and auxiliary space lower bounds. J. Comput. Syst. Sci. **65**(2), 393–414 (2002)
10. Pighizzini, G.: Strongly limited automata. Fundam. Inform. **148**(3–4), 369–392 (2016). http://dx.doi.org/10.3233/FI-2016-1439
11. Pighizzini, G., Pisoni, A.: Limited automata and regular languages. Int. J. Found. Comput. Sci. **25**(7), 897–916 (2014). http://dx.doi.org/10.1142/S0129054114400140
12. Pighizzini, G., Pisoni, A.: Limited automata and context-free languages. Fundam. Inf. **136**(1–2), 157–176 (2015). http://dx.doi.org/10.3233/FI-2015-1148
13. Sloane, N.J.A.: The on-line encyclopedia of integer sequences. http://oeis.org/A007814
14. Wagner, K.W., Wechsung, G.: Computational Complexity. D. Reidel Publishing Company, Dordrecht (1986)

A Characterization of Infinite LSP Words

Gwenaël Richomme[1,2]($\boxtimes$)

[1] Univ. Paul-Valéry Montpellier 3, UFR 6, Dpt MIAp, Case J11,
Rte de Mende, 34199 Montpellier Cedex 5, France
[2] LIRMM (CNRS, Univ. Montpellier), UMR 5506 - CC 477,
161 rue Ada, 34095 Montpellier Cedex 5, France
gwenael.richomme@lirmm.fr

Abstract. G. Fici proved that a finite word has a minimal suffix automaton if and only if all its left special factors occur as prefixes. He called LSP all finite and infinite words having this latter property. We characterize here infinite LSP words in terms of S-adicity. More precisely we provide a finite set of morphisms S and an automaton $\mathcal{A}$ such that an infinite word is LSP if and only if it is S-adic and all its directive words are recognizable by $\mathcal{A}$.

Keywords: Generalizations of Sturmian words · Morphisms · S-adicity

1 Introduction

Extending an initial work by M. Sciortino and L.Q. Zamboni [15], G. Fici investigated relations between the structure of the suffix automaton built from a finite word w and the combinatorics of this word [8]. He proved that words having their associated automaton with a minimal number of states (with respect to the length of w) are the words having all their left special factors as prefixes. G. Fici asked in the conclusion of his paper for a characterization of the set of words having the previous property, that he called the LSP property, both in the finite and the infinite case. We provide such a characterization for infinite words in the context of S-adicity.

We assume that readers are familiar with combinatorics on words; for omitted definitions (as for instance, factor, prefix, ...) see, *e.g.*, [6,13,14]. Given an alphabet A, A^* is the set of all finite words over A and A^ω is the set of all infinite words over A. A finite word u is a *left special factor* of a finite or infinite word w if there exist at least two distinct letters a and b such that both au and bu occur in w. Given two alphabets A and B, a *morphism* (*endomorphism* when $A = B$) f is a map from A^* to B^* such that for all words u and v over A, $f(uv) = f(u)f(v)$. Morphisms extend naturally to infinite words.

Let S be a set of morphisms. An infinite word $\mathbf{w}$ is said S-adic if there exists a sequence $(f_n)_{n\geq 1}$ of morphisms in S and a sequence of letters $(a_n)_{n\geq 1}$ such that $\lim_{n\to+\infty} |f_1 f_2 \cdots f_n(a_{n+1})| = +\infty$ and $\mathbf{w} = \lim_{n\to+\infty} f_1 f_2 \cdots f_n(a_{n+1}^\omega)$. The sequence $(f_n)_{n\geq 1}$ is called the *directive word* of $\mathbf{w}$. We consider here S-adicity in a rather larger way: a word $\mathbf{w}$ is S-*adic* with directive word $(f_n)_{n\geq 1}$ if

© Springer International Publishing AG 2017
É. Charlier et al. (Eds.): DLT 2017, LNCS 10396, pp. 320–331, 2017.
DOI: 10.1007/978-3-319-62809-7_24

there exists an infinite sequence of infinite words $(\mathbf{w}_n)_{n \geq 1}$ such that $\mathbf{w}_1 = \mathbf{w}$ and $\mathbf{w}_n = f_n(\mathbf{w}_{n+1})$ for all $n \geq 1$. Denoting $w_k = f_k f_{k+1} \cdots f_n(a_{n+1}^\omega)$ shows that if the former definition is verified, the latter is also verified. This second definition may include degenerate cases as, for instance, the word a^ω that is $\{Id\}$-adic with Id the morphism mapping a on a. For more information on S-adic systems, readers can consult, $e.g.$, papers [2,3] and their references.

Let $p_\mathbf{w}$ be the *factor complexity* of the infinite word $\mathbf{w}$, that is the function that counts the number of different factors of $\mathbf{w}$. If $\mathbf{w}$ is an infinite LSP word, by definition, it has at most one left special factor of each length. Thus it is well-known that $p_\mathbf{w}(n+1) - p_\mathbf{w}(n) \leq \#A - 1$ (where for any set X, $\#X$ denotes the cardinality of X). We let readers verify that all infinite LSP words are uniformly recurrent (all factors occur infinitely many times with bounded gaps). By a result of S. Ferenczi [7] (see also [9–11]), there exists a finite set S of morphisms such that all infinite LSP words are S-adic. But this general result does not provide a characterization of infinite LSP words.

Our characterization is twofold. First we exhibit an adapted finite set of morphisms S_{bLSP}. Second we show that there exists an automaton that recognizes the set of directive words of infinite LSP words. In the binary case, our result can be seen as a version for infinite words of a result of M. Sciortino and L.Q. Zamboni [15] (see the conclusion). In the ternary case, morphisms in S_{bLSP} are the mirror morphisms of Arnoux-Rauzy-Poincaré morphisms (here f is a *mirror morphism* of g if $f(a)$ is the mirror image or reversal of $g(a)$ for all letters a). These morphisms were used by V. Berthé and S. Labbé [5] to provide an S-adic system recognizing sequences arising from the study of the Arnoux-Rauzy-Poincaré multidimensional continued fraction algorithm. For alphabets of cardinality at least 4, new morphisms appear.

The paper is organized as follows. After introducing in Sect. 2 our basis of morphisms S_{bLSP}, in Sect. 3, we show that all infinite LSP words are S_{bLSP}. Section 4 introduces a property of infinite LSP words and a property of morphisms in S_{bLSP} that allow to explain why the LSP property is lost when applying a LSP morphism to an infinite LSP word. Section 5 allows to trace the origin of the previous property of infinite LSP words. Based on this information, Sect. 6 defines our automaton and Sect. 7 proves our characterization of infinite LSP words. We end with a few words on characterizations of finite LSP words.

2   Some Basic Morphisms

We call *basic LSP morphism* on an alphabet A, or $bLSP$ in short, any endomorphism f of A^* that verifies:

- there exists a letter α such that $f(\alpha) = \alpha$, and
- for all letters $\beta \neq \alpha$, there exists a letter γ such that $f(\beta) = f(\gamma)\beta$

We let $S_{\mathrm{bLSP}}(A)$ (or shortly S_{bLSP} when A is clear) denote the set of all bLSP morphisms over the alphabet A. Observe that for any bLSP morphism f, there exists a unique letter α such that $f(\alpha) = \alpha$. We let first(f) denote this letter as it is also the first letter of $f(\beta)$ for any letter β. We also let $[u_1, u_2, \ldots]$ denote the morphism defined by $a \mapsto u_1$, $b \mapsto u_2$, ... For instance, $[a, ab, abc, abcd, abcde]$ defines the morphism f such that $f(a) = a$, $f(b) = ab$, $f(c) = abc$, $f(d) = abcd$, $f(e) = abcde$.

Remark 1. By definition of bLSP morphisms, given an alphabet A, there is a bijection between $S_{\text{bLSP}}(A)$ and the set of labeled rooted trees with label in A (all labels are on vertices and distinct vertices have distinct labels). Given a labeled rooted tree $T = (A, E)$, the associated bLSP morphism f is the one such that, for all letters β, $f(\beta)$ is the word obtained concatenating vertices on the path in T from the root to β. For instance, the rooted trees associated with morphisms $[a, ab, abc, abcd]$, $[a, ab, abc, abd]$, $[a, ab, abc, ad]$ and $[a, ab, ac, ad]$ are given in Fig. 1.

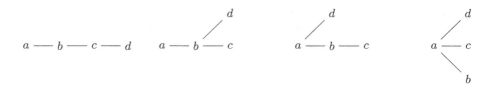

Fig. 1. Rooted trees associated with bLSP morphisms

The previous remark allows to enumerate bLSP morphisms (see Sequence A000169 in The On-Line Encyclopedia of Integer Sequences (https://oeis.org/A000169) whose first values are 1, 2, 9, 64, 625, 7776, 117649, 2097152).

Here follows some examples of bLSP morphisms.

- $S_{\text{bLSP}}(\{a, b\}) = \{[a, ab], [ba, b]\}$. These morphisms are well-known in the context of Sturmian words. They are denoted τ_a and τ_b in [4] from which it can be seen that standard Sturmian words are non-periodic $\{\tau_a, \tau_b\}$-adic words (see also [12]).
- $S_{\text{bLSP}}(\{a, b, c\}) = \{[a, ab, abc], [a, ab, ac], [a, acb, ac], [ba, b, bac], [ba, b, bc], [bca, b, bc], [ca, cb, c], [ca, cab, c], [cba, ca, c]\} = \{p^{-1} \circ [a, ab, abc] \circ p, p^{-1} \circ [a, ab, ac] \circ p \mid p \in perm(A)\}$ where $perm(A)$ is the set of all endomorphisms of A^* whose restriction to the set of letters is a permutation of the alphabet. As mentioned in the introduction, these sets $S_{\text{bLSP}}(\{a, b, c\})$ is also the set of mirror morphisms considered in [5], that is mirrors of the Poincaré substitutions (defined for $\{i, j, k\} = \{a, b, c\}$ by $i \mapsto ijk$, $j \mapsto jk$, $k \mapsto k$) and the Arnoux-Rauzy substitutions (defined for $\{i, j, k\} = \{a, b, c\}$ by $i \mapsto ik$, $j \mapsto jk$, $k \mapsto k$).
- The set $S_{\text{bLSP}}(\{a, b, c, d\})$ is the set of all morphisms on the form $p^{-1} \circ f \circ p$ for $p \in perm(A)$, and f being one of the following morphisms: $[a, ab, abc, abcd]$, $[a, ab, abc, abd]$, $[a, ab, abc, ad]$ and $[a, ab, ac, ad]$.

We end this section with some basic properties of bLSP morphisms that follow directly from the definition. For a non-empty word u, let first(u) denote its first letter, last(u) its last letter and alph(u) its set of letters.

Property 2. Let f be a bLSP morphism over the alphabet A.

1. there exists a unique letter $\alpha \in A$ such that for all $\beta \in A$, first$(f(\beta)) = \alpha$;
2. for all $\beta \in A$, last$(f(\beta)) = \beta$;
3. there exists a unique letter $\alpha \in A$ such that $f(\alpha) = \alpha$: $\alpha = $ first(f);
4. $f(A)$ is a suffix code (no word of $f(A)$ is a suffix of another word in $f(A)$);
5. f is injective both on the set of finite words and the set of infinite words;
6. for all $\beta \in A$, $x, y \in A^*$, if $|x| = |y|$ and if $x\beta$ and $y\beta$ are factors of words in $f(A)$, then $x = y$;
7. for all letters β, γ, $|f(\beta)|_\gamma \leq 1$.

3 S_{bLSP}-Adicity of Infinite LSP Words

Proposition 3. *Any infinite LSP word is S_{bLSP}-adic.*

Given a set S of morphisms, in order to prove that infinite words verifying a property P are S-adic, it suffices to prove that for all infinite words $\mathbf{w}$ verifying P that:

1. there exists $f \in S$ and an infinite word $\mathbf{w}'$ such that $\mathbf{w} = f(\mathbf{w}')$, and
2. if $\mathbf{w} = f(\mathbf{w}')$ with $f \in S$, then $\mathbf{w}'$ verifies Property P.

Hence Proposition 3 is a direct consequence of the next two lemmas.

Lemma 4. *Given any finite or infinite LSP word $\mathbf{w}$, there exist a bLSP morphism f on alph$(\mathbf{w})$ and an infinite word $\mathbf{w}'$ such that $\mathbf{w} = f(\mathbf{w}')$.*

Proof. Let $\mathbf{w}$ be a non-empty finite or infinite LSP word and let α be its first letter. Let X be the set of words over alph$(w)\backslash\{\alpha\}$ such that w can be factorized over $\{\alpha\} \cup X$. Let G be the graph (alph$(w), E$) with E the set of edges (β, γ) such that $\beta\gamma$ is a factor of a word αu with $u \in X$. By LSP Property of $\mathbf{w}$, each letter occurring in a word of X is not left special in $\mathbf{w}$. Hence G is a rooted tree with α as root, that is, for any letter β, there exists a unique path from α to β. We let u_β denote the word obtained by concatenating the letters occurring in the path. Let f be the morphism defined by $f(\beta) = u_\beta$. By construction, f is bLSP and $\mathbf{w} = f(\mathbf{w}')$ for a word $\mathbf{w}'$. □

Remark 5. The word $\mathbf{w}'$ in Lemma 4 is unique. The morphism f is not unique but its restriction to alph$(\mathbf{w}')$ is. It can also be observed that this restriction is entirely defined by the first letter of $\mathbf{w}$ and the factors of length two of $\mathbf{w}$.

Lemma 6. *For any bLSP morphism f and any infinite word $\mathbf{w}$, if $f(\mathbf{w})$ is LSP then $\mathbf{w}$ is LSP.*

Proof. Assume by contradiction that **w** is not LSP. This means that **w** has (at least) one left special factor that is not one of its prefixes. Considering such a factor of minimal length, there exist a word u and letters a, b, β, γ such that $a \neq b$, $\beta \neq \gamma$, ua is a prefix of **w**, βub and γub are factors of **w**. Recall that f is a bLSP morphism: let $\alpha = \text{first}(f)$. The word $f(u)f(a)\alpha$ is a prefix of $f(\mathbf{w})$. Moreover by Property 2(2), words $\beta f(u)f(b)\alpha$ and $\gamma f(u)f(b)\alpha$ are factors of **w** (here the fact that **w** is infinite is needed: each factor is followed by a letter whose image begins with α). As $f(a) \neq f(b)$ and as the letter α occurs only as a prefix in $f(a)$ and $f(b)$, $f(a)\alpha$ is not a prefix of $f(b)\alpha$ and, conversely, $f(b)\alpha$ is not a prefix of $f(a)\alpha$. Hence there exist a word v and letters α', β' such that $\alpha' \neq \beta'$, $v\alpha'$ and $v\beta'$ are respectively prefixes of $f(a)\alpha$ and $f(b)\alpha$. It follows that $f(u)v\alpha'$ is a prefix of $f(\mathbf{w})$ while $\beta f(u)v\beta'$ and $\gamma f(u)v\beta'$ are factors of $f(\mathbf{w})$: $f(\mathbf{w})$ is not LSP. $\qquad\square$

Observe that Lemma 6 does not hold for finite words. For instance the word baa is not LSP while its image $abaa$ by the morphism $[a, ab]$ is LSP.

To end this section let us mention (without proof by lack of space) that in the binary case the converses of Lemma 6 and Proposition 3 hold.

Proposition 7. *If* **w** *is a binary LSP infinite word and if f belongs to the set* $\{[a, ab], [ba, b]\}$ *then $f(\mathbf{w})$ is also LSP. Consequently a binary word is LSP if and only if it is $\{[a, ab], [ba, b]\}$-adic.*

4   Fragility of Infinite LSP Words

For alphabets of cardinality at least 3, the converse of Lemma 6 is false: there exist an infinite LSP word **w** and a bLSP morphism f such that $f(\mathbf{w})$ is not LSP. For instance, let **F** be the well-known Fibonacci word (the fixed point of the endomorphism $[ab, a]$), and let g be the bLSP morphism $[a, acb, ac]$. The word $g^2(\mathbf{F})$ begins with the word $g^2(ab) = g(aacb) = aaacacb$ that contains the factor ac which is left special but not a prefix of the word. Hence the word $g^2(\mathbf{F})$ is not LSP while **F** is LSP and g is bLSP (actually one can prove, using Lemma 10 below, that $g(\mathbf{F})$ is LSP).

In what follows, we introduce some properties of LSP words and morphisms that explain in which context a (breaking) bLSP morphism can map a (fragile) infinite LSP word on a non-LSP word.

Definition 8. *Let a, b, c be three pairwise distinct letters. An infinite word* **w** *is (a, b, c)-fragile if there exist a word u and distinct letters α and β such that the word ua is a prefix of* **w** *and the words αub and βuc are factors of* **w**. *We will also say that* **w** *is (a, b, c, α, β)-fragile when we need letters α and β. The word u is also called an (a, b, c, β, γ)-fragility of* **w**.

For instance, the empty word ε is an (a, b, c, c, a)-fragility of $g(\mathbf{F})$: εa is a prefix of $g(\mathbf{F}) = aacb \cdots$ while $c\varepsilon b$ and $a\varepsilon c$ are factors of $g(\mathbf{F})$. More generally any factor abc or acb in an infinite word starting with the letter a (and with

$a \neq b \neq c \neq a$) produces an (a, b, c)-fragility. One can also observe that, by symmetry of the definition, any (a, b, c)-fragile word is also (a, c, b)-fragile. Finally let us note that no fragility exists in words over two letters (as the definition needs three pairwise distinct letters).

The main idea of introducing the previous notion is that for any (a, b, c)-fragile LSP word $\mathbf{w}$, there exists a bLSP morphism such that $f(\mathbf{w})$ is not LSP. For instance, if $u, \alpha, \beta, \mathbf{w}$ are as in Definition 8, the word $g(u)aa$ is a prefix of $g(\mathbf{w})$ whereas words $\alpha g(u)acb$ and $\beta g(u)ac$ are factors of $g(\mathbf{w})$, so that $g(\mathbf{w})$ is not LSP since $g(u)ac$ is left special but not a prefix of $g(\mathbf{w})$.

Definition 9. Let a, b, c be three pairwise distinct letters. A morphism f is LSP (a, b, c)-*breaking*, if for all (a, b, c)-fragile infinite LSP word $\mathbf{w}$, $f(\mathbf{w})$ is not LSP.

For instance, the morphism $g = [a, acb, ac]$ is (a, b, c)-breaking.

Lemma 10. *Let $\mathbf{w}$ be an infinite LSP word and let f be a bLSP morphism. The following assertions are equivalent:*

1. *The word $f(\mathbf{w})$ is not LSP;*
2. *There exist some pairwise distinct letters a, b, c such that $\mathbf{w}$ is (a, b, c)-fragile and the longest common prefix of $f(b)$ and $f(c)$ is strictly longer than the longest common prefix of $f(a)$ and $f(b)$;*
3. *There exist some pairwise distinct letters a, b, c, such that $\mathbf{w}$ is (a, b, c)-fragile and f is LSP (a, b, c)-breaking.*

Proof. $1 \Rightarrow 2$. Assume first that $f(\mathbf{w})$ is not LSP. There exists a left special factor V of $f(\mathbf{w})$ which is not a prefix of $f(\mathbf{w})$. Let v be the longest common prefix of V and $f(\mathbf{w})$. Let a', b' be the letters such that va' is a prefix of $f(\mathbf{w})$ and vb' is a prefix of V: by construction $a' \neq b'$. Let also β, γ be distinct letters such that βV and γV are factors of $f(\mathbf{w})$ (also $\beta vb'$ and $\gamma vb'$ are factors of $f(\mathbf{w})$).

By Property 2, the letter $\alpha = \text{first}(f)$ is the unique letter that can be left special in $f(\mathbf{w})$. This implies $v \neq \varepsilon$ and $\text{first}(v) = \text{first}(f)$. As α occurs exactly at the first position in all images of letters, occurrences of α mark the beginning of images of letters in $f(\mathbf{w})$. Considering the last occurrence of α in v, we can write $v = f(u)\alpha x$ with $|x|_\alpha = 0$. Let a, b, c be letters such that:

- ua is a prefix of $\mathbf{w}$, and, $va' = f(u)\alpha xa'$ is a prefix of $f(ua)$ when $a' \neq \alpha$ or $v = f(ua)$ when $a' = \alpha$;
- βub is a factor of $\mathbf{w}$, and, $\beta vb'$ is a prefix of $\beta f(ub)$ when $b' \neq \alpha$ or $v = f(ub)$ when $b' = \alpha$;
- γuc is a factor of $\mathbf{w}$, and, $\gamma vb'$ is a prefix of $\gamma f(uc)$ when $b' \neq \alpha$ or $v = f(uc)$ when $b' = \alpha$.

As $a' \neq b'$, we have $a \neq b$ and $a \neq c$. Observe that until now we did not use the fact that $\mathbf{w}$ is LSP. This implies $b \neq c$ (and so $b' \neq \alpha$). Indeed otherwise ub would be a left special factor of $\mathbf{w}$ without being one of its prefixes: a contradiction with the fact that $\mathbf{w}$ is an LSP word. Thus $\mathbf{w}$ is (a, b, c)-fragile.

This ends the proof of Part $1 \Rightarrow 2$ as $\alpha x b'$ is a common prefix of $f(b)$ and $f(c)$ and αx is the longest common prefix of $f(a)$ and $f(b)$.

$2 \Rightarrow 3$. By hypothesis, $f(a) = v\delta w_1$, $f(b) = v\gamma w_2$ and $f(c) = v\gamma w_3$ for letters δ, γ and words w_1, w_2 and w_3 with $\delta \neq \gamma$. Let $\mathbf{w}'$ be any LSP (a, b, c)-fragile infinite word. Let u', α', β' be the word and letters such that $u'a$ is a prefix of $\mathbf{w}'$ while $\alpha'u'b$ and $\beta'u'c$ are factors of $\mathbf{w}'$ with $\alpha' \neq \beta'$. The word $f(\mathbf{w}')$ has $f(u')v\delta$ as a prefix and words $\alpha'f(u')v\gamma$ and $\beta'f(u')v\gamma$ as factors. As $\delta \neq \gamma$, the word $f(\mathbf{w}')$ is not LSP. The morphism f is LSP (a, b, c)-breaking.

$3 \Rightarrow 1$. This follows the definition of (a, b, c)-fragile words and LSP (a, b, c)-breaking morphisms. □

Observe that we have also proved the next result.

Corollary 11. *A bLSP morphism is LSP (a, b, c)-breaking for pairwise distinct letters a, b and c if and only if the longest common prefix of $f(b)$ and $f(c)$ is strictly longer than the longest common prefix of $f(a)$ and $f(b)$.*

5   Origin of Fragilities

Before characterizing infinite LSP words, we need to know how fragilities in an LSP word can appear. This is explained by next result. For a set X of words, we let $\mathrm{Fact}(X)$ denote the set of factors of words in X.

Lemma 12. *Assume a word u is an (a, b, c, β, γ)-fragility of $f(\mathbf{w})$ for an infinite word $\mathbf{w}$ (not necessarily LSP) over an alphabet A and f is a bLSP morphism (by definition of fragilities, a, b, c, β, γ are letters).*

- *(New fragilities). If $u = \varepsilon$, then $a = \mathrm{first}(f)$ and βb, $\gamma c \in \mathrm{Fact}(f(\mathrm{alph}(\mathbf{w})))$.*
- *(Propagated fragilities). If $u \neq \varepsilon$, there exist letters a', b', c' in $\mathrm{alph}(\mathbf{w})$ and an $(a', b', c', \beta, \gamma)$-fragility v of $\mathbf{w}$ such that $|v| < |u|$, $f(v)$ is a proper prefix of u and words ua, βub, γuc are respectively prefixes of $f(va')\alpha$, $\beta f(vb')\alpha$, $\gamma f(vc')\alpha$ with $\alpha = \mathrm{first}(f)$.*

Proof. (New fragilities). If $u = \varepsilon$, it follows from the definition of an (a, b, c, β, γ)-fragility that $a = \mathrm{first}(\mathbf{w})$ and βb, γc are factors of $f(\mathbf{w})$. Now observe that, still by the same definition, $a \notin \{b, c\}$. Thus by definition of bLSP morphisms, $a = \mathrm{first}(f)$ and βb, γc belong to $\mathrm{Fact}(f(\mathrm{alph}(\mathbf{w})))$.

(Propagated fragilities). We assume here that u is not empty. Let $\alpha = \mathrm{first}(f)$. Considering the last occurrence of α in u, observe that the word u can be decomposed in a unique way as $u = f(v)\alpha x$ with v, x words such that $|x|_\alpha = 0$. As u is an (a, b, c, β, γ)-fragility of $f(\mathbf{w})$, there exist words w_1, w_2 and w_3 such that:

- $|w_1|_\alpha = |w_2|_\alpha = |w_3|_\alpha = 0$;
- $f(v)\alpha x w_1 \alpha$ is a prefix of $f(\mathbf{w})$ and $a = \mathrm{first}(w_1\alpha)$;
- $\beta f(v)\alpha x w_2 \alpha$ and $\gamma f(v)\alpha x w_3 \alpha$ are factors of $f(\mathbf{w})$ with $b = \mathrm{first}(w_2\alpha)$ and $c = \mathrm{first}(w_3\alpha)$.

By definition of a bLSP morphism, there exist letters a', b', c' such that $f(a') = \alpha x w_1$, $f(b') = \alpha x w_2$, $f(c') = \alpha x w_3$. These letters a', b', c' are pairwise distinct since letters $a = \text{first}(w_1\alpha)$, $b = \text{first}(w_2\alpha)$ and $c = \text{first}(w_3\alpha)$ are pairwise distinct. Moreover va' is a prefix of $\mathbf{w}$ and words $\beta v b'$ and $\gamma v c'$ are factors of $\mathbf{w}$ (remember that α marks the beginning of letters in $f(\mathbf{w})$ as f is a bLSP morphism). Hence the word v is an $(a', b', c', \beta, \gamma)$-fragility of $\mathbf{w}$. Finally let us observe that $|v| \leq |f(v)| < |u|$. $\qquad\square$

6   An Automaton to Follow Fragilities

In this section, we introduce an automaton that allows to recognize all directive words of LSP words viewed as S_{bLSP}-adic words. We will prove the converse in next section. Observe that transitions of the automaton are defined in order to follow fragilities using Lemma 12.

Definition 13. We let $\mathcal{A}_{bLSP}$ denote the non-deterministic automaton whose elements are described below.

- The alphabet of $\mathcal{A}_{bLSP}$ is the set bLSP of basic LSP morphisms.
- The set of states Q is the set $2^A \times \text{bLSP} \times 2^{A^5}$. Hence a state is the data of a sub-alphabet of A, of a bLSP morphism and of a set of 5-tuples (a, b, c, β, γ) of letters whose aim is to represent the set of fragilities of a word. For a state q, we let $\text{alph}(q)$ denote the sub-alphabet of A, by $\text{bLSP}(q)$ the morphism and by $\text{set}(q)$ the set of 5-tuples.
- The set of transitions Δ is the set of triples (q, f, q') such that
   1. $f = \text{bLSP}(q)$;
   2. $\text{alph}(q) = \text{alph}(f(\text{alph}(q')))$;
   3. if $(a, b, c, \beta, \gamma) \in \text{set}(q')$ then f is not LSP (a, b, c)-breaking;
   4. $\text{set}(q)$ is the set of all 5-tuples (a, b, c, β, γ) such that a, b, c, β, γ are letters of $\text{alph}(q)$, $a \neq b \neq c \neq a$, $\beta \neq \gamma$ and one of the following two conditions holds:
      (a) $a = \text{first}(f)$, βb, γc in $\text{Fact}(f(\text{alph}(q')))$ and $\beta \neq \gamma$;
      (b) there exist a', b', c' such that $(a', b', c', \beta, \gamma) \in \text{set}(q')$ and a word x such that $xa \in \text{pref}(f(a')\alpha)$, $xb \in \text{pref}(f(b')\alpha)$ and $xc \in \text{pref}(f(c')\alpha)$ with $\alpha = \text{first}(f)$.
- All states are initial.

Figure 2 shows this automaton when the alphabet is $\{a, b\}$. In this figure, $\tau_a = [a, ab]$ and $\tau_b = [ba, b]$. States q with $\text{set}(q) \neq \emptyset$ are not drawn since binary infinite LSP words contain no fragilities. Moreover states $(\emptyset, \tau_a, \emptyset)$ and $(\emptyset, \tau_b, \emptyset)$, $(\{a\}, \tau_b, \emptyset)$ and $(\{b\}, \tau_a, \emptyset)$ are not drawn as there are no transition leaving them.

For alphabets with at least three letters, automaton $\mathcal{A}_{bLSP}$ is too huge to be drawn even restricting to states q such that $\text{set}(q)$ is a set of fragilities of an LSP word.

An infinite word $\mathbf{f}$ over bLSP is said to be *recognized* by $\mathcal{A}_{bLSP}$ if there exists an infinite path in $\mathcal{A}_{bLSP}$ whose label is $\mathbf{f}$. The aim of $\mathcal{A}_{bLSP}$ is to recognize bLSP directive words of infinite LSP words.

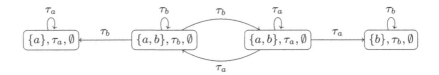

Fig. 2. $\mathcal{A}_{\mathrm{bLSP}}$ for the binary alphabet

Let **w** be an LSP word. We associate with **w** a state of $\mathcal{A}_{bLSP}$ that we let denote $q(\mathbf{w})$. This state is the state q such that:

- $\mathrm{alph}(q) = \mathrm{alph}(\mathbf{w})$:
- $\mathrm{bLSP}(q)$ is any morphism f such that $\mathbf{w} = f(\mathbf{w}')$ for some word $\mathbf{w}'$ (such a morphism exists by Lemma 4).
- $\mathrm{set}(q)$ is the set of all 5-tuples (a, b, c, β, γ) such that **w** is (a, b, c, β, γ)-fragile.

The fact that, for any LSP word **w**, any of its directive word is recognized by **w**, is a direct consequence of next lemma.

Lemma 14. *Let **w**, **w**$'$ be LSP words such that* $\mathbf{w} = f(\mathbf{w}')$ *with* $f = \mathrm{bLSP}(q(\mathbf{w}))$. *The transition* $(q(\mathbf{w}), \mathrm{bLSP}(q(\mathbf{w})), q(\mathbf{w}'))$ *exists in* $\mathcal{A}_{\mathrm{bLSP}}$.

Proof. Let $f = \mathrm{bLSP}(q(\mathbf{w}))$. Observe that $\mathrm{alph}(q(\mathbf{w})) = \mathrm{alph}(\mathbf{w})$, $\mathbf{w} = f(\mathbf{w}')$ and $\mathrm{alph}(q(\mathbf{w}')) = \mathrm{alph}(\mathbf{w}')$. Whence we have $\mathrm{alph}(q(\mathbf{w})) = \mathrm{alph}(f(\mathrm{alph}(q(\mathbf{w}'))))$.

By Lemma 12 and the definition of $q(\mathbf{w})$, a 5-tuple (a, b, c, β, γ) belongs to set $q(\mathbf{w})$ if and only if one of the following two conditions holds:

- $a = \mathrm{first}(f)$, βb, γc belong to $\mathrm{Fact}(f(\mathrm{alph}(\mathbf{w}')))$
- there exist a', b', c', β, γ in $\mathrm{alph}(\mathbf{w}')$ and an (a, b, c, β, γ)-fragility u of **w** and an $(a', b', c', \beta, \gamma)$-fragility v of **w**$'$ such that $|v| < |u|$, $f(v)$ is a proper prefix of u and words ua, βub, γuc are respectively prefixes of $f(va')\alpha$, $\beta f(vb')\alpha$, $\gamma f(vc')\alpha$ with $\alpha = \mathrm{first}(f)$.

For the second case, $u = f(v)x$ for a word x. The word xa is a prefix of $f(a')\alpha$, xb is a prefix of $f(b')\alpha$ and xc is a prefix of $f(c')\alpha$. As **w**$'$ is $(a', b', c', \beta, \gamma)$-fragile, $(a', b', c', \beta, \gamma) \in \mathrm{set}(q(\mathbf{w}'))$. Thus in both cases, Condition 4 for $(q(\mathbf{w})$, $\mathrm{bLSP}(q(\mathbf{w}), q(\mathbf{w}'))$ to be a transition of $\mathcal{A}_{\mathrm{bLSP}}$ is verified.

To end the proof we have to check Property 3 of transitions of $\mathcal{A}_{\mathrm{bLSP}}$. Assume there exists an (a, b, c, β, γ)-fragility in $\mathrm{set}(q(\mathbf{w}'))$. By definition of $q(\mathbf{w}')$, this implies that **w**$'$ has an (a, b, c, β, γ)-fragility. As $\mathbf{w} = f(\mathbf{w}')$ is LSP, $f = \mathrm{bLSP}(q(\mathbf{w}))$ is not LSP (a, b, c)-breaking. □

7   A Characterization of LSP Words

Theorem 15. *A word **w** is LSP if and only if it is S_{bLSP}-adic and all of its directive word are recognized by the automaton* $\mathcal{A}_{\mathrm{bLSP}}$.

Proof. Proposition 3 and Lemma 14 prove the only if part of Theorem 15. Let us prove the if part of Theorem 15.

Assume, by contradiction, that $\mathcal{A}_{\mathrm{bLSP}}$ recognizes a directive word $\mathbf{f}$ of a word $\mathbf{w}$ which is S_{bLSP}-adic but not LSP. Such a word contains a left special factor u that is not a prefix of $\mathbf{w}$. Among all possible triples $(\mathbf{f}, \mathbf{w}, u)$, choose one such that $|u|$ is minimal.

For $n \geq 1$, we let f_n denote the nth letter of $\mathbf{f}$ and $\mathbf{w}_n$ the word directed by $(f_k)_{k \geq n}$ ($\mathbf{w}_1 = \mathbf{w}$; $\mathbf{w}_2$ is directed by $f_2 f_3 \cdots$; $\mathbf{w}_n = f_n(\mathbf{w}_{n+1})$ for $n \geq 1$).

Step 1: $\mathbf{w}_2$ contains a fragility

First observe $|u| \geq 2$. Indeed we have $|u| \neq 0$ as the empty word is a prefix of $\mathbf{w}$. Moreover, by the structure of images of the bLSP morphism f_1, only the letter first(f_1) can be left special, whence $|u| \neq 1$.

Let $\alpha = \mathrm{first}(\mathbf{w}) = \mathrm{first}(f_1)$. Considering the last occurrence of α in u, the word u can be decomposed in a unique way $u = f_1(v)\alpha x$ with v, x words such that $|x|_\alpha = 0$.

As u is left special, there exist distinct letters β and γ such that βu and γu are factors of $\mathbf{w}$. As the letter α marks the beginning of images of letters in $\mathbf{w}$ and as for all letters δ, $f_1(\delta)$ ends with δ, we deduce that βv and γv are factors of $\mathbf{w}_2$. As $|v| < |u|$ and by choice of the triple $(\mathbf{f}, \mathbf{w}, u)$, the word v is a prefix of $\mathbf{w}_2$. Consequently $f_1(v)\alpha$ is a prefix of $\mathbf{w}$ and so $x \neq \varepsilon$.

Assume there exists a unique letter b such that $\beta v b$ is a factor of $\mathbf{w}$ and u is a prefix of $f(vb)$. Assume also that b is the unique letter c such that $\gamma v c$ is a factor of $\mathbf{w}$ and u is a prefix of $f(vc)$. As u is not a prefix of $\mathbf{w} = f_1(\mathbf{w}_2)$ and as u is a prefix of $f_1(vb)$, the word vb is not a prefix of $\mathbf{w}_2$. By choice of the triple $(\mathbf{f}, \mathbf{w}, u)$, $|vb| \geq |u|$. As $|v| < |u|$, we get $|vb| = |u| = |f_1(v)\alpha x|$. As $|f_1(v)| \geq |v|$, it follows $x = \varepsilon$: a contradiction.

From what precedes, we deduce the existence of two distinct letters b and c such that $\beta v b$ and $\gamma v c$ are factors of $\mathbf{w}_2$ with u a prefix of $f_1(vb)$ and $f_1(vc)$. As u is not a prefix of $\mathbf{w} = f_1(\mathbf{w}_2)$, the letter a that follows the prefix v of $\mathbf{w}_2$ is different from b and c. Hence the word $\mathbf{w}_2$ is (a, b, c, β, γ)-fragile and v is such a fragility.

Step 2: f_1 is LSP (a, b, c)-breaking

By definition of letters b and c at Step 1, the word αx is a common prefix of $f_1(b)$ and $f_1(c)$. Also as $u = f_1(v)\alpha x$ is not a prefix of $\mathbf{w}$ while $f_1(v)a$ is a prefix of $\mathbf{w}$, the word αx is not a prefix of $f_1(a)$. By Corollary 11, f_1 is (a, b, c)-breaking.

Step 3: Origin of fragilities of $\mathbf{w}_2$

Applying iteratively Lemma 12, we deduce the existence of an integer $n \geq 2$, a sequence of triples of pairwise distinct letters $(a_i, b_i, c_i)_{i \in \{2, \cdots, n\}}$, a sequence of $(a_i, b_i, c_i, \beta, \gamma)$-fragilities $(v_i)_{i \in \{2, \cdots, n\}}$ such that:

- v_i occurs in $\mathbf{w}_i$ for all $i \in \{2, \cdots, n\}$;
- $(a_2, b_2, c_2) = (a, b, c)$ and $v_2 = v$;
- $|v_{i+1}| < |v_i|$ for all $i \in \{2, \cdots, n-1\}$;
- $v_n = \varepsilon$.
- words $v_i a_i$, $\beta v_i b_i$, $\gamma v_i c_i$ are respectively prefixes of the words $f_i(v_{i+1}a_{i+1})\alpha_i$, $\beta f_i(v_{i+1}b_{i+1})\alpha_i$, $\gamma f_i(v_{i+1}c_{i+1})\alpha_i$ where $\alpha_i = \mathrm{first}(f_i)$ for $i \in \{2, \cdots, n-1\}$;

- $a_n = \text{first}(f_n)$;
- βb_n, γc_n belong to $\text{Fact}(f_n(\text{alph}(\mathbf{w}_{n+1})))$.

Step 4: Conclusion Let $(q_i)_{i \geq 1}$ be the sequence of states along a path recognizing **f**: for all $n \geq 1$, (q_n, f_n, q_{n+1}) is a transition of $\mathcal{A}_{\text{bLSP}}$.

At the end of Step 3, we learn that there exists an (a_n, b_n, c_n)-fragility in $f_n(\mathbf{w}_{n+1})$. Hence a_n, b_n, c_n are pairwise distinct letters. Especially as $a = \text{first}(f_n) \notin \{b_n, c_n\}$ by properties of bLSP morphisms, the words βb_n and γc_n are factors of images of letters, say b'_n and c'_n. As $\text{alph}(q_n) = \text{alph}(f_n(q_{n+1}))$, this implies that b'_n and c'_n belong to $\text{alph}(q_{n+1})$ and a_n, b_n and c_n belong to $\text{alph}(q_n)$. Moreover, as βb_n, γc_n are factors of words in $f_n(\text{alph}(q_{n+1}))$, we deduce that $(a_n, b_n, c_n, \beta, \gamma) \in \text{set}(q_n)$.

By backward induction, we can show that for all i, $2 \leq i \leq n$, $(a_i, b_i, c_i, \beta, \gamma) \in \text{set}(q_n)$. Especially $(a_2, b_2, c_2, \beta, \gamma) \in \text{set}(q_2)$. As, by Step 2, f_1 is LSP (a_2, b_2, c_2)-breaking and (q_1, f_1, q_2) is a transition of $\mathcal{A}_{\text{bLSP}}$, we get our final contradiction. □

8   Conclusion

Recall that G. Fici [8] asked for a characterization of both finite and infinite words. Observe that it can be proved that any non-empty finite LSP word w is right extendable to a longer LSP word (That is there exists a letter a occurring in w such that wa is a LSP). As a consequence one can prove:

Lemma 16. *A finite word is LSP if and only if it is a prefix of an infinite LSP word.*

This result shows that any characterization of infinite LSP words provides naturally a characterization of finite LSP words (adding "is a prefix of" before the characterization of infinite LSP words). For instance in the binary case, this allows to find back M. Sciortino and L.Q. Zamboni's result [15]: "binary words having suffix automaton with the minimal possible numbers of states are exactly the finite prefixes of standard Sturmian words" (that can be reformulated after G. Fici's work: "finite binary LSP words are exactly the finite prefixes of standard Sturmian words"). For this purpose, one can first see from Theorem 15 and Fig. 2 that directive words of binary infinite LSP words are ultimately τ_a or ultimately τ_b or ultimately contain both τ_a and τ_b. By classical results (see, *e.g.*, [1]) it can be deduced that an infinite LSP word is an infinite repetition of a finite standard word or is an infinite standard word. As any power of a finite standard word is a prefix of an infinite standard word (see [14, Chap. 2] for instance), we get M. Sciortino and L.Q. Zamboni's result.

We end this paper mentioning natural questions arising from this work. Can a smaller automaton than $\mathcal{A}_{\text{bLSP}}$ can be found for recognizing directive words of LSP infinite words? Can a similar S-adicity system can be found for infinite words having at most one left special factor? Does there exist a finite or infinite set S of morphisms such that an infinite word is LSP if and only if it S-adic (as it occurs for infinite balanced binary words)?

Acknowledgements. Many thanks to referees for their careful readings and their interesting suggestions and questions.

References

1. Berstel, J., Séébold, P.: Sturmian words. In: Lothaire, M. (ed.) Algebraic Combinatorics on Words, Encyclopedia of Mathematics and its Applications, vol. 90, pp. 45–110. Cambridge University Press, Cambridge (2002)
2. Berthé, V.: S-adic expansions related to continued fractions. In: Akiyama, S. (ed.) Natural Extension of Arithmetic Algorithms and S-adic System. RIMS Kôkyûroku Bessatsu, vol. B58, pp. 61–84 (2016)
3. Berthé, V., Delecroix, V.: Beyond substitutive dynamical systems: S-adic expansions. In: Akiyama, S. (ed.) Numeration and Substitution 2012. RIMS Kôkyûroku Bessatsu, vol. B46, pp. 81–123 (2014)
4. Berthé, V., Holton, C., Zamboni, L.Q.: Initial powers of Sturmian sequences. Acta Arith. **122**, 315–347 (2006)
5. Berthé, V., Labbé, S.: Factor complexity of S-adic words generated by the Arnoux-Rauzy-Poincaré algorithm. Adv. App. Math. **63**, 90–130 (2015)
6. Berthé, V., Rigo, M. (eds.): Combinatorics, Automata and Number Theory, Encyclopedia of Mathematics and its Applications, vol. 135. Cambridge University Press, Cambridge (2010)
7. Ferenczi, S.: Rank and symbolic complexity. Ergod. Theor. Dyn. Syst. **16**, 663–682 (1996)
8. Fici, G.: Special factors and the combinatorics of suffix and factor automata. Theor. Comput. Sci. **412**, 3604–3615 (2011)
9. Leroy, J.: Contribution à la résolution de la conjecture S-adique. Doctoral thesis, Université de Picardie Jules Verne (2012)
10. Leroy, J.: An S-adic characterization of minimal subshifts with first difference of complexity $p(n+1)-p(n) \leq 2$. Discrete Math. Theor. Comput. Sci. **16**(1), 233–286 (2014)
11. Leroy, J., Richomme, G.: A combinatorial proof of S-adicity for sequences with linear complexity. Integers **13**, 19 (2013). Article #A5
12. Levé, F., Richomme, G.: Quasiperiodic Sturmian words and morphisms. Theor. Comput. Sci. **372**(1), 15–25 (2007)
13. Lothaire, M.: Combinatorics on Words, Encyclopedia of Mathematics and its Applications, vol. 17. Addison-Wesley, Reading (1983). Reprinted in the Cambridge Mathematical Library. Cambridge University Press, UK (1997)
14. Lothaire, M.: Algebraic Combinatorics on Words, Encyclopedia of Mathematics and its Applications, vol. 90. Cambridge University Press, Cambridge (2002)
15. Sciortino, M., Zamboni, L.Q.: Suffix automata and standard Sturmian words. In: Harju, T., Karhumäki, J., Lepistö, A. (eds.) DLT 2007. LNCS, vol. 4588, pp. 382–398. Springer, Heidelberg (2007). doi:10.1007/978-3-540-73208-2_36

On Computational Complexity of Set Automata

Alexander A. Rubtsov[1,2]([✉]) and Mikhail N. Vyalyi[1,2,3]

[1] National Research University Higher School of Economics, Moscow, Russia
`rubtsov99@gmail.com, vyalyi@gmail.com`
[2] Moscow Institute of Physics and Technology, Dolgoprudny, Russia
[3] Dorodnicyn Computing Centre, FRC CSC RAS, Moscow, Russia

Abstract. We consider a computational model which is known as set automata. The set automata are one-way finite automata with an additional storage—the set. There are two kinds of set automata—the deterministic and the nondeterministic ones. We denote them as DSA and NSA respectively. The model was introduced by M. Kutrib, A. Malcher, M. Wendlandt in 2014 in [3,4]. It was shown that DSA-languages look similar to DCFL due to their closure properties and NSA-languages look similar to CFL due to their undecidability properties.

In this paper we show that this similarity is natural: we prove that languages recognizable by NSA form a rational cone, so as CFL. The main topic of this paper is computational complexity: we prove that languages recognizable by DSA belong to **P**, and the word membership problem is **P**-complete for DSA without ε-loops; languages recognizable by NSA are in **NP**, and there are **NP**-complete languages among them. Also we prove that the emptiness problem is **PSPACE**-hard for DSA.

Keywords: Set automata · Automata theory · Formal languages · Rational cone · Computational complexity · Membership problem

1 Introduction

We consider a computational model which is known as set automata. A set automaton is a one-way finite automaton equipped with an additional storage—the set $\mathbb{S}$—which is accessible through the work tape. On processing of a word, the set automaton can write a query z on the work tape and perform one of the following operations: the operation **in** inserts the word z into the set $\mathbb{S}$, the operation **out** removes the word z from the set $\mathbb{S}$ if $\mathbb{S}$ contains z, and the operation **test** is the query that verifies whether z belongs to $\mathbb{S}$. After the query the work tape is erased.

There are two kinds of set automata—the deterministic and the nondeterministic ones. We denote them as DSA and NSA respectively.

If determinism or nondeterminism of an automaton is not significant, we use abbreviation SA, and we refer to the class of languages recognizable by (N)SA as SA. We denote as DSA the class of languages recognizable by DSA.

Supported in part by RFBR grant 17–01–00300. The study has been funded by the Russian Academic Excellence Project '5–100'.

E. Charlier et al. (Eds.): DLT 2017, LNCS 10396, pp. 332–344, 2017.
DOI: 10.1007/978-3-319-62809-7_25

1.1   The Definition, Known Properties and Examples

We start with formal definitions. A set automaton M is defined by a tuple

$$M = \langle S, \Sigma, \Gamma, \lhd, \delta, s_0, F \rangle, \text{where}$$

- S is the finite set of states;
- Σ is the finite alphabet of the input tape;
- Γ is the finite alphabet of the work tape;
- $\lhd \notin \Sigma$ is the right endmarker;
- $s_0 \in S$ is the initial state;
- $F \subseteq S$ is the set of accepting states;
- δ is the transition relation:

$$\delta \subseteq S \times (\Sigma \cup \{\varepsilon, \lhd\}) \times [S \times (\Gamma^* \cup \{\mathbf{in}, \mathbf{out}\}) \cup S \times \{\mathbf{test}\} \times S].$$

In the deterministic case δ is the function

$$\delta : S \times (\Sigma \cup \{\varepsilon, \lhd\}) \to [S \times (\Gamma^* \cup \{\mathbf{in}, \mathbf{out}\}) \cup S \times \{\mathbf{test}\} \times S].$$

As usual, if $\delta(s, \varepsilon)$ is defined, then $\delta(s, a)$ is not defined for every $a \in \Sigma$. A *configuration* of M is a tuple $(s, v, z, \mathbb{S})$ consisting of the state $s \in S$, the unprocessed part of the input tape $v \in \Sigma^*$, the content of the work tape $z \in \Gamma^*$, and the content of the set $\mathbb{S} \subset \Gamma^*$. The transition relation determines the action of M on configurations. We use $\vdash$ notation for this action. It is defined as follows

$$
\begin{array}{llr}
(s, xv, z, \mathbb{S}) \vdash (s', v, zz', \mathbb{S}) & \text{if } (s, x, (s', z')) \in \delta, & z' \in \Gamma^*; \\
(s, xv, z, \mathbb{S}) \vdash (s', v, \varepsilon, \mathbb{S} \cup \{z\}) & \text{if } (s, x, (s', \mathbf{in})) \in \delta; & \\
(s, xv, z, \mathbb{S}) \vdash (s', v, \varepsilon, \mathbb{S} \setminus \{z\}) & \text{if } (s, x, (s', \mathbf{out})) \in \delta; & \\
(s, xv, z, \mathbb{S}) \vdash (s_+, v, \varepsilon, \mathbb{S}) & \text{if } (s, x, (s_+, \mathbf{test}, s_-)) \in \delta, & z \in \mathbb{S}; \\
(s, xv, z, \mathbb{S}) \vdash (s_-, v, \varepsilon, \mathbb{S}) & \text{if } (s, x, (s_+, \mathbf{test}, s_-)) \in \delta, & z \notin \mathbb{S}.
\end{array}
$$

We call a configuration *accepting* if the state of the configuration belongs to F and the word is processed till the endmarker. So the accepting configuration has the form $(s_f, \varepsilon, z, \mathbb{S})$, where $s_f \in F$.

The set automaton accepts a word w if there exists a run from the initial configuration $(q_0, w \lhd, \varepsilon, \varnothing)$ to some accepting one.

Set automata were presented by M. Kutrib, A. Malcher, M. Wendlandt in 2014 in [3,4]. The results of these conference papers are covered by the journal paper [5], so we give references to the journal variant further.

We recall briefly results from [5] about structural and decidability properties of DSA. They are presented in the tables, see Fig. 1. In the first table we list decidability problems: emptiness, regularity, equality to a regular language and finiteness. In the tables, R denotes an arbitrary regular language. The second table describes the structural properties: L, L_1 and L_2 are languages from the corresponding class; we write $+$ in a cell if the class is closed under the operation, otherwise we write $-$.

	DSA	CFL	DCFL
$L \stackrel{?}{=} \varnothing$	+	+	+
$L \stackrel{?}{\in} \mathsf{REG}$	+	−	+
$L \stackrel{?}{=} R$	+	−	+
$\|L\| \stackrel{?}{<} \infty$	+	+	+

	DSA	CFL	DCFL
$L_1 \cdot L_2$	−	+	−
$L_1 \cup L_2$	−	+	−
$L_1 \cap L_2$	−	−	−
$\Sigma^* \setminus L$	+	−	+
$L \cup R$	+	+	+
$L \cap R$	+	+	+

Fig. 1. Structural and decidability properties

From Fig. 1 one can see that DSA languages look similar to DCFL. Let us consider an example of a DSA-recognizable language that is not a DCFL (and not even a CFL).

Example 1. Denote $\Sigma_k = \{0, 1, \ldots k - 1\}$. We define $\mathrm{Per}_k = \{(w\#)^n \mid w \in \Sigma_k^*, n \in \mathbb{N}\}$ to be the language of repetitions of words over Σ_k separated by the delimiter $\#$. For any k there exists DSA M recognizing Per_k.

Proof. Firstly M copies the letters from Σ_k on the work tape until meets $\#$ and performs the operation **in** on $\#$. So, after processing of the prefix $w\#$, the set $\mathbb{S}$ contains w. Then, M copies letters from Σ_k on the work tape and performs the operation **test** on each symbol $\#$ until reaches $\triangleleft$. DSA M accepts the input iff all tests are positive; an accepted input looks like $w\#w\#\ldots w\# \in \mathrm{Per}_k$.   □

One can naturally assume that DSA class contains DCFL (or even CFL), but this assumption is false. As was shown in [5], the language $\{w\#w^R \mid w \in \Sigma^*, |\Sigma| \geqslant 2\}$ is not recognizable by any DSA.

Theorem 2. ([5]). *The classes* DCFL *and* DSA *are incomparable.*

The undecidability results for NSA are based on the fact that NSA can accept the set of invalid computations of a Turing machine.

Theorem 3. ([5]). *For NSA the questions of universality, equivalence with regular sets, equivalence, inclusion, and regularity are not semi-decidable. Furthermore, it is not semi-decidable whether the language accepted by some NSA belongs to* DSA.

It is worth to mention a quite similar model presented by K.-J. Lange and K. Reinhardt in [7]. We refer to this model as L-R-SA. In this model there are no **in** and **out** operations; in the case of **test−** result the tested word is added to the set after the query; also L-R-SA have no ε-moves. The results from [7] on computational complexity for L-R-SA are similar to ours: the membership problem is **P**-complete for L-R-DSA, and **NP**-complete for L-R-NSA.

1.2   Rational Transductions

Despite the classes DCFL and DSA are incomparable, their structural and decidability properties are similar. In this paper we show that this similarity is natural: we prove that the class of languages recognizable by NSA has the same structure as context-free languages.

Both classes are principal rational cones, and to define this structural property we need auxiliary notions.

A finite state transducer is a nondeterministic finite automaton with the output tape. Let T be a FST. The set $T(u)$ consists of all words v that T outputs on runs from the initial state to a final state while processing of u. So, FST T defines a *rational transduction* $T(\cdot)$. We also define $T(L) = \bigcup_{u \in L} T(u)$.

The *rational dominance* relation $A \leqslant_{\text{rat}} B$ holds if there exists a FST T such that $A = T(B)$, here A and B are languages.

A *rational cone* is a family of languages $\mathbf{C}$ closed under the rational dominance relation: $A \leqslant_{\text{rat}} B$ and $B \in \mathbf{C}$ imply $A \in \mathbf{C}$. If there exists a language $F \in \mathbf{C}$ such that $L \leqslant_{\text{rat}} F$ for any $L \in \mathbf{C}$, then $\mathbf{C}$ is a *principal* rational cone generated by F; we denote it as $\mathbf{C} = \mathcal{T}(F)$.

Rational transductions for context-free languages were thoroughly investigated in the 70s, particularly by the French school. The main results of this research are published in J. Berstel's book [2]. As described in [2], it follows from the Chomsky-Schützenberger theorem that CFL is a principal rational cone: CFL $= \mathcal{T}(D_2)$, where D_2 is the Dyck language on two brackets.

1.3   Our Contribution

One can consider D_2 as the language of correct protocols for operations with the stack (the push-down memory). In Sect. 3 we show that NSA languages are generated by the language SA-PROT of correct protocols for operations with the set. The similar result for the L-R-SA model was presented in [7].

It was shown in [5] that the emptiness problem for DSA is decidable. In fact the proof doesn't depend on determinism of SA. We prove the lower bounds for the emptiness problems for NSA and DSA: the problems are **PSPACE**-hard. In the case of the unary alphabet of the work tape the emptiness problem is **NP**-hard. Our proof relies on the technique of rational cones, so we put it in Sect. 3.2.

The main topic of this paper is computational complexity: we prove that the word membership problem is **P**-complete for DSA without ε-loops, DSA $\subsetneq$ **P**, and show that languages recognizable by NSA are in **NP**, and there are **NP**-complete languages.

The main technical result of the paper is SA $\subseteq$ **NP**. This result is based on the fact that the class SA is a rational cone and on our improvement of the technique of normal forms described in [5]. Due to space limitations we present only sketches of the proofs here. The complete proofs can be found in the full-version preprint [8].

2   P and NP-Complete Languages

Lemma 4. *There exists a* **P**-*hard language recognized by a DSA.*

Sketch of the proof. We reduce the language CVP (Circuit Value Problem), which is **P**-complete [6] under log-space reductions (we denote them by $\leqslant_{\log}$), to a language SA-CVP recognizable by a DSA M. The variant of the language CVP which is convenient for our purposes consists of words that are encodings of assignment sequences. Each variable P_i is encoded by a binary string, the operation basis is $\wedge, \vee, \neg, 1, 0$. An assignment has the form $P_i := P_j \, \mathrm{op} \, P_k$, where $\mathrm{op} \in \{\wedge, \vee\}$, $P_i := 1$, $P_i := 0$, $P_i := \neg P_j$. An assignment sequence belongs to CVP iff the last assignment is equal to one.

The language SA-CVP consists of words that are encodings of sequences of reversed assignments. A reversed assignment has the form $P_j \, \mathrm{op} \, P_k =: P_i$, where $\mathrm{op} \in \{\wedge, \vee\}$, $1 =: P_i$, $0 =: P_i$, $\neg P_k =: P_i$. Initially each variable is assigned to zero. Unlike CVP, reassignments of variables are allowed. A word w belongs to SA-CVP if the last assignment (the value of the reversed CVP-program) is equal to one. So, a word from SA-CVP looks like

$$w = \#\langle P_1\rangle\# \wedge \#\langle P_2\rangle\#\langle P_3\rangle\#1\#\langle P_4\rangle\# \ldots \#\langle P_j\rangle\# \, \mathrm{op} \, \#\langle P_k\rangle\#\langle P_i\rangle\#.$$

A reversed CVP-program is constructed from a regular one in log space. So, CVP $\leqslant_{\log}$ SA-CVP. DSA M verifies the correctness of a reversed CVP-program in a natural way: it stores the description of a variable $\langle P_i\rangle$ in the set iff the current assignment sets $P_i = 1$ and computes the value of an assignment $P_j \, \mathrm{op} \, P_k =: P_i$ using results of the tests $\langle P_j\rangle \in \mathbb{S}$ and $\langle P_k\rangle \in \mathbb{S}$. □

Now we show that DSA $\subseteq$ **P**.

As usually, ε-loops cause difficulties in analysis of deterministic models with ε-transitions. We say that DSA M *has an ε-loop* if there is a chain of ε-transitions from a state q to itself.

Proposition 5. *For any DSA M there is an equivalent DSA M' without ε-loops.*

Proof. There are two kinds of ε-loops: during the loops of the first kind M only writes letters on the work tape and during the loops of the second kind M performs queries. The former are evidently useless, so we simply remove them. The latter are significant: the behavior of M on these loops depends on the set's content and sometimes M goes to an infinite loop and sometimes not.

Since M is a deterministic SA, the set of words $\{u_1, \ldots, u_m\}$, that M writes on all ε-paths, is finite. We build DSA M' by M as follows. Each state of M' is marked by a vector $\boldsymbol{a} = (a_1, \ldots, a_m)$ of zeroes and ones. A content of the set is compatible with a vector $\boldsymbol{a}$ if $a_i = 1$ is equivalent to $u_i \in \mathbb{S}$. Each state of M' has the form $\langle s, \boldsymbol{a}, \mathsf{aux}\rangle$, where s is a state of M and aux is an auxiliary part of finite memory that M' uses to maintain $\boldsymbol{a}$ correctly.

To define an ε-transition $\delta(\langle s, \boldsymbol{a}, \mathsf{aux}\rangle, \varepsilon)$ of M' we follow an ε-path of M from the state s with the set content compatible with $\boldsymbol{a}$. If M goes to an ε-loop, then the ε-transition of M' is undefined. If the ε-path finishes at a state s', then $\delta(\langle s, \boldsymbol{a}, \mathsf{aux} \rangle, \varepsilon) = (\langle s', \boldsymbol{a}', \mathsf{aux}' \rangle, z)$, where $\boldsymbol{a}'$ is the vector compatible with the M's set content and z is the content of M's work tape at the end of the ε-path. □

Theorem 6. DSA $\subseteq$ **P**. *The membership problem for DSA without ε-loops is* **P**-*complete.*

Proof. The input of the membership problem is an encoding of DSA M without ε-loops and a word w. It is easy to emulate a DSA by a Turing machine with 3 tapes.

The first tape is for the input tape of the DSA, the second one is for the work tape and the third one (the storage tape) is used to maintain the set.

Let M process a subword u of the input between two queries to the set. There are no ε-loops. Therefore during this processing the automaton writes at most $c|u|$ symbols on the work tape, where c is a constant depending only on the description of the set automaton. It implies that $c|w|$ space on the storage tape is sufficient to maintain the set content on processing of the input word w. Thus each query to the set can be performed in polynomial time of the input size.

So, the membership problem for DSA without ε-loops belongs to **P** and due to Lemma 4 it is **P**-complete. By excluding ε-loops due to Proposition 5 we get that every language recognizable by DSA belongs to **P**. □

Remark 7. Proposition 5 implies an exponential upper bound on the number of M's steps during ε-moves. Now we describe a DSA M that performs $2^{c|\langle M \rangle|}$ steps on the empty input. Let $\Gamma = \{0, \ldots, n-1\}$. For $i \in 0..n-1$ M verifies whether $i \in \mathbb{S}$. If not, M puts i in the set, excludes each $j < i$ and restarts from $i = 0$. It is easy to see that M performs at least 2^n moves since each subset of Γ occurs in the set at some moment.

Now we present an NSA recognizing an **NP**-complete language. This result was also proved independently by M. Kutrib, A. Malcher, M. Wendlandt (private communication with M. Kutrib). We construct an **NP**-complete language that we call SA-SAT and reduce 3-SAT to this language. The language SA-SAT is quite similar to the language in [7] for the corresponding result for L-R-SA.

Let words $x_i \in \{0, 1\}^*$ encode variables and words $\langle \varphi \rangle \in \{0, 1\}^*$ encode 3-CNFs. SA-SAT contains the words of the form $x_1 \# x_2 \# \ldots \# x_n \# \# \langle \varphi \rangle$ such that an auxiliary 3-CNF φ' is satisfiable. The 3-CNF φ' is derived from φ as follows. For each variable x_i that appears in the list $x_1 \# x_2 \# \ldots \# x_n \# \#$ at least twice, remove all clauses containing x_i from φ and get the reduced 3-CNF φ''. Set each variable x of φ'' that is not in the list $x_1 \# x_2 \# \ldots \# x_n \# \#$ to zero, simplify φ'' and obtain as a result the 3-CNF φ'.

Lemma 8. *The language* SA-SAT *is* **NP**-*complete and it is recognized by an NSA.*

Proof. We describe an NSA M that recognizes the language SA-SAT. Firstly, M processes the prefix $x_1\#x_2\# \ldots \#x_n\#\#$, guesses the values b_i of x_i (0 or 1) and puts the pairs (x_i, b_i) in the set. If a variable x_i appears more than once in the prefix, then the NSA M nondeterministically guesses this event and puts both $(x_i, 0)$ and $(x_i, 1)$ in the set. On processing of the suffix $\langle \varphi \rangle$, for every clause of φ the NSA M nondeterministically guesses a literal that satisfies the clause and verifies whether the set contains the required value of the corresponding variable. It is easy to see that all tests are satisfied iff 3-CNF φ' is satisfiable.

The language 3-SAT is reduced to SA-SAT in a straightforward way. Also it is easy to see that SA-SAT $\in$ **NP**.   □

3   Structural Properties of SA-Languages

In this section we show that the class SA is a principal rational cone generated by the language of correct protocols. We also prove the lower bounds for the emptiness problem since they directly follow from the fact that SA is a rational cone.

To simplify arguments, hereinafter we assume that an NSA satisfies the following requirements:

(i)   the alphabet of the work tape is the binary alphabet, say, $\Gamma = \{a, b\}$;
(ii) it doesn't use the endmarker $\lhd$ on the input tape and the initial configuration is $(q_0, w, \varepsilon, \varnothing)$;
(iii) it accepts a word only if the last transition was a query.

It is easy to see that any NSA can be converted to an equivalent one satisfying these requirements.

3.1   Protocols

A *protocol* is a word $p = \#u_1\# \mathrm{op}_1 \#u_2\# \mathrm{op}_2\# \cdots \#u_n\# \mathrm{op}_n$, where $u_i \in \Gamma^*$, $\# \notin \Gamma$ and $\mathrm{op}_i \in \{\mathbf{in}, \mathbf{out}, \mathbf{test+}, \mathbf{test-}\}$. *Query words* are the words of the form $\#u\# \mathrm{op}$. So a protocol is a concatenation of query words.

We say that p is *a correct protocol for SA M on an input w in $L(M)$*, if there exists a run of M on the input w such that M performs the operation op_1 with the word u_1 on the work tape at first, then performs op_2 with u_2 on the work tape, and so on. In the case of a test, op_i indicates the result of the test: **test+** or **test−**.

We call p *a correct protocol for SA M* if there exists a word w such that p is a correct protocol for SA M on the input w. And finally, we say that p is a *correct protocol* if there exists an SA M such that p is a correct protocol for M.

We define SA-PROT to be the language of all correct protocols over the alphabet of the work tape $\Gamma = \{a, b\}$.

Proposition 9. *The language SA-PROT is recognizable by DSA.*

The proof is straightforward. We omit it due to the space limits.

We use a notation $q \xrightarrow[v]{u} p$ to express the fact that a transducer T has a run from the state q to the state p such that T reads u on the input tape and writes v on the output tape. The notation is also applied to a single transition. A rational transduction $T(u)$ mentioned in the introduction is defined with this notation as $\{v \mid q_0 \xrightarrow[v]{u} q_f, q_f \in F\}$. Recall that $T(L) = \bigcup_{u \in L} T(u)$.

We define $T^{-1}(y) = \{x \mid y \in T(x)\}$ and $T^{-1}(A) = \{x \mid T(x) \cap A \neq \varnothing\}$. It is well-known (e.g., see [2]) that for every FST T there exists a FST T' such that $T'(u) = T^{-1}(u)$.

We will prove that the class SA is a principal rational cone generated by the language SA-PROT. It means that for every SA-recognizable language L there exists a FST T such that $L = T(\text{SA-PROT})$.

Let us describe our plan. Firstly, we prove that for each SA M there is a FST T_M such that $w \in L(M)$ iff $T_M(w) \cap \text{SA-PROT} \neq \varnothing$. We call such FST T_M an *extractor* (of a protocol) for M. Then we show that the required FST T is T_M^{-1}.

Lemma 10. *For any SA* $M = \langle S, \Sigma, \Gamma, \lhd, \delta_M, s_0, F \rangle$ *there exists an extractor* $T_M = \langle S \cup \{s_0'\}, \Sigma, \Gamma, \delta, s_0', F \rangle$.

Sketch of the proof. The behavior of the extractor is similar to the behavior of the SA. When the SA writes something on the work tape, the FST writes the same on the output tape. When the SA makes a query, the FST writes a word $\# \text{op} \#$. The only difference is that the SA knows the results of the performed tests. But a nondeterministic extractor can guess the results of the tests to produce a correct output. $\qquad\square$

Theorem 11. SA *is a principal rational cone generated by* SA-PROT.

Proof. We shall prove that for every SA M there exists a FST T such that $T(\text{SA-PROT}) = L(M)$. Take $T = T_M^{-1}$, where T_M is the extractor for M. By the definition of extractor $w \in L(M)$ iff $T(w) \cap \text{SA-PROT} \neq \varnothing$. So, if $w \in L(M)$, then there is at least one correct protocol in $T_M(w)$. Therefore $w \in T_M^{-1}(\text{SA-PROT})$. In the other direction: if $w \in T_M^{-1}(\text{SA-PROT})$, then $T_M(w) \cap \text{SA-PROT} \neq \varnothing$. So, $w \in L(M)$ by the definition of extractor. $\qquad\square$

3.2 Lower Bounds for the Emptiness Problem

We present an application of the rational cones technique. It is a lower bound on complexity of the emptiness problem for NSA. We show that the emptiness problem for NSA is equivalent to the regular realizability (NRR) problem for SA-PROT and use hardness results for NRR problems.

The problem NRR(F) for a language F (a parameter of the problem) is to decide on the input nondeterministic finite automaton (NFA) $\mathcal{A}$ whether the intersection $L(\mathcal{A}) \cap F$ is nonempty.

Lemma 12

$$(L(M) \overset{?}{=} \varnothing) \leqslant_{\log} \text{NRR}(\textit{SA-PROT}) \text{ and } \text{NRR}(\textit{SA-PROT}) \leqslant_{\log} (L(M) \overset{?}{=} \varnothing).$$

Proof. It is easy to see that one can construct the extractor T_M by SA M in log space. So, by the definition of extractor, we get that $L(M) = \varnothing$ iff $T_M(\Sigma^*) \cap$ SA-PROT $= \varnothing$. Note that rational transductions preserve regularity [2]: $T_M(\Sigma^*) \in$ REG. It is shown in [9] that an NFA recognizing $T_M(\Sigma^*)$ is log-space constructible by the description of T_M. So $(L(M) \overset{?}{=} \varnothing) \leqslant_{\log}$ NRR(SA-PROT).

It was shown in [5] that for any regular language $R \subseteq \Gamma^*$ there exists SA M_R recognizing SA-PROT $\cap R$. This SA is also constructible in log space by the description of an NFA recognizing R: the proof is almost the same as for the aforementioned NFA by FST construction. Thus NRR(SA-PROT) $\leqslant_{\log} (L(M) \overset{?}{=} \varnothing)$.   □

Also we need the following relation between RR-problems and rational transductions.

Proposition 13 ([9]). *If $A \leqslant_{\mathrm{rat}} B$, then* NRR$(A) \leqslant_{\log}$ NRR(B).

We use the hardness of NRR problem for languages Per$_k$ from Example 1.

Theorem 14 ([1,10]). *The problem* NRR(Per$_1$) *is* **NP**-*complete and the problem* NRR(Per$_2$) *is* **PSPACE**-*complete.*

From these facts we derive the **PSPACE** lower bound.

Theorem 15. *The emptiness problem for NSA is* **PSPACE**-*hard. For NSA with the unary alphabet of the work tape, the emptiness problem is* **NP**-*hard.*

Proof. The language Per$_2$ is recognizable by DSA (see Example 1). By Theorem 11 we get that Per$_2 \leqslant_{\mathrm{rat}}$ SA-SAT. Thus NRR(Per$_2$) $\leqslant_{\log}$ NRR(SA-SAT) by Proposition 13 and therefore the problem NRR(SA-SAT) is **PSPACE**-hard. So, by Lemma 12, the emptiness problem for NSA is **PSPACE**-hard too.   □

Corollary 16. *Theorem 15 also holds for DSA.*

We describe briefly a proof idea. The emptiness problem for NSA is reduced to the emptiness problem for DSA. For an NSA M we build a DSA M' such that $L(M') = \varnothing$ iff $L(M) = \varnothing$. Generally, DSA M' acts as M. But the alphabet M' includes two auxiliary symbols $\#, \$ \notin \Sigma_M$. Transitions of M are uniquely encoded by words in the form $\#\$^i\#$. The codes of transitions form a prefix code. Thus they can be used to simulate nondeterministic transitions of M.

4 SA ⊆ NP

We come to the main result of the paper. It is technically hard, so we provide only a sketch of the proof here. We still assume that Requirements (i-iii) from Sect. 3 hold for NSA.

The general idea is to prove that for any $w \in L(M)$ there exists a short (polynomial in $|w|$) protocol p for w. Thus an NP-algorithm guesses the run of M on w corresponding to p and verifies that M accepts w on that run.

We start with auxiliary notions.

Consider a protocol $p = \#u_0\# \operatorname{op}_0 \#u_1\# \operatorname{op}_1\# \cdots \#u_{n-1}\# \operatorname{op}_{n-1}$. By a *segment* p_i of p we mean an occurrence of a query word $\#u_i\# \operatorname{op}_i$ in the protocol. Different segments may coincide as words. We say that a segment p_i *supports* segment p_j if $u_i = u_j$ and $\operatorname{op}_i = \textbf{in}$, $\operatorname{op}_j = \textbf{test+}$ or $\operatorname{op}_i = \textbf{out}$, $\operatorname{op}_j = \textbf{test−}$ and there is no segment p_k such that $\operatorname{op}_k \in \{\textbf{in}, \textbf{out}\}$, $u_k = u_i$ and $i < k < j$.

We say that segments $v_1, v_2, \ldots, v_k$ form a *chain* C if v_1 supports $v_i, i \in 2..k$. *Standalone queries* are the segments $\#u\#\textbf{in}$ and $\#u\#\textbf{out}$ that support no segments and the segments $\#u\#\textbf{test−}$ having no support. Each standalone query forms a *standalone chain*. Let $v_i = \#u\# \operatorname{op}_i$ be the segments of a chain. We denote the chain by $C(u)$ and call u the *pivot* of $C(u)$. Also we denote the chain C^+ if $\operatorname{op}_1 = \textbf{in}$ and C^- if $\operatorname{op}_1 \in \{\textbf{out}, \textbf{test−}\}$.

The following lemma immediately follows from the definitions.

Lemma 17. *A protocol p is correct iff each segment $\#u_i\#\textbf{test+}$ is supported and each segment $\#u_i\#\textbf{test−}$ is either supported or standalone.*

By definition, there is a correspondence between protocols and runs of SA. Segments correspond to the following parts of a run. Suppose that SA M starts from a state s with the blank work tape, writes u on the work tape on processing of x on the input tape and performs a query at a state q. We call this sequence of operations a *query run* and denote it as $s \xrightarrow[u]{x} q$. Thus an accepting run of SA M on word $w = x_0 x_1 \cdots x_{n-1}$, $x_i \in \Sigma^*$, has the following *partition* into query runs:

$$s_0 \xrightarrow[u_0]{x_0} s_1 \xrightarrow[u_1]{x_1} s_2 \xrightarrow[u_2]{x_2} \cdots \xrightarrow[u_{n-1}]{x_{n-1}} s_n, \quad s_n \in F. \tag{1}$$

We also say that a sequence of query runs $s_i \xrightarrow[u_i]{x_i} s_{i+1} \xrightarrow[u_{i+1}]{x_{i+1}} \cdots \xrightarrow[u_{j-1}]{x_{j-1}} s_j$ is a *segment of a run*, we denote it as $s_i \xrightarrow{x_i x_{i+1} \cdots x_j} s_j$.

Note that $x_i = \varepsilon$ is possible and it is the main obstacle to prove an existence of a short protocol: a segment $s_i \xrightarrow{\varepsilon} s_j$ may contain a lot of queries that support the tests on the nonempty x_ks. So the total number of queries may be arbitrary large. If we had no ε-transitions in NSA, we would have had the same proof as for the result L-R-NSA $\in$ **NP** in [7]. But ε-transitions require more careful consideration.

To overcome this difficulty we will use a special form of NSA that we call *Atomic Action Normal Form* (AANF). AANF is a refinement of the infinite action normal form from [5] developed for DSA.

We also use the notation $s \xrightarrow[u]{x} q$ for the relation that holds if there is a query run that starts from the state s, reads the word x, writes the word u and makes a query at the state q. We always indicate the meaning of the notation: a query or a relation. A relation $s \xrightarrow[U]{X} q$ means that $U = \{u \mid \exists x \in X : s \xrightarrow[u]{x} q\}$.

Definition 18. *We say that an NSA M is in AANF if the following conditions hold.*

- *Requirements* (i-iii) *from Sect. 3 hold.*
- $S = \{s_0\} \cup S_{test+} \cup S_{test-} \cup S_{in} \cup S_{out} \cup S_{\varepsilon\text{-write}} \cup S_{\not{\varepsilon}\text{-write}},\ S_{op} \cap S_{op'} = \varnothing$
  if op $\neq$ op': *each state q such that the relation* $s \xrightarrow[u]{x} q$ *holds is marked by the operation of the query; states that occur while writing u are marked as* ε**-write** *if* $x = \varepsilon$ *and as* $\not{\varepsilon}$**-write** *if* $x \neq \varepsilon$.
- *There exists a finite family of regular languages* $\mathscr{L}_M$ *such that*
  - *for all* $A_\alpha, A_\beta \in \mathscr{L}_M : A_\alpha \cap A_\beta = \varnothing$ *if* $\alpha \neq \beta$;
  - *if* $s \xrightarrow[U]{\Sigma^*} q$, $s' \xrightarrow[U']{\Sigma^*} q'$, *then either* $U \cap U' = \varnothing$ *or* $U = U' \in \mathscr{L}_M$;
  - $|A_\alpha| = \infty$ *for every* $A_\alpha \in \mathscr{L}_M$.

Informally, AANF implies that each query run $s \xrightarrow[u]{x} q$ belongs to an equivalence class corresponding to $A_\alpha \in \mathscr{L}_M$, $u \in A_\alpha$, and the number of such classes is finite. Moreover, in the case of $x = \varepsilon$ the relation $s \xrightarrow[A_\alpha]{\varepsilon} q$ holds, and it means that one can replace u (in the segment) by any word $u_\alpha \in A_\alpha$ and obtain a run for w again. But this run may be incorrect, so later we will carefully choose u_α to maintain the correctness of the run. Thus, AANF helps us to struggle with ε-queries as well as to make non-ε-segments short.

Lemma 19. *For any NSA M there exists an NSA M' in AANF such that* $L(M) = L(M')$.

From now on we fix an NSA M in AANF. We say that p_i is an ε-*segment* of a protocol if the corresponding query run has the form $s_i \xrightarrow[u_i]{\varepsilon} s_{i+1}$. We say that an ε-segment p_i has type α if the relation $s_i \xrightarrow[A_\alpha]{\varepsilon} s_{i+1}$ holds for $A_\alpha \in \mathscr{L}_M$. We call a chain $C(u)$ of ε-segments an ε-*chain*. We denote ε-chain C_α if all segments in the chain have type α. Different segments of an ε-chain can't have different types due to AANF definition.

Now we are ready to describe the main steps of the proof. Each step is a transformation of an accepting run to some other accepting run of M on w. We call such transformations *valid*. We call a word u *short* if $|u|$ is polynomial in $|w|$. We call a run *short* if the corresponding protocol is short.

Lemma 20. *For any* $A_\alpha \in \mathscr{L}_M$ *there exist short words* $u_\alpha^+ \in A_\alpha$ *and* $u_\alpha^- \in A_\alpha$ *such that for any run the simultaneous replacement of the words u in all segments of* ε-*chains* $C_\alpha^+(u)$ *and* $C_\alpha^-(u)$ *by* u_α^+ *and* u_α^- *respectively is a valid transformation of the run. Moreover, during the transformed run*

- *M never adds the words* u_α^- *to* $\mathbb{S}$;
- *M never removes the words* u_α^+ *from* $\mathbb{S}$;
- *all segments of* ε-*chains* C^- *become standalone;*
- *the length of any word* $u_\alpha^\pm$ *is* $O(|w|)$.

From now on we assume w.l.o.g. that all runs into consideration have the properties described in Lemma 20.

Lemma 21. *If in each query run $s_i \xrightarrow[u_i]{x_i} s_{i+1}$, $x_i \neq \varepsilon$, of a run the word u_i is short, then there is a valid transformation of the run to a short one.*

Lemma 22. *Assume that $C(u)$ is a non-ε-chain of segments $v_1, \ldots v_k$, $v_i = \#u\# \mathrm{op}_i$. Then there is a short word u' such that the replacement of v_i by $\#u'\# \mathrm{op}_i$ is a valid transformation of the run.*

Theorem 23. *There is an NP-algorithm verifying for an input word w whether $w \in L(M)$, where M is an NSA.*

Proof. W.l.o.g. M is NSA in AANF. We will show that for any $w \in L(M)$ there exists a short accepting run. Thus, an NP-algorithm guesses a short run and checks its correctness.

At first, we apply Lemma 20 to an accepting run on the input w to get a run satisfying the conditions of the lemma. After that we shall transform all non-ε-segments to short ones. Note that there is no more than $|w|$ non-ε-segments in any run and therefore there is no more than $|w|$ non-ε-chains. We apply Lemma 22 no more than $|w|$ times to get a run in which all query runs $s_i \xrightarrow[u_i]{x_i} s_{i+1}$, $x_i \neq \varepsilon$, are short. Finally, we apply Lemma 21 and get a short run for w. $\square$

Acknowledgements. We thank Dmitry Chistikov for the feedback and discussion of the text's results and suggestion for improvements and anonymous referees for helpful comments.

References

1. Anderson, T., Loftus, J., Rampersad, N., Santean, N., Shallit, J.: Special issue: LATA 2008 detecting palindromes, patterns and borders in regular languages. Inf. Comput. **207**(11), 1096–1118 (2009)
2. Berstel, J.: Transductions and Context-Free Languages. Teubner, Stuttgart (1979)
3. Kutrib, M., Malcher, A., Wendlandt, M.: Deterministic set automata. In: Shur, A.M., Volkov, M.V. (eds.) DLT 2014. LNCS, vol. 8633, pp. 303–314. Springer, Cham (2014). doi:10.1007/978-3-319-09698-8_27
4. Kutrib, M., Malcher, A., Wendlandt, M.: Regularity and size of set automata. In: Jürgensen, H., Karhumäki, J., Okhotin, A. (eds.) DCFS 2014. LNCS, vol. 8614, pp. 282–293. Springer, Cham (2014). doi:10.1007/978-3-319-09704-6_25
5. Kutrib, M., Malcher, A., Wendlandt, M.: Set automata. Int. J. Found. Comput. Sci. **27**(02), 187–214 (2016)
6. Ladner, R.E.: The circuit value problem is log space complete for P. SIGACT News **7**(1), 18–20 (1975)
7. Lange, K.J., Reinhardt, K.: Set automata. In: Combinatorics, Complexity and Logic, Proceedings of the DMTCS 1996. pp. 321–329. Springer (1996)

8. Rubtsov, A.A., Vyalyi, M.N.: On computational complexity of Set Automata. ArXiv e-prints https://arxiv.org/pdf/1704.03730, April 2017
9. Rubtsov, A., Vyalyi, M.: Regular realizability problems and context-free languages. In: Shallit, J., Okhotin, A. (eds.) DCFS 2015. LNCS, vol. 9118, pp. 256–267. Springer, Cham (2015). doi:10.1007/978-3-319-19225-3_22
10. Vyalyi, M.N.: On the models of nondeterminizm for two-way automata. In: Proceedings of VIII International Conference on Discrete Models in the Theory of Control Systems, pp. 54–60 (2009). (in Russian)

On the Number of Rich Words

Josef Rukavicka[(✉)]

Department of Mathematics,
Faculty of Nuclear Sciences and Physical Engineering,
Czech Technical University in Prague,
Trojanova 13, 120 01 Prague 2, Czech Republic
josef.rukavicka@seznam.cz

Abstract. Any finite word w of length n contains at most $n+1$ distinct palindromic factors. If the bound $n+1$ is reached, the word w is called rich. The number of rich words of length n over an alphabet of cardinality q is denoted $R_q(n)$. For binary alphabet, Rubinchik and Shur deduced that $R_2(n) \leq c1.605^n$ for some constant c. In addition, Guo, Shallit and Shur conjectured that the number of rich words grows slightly slower than $n^{\sqrt{n}}$. We prove that $\lim\limits_{n \to \infty} \sqrt[n]{R_q(n)} = 1$ for any q, i.e. $R_q(n)$ has a subexponential growth on any alphabet.

Keywords: Rich words · Enumeration · Palindromes · Palindromic factorization

1 Introduction

The study of palindromes is a frequent topic and many diverse results may be found. In recent years, a number of articles deal with so-called *rich* words, or also words having *palindromic defect* 0. They are words having the maximum number of palindromic factors. As noted by [6], a finite word w contains at most $|w| + 1$ distinct palindromic factors with $|w|$ being the length of w. The rich words are exactly those that attain this bound. It is known that on a binary alphabet the set of rich words contains factors of Sturmian words, factors of complementary symmetric Rote words, factors of the period-doubling word, etc., see [1,4,6,13]. On a multiliteral alphabet, the set of rich words contains for example factors of Arnoux—Rauzy words and factors of words coding symmetric interval exchanges.

Rich words can be characterized using various properties, see for instance [2,5,8]. The concept of rich words can also be generalized to respect so-called pseudopalindromes, see [10]. In this paper we focus on an unsolved question of computing the number of rich words of length n over an alphabet with $q > 1$ letters. This number is denoted $R_q(n)$.

This question is investigated in [15], where J. Vesti gives a recursive lower bound on the number of rich words of length n, and an upper bound on the number of binary rich words. Both these estimates seem to be very rough. In [9],

© Springer International Publishing AG 2017
É. Charlier et al. (Eds.): DLT 2017, LNCS 10396, pp. 345–352, 2017.
DOI: 10.1007/978-3-319-62809-7_26

C. Guo, J. Shallit and A.M. Shur construct for each n a large set of rich words of length n. Their construction gives, currently, the best lower bound on the number of binary rich words, namely $R_2(n) \geq \frac{C^{\sqrt{n}}}{p(n)}$, where $p(n)$ is a polynomial and the constant $C \approx 37$. On the other hand, the best known upper bound is exponential. As mentioned in [9], a calculation performed recently by M. Rubinchik provides the upper bound $R_2(n) \leq c1.605^n$ for some constant c, see [11].

Our main result stated as Theorem 10 shows that $R_q(n)$ has a subexponential growth on any alphabet. More precisely, we prove that

$$\lim_{n \to \infty} \sqrt[n]{R_q(n)} = 1.$$

In [14], Shur calls languages with the above property *small*. Our result is an argument in favor of a conjecture formulated in [9] saying that for some infinitely growing function $g(n)$ the following holds true $R_2(n) = \mathcal{O}\left(\frac{n}{g(n)}\right)^{\sqrt{n}}$.

To derive our result we consider a specific factorization of a rich word into distinct rich palindromes, here called UPS-factorization (Unioccurrent Palindromic Suffix factorization), see Definition 2. Let us mention that another palindromic factorizations have already been studied, see [3,7]: *Minimal* (minimal number of palindromes), *maximal* (every palindrome cannot be extended on the given position) and *diverse* (all palindromes are distinct). Note that only the *minimal* palindromic factorization has to exist for every word.

The article is organized as follows: Sect. 2 recalls notation and known results. In Sect. 3 we study a relevant property of UPS-factorization. The last section is devoted to the proof of our main result.

2   Preliminaries

Let us start with a couple of definitions: Let A be an alphabet of q letters, where $q > 1$ and $q \in \mathbb{N}$ ($\mathbb{N}$ denotes the set of nonnegative integers). A finite sequence $u_1 u_2 \cdots u_n$ with $u_i \in A$ is a *finite word*. Its length is n and is denoted $|u_1 u_2 \cdots u_n| = n$. Let A^n denote the set of words of length n. We define that A^0 contains just the empty word. It is clear that the size of A^n is equal to q^n. Given $u = u_1 u_2 \cdots u_n \in A^n$ and $v = v_1 v_2 \cdots v_k \in A^k$ with $0 \leq k \leq n$, we say that v is a *factor* of u if there exists i such that $0 \leq i$, $i + k \leq n$ and $u_{i+1} = v_1$, $u_{i+2} = v_2, \ldots, u_{i+k} = v_k$.

A word $u = u_1 u_2 \cdots u_n$ is called a *palindrome* if $u_1 u_2 \cdots u_n = u_n u_{n-1} \cdots u_1$. The empty word is considered to be a palindrome and a factor of any word.

A word u of length n is called *rich* if u has $n+1$ distinct palindromic factors. Clearly, $u = u_1 u_2 \cdots u_n$ is rich if and only if its *reversal* $u_n u_{n-1} \cdots u_1$ is rich as well.

Any factor of a rich word is rich as well, see [8]. In other words, the language of rich words is factorial. In particular it means that $R_q(n) R_q(m) \leq R_q(n + m)$ for any $m, n, q \in \mathbb{N}$. Therefore, the Fekete's lemma implies existence of the limit of $\sqrt[n]{R_q(n)}$ and moreover

$$\lim_{n \to \infty} \sqrt[n]{R_q(n)} = \inf \left\{ \sqrt[n]{R_q(n)} : n \in \mathbb{N} \right\}.$$

For a fixed n_0, one can find the number of all rich words of length n_0 and obtain an upper bound on the limit. Using a computer Rubinchik counted $R_2(n)$ for $n \le 60$, (see the sequence A216264 in OEIS). As $\sqrt[60]{R_2(60)} < 1.605$, he obtained the upper bound given in Introduction.

As shown in [8], any rich word u over an alphabet A is richly prolongable, i.e., there exist letters $a, b \in A$ such that aub is also rich. Thus a rich word is a factor of an arbitrarily long rich word. But the question whether two rich words can appear simultaneously as factors of a longer rich word may have a negative answer. It means that the language of rich words is not recurrent. This fact makes the enumeration of rich words hard.

3   Factorization of Rich Words into Rich Palindromes

Let us recall one important property of rich words [6, Definition 4 and Proposition 3]: The longest palindromic suffix of a rich word w has exactly one occurrence in w (we say that the longest palindromic suffix of w is *unioccurrent* in w). It implies that $w = w^{(1)}w_1$, where w_1 is a palindrome which is not a factor of $w^{(1)}$. Since every factor of a rich word is a rich word as well, it follows that $w^{(1)}$ is a rich word and thus $w^{(1)} = w^{(2)}w_2$, where w_2 is a palindrome which is not a factor of $w^{(2)}$. Obviously $w_1 \neq w_2$. We can repeat the process until $w^{(p)}$ is the empty word for some $p \in \mathbb{N}$, $p \ge 1$. We express these ideas by the following lemma:

Lemma 1. *Let w be a rich word. There exist distinct non-empty palindromes $w_1, w_2, \ldots, w_p$ such that*

$$w = w_p w_{p-1} \cdots w_2 w_1 \text{ and } w_i \text{ is the longest palindromic suffix of}$$
$$w_p w_{p-1} \cdots w_i \text{ for } i = 1, 2, \ldots, p. \quad (1)$$

Definition 2. *We define UPS-factorization (Unioccurrent Palindromic Suffix factorization) to be the factorization of a rich word w into the form (1).*

Since the w_i in the factorization (1) are non-empty, it is clear that $p \le n = |w|$. From the fact that the palindromes w_i in the factorization (1) are distinct we can derive a better upper bound on p. The aim of this section is to prove the following theorem:

Theorem 3. *There is a constant $c > 1$ such that for any rich word w of length n the number p of palindromes in the UPS-factorization of $w = w_p w_{p-1} \cdots w_2 w_1$ satisfies*

$$p \le c \frac{n}{\ln n}. \quad (2)$$

Before proving the theorem, we show two auxiliary lemmas:

Lemma 4. *Let* $q, n, t \in \mathbb{N}$ *such that*

$$\sum_{i=1}^{t} iq^{\lceil \frac{i}{2} \rceil} \geq n. \tag{3}$$

The number p of palindromes in the UPS-factorization $w = w_p w_{p-1} \cdots w_2 w_1$ of any rich word w with $n = |w|$ satisfies

$$p \leq \sum_{i=1}^{t} q^{\lceil \frac{i}{2} \rceil}. \tag{4}$$

Proof. Let $f_1, f_2, f_3, \ldots$ be an infinite sequence of all non-empty palindromes over an alphabet A with $q = |A|$ letters, where the palindromes are ordered in such a way that $i < j$ implies that $|f_i| \leq |f_j|$. Therefore, the palindromes $f_1, \ldots, f_q$ are of length 1, the palindromes $f_{q+1}, \ldots, f_{2q}$ are of length 2, etc. Since $w_1, \ldots, w_p$ are distinct non-empty palindromes we have $\sum_{i=1}^{p} |f_i| \leq \sum_{i=1}^{p} |w_i| = n$. The number of palindromes of length i over the alphabet A with q letters is equal to $q^{\lceil \frac{i}{2} \rceil}$ (just consider that the "first half" of a palindrome determines its second half). The number $\sum_{i=1}^{t} iq^{\lceil \frac{i}{2} \rceil}$ equals the length of a word obtained as concatenation of all palindromes of length less than or equal to t. Since $\sum_{i=1}^{p} |f_i| \leq n \leq \sum_{i=1}^{t} iq^{\lceil \frac{i}{2} \rceil}$, it follows that the number of palindromes p is less than or equal to the number of all palindromes of length at most t; this explains the inequality (4). $\square$

Lemma 5. *Let* $N \in \mathbb{N}$, $x \in \mathbb{R}$, $x > 1$ *such that* $N(x-1) \geq 2$. *We have*

$$\frac{Nx^N}{2(x-1)} \leq \sum_{i=1}^{N} ix^{i-1} \leq \frac{Nx^N}{(x-1)}. \tag{5}$$

Proof. The sum of the first N terms of a geometric series with the quotient x is equal to $\sum_{i=1}^{N} x^i = \frac{x^{N+1}-x}{x-1}$. Taking the derivative of this formula with respect to x with $x > 1$ we obtain: $\sum_{i=1}^{N} ix^{i-1} = \frac{x^N(N(x-1)-1)+1}{(x-1)^2} = \frac{Nx^N}{x-1} + \frac{1-x^N}{(x-1)^2}$. It follows that the right inequality of (5) holds for all $N \in \mathbb{N}$ and $x > 1$. The condition $N(x-1) \geq 2$ implies that $\frac{1}{2}N(x-1) \leq N(x-1) - 1$, which explains the left inequality of (5). $\square$

We can start the proof of Theorem 3:

Proof (Proof of Theorem 3). Let $t \in \mathbb{N}$ be a minimal nonnegative integer such that the inequality (3) in Lemma 4 holds. It means that:

$$n > \sum_{i=1}^{t-1} iq^{\lceil \frac{i}{2} \rceil} \geq \sum_{i=1}^{t-1} iq^{\frac{i}{2}} = q^{\frac{1}{2}} \sum_{i=1}^{t-1} iq^{\frac{i-1}{2}} \geq \frac{(t-1)q^{\frac{t}{2}}}{2(q^{\frac{1}{2}} - 1)}, \tag{6}$$

where for the last inequality we exploited (5) with $N = t - 1$ and $x = q^{\frac{1}{2}}$. If $q \geq 9$, then the condition $N(x - 1) = (t - 1)(q^{\frac{1}{2}} - 1) \geq 2$ is fulfilled (it is the condition from Lemma 5) for any $t \geq 2$. Hence let us suppose that $q \geq 9$ and $t \geq 2$. From (6) we obtain:

$$\frac{q^{\frac{t}{2}}}{q^{\frac{1}{2}} - 1} \leq \frac{2n}{t - 1} \leq \frac{4n}{t}. \tag{7}$$

Since t is such that the inequality (3) holds and $i \leq q^{\frac{i+1}{2}}$ for any $i \in \mathbb{N}$ and $q \geq 2$, we can write:

$$n \leq \sum_{i=1}^{t} i q^{\frac{i+1}{2}} \leq \sum_{i=1}^{t} q^{i+1} = q^2 \frac{q^t - 1}{q - 1} \leq \frac{q^2}{q - 1} q^t \leq q^{2t}. \tag{8}$$

We apply the logarithm on the previous inequality:

$$\ln n \leq 2t \ln q. \tag{9}$$

An upper bound on the number of palindromes p in UPS-factorization follows from (4), (7), and (9):

$$p \leq \sum_{i=1}^{t} q^{\lceil \frac{i}{2} \rceil} \leq \sum_{i=1}^{t} q^{\frac{i+1}{2}} \leq q^{\frac{3}{2}} \frac{q^{\frac{t}{2}}}{q^{\frac{1}{2}} - 1} \leq q^{\frac{3}{2}} \frac{4n}{t} \leq q^{\frac{3}{2}} 8 \ln q \frac{n}{\ln n}. \tag{10}$$

The previous inequality requires that $q \geq 9$ and $t \geq 2$. If $t = 1$ then we can easily derive from (3) that $n \leq q$ and consequently $p \leq n \leq q$. Thus the inequality $p \leq q^{\frac{3}{2}} 8 \ln q \frac{n}{\ln n}$ holds as well for this case. Since every rich word over an alphabet with the cardinality $q < 9$ is also a rich word over the alphabet with the cardinality 9, the estimate (2) in Theorem 3 holds if we set the constant c as follows: $c = \max\{8q^{\frac{3}{2}} \ln q, 8 \cdot 9^{\frac{3}{2}} \ln 9\}$.

Remark 6. Note that in [12] it is shown that most of palindromic factors of a random word of length n are of length close to $\ln(n)$ (compare to Theorem 3).

4   Rich Words Form a Small Language

Recall the definition of a *small* language; the aim of this section is to show that the set of rich words forms a small language, see Theorem 10.

We present a recurrent inequality for $R_q(n)$. To ease our notation we omit the specification of the cardinality of alphabet and write $R(n)$ instead of $R_q(n)$.

Let us define

$$\kappa_n = \left\lceil c \frac{n}{\ln n} \right\rceil,$$

where c is the constant from Theorem 3 and $n \geq 2$.

Theorem 7. *If $n \geq 2$, then*

$$R(n) \leq \sum_{p=1}^{\kappa_n} \sum_{\substack{n_1, n_2, \ldots, n_p \geq 1 \\ n_1 + n_2 + \cdots + n_p = n}} R\left(\left\lceil \frac{n_1}{2} \right\rceil\right) R\left(\left\lceil \frac{n_2}{2} \right\rceil\right) \ldots R\left(\left\lceil \frac{n_p}{2} \right\rceil\right). \quad (11)$$

Proof. Given $p, n_1, n_2, \ldots, n_p$, let $R(n_1, n_2, \ldots, n_p)$ denote the number of rich words with UPS-factorization $w = w_p w_{p-1} \ldots w_1$, where $|w_i| = n_i$ for $i = 1, 2, \ldots, p$. Note that any palindrome w_i is uniquely determined by its prefix of length $\left\lceil \frac{n_i}{2} \right\rceil$; obviously this prefix is rich. Hence the number of words that appear in the UPS-factorization as w_i cannot be larger than $R(\left\lceil \frac{n_i}{2} \right\rceil)$. It follows that $R(n_1, n_2, \ldots, n_p) \leq R(\left\lceil \frac{n_1}{2} \right\rceil) R(\left\lceil \frac{n_2}{2} \right\rceil) \ldots R(\left\lceil \frac{n_p}{2} \right\rceil)$. The sum of this result over all possible p (see Theorem 3) and $n_1, n_2, \ldots, n_p$ completes the proof.

Proposition 8. *Let $h > 1$, $K \geq 1$ and $\beta_n = \Theta\left(\frac{n}{\ln n}\right)$. If $\Gamma(n)$ is a sequence of positive integers such that $\Gamma(n) \leq K^{\beta_n} h^{\frac{n+\beta_n}{2}} \left(\frac{en}{\beta_n}\right)^{\beta_n}$, then $\lim_{n \to \infty} \sqrt[n]{\Gamma(n)} \leq \sqrt{h}$.*

Proof. For any constant α we have $\lim_{n \to \infty} \alpha^{\frac{\beta_n}{n}} = 1$. Moreover, $\lim_{n \to \infty} \left(\frac{n}{\beta_n}\right)^{\frac{\beta_n}{n}} = 1$. Let us suppose that $\Gamma(n) = K^{\beta_n} h^{\frac{n+\beta_n}{2}} \left(\frac{en}{\beta_n}\right)^{\beta_n}$. Using these two equalities we obtain $\lim_{n \to \infty} K^{\frac{\beta_n}{n}} h^{\frac{n+\beta_n}{2n}} \left(\frac{en}{\beta_n}\right)^{\frac{\beta_n}{n}} = \lim_{n \to \infty} h^{\frac{1}{2} + \frac{\beta_n}{2n}} = \sqrt{h}$. Since $\sqrt[n]{\Gamma(n)} \leq K^{\frac{\beta_n}{n}} h^{\frac{n+\beta_n}{2n}} \left(\frac{en}{\beta_n}\right)^{\frac{\beta_n}{n}}$, we conclude that $\lim_{n \to \infty} \sqrt[n]{\Gamma(n)} \leq \sqrt{h}$.

Next, we show that $R(n)$ satisfies the conditions of Proposition 8 with $\beta_n = \kappa_n$.

Proposition 9. *If $h > 1$ and $K \geq 1$, then $R(n) \leq K^{\kappa_n} h^{\frac{n+\kappa_n}{2}} \left(\frac{en}{\kappa_n}\right)^{\kappa_n}$.*

Proof. For any integers $p, n_1, \ldots, n_p \geq 1$, the assumption implies that $R(\left\lceil \frac{n_1}{2} \right\rceil)$ $R(\left\lceil \frac{n_2}{2} \right\rceil) \cdots R(\left\lceil \frac{n_p}{2} \right\rceil) \leq K^p h^{\frac{n_1+1}{2}} h^{\frac{n_2+1}{2}} \cdots h^{\frac{n_p+1}{2}} \leq K^p h^{\frac{n+p}{2}}$. Using (11) we obtain:

$$R(n) \leq K^{\kappa_n} h^{\frac{n+\kappa_n}{2}} \sum_{p=1}^{\kappa_n} \sum_{\substack{n_1, n_2, \ldots, n_p \geq 1 \\ n_1 + n_2 + \cdots + n_p = n}} 1. \quad (12)$$

The sum

$$S_n = \sum_{\substack{n_1 + n_2 + \cdots + n_p = n \\ n_1, n_2, \ldots, n_p \geq 1}} 1$$

can be interpreted as the number of ways how to distribute n coins between p people in such a way that everyone has at least one coin. That is why $S_n = \binom{n-1}{p-1}$.

It is known (see Appendix for a proof) that

$$\sum_{i=0}^{L} \binom{N}{i} \leq \left(\frac{eN}{L}\right)^L, \text{ for any } L, N \in \mathbb{Z}^+ \text{ and } L \leq N. \quad (13)$$

From (12) we can write: $R(n) \leq K^{\kappa_n} h^{\frac{n+\kappa_n}{2}} \left(\frac{en}{\kappa_n}\right)^{\kappa_n}$.

The main theorem of this article is a simple consequence of the previous proposition.

Theorem 10. *Let $R(n)$ denote the number of rich words of length n over an alphabet with q letters. We have $\lim_{n \to \infty} \sqrt[n]{R(n)} = 1$.*

Proof. Let us suppose that $\lim_{n \to \infty} \sqrt[n]{R(n)} = \lambda > 1$. Let $\epsilon > 0$ be such that $\lambda + \epsilon < \lambda^2$. The definition of a limit implies that there is n_0 such that $\sqrt[n]{R(n)} < \lambda + \epsilon$ for any $n > n_0$, i.e. $R(n) < (\lambda + \epsilon)^n$. Let $K = \max\{R(1), R(2), \ldots, R(n_0)\}$. It holds for any $n \in \mathbb{N}$ that $R(n) \leq K(\lambda + \epsilon)^n$. Using Propositions 8 and 9 we obtain $\lim_{n \to \infty} \sqrt[n]{R(n)} \leq \sqrt{\lambda + \epsilon} < \lambda$, and this is a contradiction to our assumption that $\lim_{n \to \infty} \sqrt[n]{R(n)} = \lambda > 1$, it follows that $\lambda = 1$ (obviously $\lambda \geq 1$ since it holds that $R(n + 1) \geq R(n) \geq 1$ for all $n > 0$).

Acknowledgments. The author wishes to thank Edita Pelantová and Štěpán Starosta for their useful comments. The author acknowledges support by the Czech Science Foundation grant GAČR 13-03538S and by the Grant Agency of the Czech Technical University in Prague, grant No. SGS14/205/OHK4/3T/14.

Appendix

For the reader's convenience, we provide a proof of the well-known inequality we used in the proof of Proposition 9.

Lemma 11. $\sum_{k=0}^{L} \binom{N}{k} \leq \left(\frac{eN}{L}\right)^L$, *where $L \leq N$ and $L, N \in \mathbb{Z}^+$ ($\mathbb{Z}^+$ denotes the set of positive integers).*

Proof. Consider $x \in (0, 1]$. The binomial theorem states that

$$(1 + x)^N = \sum_{k=0}^{N} \binom{N}{k} x^k \geq \sum_{k=0}^{L} \binom{N}{k} x^k.$$

By dividing by the factor x^L we obtain

$$\sum_{k=0}^{L} \binom{N}{k} x^{k-L} \leq \frac{(1 + x)^N}{x^L}.$$

Since $x \in (0, 1]$ and $k - L \leq 0$, then $x^{k-L} \geq 1$, it follows that

$$\sum_{k=0}^{L} \binom{N}{k} \leq \frac{(1 + x)^N}{x^L}.$$

Let us substitute $x = \frac{L}{N} \in (0, 1]$ and let us use the inequality $1 + x < e^x$, that holds for all $x > 0$:

$$\frac{(1 + x)^N}{x^L} \leq \frac{e^{xN}}{x^L} = \frac{e^{\frac{L}{N}N}}{(\frac{L}{N})^L} = \left(\frac{eN}{L}\right)^L.$$

References

1. Balková, L.: Beta-integers and Quasicrystals, Ph. D. thesis, Czech Technical University in Prague and Université Paris Diderot-Paris 7 (2008)
2. Balková, L., Pelantová, E., Starosta, Š.: Sturmian jungle (or garden?) on multiliteral alphabets. RAIRO Theor. Inf. Appl. **44**, 443–470 (2010)
3. Bannai, H., Gagie, T., Inenaga, S., Kärkkäinen, J., Kempa, D., Piątkowski, M., Puglisi, S.J., Sugimoto, S.: Diverse palindromic factorization is NP-complete. In: Potapov, I. (ed.) DLT 2015. LNCS, vol. 9168, pp. 85–96. Springer, Cham (2015). doi:10.1007/978-3-319-21500-6_6
4. Massé, A.B., Brlek, S., Labbé, S., Vuillon, L.: Palindromic complexity of codings of rotations. Theor. Comput. Sci. **412**, 6455–6463 (2011)
5. Bucci, M., De Luca, A., Glen, A., Zamboni, L.Q.: A new characteristic property of rich words. Theor. Comput. Sci. **410**, 2860–2863 (2009)
6. Droubay, X., Justin, J., Pirillo, G.: Episturmian words and some constructions of de Luca and Rauzy. Theor. Comput. Sci. **255**, 539–553 (2001)
7. Frid, A., Puzynina, S., Zamboni, L.: On palindromic factorization of words. Adv. Appl. Math. **50**, 737–748 (2013)
8. Glen, A., Justin, J., Widmer, S., Zamboni, L.Q.: Palindromic richness. Eur. J. Comb. **30**, 510–531 (2009)
9. Guo, C., Shallit, J., Shur, A.M.: Palindromic rich words and run-length encodings. Inform. Process. Lett. **116**, 735–738 (2016)
10. Pelantová, E., Starosta, Š.: Palindromic richness for languages invariant under more symmetries. Theor. Comput. Sci. **518**, 42–63 (2014)
11. Rubinchik, M., Shur, A.M.: EERTREE: an efficient data structure for processing palindromes in strings. In: Lipták, Z., Smyth, W.F. (eds.) IWOCA 2015. LNCS, vol. 9538, pp. 321–333. Springer, Cham (2016). doi:10.1007/978-3-319-29516-9_27
12. Rubinchik, M., Shur, A.M.: The number of distinct subpalindromes in random words. Fund. Inf. **145**, 371–384 (2016)
13. Schaeffer, L., Shallit, J.: Closed, palindromic, rich, privileged, trapezoidal, and balanced words in automatic sequences. Electr. J. Comb. **23**, P1.25 (2016)
14. Shur, A.M.: Growth properties of power-free languages. Comput. Sci. Rev. **6**, 187–208 (2012)
15. Vesti, J.: Extensions of rich words. Theor. Comput. Sci. **548**, 14–24 (2014)

One-Way Bounded-Error Probabilistic Pushdown Automata and Kolmogorov Complexity
(Preliminary Report)

Tomoyuki Yamakami[✉]

Faculty of Engineering, University of Fukui, 3-9-1 Bunkyo, Fukui 910-8507, Japan
TomoyukiYamakami@gmail.com

Abstract. One-way probabilistic pushdown automata (or ppda's) are a simple model of randomized computation with last-in first-out memory device known as stacks and, when error probabilities are bounded away from 1/2, ppda's can characterize a family of bounded-error probabilistic context-free languages (BPCFL). We resolve a fundamental question raised by Hromkovič and Schnitger [Inf. Comput. 208 (2010) 982–995] concerning the limitation of the language recognition power of bounded-error ppda's. More specifically, we prove that a well-known language—the set of palindromes—cannot be recognized by any bounded-error ppda; in other words, this language stays outside of BPCFL. Furthermore, we show that, with bounded-error probability, no ppda can determine whether the center bit of input string is 1 (one). For those impossibility results, we utilize a complexity measure of algorithmic information known as Kolmogorov complexity. In our proofs, we first transform ppda's into an ideal shape and then lead to a key lemma by employing a Kolmogorov complexity argument.

Keywords: Probabilistic pushdown automata · Bounded error probability · BPCFL · Palindromes · Kolmogorov complexity

1 Overview: Challenges and Solutions

A *pushdown automaton* is a fundamental, mathematical model of computation, equipped with a single tape and a stack—a memory device that stores a series of symbols in the last-come fast-served (or the last-in first-out) manner. Unlike a standard model of Turing machine, the pushdown automata are generally allowed to make λ-*moves* (or λ-*transitions*), by which we can conduct stack operations without reading any input symbol. In particular, one-way nondeterministic pushdown automata are known to characterize context-free languages.

This work was done at the University of Toronto from August 2016 to March 2017 and supported by Natural Sciences and Engineering Research Council of Canada.

© Springer International Publishing AG 2017
É. Charlier et al. (Eds.): DLT 2017, LNCS 10396, pp. 353–364, 2017.
DOI: 10.1007/978-3-319-62809-7_27

A pushdown automaton that chooses the next move (or transition) by flipping a fair coin is a *probabilistic pushdown automaton*. Its extraordinary computational power was discovered in late 1970 s through early 1980s. In this paper, we wish to continue studying the computational complexity of languages recognized by one-way probabilistic pushdown automata with error probability bounded away from $1/2$.

1.1 One-Way Probabilistic Pushdown Automata

Concerning the language-recognition power of *one-way probabilistic pushdown automata* (or *ppda's*, in short), early studies unearthed ppda's surprising power by exploiting random selection of their moves. Freivalds (cited in [5]), for instance, demonstrated that a ppda can recognize a non-context-free language $kEqual = \{w \in \{a_1, a_2, \ldots, a_k, b_1, b_2, \ldots, b_k\}^* \mid \forall i \in \{1, 2, \ldots, k\}\, [\#_{a_i}(w) = \#_{b_i}(w)]\}$ for each index $k \geq 3$ with arbitrarily small two-sided error probability, where $\#_a(x)$ means the number of all occurrences of symbol a in x. A much simpler language $L_{keq} = \{a_1^n a_2^n \cdots a_k^n \mid n \geq 0\}$ has one-sided error probability. In sharp contrast, as Kaņeps, Geidmanis, and Freivalds [5] claimed, all tally languages (i.e., one-letter languages) recognized by ppda's with two-sided bounded-error probability are no more complex than regular languages. Unlike polynomial-time probabilistic Turing machines, ppda's cannot in general amplify their success probability; as an example, Hromkovič and Schnitger [4] presented a language that is recognized by a certain ppda with error probability exactly $1/3$ but cannot be recognized by any ppda with error probability at most $\frac{1}{3} - 2^{-n/8 + c\log n}$ for a certain absolute constant $c > 0$.

Conventionally, the families of languages recognized by ppda's with bounded-error probability and unbounded-error probability are respectively denoted by BPCFL and PCFL, analogous to BPP and PP in the polynomial-time setting, whereas *one-way deterministic pushdown automata* (or *dpda's*) and *one-way nondeterministic pushdown automata* (or *npda's*) respectively define the language families DCFL and CFL.

Concerning PCFL, Macarie and Ogihara [9] showed that PCFL properly contains CFL. Moreover, any language L that is log-space many-one reducible to languages in PCFL (notationally, LOGPCFL) is characterized as $\{x \mid f(x) > g(x)\}$ for two functions f, g in $\#SAC^1$; hence, the language falls into TC^1.

As for BPCFL, Hromkovič and Schnitger [4] proved that BPCFL and CFL are incomparable, namely, CFL $\not\subseteq$ BPCFL and BPCFL $\not\subseteq$ CFL. To show this incomparability, they claimed that L_{3eq} belongs to BPCFL but a context-free language $IP = \{xy \mid x, y \in \{0,1\}^*, |x| = |y|, x^R \odot y \equiv 1 \pmod 2\}$ cannot belong to BPCFL, where $\odot$ denotes the bitwise binary inner product. The latter claim was further expanded by Yamakami (arXiv version of [14]) to the advised context-free language family CFL/n (which was introduced in [12]) as BPCFL $\not\subseteq$ CFL/n by a simple application of the *swapping lemma for context-free languages* [11] (which was re-proven in [15, Corollary 4.2]). Unlike BPP $\subseteq \Sigma_2^p$, it seems unlikely to hold that BPCFL $\subseteq \Sigma_2^{CFL}$, because there is an oracle A for which BPCFL$^A \not\subseteq \Sigma_2^{CFL, A}$ (arXiv version of [14]), where Σ_2^{CFL} is the second level of the *CFL hierarchy* [14], which is built over CFL by allowing access to oracles using write-only query tapes.

1.2   Unsolved Questions on BPCFL

Although the language IP does not belong to BPCFL, its "marked" language $IP_\# = \{u\#w \mid |u| = |w|, uw \in IP\}$ (with a separator $\# \notin \{0,1\}$) is in DCFL and thus it must belong to BPCFL. In the proof of $IP \notin$ BPCFL, Hromkovič and Schnitger [4] started a contradictory assumption of the existence of a ppda M for IP and then transformed M into a two-party protocol of *two-trial randomized communication*. This model is quite different from a standard communication model used in communication complexity theory. At the next step, they transformed it into a one-way communication model by fixing a part of inputs and then applied a discrepancy method to lead to a desired contradiction. The discrepancy methodology was occasionally used in automata theory (see, e.g., [13,15]). Their proof strategy, however, has a severe limitation and does not seem to be applicable to various other languages. In particular, Hromkovič and Schnitger left the following membership question unsolved in [4, Sects. 4.1 and 5]. (1) Does $Pal = \{w \in \{0,1\}^* \mid w = w^R\}$ (palindromes) belong to BPCFL? This language is recognized by an appropriate npda, and thus it is a context-free language. Related to the first question, we may ask one more natural question regarding a subset of Pal. (2) Is $Pal_{even} = \{ww^R \mid w \in \{0,1\}^*\}$ in BPCFL? The both languages Pal and Pal_{even} stem from an intuition that finding the center bit location of an input string may be difficult for ppda's. An npda, in stark contrast, has an ability to "guess" (i.e., nondeterministically choose) the center point of each input string and utilize that information to check whether the input is of the form ww^R. From this observation, we ask one more question. (3) Does $Center = \{u1w \mid u, w \in \{0,1\}^*, |u| = |w|\}$ belong to BPCFL? The language $Center$ is also context-free; in comparison, a similar language $L_{center} = \{0^n 10^n \mid n \geq 0\}$ already belongs to DCFL.

To resolve all the questions, we definitely need a new proof technique, involving new technical tools in order to analyze the behaviors of ppda's.

1.3   Our Solutions

Here, we use a standard model of ppda's, which are a randomized version of deterministic pushdown automata (or dpda's). A *one-way probabilistic pushdown automaton* (or a *ppda*, in short) M is similar to a one-way nondeterministic pushdown automaton (or an npda) except that, instead of making nondeterministic choices, M flips a *fair coin* at each step and branches out into two possible transitions. The *probability* of each computation path is determined by the number of coin flips. The ppda has a read-only input tape, on which an input string is initially placed, surrounded by blank symbols[1] and the ppda is assumed to halt

---

[1] There is another well-known model in which an input tape has two endmarkers and a ppda halts after reading the right endmarker (see, e.g. [6]).

after reading all non-blank symbols. Ppda's can probabilistically choose either λ-moves or non-λ-moves, or both at any moment. Those behaviors are quite different from npda's, in which we can eliminate all λ-moves and thus we can make them halt exactly $n + 1$ steps just after reading off inputs of length n.

It is, however, possible to transform the original model of ppda's in several different ways; here, we will convert every ppda into an "ideal form" in Sect. 2.3.

Recall three open questions raised in Sect. 1.2. In this paper, we resolve negatively all the questions and obtain the following main theorem.

Theorem 1 (Main Theorem). *The following languages are all located outside of* BPCFL: *(1)* Pal *and* Pal$_{even}$, *and (2)* Center.

To prove Theorem 1, we need to explore the properties of ppda's. First of all, ppda's lack handy tools, such as the pumping lemma [1] and the swapping lemma [11] (re-proven in [15]) with which npda's can naturally provide. As noted in Sect. 1.2, the existing tool of [4] is not useful to prove Theorem 1, either. We thus wish to seek a different approach, known as *Kolmogorov complexity*. Roughly speaking, the conditional Kolmogorov complexity $C(x|y)$ of string x conditioned to string y is the minimal amount of information necessary to produce x algorithmically from y. The Kolmogorov complexity $C(x)$ of x is simply $C(x|\lambda)$, where λ is the empty string. This notion of Kolmogorov complexity was used to describe basic properties of formal languages. Notably, Li and Vitányi [7] and Glier [2] proposed Kolmogorov-complexity versions of the pumping lemmas for deterministic finite automata and dpda's, respectively. Here, we will propose a similar lemma for bounded-error ppda's.

Since dpda's are also bounded-error ppda's, our lemma must supersede (the corollary of) KC-DCFL Lemma in [2]. To describe the desired lemma of ours, we need to introduce basic terminology. Given two languages L_1 and L_2, we say that a program p *decides* L_1 *for* L_2 if, for any string $x \in L_2$, p can determine whether x is in L_1 (for other inputs, we do not require any condition). Given a sequence $(v_1, v_2, \ldots, v_j)$ of j strings, we write $v_{[j]}$ to denote the concatenated string $v_1 v_2 \cdots v_j$. We also set $v_{[0]} = \lambda$. For two strings x and y, $y^\infty x$ and xy^∞ are a left-infinite string and a right-infinite string of the forms $\cdots yyyx$ and $xyyy \cdots$, respectively. Moreover, $Pref(x)$ is composed of all prefixes of string x. With respect to a language L over alphabet Σ, L_x expresses the set $\{y \in \Sigma^* \mid xy \in L\}$. For each positive integer n, $[n]$ denotes the set $\{1, 2, \ldots, n\}$.

Lemma 2 (KC-BPCFL Lemma). *Let L be a language in* BPCFL *over alphabet Σ with $|\Sigma| \geq 2$. Let $x_i, y_i \in \Sigma^+$ for every $i \in \mathbb{N}$. For every constant $c_0 \geq 1$, there exist three constants $c_1, s, t \geq 1$ that satisfy the following: for any $u, v_1, \ldots, v_t, w \in \Sigma^*$ with $u \sqsubseteq_{suf} y_0^\infty x_0$ and $v_i \sqsubseteq_{pref} x_i y_i^\infty$ for all $i \in [t]$, if the following conditions (1)–(3) hold, then $C(w) \leq c_1 (\log \log |u|)^s + c_1$ holds.*

(1) Let $j \in [t]$. For any program p and any splitting $v_j = v_j' v_j''$, if p decides $L_{uv_{[j-1]}v_j'}$ for $\mathrm{Pref}(v_j'')$, then $C(v_j''|p) \leq c_0$.

(2) For any index $j \in [t]$, $C(v_j) \geq 2 \log \log |u|$.

(3) For any program p, if p lists all elements in $L_{uv_{[t]}}$ in the lexicographic order, then $C(w|p) \leq c_0$.

We will use this lemma to prove Theorem 1 in Sect. 3. The lemma itself will
be proven in Sect. 4.

2   Preparations for Our Exposition

2.1   Numbers, Alphabets, and Strings

Let $\mathbb{Z}$ denote the set of all *integers*, let $\mathbb{N}$ be the set of all *natural numbers*
(i.e., nonnegative integers), and set $\mathbb{N}^+ = \mathbb{N} - \{0\}$. Given two integers m, n with
$m \leq n$, the notation $[m, n]_{\mathbb{Z}}$ expresses an integer interval $\{m, m+1, m+2, \ldots, n\}$
and we abbreviate $[1, n]_{\mathbb{Z}}$ for $n \in \mathbb{N}^+$ as $[n]$ for simplicity. All logarithms are taken
to base 2 and $\log 0$ is conveniently identified with 0.

Let Σ be any alphabet (i.e., a nonempty finite set). For two strings $x, y \in \Sigma^*$,
the notation $y \sqsubseteq_{suf} x$ (resp., $y \sqsubseteq_{pref} x$) indicates that y is a *suffix* (resp., a
prefix) of x, namely, $x = wy$ (resp., $x = yw$) for a certain string $w \in \Sigma^*$. For a
language L and a string x, the notation L_x denotes the set $\{y \in \Sigma^* \mid xy \in L\}$.
The *empty string* is always denoted by λ.

An infinite sequence of symbols in Σ is called an *infinite string* for convenience. In particular, for two strings $x, y \in \Sigma^*$, the notation $y^\infty x$ (resp., xy^∞)
refers to a left-infinite (resp., a right-infinite) string of the form $\cdots yyyx$ (resp.,
$xyyy \cdots$). We can extend $\sqsubseteq_{pref}$ and $\sqsubseteq_{suf}$ to infinite sequences in an obvious
way.

2.2   Our Machine Model: Ppda's

Formally, a *one-way probabilistic pushdown automaton* (or a *ppda*, in short) M
is a tuple $(Q, \Sigma, \Gamma, \Theta_\Gamma, \delta, q_0, Z_0, F)$, where Q is a finite set of (inner) states, Σ
is an input alphabet, Γ is a stack alphabet, Θ_Γ is a finite subset of Γ^* with
$\lambda \in \Theta_\Gamma$, $\delta : Q \times \check{\Sigma} \times \Gamma \times Q \times \Theta_\Gamma \to [0,1]$ is a *probabilistic transition function*
(where $\check{\Sigma} = \Sigma \cup \{\lambda\}$), $q_0 \in Q$ is an initial state, $Z_0 \in \Gamma$ is a bottom marker, and
$F \subseteq Q$ is a set of accepting states (whereas $Q - F$ is a set of rejecting states).
Purely for clarity reason, we express $\delta(q, \sigma, a, p, u)$ as $\delta(q, \sigma, a|p, u)$. This value
$\delta(q, \sigma, a|p, u)$ indicates the probability that, when M scans σ on the input tape
and a in the top of the stack in inner state q, M changes its internal state to p,
moves its input tape head to the right if $\sigma \in \Sigma$, and rewrites a by u. Note that
the tape head of M moves only in one direction: from the left to the right. After
reading an input $x \in \Sigma^*$, however, the tape head is considered to move off the
input region occupied by x. A *stack content* is expressed as $a_1 a_2 \cdots a_k$, where
a_1 is the topmost stack symbol and a_k is the bottom symbol. For simplicity, we
always demand that M neither removes Z_0 nor stores more than two Z_0's.

Define $\delta[q, \sigma, a] = \sum_{(p,u) \in Q \times \Theta_\Gamma} \delta(q, \sigma, a|p, u)$ with $\sigma \in \check{\Sigma}$. When $\sigma = \lambda$, we
call its transition a λ-*move* (or a λ-*transition*) and the tape head must stay
still. Using the λ-moves, we can assume without loss of generality that M makes
the next move with probability exactly $1/2$. Note that, at any point, M can
probabilistically select either a λ-move or a non-λ-move, or both. This is formally
stated as $\delta[q, \sigma, a] + \delta[q, \lambda, a] = 1$ for any given triplet $(q, \sigma, a) \in Q \times \Sigma \times \Gamma$.

When we say that x *is completely read*, we mean that M scans and processes all symbols in x, makes all possible λ-moves that follow the scanning of the rightmost symbol of x, and moves its tape head to the next cell. We always assume, for simplicity, that M reads inputs completely. Therefore, a ppda *halts* when it reads input completely.

Generally, a ppda may produce an extremely long computation path or even an infinite computation path; however, we restrict our attention to ppda's whose *computation paths halt within $O(n)$ steps for input size n*. In what follows, we assume that all ppda's should satisfy this requirement. A standard definition of dpda's and npda's does not require such a runtime bound, because we can easily convert those machines to ones that halt within $O(n)$ time (e.g., [3]).

The *acceptance probability* of M on input x is the sum of all probabilities of accepting computation paths of M starting with x. We express by $p_{M,acc}(x)$ the acceptance probability of M on x. Similarly, we define $p_{M,rej}(x)$ to be the *rejection probability* of M on x. If M is clear from the context, we often omit script "M" entirely and write, e.g., $p_{acc}(x)$ instead of $p_{M,acc}(x)$. Given a ppda M, we simply say that M *accepts* x if $p_{acc}(x) > 1/2$ and *rejects* x otherwise. Since all computation paths are assumed to halt in linear time, for any given string x, either M accepts it or M rejects it. In this case, we conventionally say that M makes *unbounded error*.

It is useful to consider a modification of the standard notion of configuration. A *skew-configuration* of a ppda M is a triplet (q, x, r) for which M has already read x completely, it is now in state q, and its stack consists of r. Let $input(q, x, r) = x$ and $stack(q, x, r) = r$. Define $sCONF = Q \times \Sigma^* \times (\Gamma^{(-)})^* Z_0$, where $\Gamma^- = \Gamma - \{Z_0\}$. Given $c, c' \in sCONF$, $\mu(c, c')$ expresses a probability of obtaining c' from c by M in a single step. For a fixed $c_0 \in sCONF$, a *possibility space* $\Omega(c_0)$ of M starting at c_0 is $\{(c_0, c_1, \ldots) \mid \forall i \in \mathbb{N}^+ \ (c_i \in sCONF \ \& \ \mu(c_{i-1}, c_i) > 0)\}$. An *initial segment* of $(c_0, c_1, \ldots)$ is of the form $(c_0, c_1, \ldots, c_k)$ for a certain $k \in \mathbb{N}$. All initial segments of ω in $\Omega(c_0)$ form a set $\Omega_{fin}(c_0)$. In the case of $c_0 = (q_0, \lambda, Z_0)$, we write Ω and Ω_{fin} for $\Omega(c_0)$ and $\Omega_{fin}(c_0)$, respectively. For an $\omega_k = (c_0, c_1, \ldots, c_k) \in \Omega_{fin}(c_0)$, $E(\omega_k)$ denotes an event $\{\omega \in \Omega \mid \omega_k \sqsubseteq_{pref} \omega\}$ with probability $\mathrm{Prob}_\mu[E(\omega_k)] = \prod_{i=0}^{k-1} \mu(c_i, c_{i+1})$.

A *stack-configuration* of M is a pair of the form (q, r) in $Q \times \Gamma^*$. We consider a probability distribution D over all stack-configurations, namely, a function $D : Q \times \Gamma^* \to [0, 1]$ satisfying $\sum_{(q,r) \in Q \times \Gamma^*} D(q, r) = 1$. In contrast, a *partial probability distribution* $D : Q \times \Gamma^* \to [0, 1]$ requires only $\sum_{(q,r) \in Q \times \Gamma^*} D(q, r) \leq 1$. For such a D, let $|D| = \sum_{(q,r) \in Q \times \Gamma^*} D(q, r)$. Let D_x denote a probability distribution produced by M after reading x completely.

2.3   Ppda's in an Ideal Shape

For every npda, it is always possible to eliminate all λ-moves and limit the set Q of inner states to $Q_{simple} = \{q_0, q, q_{acc}\}$ by first transforming an npda M to its recognition-equivalent context-free grammar, converting it to Greibach Normal Form, and then translating it back to its recognition-equivalent npda (see, e.g., [3]). For ppda's, in contrast, we can neither eliminate λ-moves nor limit Q to Q_{simple}. Furthermore, we cannot control the number of consecutive λ-moves. Despite all those difficulties, we can still abridge certain behaviors of ppda's.

First of all, we can restrict ppda's so that they make only the following transitions. (1) Scanning $\sigma \in \Sigma$, preserve a topmost stack symbol (called a *stationary operation*). (2) Scanning $\sigma \in \Sigma$, push a new symbol u ($\in \Gamma - \{Z_0\}$) without changing any other symbols in the stack. (3) Scanning $\sigma \in \Sigma$, pop a topmost stack symbol. (4) Without scanning any input symbol (i.e., making a λ-move), pop a topmost stack symbol. Additionally, Step (4) comes only after Steps (3) or (4). A ppda that satisfies these conditions is said to be *in an ideal shape*. We can state those conditions more formally. A ppda N is in an ideal shape if it satisfies the following conditions. If $\delta(q, \sigma, a|p, u) \neq 0$, then (i) $\sigma = \lambda$ implies $u = \lambda$ and (ii) $\sigma \neq \lambda$ implies $u \in \{\lambda, ba, a\}$ for a certain $b \in \Gamma - \{Z_0\}$. Moreover, for any (q, σ, a) with $\sigma \neq \lambda$, (iii) if $\delta(q, \sigma, a|p, ba) \neq 0$ with $b \in \Gamma$, then $\delta[p, \lambda, b] = 0$ and (iv) if $\delta(q, \sigma, a|p, a) \neq 0$, then $\delta[p, \lambda, a] = 0$.

The next lemma will show that any ppda can be converted into its "equivalent" ppda in an ideal shape.

Lemma 3 (Ideal Shape Lemma). *For each ppda M, there exists another ppda N in an ideal shape for which $p_{M,acc}(x) = p_{N,acc}(x)$ for all x.*

The proof of Lemma 3 is similar in nature to one in [10]. In the remaining sections, ppda's are implicitly assumed to be in an ideal shape.

2.4   BPCFL and Closure Properties

Let M be a ppda with input alphabet Σ. The notation $L(M)$ stands for the set of all strings $x \in \Sigma^*$ accepted by M; that is, $L(M) = \{x \in \Sigma^* \mid p_{M,acc}(x) > 1/2\}$. Given a language L, we say that M *recognizes* L if $L = L(M)$. We write PCFL for the family of all languages recognized by unbounded-error ppda's. In contrast, a ppda M is said to make *bounded error* if there exists a constant $\varepsilon \in [0, 1/2)$ (called an *error bound*) such that, for every input x, either $p_{M,acc}(x) \geq 1 - \varepsilon$ or $p_{M,rej}(x) \geq 1 - \varepsilon$. As noted in Sect. 1, BPCFL is the language family consisting of all languages recognized by bounded-error ppda's. We further define $\mathrm{BPCFL}_\varepsilon$ as a variant obtained from BPCFL by fixing an error bound to be ε for underlying ppda's. Note that $\mathrm{BPCFL}_0 = \mathrm{DCFL}$, $\mathrm{BPCFL} = \bigcup_{0 \leq \varepsilon < 1/2} \mathrm{BPCFL}_\varepsilon$, and $\mathrm{DCFL} \subsetneq \mathrm{BPCFL} \subsetneq \mathrm{PCFL}$ [4].

We briefly state several closure properties of BPCFL under natural Boolean operations. By exchanging the roles of Q_{acc} and Q_{rej}, we can easily verify that BPCFL is closed under complementation.[2] Nevertheless, other Boolean closure properties, such as union and intersection, hold for BPCFL in a limited form.

---

[2] This argument implicitly utilizes the fact that our ppda's have a runtime bound of $O(n)$ for all computation paths.

Lemma 4. BPCFL *is closed under union and intersection with regular languages; namely,* $\{A \circ B \mid A \in \text{BPCFL}, B \in \text{REG}\} \subseteq \text{BPCFL}$ *for any operator* $\circ \in \{\cup, \cap\}$.

Lemma 5. *Let* M_1 *and* M_2 *be two ppda's with the same input alphabet and assume that* M_1 *and* M_2 *make error with probability at most* ε_1 *and* ε_2, *respectively. If* $0 \leq \varepsilon_1 + \varepsilon_2 < 1/3$, *then* $L(M_1) \cup L(M_2)$ *as well as* $L(M_1) \cap L(M_2)$ *can be recognized by suitable ppda's with error probability at most* $\frac{2+3\varepsilon_1+3\varepsilon_2}{6}$.

We omit the proofs of the above two lemmas. The upper bound of Lemma 5 is *tight* when $\varepsilon_1 = \varepsilon_2 = 0$, because, as noted in Sect. 1.1, Hromkovič and Schnitger [4] exhibited two languages that satisfy the premise of the lemma but their union cannot be recognized by any ppda with error probability at most $\frac{1}{3} - 2^{-n/8+c\log n}$ for a constant $c > 0$.

2.5 Kolmogorov Complexity

For simplicity, our alphabet Σ contains at least 0 and 1. For any string x over Σ, the Kolmogorov complexity of x is roughly the size of the "minimal" program such that, when it runs with no input, it eventually produces x and halts.

Given a number $n \in \mathbb{N}^+$, $bin(n)$ expresses the binary representation of n; e.g., $bin(1) = 1$, $bin(2) = 10$, and $bin(3) = 11$. Given a string $x = x_1 x_2 \cdots x_n$ of length n in Σ^*, the *self-delimiting code* of x is $\overline{x} = 1^{|bin(n)|} 0 bin(n) x$. We fix a universal Turing machine U, which can simulate the behaviors of any given program p on any given input y. The *conditional Kolmogorov complexity of x conditional to y* is $C(x|y) = \min\{|p| \mid U(\overline{p}y) = x, p \in \Sigma^*\}$ and the *Kolmogorov complexity of x* is $C(x) = C(x|\lambda)$. It is known that (1) $C(x|y) \leq C(x) \leq |x| + O(1)$, (2) for any totally recursive function $f : \Sigma^* \to \Sigma^*$, $C(f(x)|y) \leq C(x|y) + O(1)$, and (3) $C(x) \leq C(x|y) + C(y) + O(\min\{\log|x|, \log|y|\})$ (see, e.g., [8] for details). A string x is called *compressible* if $C(x) < |x|$ and *incompressible* otherwise. Similarly, a number $n \in \mathbb{N}^+$ is *compressible* if $C(bin(n)) < \log n$ and *incompressible* otherwise.

3   Proof of the Main Theorem

Using KC-BPCFL Lemma (Lemma 2), we will demonstrate how to prove the main theorem (Theorem 1). Due to page limit, we will prove only Theorem 1(1) for *Pal* and *Pal*$_{even}$. The first goal of ours is to verify by way of contradiction that no ppda can recognize *Pal*$_{even}$ with bounded-error probability. For this purpose, we will start with a contradictory assumption that *Pal*$_{even}$ belongs to BPCFL, and we will then apply KC-BPCFL Lemma to lead to the desired contradiction.

Proof of Theorem 1(1). Here, we will first prove that *Pal*$_{even} \notin$ BPCFL. To lead to a contradiction, we assume that *Pal*$_{even} \in$ BPCFL. For convenience, we write L for Pal_{even}. To apply Lemma 2, we need to show that Conditions (1)–(3) of the lemma are satisfied. We set $x = 1$, $\tilde{x} = 11$, and $y = 0$.

As the starting point, we make c_0 large enough and then take constants $c_1, s, t \geq 1$ given by the lemma. We also set n to be a sufficiently large incompressible even number. This implies that $C(n) \geq \log n$. We define $u = 0^n1$, $v_1 = 10^n$, $v_j = 110^n$ for each index $j \in [2, t]_{\mathbb{Z}}$, and $w = 110^n$. Note that $u \sqsubseteq_{suf} y^\infty x$, $v_1 \sqsubseteq_{pref} xy^\infty$, and $v_j \sqsubseteq_{pref} \tilde{x}y^\infty$ for any $j \in [2, t]_{\mathbb{Z}}$. It follows from these definitions that $uv_1 \in L$, $uv_1v_2 \in L$, $uv_{[3]} \in L$, ..., $uv_{[t]} \in L$, and $uv_{[t]}w \in L$.

(1) We want to see that Condition (1) is met. Let $j \in [t]$. There are several cases to consider separately. Here, we consider only the case of $j \geq 2$. Let $v_j = v_j'v_j''$ be any splitting. Let p be a program that decides $L_{uv_{[j-1]}v_j'}$ for $Pref(v_j'')$. Let us examine the case where $v_j' = 110^r$ and $v_j'' = 0^{n-r}$ for a certain index $r \in [0, t]_{\mathbb{Z}}$. Note that $L_{uv_{[j-1]}v_j'} = \{x \mid 0^n1(10^n1)^{j-1}10^rx \in L\}$ and $Pref(v_j'') = \{0^i \mid 0 \leq i \leq n - r\}$. It is easy to see that v_j'' is lexicographically the first element in $L_{uv_{[j-1]}v_j'}$. We define an algorithm that runs p consecutively on different inputs $\lambda, 0, 0^2, 0^3, \ldots$ until p accepts for the first time, and then outputs the first accepted input as v_j''. It then follows that $C(v_j''|p)$ is upper-bounded by a certain constant, which is not depending on (j, n). The other cases, such as $v_j' = 1$ and $v_j'' = 10^n$, are similarly treated.

(2) For each index $j \in [t]$, since n is incompressible, we obtain $C(v_j) \geq C(bin(n)) \geq \log n$. Thus, Condition (2) is satisfied.

(3) Take a program p that lists all elements in $L_{uv_{[t]}}$ in the lexicographic order. Let us consider another program that, with feeding no input, runs p and outputs lexicographically the first element in $L_{uv_{[t]}}$. Since $L_{uv_{[t]}} = \{x \mid 0^n1(10^n1)^{t-1}1x \in L\}$, w is lexicographically the first string in $L_{uv_{[k]}}$. Therefore, this new program indeed produces w. As a consequence, $C(w|p)$ is upper-bounded by a certain constant. Condition (3) is therefore met.

From the above (1)–(3), L satisfies all the premises of KC-BPCFL Lemma, and therefore we obtain $C(w) \leq c_1(\log\log|u|)^s + c_1 < \log n$ since n is sufficiently large. On the contrary, it follows that $C(w) \geq C(bin(n)) \geq \log n$ because n is incompressible. This brings us a clear contradiction. Therefore, we conclude that Pal_{even} is not in BPCFL.

It is easy to prove by Lemma 4 that $Pal \in$ BPCFL implies $Pal_{even} \in$ BPCFL. Since $Pal_{even} \notin$ BPCFL, we immediately obtain $Pal \notin$ BPCFL.   □

4 Proof of the KC-BPCFL Lemma

4.1 A Technical Proposition

Finally, we are ready to show KC-BPCFL Lemma (Lemma 2). Firstly, we intend to prove a technical assertion, Proposition 6, which describes the behaviors of partial probability distributions of stack-configurations of a given ppda.

Let us introduce necessary terminology to describe the proposition. Take a bounded-error ppda $M = (Q, \Sigma, \Gamma, \Theta_\Gamma, \delta, q_0, Z_0, F)$ and denote its error bound in $[0, 1/2)$ by ε. Suppose that a sequence $k = (k_1, k_2, \ldots, k_j)$ on $\mathbb{N}^+$ is fixed. Given strings $u, v_1, v_2, \ldots, v_j$ and a stack-configuration (q, r) of M, we denote by $D^{(k)}_{u|v_{[j-1]}|v_j}(q, r)$ the probability of generating (q, r) by M after reading $uv_{[j-1]}v_j$ completely under the condition that, while reading each v_i for $i \in [j-1]$, M's stack height keeps above k_i. Moreover, we define $D^{(k)}_{u|v_{[j-1]}|v_j, \leq}(q, r)$ to be the probability of generating (q, r) by M after reading $uv_{[j-1]}v_j$ completely such that, while reading each v_i ($i \in [j-1]$), M's stack height is more than k_i and, at a certain point of reading v_j, its stack height becomes less than or equal to k_j.

Given $c_0 \in sCONF$, $z \in \Sigma^*$, and $k \in \mathbb{N}$, an event $E_{>k}(c_0, z)$ consists of all $(c_0, c_1, \ldots) \in \Omega(c_0)$ satisfying that there is an $i \in \mathbb{N}$ with $input(c_i) = z$ and, for any such i and any $j \in [i]$, $|stack(c_j)| > k$. A biased transition probability $\mu_{>k}(c, c')$ is set to be $\mu(c, c')/\mu_{>k}[c]$ if $|stack(c)|, |stack(c')| > k$ and $\mu_{>k}[c] > 0$, and 0 otherwise, where $\mu_{>k}[c] = \mathrm{Prob}_\mu[\{(c, c') \mid c' \in sCONF, |stack(c')| > k\}]$. For any partial probability distribution D, we define a new partial probability distribution $H_{z, >k}D$ as $H_{z, >k}D(q, r) = \sum_{c_0} D(c_0)\mathrm{Prob}_{\mu_{>k}}[E_{>k}(c_0, z)]$. Write $H_z D$ for $H_{z, >0}D$. Similarly, $(c_0, c_1, \ldots)$ is in $E_{\leq k}(c_0, z)$ if there are indices $i \in \mathbb{N}$ and $j \in [i]$ for which $input(c_i) = z$ and $|stack(c_j)| \leq k$. We then define $\mu_{\leq k}(c, c') = \mu(c, c')/(1 - \mu_{>k}[c])$ if $|stack(c')| \leq k$ and $\mu_{>k}[c] < 1$, and 0 otherwise. Finally, set $H_{z, \leq k}D(q, r) = \sum_{c_0} D(c_0)\mathrm{Prob}_{\mu_{\leq k}}[E_{\leq k}(c_0, z)]$.

Let $\varepsilon' \in [0, 1/2)$ and let D and D' be two partial probability distributions on $sCONF$. We denote by $p_{acc}(D)$ (resp., $p_{rej}(D)$) the acceptance (resp., rejection) probability of M at D; namely, the sum of probabilities $D(q, r)$ for all $q \in F$ (resp., $q \in Q - F$) and all $r \in (\Gamma^{(-)})^* Z_0$. Now, let $H'_\sigma \in \{H_\sigma, H_{\sigma, >k}, H_{\sigma, \leq k}\}$, where σ refers to a "symbolic" variable. We say that (D', H'_σ) is ε'-mimics (D, H_σ) on x if, for all strings $z \in Pref(x)$ and any sign $\tau \in \{acc, rej\}$, $p_\tau(H'_z D') \geq \varepsilon'$ iff $p_\tau(H_z D) > 1/2$.

The proposition given below is quite technical but the meaning of each condition will become clear in the proof of Lemma 2 in Sect. 4.2.

Proposition 6. *Let Σ be an alphabet with $0, 1 \in \Sigma$. Let M be a ppda that recognizes a language with error probability at most ε with stack alphabet Γ. Let $x_i, y_i \in \Sigma^+$ be any strings for each $i \in \mathbb{N}$. Given each constant $c_0 \geq 1$, there always exist constants $c_1, s, t, k_1, \ldots, k_t \geq 1$ that satisfy the following statement. For any strings $u, v_1, \ldots, v_t, w \in \Sigma^*$ with $u \sqsubseteq_{suf} y_0^\infty x_0$ and $v_i \sqsubseteq_{pref} x_i y_i^\infty$ for all $i \in [t]$, if the following three conditions (1)–(3) hold, then $C(w) \leq c_1(\log\log|u|)^s + c_1$. Let $\bar\varepsilon$ be $1 - 2\varepsilon$ if $\varepsilon > \frac{1}{3}$ and 2ε otherwise.*

(1) Let j be any number in $[t]$ and let $H'_\sigma \in \{H_\sigma, H_{\sigma, >k_j}, H_{\sigma, \leq k_j}\}$. Consider any splitting $v_j = v'_j v''_j$ and any partial probability distribution D_j on $sCONF$. Assume that (D_j, H'_σ) ε-mimics $(D_{uv_{[j-1]}v'_j}, H_\sigma)$ on v''_j. Letting $A_{v'_j} = \{(q, r, \eta) \mid \eta = D_j(q, r)/|D_j| > 0\}$, it then follows that $C(v''_j|\langle A_{v'_j}, |D_j|, \varepsilon\rangle) \leq c_0$, where $\langle \cdot, \cdot, \cdot \rangle$ expresses an appropriate effective encoding.

(2) For any index $j \in [t]$, $C(v_j) \geq 2\log\log|u|$.

(3) Finally, let D_{t+1} be any partial probability distribution on $sCONF$. Assume that (D_{t+1}, H_σ) $\bar{\varepsilon}$-mimics $(D_{uv_{[t]}}, H_\sigma)$ on w. Letting $B_{t+1} = \{(q, r, \eta) \mid \eta = D_{t+1}(q, r)/|D_{t+1}| > 0\}$, it then follows that $C(w|\langle B_{t+1}, |D_{t+1}|, \varepsilon \rangle) \leq c_0$.

Since the proof of Proposition 6 is lengthy, we leave it to a journal version of this paper.

4.2 Proof of Lemma 2

Now, we are ready to give the proof of Lemma 2 based on Proposition 6.

Proof of Lemma 2. Let $L \in \text{BPCFL}$ and take an appropriate ideal-shape ppda M that recognizes L with error probability at most $\varepsilon \in [0, 1/2)$. Assuming Conditions (1)–(3) of Lemma 2, hereafter, we want to derive the conclusion of $C(w) \leq c_1(\log\log |u|)^s + c_1$. To obtain this consequence, it suffices to show that Conditions (1)–(3) of Proposition 6 are satisfied because the proposition ensures the same conclusion. In what follows, we will discuss each of the three conditions in Proposition 6 separately. Let $u, v_1, \ldots, v_t, w$ denote arbitrary strings.

To simplify our argument below, we intend to ignore an approximation of each partial probability distribution produced at any moment in a computation of M.

(1) We first target Condition (1) of Proposition 6. Let $j \in [t]$. Let $v_j = v_j' v_j''$ be any splitting. Let $D_j = D_{uv_{[j-1]}v_j'}$ and $A_{v_j'} = \{(q, r, \eta) \mid \eta = D_j(q, r)/|D_j| > 0\}$.

Let us consider the following program p. Taking input x and auxiliary input $(A_{v_j'}, |D_j|, \varepsilon')$, recover D_j from $(A_{v_j'}, |D_j|, \varepsilon)$ and then run M on x starting with D_j. Accept if M's acceptance probability is more than $1 - \varepsilon$. Notice that we simulate M in a deterministic fashion. Here, we claim the following.

Claim 7. *The above program p decides $L_{uv_{[j-1]}v_j'}$ for $Pref(v_j'')$.*

Proof. Let $z \in Pref(v_j'')$. Note that, by $L = L(M)$, $uv_{[j-1]}v_j'z \in L$ iff $p_{acc}(z|D_j) \geq 1 - \varepsilon$. By the definition of p, whenever $uv_{[j-1]}v_j'z \in L$, p correctly accepts z with probability at least $1 - \varepsilon$. Hence, we can determine whether $z \in L_{uv_{[j-1]}v_j'}$. □

Condition (1) of Lemma 2 then ensures that $C(v_j''|p) \leq c_0$. Clearly, p depends on $(A_{v_j'}, |D_j|, \varepsilon')$ as well as a constant amount of information on M. Hence, $C(v_j''|\langle A_{v_j'}, |D_j|, \varepsilon' \rangle) \leq c_0'$ for an appropriate constant $c_0' \geq c_0$. This proves the validity of Condition (1) of the proposition.

(2) Condition (2) of the proposition obviously holds.

(3) Finally, we want to assert the validity of Condition (3) of Proposition 6. Let $D_{t+1} = D_{uv_{[t]}}$ and let $B_{t+1} = \{(q, r, \eta) \mid \eta = D_{t+1}(q, r)/|D_{t+1}| > 0\}$.

Here, we consider the following program p. At each round, recursively generate a string x in the lexicographic order. From $\langle B_{t+1}, |D_{t+1}|, \varepsilon \rangle$, generate D_{t+1} and run M on x starting with D_{t+1}. Whenever x is accepted with probability at least $1 - \varepsilon$, output x. Continue to the next round.

Claim 8. *The above program p lists all elements in $L_{uv_{[t]}}$.*

Thus, Condition (3) of the lemma implies $C(w|p) \leq c_0$. Since p depends on $(B_{t+1}, |D_{t+1}|, \varepsilon)$ and M, we obtain $C(w|\langle B_{t+1}, |D_{t+1}|, \varepsilon \rangle) \leq c_0''$ for a certain constant $c_0'' \geq c_0$.

Since all the premises of Proposition 6 are satisfied for $\max\{c_0', c_0''\}$, the proposition implies $C(w) \leq c_1(\log \log |u|)^s + c_1$ for certain constants $c_1, s \geq 1$. This is also the conclusion of Lemma 2. Therefore, the lemma is true.

In the end of this proof, we remark that, in general, we cannot deal with arbitrary real numbers that appear in any partial probability distribution in the above argument. Hence, if we want to handle encoding $\langle A_{v_j'}, |D_j|, \varepsilon \rangle$, then we need to approximate each real value $D_j(q, r)$ with precision of ℓ bits, where $\ell = \lceil \log(1/\eta_0) \rceil$ with $\eta_0 = (\frac{1}{2} - \varepsilon) \cdot \frac{1}{|Q||\Gamma^{\leq k_0}|}$. We leave to the reader the justification of the above argument by using this approximation.   □

References

1. Bar-Hillel, Y., Perles, M., Shamir, E.: On formal properties of simple phrase-structure grammars. Z. Phonetik Sprachwiss. Kommunikationsforsch **14**, 143–172 (1961)
2. Glier, O.: Kolmogorov complexity and deterministic context-free languages. SIAM J. Comput. **32**, 1389–1394 (2003)
3. Hopcroft, J.E., Ullman, J.D.: Introduction to Automata Theory, Languages, and Computation. Addison-Wesley, Reading (1979)
4. Hromkovič, J., Schnitger, G.: On probabilistic pushdown automata. Inf. Comput. **208**, 982–995 (2010)
5. Kaņeps, J., Geidmanis, D., Freivalds, R.: Tally languages accepted by Monte Carlo pushdown automata. In: Rolim, J. (ed.) RANDOM 1997. LNCS, vol. 1269, pp. 187–195. Springer, Heidelberg (1997). doi:10.1007/3-540-63248-4_16
6. Lewis, H.R., Papadimitriou, C.H.: Elements of the Theory of Computation, 2nd edn. Prentice-Hall, Englewood Cliffs (1998)
7. Li, M., Vitányi, P.: A new approach to formal language theory by Kolmogorov complexity. SIAM J. Comput. **24**, 398–410 (1995)
8. Li, M., Vitányi, P.: Kolmogorov Complexity and Its Applications, 2nd edn. Springer, New York (1997)
9. Macarie, I.I., Ogihara, M.: Properties of probabilistic pushdown automata. Theor. Comput. Sci. **207**, 117–130 (1998)
10. Pighizzini, G., Pisoni, A.: Limited automata and context-free languages. Fundam. Inf. **136**, 157–176 (2015)
11. Yamakami, T.: Swapping lemmas for regular and context-free languages. arXiv:0808.4122 (2008)
12. Yamakami, T.: The roles of advice to one-tape linear-time turing machines and finite automata. Int. J. Found. Comput. Sci. **21**, 941–962 (2010)
13. Yamakami, T.: Immunity and pseudorandomness of context-free languages. Theor. Comput. Sci. **412**, 6432–6450 (2011)
14. Yamakami, T.: Oracle pushdown automata, nondeterministic reducibilities, and the hierarchy over the family of context-free languages. In: Geffert, V., Preneel, B., Rovan, B., Štuller, J., Tjoa, A.M. (eds.) SOFSEM 2014. LNCS, vol. 8327, pp. 514–525. Springer, Cham (2014). doi:10.1007/978-3-319-04298-5_45. A complete version arXiv:1303.1717v2 under a slightly different title
15. Yamakami, T.: Pseudorandom generators against advised context-free languages. Theor. Comput. Sci. **613**, 1–27 (2016)

Classifying Non-periodic Sequences
by Permutation Transducers

Hans Zantema[1,2(✉)] and Wieb Bosma[2,3]

[1] Department of Computer Science, TU Eindhoven,
P.O. Box 513, 5600 MB Eindhoven, The Netherlands
h.zantema@tue.nl
[2] Radboud University Nijmegen,
P.O. Box 9010, 6500 GL Nijmegen, The Netherlands
w.bosma@math.ru.nl
[3] Centrum voor Wiskunde en Informatica, Amsterdam, The Netherlands

Abstract. Transducers order infinite sequences into natural classes, but permutation transducers provide a finer classification, respecting certain changes to finite segments. We investigate this hierarchy for non-periodic sequences over $\{0,1\}$ in which the groups of 0s and 1s grow according to simple functions like polynomials. In this hierarchy we find infinite strictly ascending chains of sequences, all being equivalent with respect to ordinary transducers.

1 Introduction

Equivalence under transducers organizes infinite sequences into a hierarchy with interesting properties, as ongoing research is revealing, see for example [3,6] and the conference paper [5] at DLT 2016. In this setting the main definition is that for two sequences σ, τ we have that $\sigma \geq \tau$ if and only if there exists a transducer T that produces τ when consuming σ. Here a transducer is a deterministic automaton producing output strings on every transition. Two sequences σ, τ are called equivalent, notation $\sigma \sim \tau$ if both $\sigma \geq \tau$ and $\tau \geq \sigma$. A straightforward construction shows that $\sigma \sim u\sigma$ for any sequence σ and any finite string u, so prepending or removing a finite initial word remains inside the class. The pre-order $\geq$ gives rise to an order on the equivalences classes of $\sim$; the bottom element in this order consists of the class of ultimately periodic sequences.

In the current paper we investigate a more fine-grained hierarchy on sequences based on an alternative pre-order $\geq_p$. Here prepending or removing initial segments may change the class, but other basic properties are kept, like $\sigma \geq_p h(\sigma)$ for any morphism h. The idea is that we add the requirement that transducers should be *permutation transducers*. This means that not only for every state and symbol there is exactly one outgoing arrow (as is required by determinism), but also exactly one incoming arrow: it will thus be a permutation automaton (see [1,7,8]) with output, just like a finite state transducer is a DFA with output. Our original motivation for permutation transducers was to be able

© Springer International Publishing AG 2017
É. Charlier et al. (Eds.): DLT 2017, LNCS 10396, pp. 365–377, 2017.
DOI: 10.1007/978-3-319-62809-7_28

to compare and classify two-sided sequences as was elaborated in [2]. There we already made some first investigations on ordering (one-sided) sequences by permutation transducers, raising several issues that we worked out in the current paper.

So we define $\sigma \geq_p \tau$ if and only if a permutation transducer P exists such that $P(\sigma) = \tau$, and $\sigma \sim_p \tau$ if and only if both $\sigma \geq_p \tau$ and $\tau \geq_p \sigma$. In [2] we already showed that $0^\omega \not\sim_p 10^\omega$, a clear illustration that initial segments matter in this context. Again the pre-order $\geq_p$ on sequences gives rise to an order on $\sim_p$-equivalence classes; here the bottom element is the class of all periodic sequences. In [2] we showed that the ultimately periodic sequences that are not periodic form an atomic class. Here the focus is on sequences that are not ultimately periodic. In particular, we look at sequences of the shape

$$\langle f \rangle = 10^{f(0)}10^{f(1)}10^{f(2)} \cdots \quad \text{and} \quad [\![f]\!] = 0^{f(0)}1^{f(1)}0^{f(2)} \cdots ,$$

for various functions $f : \mathbb{N} \to \mathbb{N}$, in particular polynomials. Based on ordinary transducers one has $\langle f \rangle \sim [\![f]\!]$ if $f(n) > 0$ for all $n \in \mathbb{N}$, and for all linear f, g it holds $\langle f \rangle \sim \langle g \rangle$. A main result of [4] states that the class containing the sequences of the shape $\langle f \rangle$ for f linear is *atomic*, that is, if $\langle f \rangle \geq \sigma$ for f linear, then either $\sigma \geq \langle f \rangle$ or σ is ultimately periodic. In [3] it was shown that a similar result holds for quadratic functions, while in [5,6] it was shown that for higher degree it does not hold. Here we are interested in considering $\geq_p$ and $\sim_p$ instead.

In [2] we already showed that $\langle f \rangle \sim_p \langle g \rangle$ for f, g linear. Here we show that the corresponding class is *not* atomic: we show that for ascending f we have $\langle f \rangle \geq_p [\![f]\!]$ but not the other way around, and we even show that the class containing $\langle f \rangle$ for linear f is an upper bound of infinitely many distinct classes, in particular

$$[\![n]\!] <_p [\![n+2]\!] <_p [\![n+4]\!] <_p [\![n+8]\!] <_p [\![n+16]\!] <_p \cdots \leq_p \langle n \rangle.$$

We write $\sigma <_p \tau$ if $\tau \geq_p \sigma$ but not $\sigma \geq_p \tau$, and use $[\![f(n)]\!]$ as shorthand for $[\![n \mapsto f(n)]\!]$, and similarly for $\langle \cdot \rangle$. While all $\langle f \rangle$ for f quadratic are equivalent under ordinary transduction, we show that this does not hold for $\sim_p$; in particular, we obtain the infinite ascending chain

$$\langle (n+1)^2 \rangle <_p \langle n^2 \rangle <_p \langle (n-1)^2 \rangle <_p \langle (n-2)^2 \rangle <_p \langle (n-4)^2 \rangle <_p \langle (n-8)^2 \rangle <_p \cdots .$$

Typically, for proving $\sigma <_p \tau$ the easier part is giving an explicit permutation transducer P satisfying $P(\tau) = \sigma$. The hard part is showing that a permutation transducer for the other way around does not exist. For instance, the easy part of showing $[\![n]\!] <_p [\![n + 2]\!]$ can be done using the following permutation transducer, proving $[\![n + 2]\!] \geq_p [\![n]\!]$. In presenting a transducer by a picture an arrow labeled by $a|u$ means that an input symbol a is consumed and the string u is

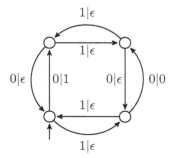

produced as output. The initial state is indicated by an incoming arrow not starting in a state. When consuming $[\![n+2]\!] = 0^2 1^3 0^4 1^5 \cdots$ by this permutation transducer indeed $[\![n]\!] = 10^2 1^3 0^4 \cdots$ is produced. Here the two 1-arrows between the two top states are never used, and may also produce anything else.

An even simpler ordinary transducer doing the same job, but which is not a permutation transducer, is easily found, as $[\![n]\!]$ can also be obtained from $[\![n+2]\!]$ by simply putting a single symbol 1 in front.

The remaining proof obligation, that no permutation transducer exists transforming $[\![n]\!]$ to $[\![n+2]\!]$, is much harder. For doing this we investigate the pattern of any sequence that can be obtained by applying a permutation transducer to $[\![n]\!]$, and then prove that $[\![n+2]\!]$ does not satisfy this pattern. For all other claims in this paper containing '$<_p$' we give similar arguments all being instances of the following three cases. The first case investigates the creation of isolated 1s, leading to $[\![f]\!] <_p \langle f \rangle$ for ascending f. The second one investigates transducts of $\langle f \rangle$ and $[\![f]\!]$ for those f (such as $f(n) = n!$) for which for every m there exists N such that $f(n) \equiv 0 \bmod m$ whenever $n > N$. The third one investigates transducts of $\langle f \rangle$ and $[\![f]\!]$ for f, such as polynomials, for which $n \mapsto f(n) \bmod m$ is periodic for every m.

We consider four basic ways to transform functions $f : \mathbb{N} \to \mathbb{N}$: transforming $f(n)$ to $f(n) + k$ and to $f(n + k)$ for any $k \geq 1$, and to $kf(n)$ and to $f(kn)$ for any $k > 1$. For all of them we investigate how $\langle f(n) \rangle$ and $[\![f(n)]\!]$ relate to their transformed variants, both with respect to ordinary transducers and permutation transducers.

The paper is organized as follows. We start by preliminaries in Sect. 2. In Sect. 3 we classify permutation transducts of particular sequences σ, in order to be able to prove $\sigma \not\geq_p \tau$ for certain τ. In Sect. 4 we investigate the effect of transforming f in the above given four ways. In Sect. 5 we investigate $[\![f]\!]$ and $\langle f \rangle$ for linear functions f; in particular, we give an infinite strictly ascending chain of them. In Sect. 6 we investigate polynomials of higher degree; in particular, we give an infinite strictly ascending chain of sequences $\langle f \rangle$ for quadratic polynomials f. Due to lack of space some proofs are omitted in this paper; they can be found in the full version [9].

2   Preliminaries

In the following we assume $\Sigma = \{0, 1\}$.

Definition 1. *A finite state transducer $T = (Q, q_0, \delta, \lambda)$ consists of a finite set Q, $q_0 \in Q$, $\delta : Q \times \Sigma \to Q$, $\lambda : Q \times \Sigma \to \Sigma^*$. For $\sigma : \mathbb{N} \to \Sigma$ we define $T(\sigma) = \lambda(q_0, \sigma(0))\lambda(q_1, \sigma(1))\lambda(q_2, \sigma(2)) \cdots$ for q_i defined by $q_{i+1} = \delta(q_i, \sigma(i))$ for $i \geq 0$.*

A permutation transducer over Σ is a finite state transducer $T = (Q, q_0, \delta, \lambda)$ with the additional requirement that for every $a \in \Sigma$ the function $q \mapsto \delta(q, a)$ is a bijection from Q to Q.

For $\sigma, \tau : \mathbb{N} \to \Sigma$ we define $\geq_p$, $\sim_p$ and $>_p$ by

$$\sigma \geq_p \tau \iff \exists \text{ permutation transducer } T : \tau = T(\sigma),$$

$$\sigma \sim_p \tau \iff \sigma \geq_p \tau \wedge \tau \geq_p \sigma, \quad \sigma >_p \tau \iff \sigma \geq_p \tau \wedge \neg(\tau \geq_p \sigma).$$

In drawing pictures for transducers we write an arrow from p to q labeled by $a|u$ if $\delta(p, a) = q$ and $\lambda(p, a) = u$. We use $\geq, \sim, >$ for the similar relations on sequences based on ordinary finite state transducers, that is, without the additional bijectivity requirement. These were studied extensively in [3–6]. To see the effect of the additional requirement of permutation transducers, throughout the paper in presenting properties of $\geq_p, \sim_p, >_p$ we often present the corresponding properties of $\geq, \sim, >$.

For a homomorphism $h : \Sigma \to \Sigma^+$ the transducer $T_h = (\{q_0\}, q_0, \delta, \lambda)$ defined by $\delta(q_0, a) = q_0$ and $\lambda(q_0, a) = h(a)$ for all $a \in \Sigma$ is a permutation transducer satisfying $T_h(\sigma) = h(\sigma)$ for all σ, proving that $\sigma \geq_p h(\sigma)$. In particular, for choosing h to be the identity we obtain that $\geq_p$ is reflexive. A straightforward construction given in [2] shows that $\geq_p$ is transitive. Hence $\geq_p$ is a pre-order, yielding a partial order on equivalence classes with respect to the equivalence relation $\sim_p$. By defining $h(a) = 0$ for all $a \in \Sigma$ we obtain $T_h(\sigma) = 0^\omega$ for every σ. Hence the equivalence class of 0^ω is the bottom element in this order; it consists of all (purely) periodic sequences as was shown in [2].

A *partial permutation transducer* $T = (Q, q_0, \delta, \lambda)$ consists of a finite set Q and initial state $q_0 \in Q$, together with a partial function $\delta : Q \times \Sigma \to Q$ such that for every $q \in Q$, $a \in \Sigma$ there is at most one $q' \in Q$ such that $\delta(q', a) = q$, and $\lambda : Q \times \Sigma \to \Sigma^*$ is a partial function that is defined on the same pairs that δ is defined for. Thus, in a permutation transducer for every symbol $a \in \Sigma$ there is exactly one incoming and exactly one outgoing a-arrow for every state $q \in Q$, but in a partial permutation transducer 'exactly one' is weakened to 'at most one'. As observed in [2], just like every partial permutation of a set can be extended to a permutation, every partial permutation transducer can be extended to a permutation transducer. Sometimes we will present a permutation transducer by only giving a partial permutation transducer and leaving the extension implicit.

From the introduction recall the definitions

$$\langle f \rangle = \sigma_f = 10^{f(0)}10^{f(1)}10^{f(2)} \cdots \quad \text{and} \quad [\![f]\!] = 0^{f(0)}1^{f(1)}0^{f(2)} \cdots$$

for any $f : \mathbb{N} \to \mathbb{N}$. For the latter it is natural to require $f(n) > 0$ for all $n > 0$, to avoid collapsing groups; we will say that f is *positive* if it satisfies this property. Note that every sequence σ that is not eventually constant has a natural representation $[\![f]\!]$ for some (positive) function f. The same is true for $\langle f \rangle$ if $\sigma(0) = 1$, with f usually not positive.

Writing $\langle f(n) \rangle$ for $\langle f \rangle$, we obtain $\langle n \rangle = 11010010001 \cdots$, and $\langle n + 1 \rangle = 101001000 \cdots$, so $\langle n + 1 \rangle = \text{tail}(\langle n \rangle)$. Using similar shorthand notation, $[\![n]\!] = 1^10^21^30^41^5 \cdots$, and $[\![n + 1]\!] = 0^11^20^31^40^5 \cdots$, so $T_h([\![n]\!]) = [\![n + 1]\!]$ and $T_h([\![n + 1]\!]) = [\![n]\!]$ for $h(0) = 1, h(1) = 0$, proving $[\![n]\!] \sim_p [\![n + 1]\!]$.

We continue with a fruitful lemma.

Lemma 2. *Let P be a permutation transducer over a finite alphabet Σ. Then there exists an integer $N > 0$ such that for every state q and every $u \in \Sigma^+$ it holds that $\delta(q, u^N) = q$.*

Proof. Let n be the number of states. Then $q \mapsto \delta(q, u)$ is a permutation on the n states. Choose N to be the least common multiple of all k with $k \leq n$. Then $q \mapsto \delta(q, u^N) = (q \mapsto \delta(q, u))^N$ is the identity. $\qquad\square$

Proposition 3. *For every positive function $f : \mathbb{N} \to \mathbb{N}$ holds*

- $\langle f \rangle \sim \llbracket f \rrbracket$, *and*
- $\langle f \rangle \geq_p \llbracket f \rrbracket$.

Proof. This is proved by the following two transducers.

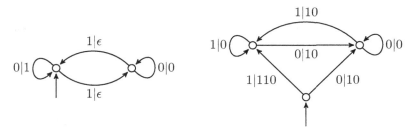

The left one is a permutation transducer replacing sequences of consecutive 0's that are demarcated by a single 1, alternatingly by the same number of 0's or 1's; hence transforming $\langle f \rangle$ to $\llbracket f \rrbracket$, showing $\langle f \rangle \geq_p \llbracket f \rrbracket$ and thus also $\langle f \rangle \geq \llbracket f \rrbracket$.

The right one is an ordinary transducer (but not a permutation transducer) transforming $\llbracket f \rrbracket$ to $\langle f \rangle$, showing that $\llbracket f \rrbracket \geq \langle f \rangle$. Together with the just observed $\langle f \rangle \geq \llbracket f \rrbracket$ this proves $\langle f \rangle \sim \llbracket f \rrbracket$. $\qquad\square$

Now we show that $\llbracket f \rrbracket \geq_p \langle f \rangle$ does not generally hold: for certain f no permutation transducer P exists transforming $\llbracket f \rrbracket$ to $\langle f \rangle$. The key idea is that the isolated 1s in $\langle f \rangle$ can not be created when the input only contains big groups of 0s and 1s as in $\llbracket f \rrbracket$. In fact we prove the following stronger result.

Theorem 4. *Let $f, g : \mathbb{N} \to \mathbb{N}$ satisfy $\lim_{n\to\infty} f(n) = \lim_{n\to\infty} g(n) = \infty$. Then no permutation transducer P such that $P(\llbracket f \rrbracket) = \langle g \rangle$.*

Proof. Assume such a $P = (Q, q_0, \delta, \lambda)$ exists. Use Lemma 2 to choose p such that $\delta(q, 0^p) = \delta(q, 1^p) = q$ for all states q. Write $u(q, 0) = \lambda(q, 0^p)$ and $u(q, 1) = \lambda(q, 1^p)$ for all states q. Since $\lim_{n\to\infty} f(n) = \infty$ a number N exists such that $f(n) > 2p$ for all $n \geq N$. Hence beyond a finite initial part, the sequence $\llbracket f \rrbracket$ is composed of strings 0^k and 1^k for $k > 2p$. For each such string the permutation transducer P produces a prefix of $u(q, i)^\omega$ that starts by $u(q, i)^2$ for some state q and $i \in \{0, 1\}$. Beyond a finite initial part, the resulting output $\langle g \rangle$ is the concatenation of such prefixes. Assume that one of the occurring strings $u(q, i)^2$ contains a symbol 1. Then it contains at least two symbols 1 at distance at most m, where m is the maximal size of all $u(q, i)$. Since $\lim_{n\to\infty} g(n) = \infty$, the total number of occurrences of $u(q, i)^2$ in $\langle g \rangle$ that contain a symbol 1, is finite. This contradicts the fact that $\langle g \rangle$ contains infinitely many 1s. $\qquad\square$

For proving more claims of the type $\sigma \not\geq_p \tau$ we typically investigate the shape of sequences $P(\sigma)$, the *transducts* of σ: then it remains to show that τ is not of the required shape. In the next section we give a number of results of this type.

3   Classifying Transducts

We want to classify permutation transducts of sequences of the shape $\langle f \rangle$ and $[\![f]\!]$ for well-known functions f, like polynomials. A key property of polynomials f that will be exploited is that the function $n \mapsto (f(n) \bmod m)$ is periodic for every $m > 0$. We start by a class of functions for which the analysis is slightly simpler, namely functions like $f(n) = n!$ for which $n \mapsto (f(n) \bmod m)$ is ultimately 0 for every $m > 0$.

Theorem 5. *Let $f : \mathbb{N} \to \mathbb{N}$ be a positive function for which for every $m > 0$ there exists $N \in \mathbb{N}$ such that $f(n) \equiv 0 \bmod m$ for all $n > N$. If $[\![f]\!] \geq_p \sigma$ then there exist $u, c, d \in \Sigma^*$ and $b, h \in \mathbb{N}$ such that*

$$\sigma = u \prod_{i=0}^{\infty} \left(c^{f(b+2i)/h} d^{f(b+2i+1)/h} \right) = u c^{f(b)/h} d^{f(b+1)/h} c^{f(b+2)/h} d^{f(b+3)/h} \cdots .$$

Proof. Let $P = (Q, q_0, \delta, \lambda)$ be a permutation transducer such that $P([\![f]\!]) = \sigma$. By Lemma 2 there exists h such that $\delta(q, 0^h) = \delta(q, 1^h) = q$ for all q. Choose b even such that $f(i) \equiv 0 \bmod h$ for all $i \geq b$. Let u be the output of P of the initial part $v = 0^{f(0)} 1^{f(1)} \cdots 1^{f(b-1)}$, and let $q = \delta(q_0, v)$. Let $c = \lambda(q, 0^h)$ and $d = \lambda(q, 1^h)$. Then the next blocks $0^{f(b)}$, $1^{f(b+1)}$, $0^{f(b+2)}$, $1^{f(b+3)}, \ldots$ produce output $c^{f(b)/h} d^{f(b+1)/h} c^{f(b+2)/h} d^{f(b+3)/h} \cdots$, exactly the pattern claimed.   □

Corollary 6. $[\![n!]\!] \not\geq_p [\![n! - 1]\!]$.

Proof. Suppose that $[\![n!]\!] \geq_p [\![n! - 1]\!]$. Then by Theorem 5 we obtain

$$[\![n! - 1]\!] = u c^{b!/h} d^{(b+1)!/h} c^{(b+2)!/h} d^{(b+3)!/h} \cdots .$$

Since in $[\![n! - 1]\!]$ only groups of 0s and 1s occur of increasing size, both c and d either consist only of 0s or only of 1s. Since $[\![n! - 1]\!]$ contains infinitely many 0s and infinitely many 1s, either c consists of 0s and d consists of 1s, or the other way around. But then the resulting consecutive groups of 0s and 1s have sizes $|c|b!/h$, $|d|(b+1)!/h$, $|c|(b+2)!/h)$, $|d|(b+3)!/h\cdots$, ultimately divisible by any number, which does not hold for the group sizes $n! - 1$, $(n+1)! - 1$, $(n+2)! - 1$, $(n+3)! - 1 \ldots$ in $[\![n! - 1]\!]$. This contradiction proves $[\![n!]\!] \not\geq_p [\![n! - 1]\!]$.   □

Corollary 7. $[\![n!]\!] \not\geq_p [\![(2n)!]\!]$.

Proof. As in the previous proof, use the form of the transducts of $[\![n!]\!]$ given by Theorem 5: again c and d both consist of copies of a single symbol, different for the two. But now it will be impossible for such transduct to equal $[\![(2n)!]\!]$ because the growth of the groups in the transduct is like a multiple of $n!$, which is much slower than that of the groups in $[\![(2n)!]\!]$.   □

For the same class of functions we now give a characterization for transducts of $\langle f \rangle$ rather than $[\![f]\!]$.

Theorem 8. *Let* $f : \mathbb{N} \to \mathbb{N}$ *be a function for which for every* $m > 0$ *there exists* $N \in \mathbb{N}$ *such that* $f(n) \equiv 0 \bmod m$ *for all* $n > N$. *If* $\langle f \rangle \geq_p \sigma$ *then there exist* $k > 0$, $a \geq 0$ *and* $u, p_0, \cdots, p_{k-1}, c_0, \cdots, c_{k-1} \in \Sigma^*$ *such that*

$$\sigma = u \prod_{j=0}^{\infty} \left(\prod_{i=0}^{k-1} p_i c_i^{f(a+i+jk)/k} \right) = u p_0 c_0^{f(a)/k} p_1 c_1^{f(a+1)/k} \cdots.$$

Proof. Assume that $P(\langle f \rangle) = \sigma$ for a permutation transducer $P = (Q, q_0, \delta, \lambda)$. Choose k by Lemma 2 such that $\delta(q, 0^k) = \delta(q, 1^k) = q$ for all $q \in Q$. By the assumption on f there exists a such that $f(n) \equiv 0 \bmod k$ for all $n \geq a$. Let $v = 10^{f(0)} 10^{f(1)} 1 \cdots 10^{f(a-1)}$, which is a prefix of $\langle f \rangle$. Let $u = \lambda(q_0, v)$, and $r_0 = \delta(q_0, v)$. Define $r_i = \delta(r_0, 1^i)$ for $i = 1, 2, \ldots, k$; since $\delta(r_0, 1^k) = r_0$ we have $r_k = r_0$. Since $f(a+i) \equiv 0 \bmod k$ we obtain $r_{i+1} = \delta(r_i, 10^{f(a+i)})$ for $i = 0, \ldots, k-1$. Write $p_i = \lambda(r_i, 1)$ and $c_i = \lambda(r_{i+1}, 0^k)$, then by using $\delta(r_{i+1}, 0^k) = r_{i+1}$ for $i = 0, \ldots, k-1$, we obtain the desired result that σ equals

$$u\lambda(r_0, 10^{f(a)})\lambda(r_1, 10^{f(a+1)})\lambda(r_2, 10^{f(a+2)}) \cdots = u \prod_{j=0}^{\infty} \left(\prod_{i=0}^{k-1} p_i c_i^{f(a+i+jk)/k} \right).$$

$\square$

Corollary 9. $\langle n! \rangle \not\geq_p \langle n! - 1 \rangle$.

Proof. Suppose that $\langle n! \rangle \geq_p \langle n! - 1 \rangle$. Then by Theorem 8 we obtain

$$\langle n! - 1 \rangle = u p_0 c_0^{a!/k} p_1 c_1^{(a+1)!/k} p_2 c_2^{(a+2)!/k} \cdots.$$

Since in $\langle n! - 1 \rangle$ only increasing groups of 0s occur between consecutive 1s, every p_i contains at most one 1 and every c_i only consists of 0s. By possibly doubling k, we may assume that two distinct p_is contain a 1; let p_g and p_h be the first two containing a 1. For $i = 0, \ldots, h - g - 1$ define d_i by $c_i = 0^{d_i}$. Then for every $j \geq 0$ the string $p_g c_g^{(a+g+jk)/k} p_{g+1} c_{g+1}^{(a+g+1+jk)/k} \cdots p_h$ is a part of $\langle n! - 1 \rangle$ containing exactly two 1s, with exactly $C + \sum_{i=0}^{h-g-1} \frac{(a+g+jk)!d_i}{k}$ separating 0s, for some constant $C \geq 0$. Choose $N > 2C + 2$. Then for j large enough all of these groups of 0s have size $C \bmod N$, contradicting the fact that after a finite part $\langle n! - 1 \rangle$ only contains groups of 0s of size $-1 \bmod N$. $\square$

Next we switch to functions f, like polynomials, for which $n \mapsto (f(n) \bmod m)$ is periodic for every $m > 0$. For transducts of $\langle f \rangle$ under permutation transducers the following characterization was given in [2].

Theorem 10. *Let* $f : \mathbb{N} \to \mathbb{N}$ *be a function for which* $n \mapsto (f(n) \bmod m)$ *is periodic for every* $m > 0$. *Then* $\langle f \rangle \geq_p \sigma$ *for* $\sigma : \mathbb{N} \to \Sigma$ *if and only if there exist* $k, h > 0$ *and* $p_0, \cdots, p_{k-1}, c_0, \cdots, c_{k-1} \in \Sigma^*$ *such that*

$$\sigma = \prod_{j=0}^{\infty} \left(\prod_{i=0}^{k-1} p_i c_i^{\lfloor f(i+jk)/h \rfloor} \right) = p_0 c_0^{\lfloor f(0)/h \rfloor} p_1 c_1^{\lfloor f(1)/h \rfloor} \cdots.$$

We give a similar description of transducts of $[\![f]\!]$.

Theorem 11. *Let $f : \mathbb{N} \to \mathbb{N}$ be a function for which $n \mapsto (f(n) \bmod m)$ is periodic for every $m > 0$. If $[\![f]\!] \geq_p \sigma$ for $\sigma : \mathbb{N} \to \Sigma$ then there exist $k, h > 0$ and $p_0, \cdots, p_{k-1}, c_0, \cdots, c_{k-1} \in \Sigma^*$ such that*

$$\sigma = \prod_{j=0}^{\infty} \left(\prod_{i=0}^{k-1} p_i(c_i p_i)^{\lfloor f(i+jk)/h \rfloor} \right) = p_0(c_0 p_0)^{\lfloor f(0)/h \rfloor} p_1(c_1 p_1)^{\lfloor f(1)/h \rfloor} \cdots,$$

with the additional constraint that $p_i = \epsilon$ if $f(i) \equiv 0 \bmod h$, and c_i is non-empty.

The proof follows the same lines as that of Theorem 8, but slightly more technical; for details we refer to the full version [9] of this paper.

4   Basic Function Operations

In this section we investigate how $\langle f(n) \rangle$ relates to $\langle f(n+k) \rangle$, $\langle f(n)+k \rangle$, $\langle kf(n) \rangle$ and $\langle f(kn) \rangle$, and similarly for $[\![\cdot]\!]$. For completeness we do not only consider $\sim_p, \geq_p, \leq_p$ based on permutation transducers, but also $\sim, \geq, \leq$ based on ordinary transducers.

Theorem 12. *When linear operations on sequences of the form $\langle f \rangle$ or $[\![f]\!]$ are performed, the general relation between the original sequence and its image by ordinary or permutation is given by an entry in the following two tables:*

$\langle f(n) \rangle \leq, \geq, \nleq_p, \geq_p \langle f(n) + k \rangle$	$[\![f(n)]\!] \leq, \geq, \nleq_p, \ngeq_p [\![f(n) + k]\!]$
$\langle f(n) \rangle \leq, \geq, \nleq_p, ? \langle f(n+k) \rangle$	$[\![f(n)]\!] \leq, \geq, \nleq_p, \ngeq_p [\![f(n+k)]\!]$
$\langle f(n) \rangle \leq, \geq, \leq_p, \geq_p \langle kf(n) \rangle$	$[\![f(n)]\!] \leq, \geq, \leq_p, \geq_p [\![kf(n)]\!]$
$\langle f(n) \rangle \nleq, \geq, \nleq_p, \geq_p \langle f(kn) \rangle$	$[\![f(n)]\!] \nleq, \geq, \nleq_p, \ngeq_p [\![f(kn)]\!]$

An entry of the form $\sigma \geq \tau$ indicates that for every f, k a transducer exists transforming σ to τ, while $\sigma \ngeq \tau$ indicates that f, k exist for which such a transducer does not exist, and similar for $\leq, \leq_p, \geq_p$. The question mark '?' states that this question is open for $\langle f(n) \rangle \geq_p \langle f(n + k) \rangle$.

The proofs of all claims can be found in the full version of this paper [9]; for all cases either explicit transducers are given or results from Sect. 3 are applied to show that the transduct does not satisfy the required pattern for a particular function f.

5   Classes of Linear Functions

For linear polynomial functions $f_{k,l}(n) = kn + l$, the relations between $\langle f_{k,l} \rangle$ were already dealt with in [2].

Theorem 13. *For all $k, l \in \mathbb{N}$ with $k \geq 1$: $\langle kn + l \rangle \sim_p \langle n \rangle$; in other words: all (non-constant) linear functions are equivalent under $\sim_p$.*

The situation is markedly different for $[\![f_{k,l}]\!]$, as we will see in the next theorem. Note that by Proposition 3 and Theorem 4 we already know $[\![n]\!] <_p \langle n \rangle$, so $\langle n \rangle$ is not atomic. We do not yet know whether $[\![n]\!]$ is atomic or not.

Theorem 14. *Under permutation transduction there is an infinite, strictly ascending sequence of equivalence classes containing $[\![f]\!]$ for linear polynomials f, in between $[\![n]\!]$ and $\langle n \rangle$; in particular*

$$[\![n]\!] <_p [\![n+2]\!] <_p [\![n+4]\!] <_p [\![n+8]\!] <_p \cdots \leq_p \langle n \rangle.$$

Before we give the proof, we state a corollary that settles a question from the previous section.

Corollary 15. *No permutation transducer P exists such that $P([\![n+1]\!]) = [\![n+2]\!]$.*

Proof. This is the content of the first strict inequality in Theorem 14, in combination with $[\![n]\!] \sim_p [\![n+1]\!]$. The latter follows from $\overline{[\![n+1]\!]} = [\![n]\!]$, in which $\overline{\sigma}$ denotes the complement of σ, obtained by permuting the two symbols $0, 1$; it will be clear that $\overline{\sigma}$ can be obtained from σ by a one-state permutation transducer. □

The proof of Theorem 14 is given in two parts. The first (existence) part is immediate from the following lemma.

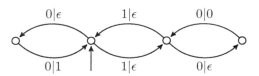

$$0|\epsilon \qquad\qquad 1|\epsilon \qquad\qquad 0|0$$

$$0|1 \qquad\qquad 1|\epsilon \qquad\qquad 0|\epsilon$$

Lemma 16. *Let $k \geq 0$ be an integer; then $[\![n+k]\!] \geq_p [\![n+\lfloor \frac{k+1}{2} \rfloor]\!]$.*

Proof. For $k = 0, 1$ the statement is trivial; so assume that $k \geq 2$. Consider the four-state permutation transducer given above. For even k it will remove any 1's from the input sequence $[\![n+k]\!]$ and alternatingly divide by 2 or divide by 2 and complement, any sequence of 0's; the result is $\overline{[\![n+\frac{k}{2}]\!]}$. For odd k it converts $\overline{[\![n+k]\!]}$ into $[\![n+\frac{k+1}{2}]\!]$. Taking complements is easily achieved by a permutation transducer, and the permutation property is transitive. □

In fact this lemma yields a stronger result than required for Theorem 14, as stated in the following corollary.

Corollary 17. *For every $a > 0$ we have $[\![n+a]\!] \geq_p [\![n]\!]$.*

Proof. Starting by $[\![n+a]\!]$ repeat applying Lemma 16 until $[\![n+1]\!]$ is obtained. Then the corollary follows from transitivity of $\geq_p$ and $[\![n+1]\!] \sim_p [\![n]\!]$. □

To complete the proof of Theorem 14 we have to prove that for none of the strict steps a backward transduction is possible. This is immediate from the following stronger result.

Proposition 18. *For $a \geq 0$, $b \geq a + 2$ no permutation transducer P exists satisfying $P(\llbracket n + a \rrbracket) = \llbracket n + b \rrbracket$.*

Proof. The permutation transducts of $\llbracket n + a \rrbracket$, according to Theorem 11, are of the form

$$\prod_{j=0}^{\infty} \left(\prod_{i=0}^{k-1} p_i(c_i p_i)^{\lfloor \frac{i+jk+a}{h} \rfloor} \right).$$

Replace the period k by a multiple (if necessary) in order for h to be a divisor of k, and write $m = \frac{k}{h}$ and $a_i = \lfloor \frac{i+a}{h} \rfloor$ for $i = 0, \ldots, k - 1$. Write $b_i = \lfloor \frac{i+a}{h} \rfloor - a_i m$ for $i = 0, \ldots, k - 1$, note that $0 \leq b_i < m$. Then the transduct is of the shape $\prod_{j=0}^{\infty} \left(\prod_{i=0}^{k-1} p_i(c_i p_i)^{b_i + (j+a_i)m} \right)$. Writing $w_i = (c_i p_i)^m$ and replacing p_i by $p_i(c_i p_i)^{b_i}$, we conclude that the transduct is of the shape $P(\llbracket n + a \rrbracket) = \prod_j \prod_{i=0}^{k-1} p_i(w_i)^{j+a_i}$.

Now suppose that this image equals $\llbracket n + b \rrbracket$, for some $b \geq a + 2$. It is impossible for w_i to contain both 0 and 1, since in that case $\prod_j \prod_i p_i w_i^{j+a_i}$ contains infinitely many pairs of the same symbol separated by a fixed number of copies of the other symbol, which $\llbracket n + b \rrbracket = 0^b 1^{b+1} 0^{b+2} 1^{b+3} \cdots$ clearly does not.

Hence each w_i consist of copies of a single symbol; if w_{i+1} consists of the same symbol or equals ϵ, we can merge $p_i w_i p_{i+1} w_{i+1}$ and reduce k. Hence without loss of generality we may assume that w_i and w_{i+1} consist of different symbols; the same then holds for p_i and p_{i+1} (but they could equal ϵ). By multiplying the period k we may assume that k is even and $k > b$.

The linear growth of $\llbracket n + b \rrbracket$ implies that each w_i will consist of exactly k symbols; since $p_i w_i$ and $p_{i+1} w_{i+1}$ are consecutive blocks of different symbols, $\#p_{i+1} \bmod k = (\#p_i + 1) \bmod k$.

Since $k > b > a$ and $a_i = \lfloor \frac{i+a}{k} \rfloor$ we obtain $a_i = 0$ for $i < k - a$. So $p_0 p_1 \cdots p_{k-a-1}$ is an initial part of $\llbracket n + b \rrbracket$, in which p_i alternatingly consist of 0s and 1s. If $p_0 = \epsilon$ then p_1 is the first group of 0s, being 0^b, contradicting $\#p_{i+1} \bmod k = (\#p_i + 1) \bmod k$. Hence $p_0 = 0^b$, and by $\#p_{i+1} \bmod k = (\#p_i + 1) \bmod k$ we obtain $\#p_i = b + i$ for $i = 0, 1, \ldots, k - a - 1$. But since $\#(c_i p_i) = k$ we obtain $\#p_i \leq k$. But then we have $b + k - a - 1 = \#p_{k-a-1} \leq k$, contradicting $b \geq a + 2$. $\quad\square$

6   Higher Degree Polynomials

Theorem 19. *Let $f, g : \mathbb{N} \to \mathbb{N}$ be two polynomials of degree $n > 1$ with the same leading coefficient such that*

- *$f - g$ is not constant, and*
- *$\lim_{x \to \infty}(f(ax) - a^n g(x)) = \infty$ for every $a > 1$.*

Then no permutation transducer P exists such that $P(\langle f \rangle) = \langle g \rangle$.

Proof. Assume that such a P exists. Then according to Theorem 10 there exist $k, h > 0$ and $p_0, \cdots, p_{k-1}, c_0, \cdots, c_{k-1} \in \Sigma^*$ such that

$$\langle g \rangle = \prod_{j=0}^{\infty} \left(\prod_{i=0}^{k-1} p_i c_i^{\lfloor \frac{f(i+jk)}{h} \rfloor} \right).$$

Since $\lim_{n \to \infty} g(n) = \infty$, for every $i = 0, \ldots, k-1$ no 1 occurs in c_i, and at most one 1 occurs in p_i. Since $\langle g \rangle$ contains symbols 1, at least one of the p_i's contains a symbol 1.

If there is only one such p_i, by doubling k we make it two.

Let p_a and p_b be the first two p_i's containing a 1. Let q be the total number of 1's in $p_0, \ldots, p_{k-1}$, and $c_i = 0^{a_i}$ for $i = 0, \ldots, k-1$. Now we count the number of 0's right after the $qj + 1$-th 1 of $\langle f \rangle$ in two ways, and obtain that there is constant $c \geq 0$ (corresponding to the number of 0's occurring in some p_i's) such that

$$c + \sum_{i=a}^{b-1} \lfloor \frac{f(i+jk)}{h} \rfloor a_i = g(jq)$$

for all $j \geq 0$.

First we consider the case $k = q$. Then $a = 0$, $b = 1$ and we have $c + a_0 \lfloor \frac{f(jk)}{h} \rfloor = g(jk)$ for all $j \geq 0$. This is only possible if $f - g$ is constant, which we assumed to be not. In the remaining case we have $0 < q < k$.

Write $A = \sum_{i=a}^{b-1} a_i$.

Using that f is ascending for sufficiently large arguments, we have $f(jk) \leq f(i+jk) \leq f((j+1)k)$ for $j > C$ for some C, and $a \leq i < b$.

Using this and $x - 1 \leq \lfloor x \rfloor \leq x$ for all x, we obtain

$$c + A(\frac{f(jk)}{h} - 1) \leq g(jq) \leq c + A\frac{f((j+1)k)}{h}$$

for all $j \geq C$. Then for $j \to \infty$ in the above inequalities we obtain $Ak^n = hq^n$. Then the left inequality yields

$$c - A + (\frac{q}{k})^n f(jk) = c + A(\frac{f(jk)}{h} - 1) \leq g(jq)$$

for all $j \geq C$. This contradicts $\lim_{x \to \infty}(f(ax) - a^n g(x)) = \infty$ for $a = \frac{k}{q} > 1$. $\square$

Corollary 20. $\langle (n+1)^2 \rangle \not\geq_p \langle n^2 \rangle$ *and* $\langle n^2 \rangle \not\geq_p \langle (n-1)^2 \rangle$.

Lemma 21. *For $k > 0$ there is no permutation transducer P such that $P : \langle (n-k)^2 \rangle \mapsto \langle (n-2k)^2 \rangle$.*

Proof. Apply Theorem 19 directly to $f = (n - k)^2$ and $g = (n - 2k)^2$. $\square$

Corollary 22. *The following provides an infinite ascending chain of quadratic polynomial functions that are non-equivalent under permutation transducers:*

$$\langle (n+1)^2 \rangle <_p \langle n^2 \rangle <_p \langle (n-1)^2 \rangle <_p \langle (n-2)^2 \rangle <_p \langle (n-4)^2 \rangle <_p \langle (n-8)^2 \rangle <_p \cdots.$$

Proof. Consider the three permutation transducers

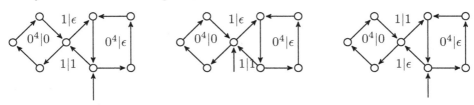

based on the principle that the left cycle reduces $10^{4m}1$ to 10^m and the right cycle $10^{4m+1}1$ to 1. Here $0^4|u$ means that all four arrows consume 0, while only one has output u, the others have empty output. It is not hard to see that the first transduces $\langle(n-2k)^2\rangle$ to $\langle(n-k)^2\rangle$ for every $k>0$, the second transduces $\langle(n-1)^2\rangle \mapsto \langle n^2\rangle$ and the third transduces $\langle n^2\rangle$ to $\langle(n+1)^2\rangle$. None of the arrows is reversible by Corollary 20 and Lemma 21. $\qquad\square$

Remark 23. Now consider the permutation transducers:

The one on the left (or the first of the three transducers in the previous picture) is easily seen to provide transition from $\langle(n+2k)^2\rangle$ to $\langle(n+k)^2\rangle$, for $k>0$, so

$$\langle(n+1)^2\rangle \leq_p \langle(n+2)^2\rangle \leq_p \langle(n+4)^2\rangle \leq_p \langle(n+8)^2\rangle \leq_p \cdots;$$

but here transductions in the opposite direction are not ruled out by Theorem 19. The other transducer shows that $\langle(n+2k-1)^2\rangle \geq_p \langle(n+k)^2\rangle$ for $k>0$, and puts $\langle(n+2k-1)^2\rangle$ in some infinite non-descending sequence. For example:

$$\langle(n+1)^2\rangle \leq_p \langle(n+2)^2\rangle \leq_p \langle(n+3)^2\rangle \leq_p \langle(n+5)^2\rangle \leq_p \langle(n+9)^2\rangle \cdots.$$

References

1. Allouche, J.P., Shallit, J.: Automatic Sequences: Theory, Applications, Generalizations. Cambridge University Press, Cambridge (2003)
2. Bosma, W., Zantema, H.: Ordering sequences by permutation transducers. Indagationes Math. **38**, 38–54 (2017)
3. Endrullis, J., Grabmayer, C., Hendriks, D., Zantema, H.: The degree of squares is an atom. In: Manea, F., Nowotka, D. (eds.) WORDS 2015. LNCS, vol. 9304, pp. 109–121. Springer, Cham (2015). doi:10.1007/978-3-319-23660-5_10
4. Endrullis, J., Hendriks, D., Klop, J.W.: Degrees of streams. Integers Electron. J. Comb. Number Theor. **11B**(A6), 1–40 (2011). Proceedings of Leiden Numeration Conference (2010)
5. Endrullis, J., Karhumäki, J., Klop, J.W., Saarela, A.: Degrees of infinite words, polynomials and atoms. In: Brlek, S., Reutenauer, C. (eds.) DLT 2016. LNCS, vol. 9840, pp. 164–176. Springer, Heidelberg (2016). doi:10.1007/978-3-662-53132-7_14

6. Endrullis, J., Klop, J.W., Saarela, A., Whiteland, M.: Degrees of transducibility. In: Manea, F., Nowotka, D. (eds.) WORDS 2015. LNCS, vol. 9304, pp. 1–13. Springer, Cham (2015). doi:10.1007/978-3-319-23660-5_1
7. Pin, J.-E.: On reversible automata. In: Simon, I. (ed.) LATIN 1992. LNCS, vol. 583, pp. 401–416. Springer, Heidelberg (1992). doi:10.1007/BFb0023844
8. Sakarovitch, J.: Elements of automata theory. Cambridge University Press, Cambridge (2009)
9. Zantema, H., Bosma, W.: Extended version of this paper (2017). http://www.win.tue.nl/~hzantema/permtr.pdf

Author Index

Printed in the United States
By Bookmasters